Deep Learning in Plant Science and Agricultural Biotechnology

CABI BIOTECHNOLOGY SERIES

Biotechnology, in particular the use of transgenic organisms, has a wide range of applications including agriculture, forestry, food, and health. There is evidence that it could make a major impact in producing plants and animals that are able to resist stresses and diseases, thereby increasing food security. There is also potential to produce pharmaceuticals in plants through biotechnology and provide foods that are nutritionally enhanced. Genetically modified organisms can also be used in cleaning up pollution and contamination. However, the application of biotechnology has raised concerns about biosafety, and it is vital to ensure that genetically modified organisms do not pose new risks to the environment or health. To understand the full potential of biotechnology and the issues that relate to it, scientists need access to information that not only provides an overview of and background to the field but also keeps them up to date with the latest research findings.

This series, which extends the scope of CABI's successful "Biotechnology in Agriculture" series, addresses all topics relating to biotechnology including transgenic organisms, molecular analysis techniques, molecular pharming, *in vitro* culture, public opinion, economics, development, and biosafety. Aimed at researchers, upper-level students, and policymakers, titles in the series provide international coverage of topics related to biotechnology, including both a synthesis of facts and discussions of future research perspectives and possible solutions.

Titles Available

1. **Animal Nutrition with Transgenic Plants**
   Edited by G. Flachowsky
2. **Plant-Derived Pharmaceuticals: Principles and Applications for Developing Countries**
   Edited by K.L. Hefferon
3. **Transgenic Insects: Techniques and Applications**
   Edited by M.Q. Benedict
4. **Bt Resistance: Characterization and Strategies for GM Crops Producing *Bacillus thuringiensis* Toxins**
   Edited by Mario Soberón, Yulin Gao, and Alejandra Bravo
5. **Plant Gene Silencing: Mechanisms and Applications**
   Edited by Tamas Dalmay
6. **Ethical Tensions from New Technology: The Case of Agricultural Biotechnology**
   Edited by Harvey James
7. **GM Food Systems and their Economic Impact**
   Tatjana Brankov and Koviljko Lovre
8. **Endophyte Biotechnology: Potential for Agriculture and Pharmacology**
   Edited by Alexander Schouten
9. **Forest Genomics and Biotechnology**
   Richard Meilan and Matias Kirst
10. **Transgenic Insects: Techniques and Applications, 2nd Edition**
    Edited by M.Q. Benedict and Maxwell J. Scott
11. **Plant Omics: Advances in Big Data Biology**
    Edited by Hajime Ohyanagi, Eiji Yamamoto, Ai Kitazumi, and Kentaro Yano

12. **Next-Generation Sequencing and Agriculture**
Edited by Phillip Bayer and Dave Edwards

13. **Aquaculture and Fisheries Biotechnology: Genetic Approaches**
Rex A. Dunham

14. **Algal Biotechnology**
Edited by Qiang Wang

15. **Genetic Engineering of Vegetable Crops**
Edited by Pritam Kalia

16. **Plant Breeding Technology: Future Trends and Challenges**
Edited by Zeba Khan

17. **Stress-Resilient Crops: Coordinated Omics-CRISPR-Nanotechnology Strategies**
Edited by Jen-Tsung Chen

18. **High-Throughput Plant Metabolomics**
Edited by Jen-Tsung Chen

19. **Plant Biotechnology and Sustainable Agriculture: Bridging the Gap for Global Food Security**
Edited by Tariq Aftab

20. **Biotechnology of Orchids and Other Ornamentals**
Edited by Pablo Bolaños-Villegas, Jian-Zhi Huang, and Teresita Amore

21. **Deep Learning in Plant Science and Agricultural Biotechnology**
Edited by Jen-Tsung Chen

Deep Learning in Plant Science and Agricultural Biotechnology

Edited by

Jen-Tsung Chen

CABI

CABI is a trading name of CAB International

CABI
Nosworthy Way
Wallingford
Oxfordshire OX10 8DE
UK

Tel: +44 (0)1491 832111
E-mail: info@cabi.org
Website: www.cabi.org

CABI
200 Portland Street
Boston
MA 02114
USA

Tel: +1 (617)682-9015
E-mail: cabi-nao@cabi.org

A catalogue record for this book is available from the British Library, London, UK.

ISBN-13: 9781836991106 (hardback)
9781836991113 (ePDF)
9781836991120 (ePub)

DOI: 10.1079/9781836991120.0000

Commissioning Editor: Ward Cooper
Editorial Assistant: Emma McCann
Production Editor: Shankari Wilford

Typeset by Straive, Pondicherry, India
Printed and bound in the UK by CPI Group (UK) Ltd, Croydon, CR0 4YY

Contents

Contributors

Banzeer Ahsan Abbasi, Department of Botany, Rawalpindi Women University, 6th Road, Satellite Town, Rawalpindi 46300, Pakistan. Email: benazirahsanabbasi786@gmail.com

Hakima Achetoui, Department of Plant Protection, Phytopathology Unit, Ecole Nationale d'Agriculture de Meknès, Km10, Rte Haj Kaddour, BP S/40, Meknès 50001, Morocco. Email: achetoui.hakima@usmba.ac.ma

Sarfraz Ahmad, Department of Genetics and Plant Breeding, College of Agriculture Mandawa, Swami Keshwanand Rajasthan Agriculture University, Bikaner, Rajasthan, India. ORCID: 0000-0002-6197-2116. Email: sarfraz_firoz@rediffmail.com

Zubaria Tul Ain, Department of Plant Sciences, Faculty of Biological Sciences, Quaid-i-Azam University, Islamabad 45320, Pakistan. Email: Zubaria@bs.qau.edu

Shaukat Ali, Department of Agronomy, Krishi Vigyan Kendra Chandgothi, Swami Keshwanand Rajasthan Agriculture University, Bikaner, Rajasthan, India. Email: shaukatagro@gmail.com

Arumugam Vijaya Anand, Department of Human Genetics and Molecular Biology, Bharathiar University, Coimbatore 641046, Tamil Nadu, India. Email: avahgmb@buc.edu.in

Muhammad Anas, Department of Plant Sciences, Faculty of Biological Sciences, Quaid-i-Azam University Islamabad 45320, Pakistan; Department of Soil Science and Plant Nutrition, University of Selcuk Campus, Konya 42079, Turkiye. Email: muhammadanas@bs.qau.edu.pk

Charan Babu Ankela, IARI Mega University, Hyderabad hub, ICAR – Central Research Institute for Dryland Agriculture, Hyderabad, Telangana 500059, India. Email: Charanbabu4927@gmail.com

K. Aruna, Professor Jayashankar Telangana Agricultural University, Hyderabad, Telangana 500030, India. Email: arunasree613@gmail.com

Sanam Ashraf, Department of Plant Sciences, Faculty of Biological Sciences, Quaid-i-Azam University, Islamabad 45320, Pakistan. Email: sanamwali47@gmail.com

Esakkimuthu Balaji, Prashant Fertility and Research Centre, Chennai 600 031, Tamil Nadu, India. Email: balaji9994651754@gmail.com

Loulaq Banday, Department of Biotechnology, University of Kashmir, Srinagar, UT J&K, India. Email: loulaqanwar@gmail.com

Madhumita Barooah, Department of Agricultural Biotechnology, Assam Agricultural University, Jorhat, Assam, India. Email: madhumita.barooah@aau.ac.in

Sana Basharat, School of Breeding and Multiplication (Sanya Institute of Breeding and Multiplication), Sanya, China; College of Tropical Agriculture and Forestry, Hainan University, Sanya 572025, China; Fang Zhiyuan Academician Team Innovation Center, Hainan, China. Email: sanabasharat143@gmail.com

Vikas Belwal, Department of Tree Improvement, College of Forestry Ranichauri, VCSG Uttarakhand University of Horticulture and Forestry, Bharsar, Uttarakhand, India. Email: vikasbelwal@hotmail.com

Vadivelu Bharathi, Department of Biochemistry, Thanthai Periyar Government College, Tiruchirappalli 620 023, Tamil Nadu, India. Email: bharathi2679@gmail.com

Amarjeet Singh Bhogal, Department of Plant Breeding and Genetics, Assam Agricultural University, Jorhat, Assam, India. Email: singhhamarjeet1820@gmail.com

Sudipta Sankar Bora, Silchar Regional Forensic Science Laboratory, Ghungur, Silchar, Assam under the Directorate of Forensic Science, Kahilipara, Guwahati, Assam, India. Email: sudip.asm@gmail.com

Dibosh Bordoloi, Department of Plant Breeding and Genetics, Assam Agricultural University, Jorhat, Assam, India. Email: dibosh.bordoloi@aau.ac.in

Wadie Chalgham, Department of Mechanical and Aerospace Engineering, University of California, Los Angeles, CA 90095, USA. Email: wadie.chalgham@ucla.edu

Rakhi Chandoliya, Department of Biotechnology, School of Bioengineering and Biosciences, Lovely Professional University, Jaladhar-Punjab, India; Fermentation and Phytofarming Technology Division, CSIR-Institute of Himalayan Bioresource Technology, Palampur 176061, India. Email: chandoliyarakhi76@gmail.com

Ankan Das, Palli Siksha Bhavana (Institute of Agriculture), Visva-Bharati, Sriniketan, West Bengal 731236, India. ORCID: 0009-0008-1410-8158. Email: rickdas.official@gmail.com

Siddhartha Das, Department of Plant Pathology, M.S. Swaminathan School of Agriculture, Centurion University of Technology and Management, Paralakhemundi, Odisha, India. ORCID: 0000-0002-7482-2555. Email: siddhartha.das10@gmail.com

Rashadin Md Bodeul Ddoza, Visva-Bharati University, Shantiniketan, Birbhum, West Bengal, India. Email: rashadinaunkon18@gmail.com

Sandip Debnath, Department of Genetics and Plant Breeding, Palli Siksha Bhavana (Institute of Agriculture), Visva-Bharati, Sriniketan, West Bengal 731236, India. ORCID: 0000-0002-0234-6633. Email: sandip.debnath@visva-bharati.ac.in

Ilham Dehbi, Department of Plant Protection, Phytopathology Unit, Ecole Nationale d'Agriculture de Meknès, Km10, Rte Haj Kaddour, BP S/40, Meknès 50001, Morocco; Laboratory of Plant Biotechnology and Molecular Biology, Faculty of Sciences, Moulay Ismail University, BP 11201, Zitoune, Meknes 50000, Morocco. Email: ilhamdehbi60@gmail.com

G. Thirumala Devi, Rythu Sadhikara Samstha, IGGAARL, Guntur, India. Email: daddychinnari@gmail.com

Thamaraiselvan Nandhini Devi, Department of Biomedical Engineering, Selvam College of Technology, Namakkal 637 003, Tamil Nadu, India. Email: bbrctry@gmail.com

Tabasum Mohi Ud Din, Department of Botany, Amar Singh College, Srinagar, UT J&K, India. Email: tabsumji@gmail.com

Pooja Erapuram, Vasantrao Naik Marathwada Krishi Vidyapeeth, Parbhani, Maharashtra 45320, India. Email: poojaerapuram@gmail.com

J. Essemine, National Key Laboratory of Plant Molecular Genetics, Chinese Academy of Science (CAS), Beijing, China. Email: jemaa.essemine@gmail.com

Xianzhong Feng, Key Laboratory of Soybean Molecular Design Breeding, Northeast Institute of Geography and Agroecology, Chinese Academy of Sciences, Changchun 130102, China. Email: fengxianzhong@iga.ac.cn

Sunil Gocher, Department of Entomology, College of Agriculture Jobner, Sri Karan Narendra Agriculture University, Jobner, Rajasthan, India. Email: sunilkdev1776@gmail.com

Dikshita Gogoi, Department of Plant Breeding and Genetics, Assam Agricultural University, Jorhat, Assam, India. Email: dikshitagogoi90@gmail.com

Rahul Gogoi, Department of Agricultural Biotechnology, Assam Agricultural University, Jorhat, Assam, India. Email: rahulgogoi.biotech@gmail.com

Subrata Goswami, Institute of Agricultural Sciences, Banaras Hindu University, Varanasi, Uttar Pradesh, India. Email: subrata.kv@gmail.com

Mokhtar Guerfel, Applied College, Shaqra University, Afif, Saudi Arabia. Email: mokhtar.guerfel@gmail.com

Chappali Harendra, ICAR-Indian Agricultural Research Institute, New Delhi 110012, India. Email: harendrakumar159007@gmail.com

Dikshita Hazarika, Department of Plant Breeding and Genetics, Assam Agricultural University, Jorhat, Assam, India. Email: dikshita.haz04@gmail.com

Gayatree Hazarika, Department of Plant Breeding and Genetics, Assam Agricultural University, Jorhat, Assam, India. Email: gayatree.hazarika58@gmail.com

Akbar Hossain, Bangladesh Wheat and Maize Research Institute, Dinajpur, Bangladesh. Email: akbarhossainwrc@gmail.com

Ishfaq Majid Hurrah, Plant Biotechnology Division, CSIR-Indian Institute of Integrative Medicine, Sanat Nagar, Srinagar, UT J&K, India. Email: eshfaqmajeed@gmail.com

Chayma Ikan, Centre d'Agrobiotechnologie et Bioingénierie, Unité de Recherche labellisée CNRST (Centre AgroBiotech-URL-CNRST-05), Université Cadi Ayyad, Marrakech 40000, Morocco. Email: ikanchaima@gmail.com

Shumaila Ijaz, Department of Burn and Plastic Surgery, Shenzhen Institute of Translational Medicine, The First Affiliated Hospital of Shenzhen University, School of Biomedical Engineering, Shenzhen University Medical School, Shenzhen University, Shenzhen 518035, China. Email: shumailaijaz1996@yahoo.com

Javed Iqbal, Department of Botany, Bacha Khan University, Charsadda 24420, Khyber Pakhtunkhwa, Pakistan. Email: javed89qau@gmail.com

Rashid Iqbal, Department of Agronomy, Faculty of Agriculture and Environment, The Islamia University of Bahawalpur, Bahawalpur 63100, Pakistan; Department of Life Sciences, Western Caspian University, Baku, Azerbaijan. Email: rashid.iqbal@iub.edu.pk

Md Mazharul Islam, School of Agriculture, University of the Virgin Islands, St Croix, VI 00850, USA. Email: mazharulislam4035@gmail.com

Shreedevi Jinnur, University of Agricultural Sciences, Bengaluru, Karnataka 560065, India. Email: shreedevijinnur@gmail.com

Rohit Joshi, Fermentation and Phytofarming Technology Division, CSIR-Institute of Himalayan Bioresource Technology, Palampur 176061, India; Academy of Scientific and Innovative Research (AcSIR), CSIR-HRDC Campus, Ghaziabad 201002, India. ORCID: 0000-0002-6524-4722. Email: rohitjoshi@ihbt.res.in; joshirohit6@gmail.com

Muhammad Kaleem, Department of Plant Sciences, Faculty of Biological Sciences, Quaid-i-Azam University Islamabad 45320, Pakistan. Email: mkaleem@bs.qau.edu.pk

Rubina Khan, Department of Genetics and Plant Breeding, College of Agriculture Jobner, Sri Karan Narendra Agriculture University, Jobner, Rajasthan, India. Email: rkrubikhan7@gmail.com

D.J. Kotresh, University of Agricultural Sciences Bangalore, Karnataka 560065, India. Email: kotreshdj21@gmail.com

Suthari Vijay Kumar, Uttar Banga Krishi Vishwavidyalaya, Pundibari, Cooch Behar, West Bengal 736165, India. Email: sutharivijay@gmail.com

Khushbu Kumari, Fermentation and Phytofarming Technology Division, CSIR-Institute of Himalayan Bioresource Technology, Palampur 176061, India; Academy of Scientific and Innovative Research (AcSIR), CSIR-HRDC Campus, Ghaziabad 201002, India. Email: rawatkhushu27@gmail.com

Santanu Kundu, ICAR-Indian Agricultural Research Institute, New Delhi 110012, India. ORCID: 0009-0001-4182-6939. Email: santanukundu120@gmail.com

Rachid Lahlali, Department of Plant Protection, Phytopathology Unit, Ecole Nationale d'Agriculture de Meknès, Km10, Rte Haj Kaddour, BP S/40, Meknès 50001, Morocco. Email: rlahlali@enameknes.ac.ma

Arafat Abdel Hamed Abdel Latef, Department of Plant Physiology, South Valley University, Qena, Egypt. Email: moawad76@gmail.com

Anandan Lokesh, Department of Genetics, University of Madras, Chennai 600113, Tamil Nadu, India. Email: lokeshanandan2001@gmail.com

Tariq Mahmood, Department of Plant Sciences, Faculty of Biological Sciences, Quaid-i-Azam University Islamabad 45320, Pakistan. Email: tmahmood@qau.edu.pk

Sweety Maity, Department of Agricultural Biotechnology, College of Agriculture, Odisha University of Agriculture and Technology, Bhubaneswa, Odisha, India. Email: maitysweety539@gmail.com

Sayanti Mandal, Department of Chemistry and Biochemistry, Sharda School of Engineering and Sciences, Sharda University, Greater Noida, Uttar Pradesh, India. Email: mandalsayanti@gmail.com

Tushar Kanti Mandal, ITH/IF, Wing "B" Building Congo, Embassy Tech Zone-SEZ, Rajiv Gandhi Infotech Park Phase II, Hinjawadi, Mercedes-Benz Research and Development India Pvt. Ltd, Bengaluru, Karnataka 560066, India. Email: tushar_kanti.mandal@mercedes-benz.com

Ramasamy Manikandan, Department of Biochemistry, Shrimati Indira Gandhi College, Tiruchirappalli 620 002, Tamil Nadu, India. Email: mani_r_trichy@yahoo.co.in

Mani Manoj, Department of Human Genetics and Molecular Biology, Bharathiar University, Coimbatore 641 046, Tamil Nadu, India. Email: manojinthravathi@gmail.com

Abdelilah Meddich, Centre d'Agrobiotechnologie et Bioingénierie, Unité de Recherche labellisée CNRST (Centre AgroBiotech-URL-CNRST-05), Université Cadi Ayyad, Marrakech, 40000, Morocco; African Sustainable Agriculture Research Institute (ASARI), Mohammed VI Polytechnic University (UM6P), Laâyoune 70000, Morocco. Email: a.meddich@uca.ma

Sucharita Mondal, Palli Siksha Bhavana (Institute of Agriculture), Visva-Bharati, Sriniketan, West Bengal 731236, India. ORCID: 0009-0003-6277-7173. Email: heavenlysucharita28@gmail.com

Isack Ibrahim Mrisho, College of Chinese Medicinal Material, Jilin Agricultural University, Changchun 130118, China. Email: issackmson98@gmail.com

Shruti Mukhopadhyay, Palli Siksha Bhavana (Institute of Agriculture), Visva-Bharati, Sriniketan, West Bengal 731236, India. ORCID: 0009-0003-0694-1290. Email: shruti03megha@gmail.com

Abdul Samad Mumtaz, Department of Plant Sciences, Faculty of Biological Sciences, Quaid-i-Azam University, Islamabad 45320, Pakistan. Email: asmumtaz@qau.edu.pk

Akhtar Munir, Department of Chemistry, Quaid-i-Azam University, Islamabad 45320, Pakistan. Email: akhtarmunir@qau.edu.pk

Ghulam Murtaza, School of Agriculture, Yunnan University, Kunming, Yunnan 650504, China. Email: murtazabotanist@gmail.com

Elshan Musazade, Key Laboratory of Soybean Molecular Design Breeding, Northeast Institute of Geography and Agroecology, Chinese Academy of Sciences, Changchun 130102, China. Email: elshan.musazade1@gmail.com

Akshita Patial, Department of Molecular Biology and Genetic Engineering, School of Bioengineering and Biosciences, Lovely Professional University, Jalandhar-Punjab, India; Fermentation and Phytofarming Technology Division, CSIR-Institute of Himalayan Bioresource Technology, Palampur, Banuri, Himachal Pradesh 176 061 India. Email: patyal.akshita17@gmail.com

Priyanka Paul, Department of Agronomy and Agroforestry, M.S. Swaminathan School of Agriculture, Centurion University of Technology and Management, Paralakhemundi, Odisha, India. Email: priyankapaul85090@gmail.com

Jeyabal Philomenathan Antony Prabhu, Department of Human Genetics and Molecular Biology, Bharathiar University, Coimbatore 641046, Tamil Nadu, India. Email: jantony.prabhu@gmail.com

Sumer Singh Punia, Department of Genetics and Plant Breeding, College of Agriculture Jobner, Sri Karan Narendra Agriculture University, Jobner, Rajasthan, India. Email: sspunia.pbg@sknau.ac.in

Mohammed Radi, Department of Plant Protection, Phytopathology Unit, Ecole Nationale d'Agriculture de Meknès, Km10, Rte Haj Kaddour, BP S/40, Meknès 50001, Morocco; Environment and

Valorization of Microbial and Plant Resources Unit, Faculty of Sciences, Moulay Ismail University, Meknes, Morocco. Email: simodefes13@gmail.com

Mohammad Ghani Raghib, Department of Horticulture, Faculty of Agriculture, Nangarhar University, Kabul-Jalalabad Highway, Daronta, Nangarhar, Afghanistan. Email: mohammadghaniraghib@gmail.com

K. Ramakrishna, Professor Jayashankar Telangana Agricultural University, Hyderabad, Telangana 500030, India. Email: palavaikrishna@gmail.com

K. Ramesh, ICAR – Indian Institute of Oilseeds Research, Hyderabad, Telangana 500030, India. Email: ramecheck@gmail.com

Ellapu Govardhana Rao, Department of Plant Breeding and Genetics, Assam Agricultural University, Jorhat, Assam, India. Email: egrpbg95@gmail.com

Iram Rasool, Plant Biotechnology Division, CSIR-Indian Institute of Integrative Medicine, Sanat Nagar, Srinagar, UT J&K, India. Email: khaniram523@gmail.com

A. Jaya Kishore Reddy, IARI Mega University, Hyderabad hub, ICAR – Central Research Institute for Dryland Agriculture, Hyderabad, Telangana 500059, India. Email: jayakishore1436@gmail.com

Challa Dinesh Kumar Reddy, Uttar Banga Krishi Viswavidyalaya, Pundibari, Cooch Behar, West Bengal 736165, India. Email: cdineshkumrreddy@gmail.com

Akarapu Sagar, IARI Mega University, ICAR-CAZRI, Jodhpur, Rajasthan 342003, India. Email: sagarakarapu11@gmail.com

Niharika Saharia, Department of Plant Breeding and Genetics, Assam Agricultural University, Jorhat, Assam, India. Email: niharikasaharia10@gmail.com

Saniya, Department of Horticulture, ICAR-National Research Centre for Grapes, Pune, Maharashtra, India. Email: saniyacg.17@gmail.com

Jhonson Raju Sankati, ICAR-Indian Agricultural Research Institute, New Delhi 110012, India. Email: santanukundu120@gmail.com

Debojit Sarma, Department of Plant Breeding and Genetics, Assam Agricultural University, Jorhat, Assam, India. Email: debojit.sarma@aau.ac.in

Muhammad Muzamil Sattar, Department of Plant Sciences, Faculty of Biological Sciences, Quaid-i-Azam University Islamabad 45320, Pakistan. Email: muzammilsattar55@gmail.com

Maksud Hasan Shah, Bidhan Chandra Krishi Viswavidyalaya, Mohanpur, Nadia, West Bengal, India. ORCID: 0000-0002-7860-9057. Email: maksudhasanshah@gmail.com

Vikas Sharma, Department of Molecular Biology and Genetic Engineering, School of Bioengineering and Biosciences, Lovely Professional University, Jalandhar, Punjab, India. Email: vikas.25269@lpu.co.in

Fatima-Ezzahra Soussani, Centre d'Agrobiotechnologie et Bioingénierie, Unité de Recherche labellisée CNRST (Centre AgroBiotech-URL-CNRST-05), Université Cadi Ayyad, Marrakech 40000, Morocco; Faculty of Sciences, Laboratory of Plant, Animal, and Agro-Industry Productions, University Ibn Toufail, Kenitra, Morocco. Email: fatimaezzahra.soussani@gmail.com

T.M. Tanjina Tarannum, Department of Agronomy, Bangladesh Agricultural University, Maimansing, Bangladesh. Email: tanjina.tarannum@gmail.com

Rajashree Mitra Thakur, Palli Siksha Bhavana (Institute of Agriculture), Visva-Bharati, Sriniketan, West Bengal 731236, India. ORCID: 0009-0004-3940-8983. Email: rajashreemitrathakur4@gmail.com

Zakir Ullah, Department of Plant Sciences, Faculty of Biological Sciences, Quaid-i-Azam University Islamabad 45320, Pakistan. Email: zakirullah@bs.qau.edu.pk

Rooma Waqar, Department of Botany, Faculty of Sciences, Mohi-ud-Din Islamic University, Nerian Sharif 12080, Azad Kashmir, Pakistan. Email: rooma.waqar667@gmail.com and rooma.waqar@miu.edu.pk

Muhammad Waseem, School of Breeding and Multiplication (Sanya Institute of Breeding and Multiplication), Sanya, China; College of Tropical Agriculture and Forestry, Hainan University, Sanya 572025, China; Fang Zhiyuan Academician Team Innovation Center, Hainan, China. Email: m.waseem.botanist@gmail.com

Preface

For decades, bioinformatics tools have made considerable contributions to the progress of biological sciences since the beginning of the post-genomic era. Their technical advancements have led to powerful approaches for investigators to accelerate knowledge exploration, particularly capable of dealing with big data, such as those resulting from multiple omics research, modeling and simulating complex biological networks, and predicting crucial players in cellular processes, such as the identification of transcription factors or hub proteins. In recent years, a subfield of bioinformatics, that is, artificial intelligence (AI), has caused a revolutionary technical breakthrough due to the rapid progress of computer science, such as machine learning (ML), deep learning (DL), generative AI, and upcoming technology of artificial general intelligence (AGI), which might play a role of game changer and are expected to make a comprehensive impact on scientific research, including biological science and, more specifically, part of plant biology.

Among AI technologies, DL is a subclass of ML, which is one of the promising algorithms for a wide range of applications. Over the years, DL models based on deep neural networks, deep belief networks, convolutional neural networks, and convolutional deep belief networks have been applied to assess their impact on advancing plant biology and agriculture, and considerable data and results have been published. However, it lacks reliable book resources for an efficient overview of this topic in the international market.

This book bridges the knowledge gap by collecting review-type chapters organized by professors and experts. It introduces the fundamentals of DL models and their current applications, followed by deeper insights into diverse subtopics within plant sciences. These include data science, functional genomics, multiple omics, stress physiology, plant pathologies (such as disease diagnosis), and plant–microbe interactions. The book also explores high-throughput technologies applied to phenotyping, synthetic biology, real-time monitoring, growth prediction, plant identification, plant classification, regulatory networks, transcription factors, and secondary metabolism, including the part of plant immunity.

This book presents DL-driven approaches for crop improvement, breeding, and smart agriculture. It covers methods and applications in image analysis, functional genomics, abiotic stress management, disease identification, detection, monitoring, and prediction, pest control strategies, unmanned aerial vehicles, genomic prediction, accelerated crop breeding, greenhouse management, weed management, tissue culture, and strategies for stress-smart and sustainable agriculture. Additionally, this book offers ways to construct environmental models and monitor plant growth for advancing crop production systems, which particularly have benefits for combating global climate change.

This book serves as an ideal reference for researchers working in plant biology, plant pathology, and agricultural sciences. It may also inspire their ideas for organizing future experiments to advance the progress of stress-smart and sustainable agriculture.

1 Deep Learning in Plant Abiotic Stress Management

Ishfaq Majid Hurrah[1], Iram Rasool[1], Tabasum Mohi Ud Din[2], Loulaq Banday[3], Tushar Kanti Mandal[4], and Sayanti Mandal[5]*

[1]Plant Biotechnology Division, CSIR-Indian Institute of Integrative Medicine, Sanat Nagar, Srinagar, UT J&K, India; [2]Department of Botany, Amar Singh College, Srinagar, UT J&K, India; [3]Department of Biotechnology, University of Kashmir, Srinagar, UT J&K, India; [4]ITH/IF, Wing "B" Building Congo, Embassy Tech Zone-SEZ, Rajiv Gandhi Infotech Park Phase II, Hinjawadi, Mercedes-Benz Research and Development India Pvt. Ltd, Pimpri-Chinchwad Maharashtra, India; [5]Department of Chemistry and Biochemistry, Sharda School of Engineering and Sciences, Sharda University, Greater Noida, Uttar Pradesh, India

Abstract

In recent decades, abiotic stress factors, including drought, salinity, heat, and nutrient deficiencies, have emerged as major challenges in global agriculture, significantly reducing crop yields and threatening food security. The advent of deep learning (DL) technologies presents a transformative opportunity for improving the detection, prediction, and management of these stresses. This chapter explores the application of DL in abiotic stress management, focusing on various architectures, such as convolutional neural networks and recurrent neural networks, for stress detection in crops using high-dimensional data from sensors, imagery, and environmental inputs. This chapter reviews case studies and real-world implementations where DL has been successfully applied to identify early stress indicators, optimize resource use, and support precision agriculture practices. Furthermore, future directions in DL, including the integration of multimodal data, edge artificial intelligence (AI) for real-time processing, climate-adaptive models, and explainable AI, are discussed. By enhancing early detection and enabling targeted interventions, DL has the potential to revolutionize how we manage abiotic stress and build climate-resilient, sustainable agricultural systems.

Keywords: agricultural sustainability, autonomous systems, climate change adaptation, convolutional neural networks (CNNs), crop stress management, deep learning (DL), edge computing, explainable AI (XAI), multimodal data integration, real-time processing, recurrent neural networks (RNNs)

1.1 Introduction

In recent decades, global agriculture has been increasingly challenged by a variety of environmental stressors that adversely impact plant development, physiological functioning, and, ultimately, crop productivity. The continuous rise in the global population, coupled with shifting climatic conditions, has placed unprecedented pressure on food systems to produce more from

*E-mail: mandalsayanti@gmail.com

DOI: 10.1079/9781836991120.0001

fewer resources while maintaining ecological sustainability (Jalil and Ansari, 2020; Prajapati *et al.*, 2024). Among the various environmental constraints, abiotic stresses, including drought, salinity, heat, cold, nutrient deficiency, and waterlogging, stand out as major limiting factors that significantly reduce plant growth and yield. These stressors provoke a wide range of physiological and biochemical responses in plants, such as impaired photosynthesis, oxidative stress, reduced water uptake, and metabolic imbalances (Habib-ur-Rahman *et al.*, 2022; Kumari *et al.*, 2022). If left unmanaged, they can lead to yield reductions, stunted growth, and, in extreme cases, total crop failure. The problem is exacerbated by climate change, which has intensified the frequency, severity, and unpredictability of abiotic stress events. Irregular precipitation patterns, increasing global temperatures, and extreme weather phenomena now occur more frequently, further compromising agricultural stability. According to the Food and Agriculture Organization (FAO), abiotic stresses are responsible for over 50% of yield losses in major crops annually, highlighting the urgent need for innovative monitoring and management strategies to safeguard food security (Oshunsanya *et al.*, 2019; Kopecká *et al.*, 2023).

Historically, the detection and management of abiotic stress in plants have relied on traditional techniques such as manual field scouting for visual symptom identification, laboratory-based physiological and biochemical assays, remote sensing technologies (satellite or unmanned aerial vehicle (UAV)-based imagery), and direct physiological measurements, such as chlorophyll fluorescence or leaf water potential. While these methods provide valuable information, they are often labor-intensive, time-consuming, costly, and spatially limited (Guo *et al.*, 2022; Zhang *et al.*, 2024). Manual inspections are subjective and prone to human error, particularly when differentiating between abiotic and biotic stresses, which can exhibit similar visible symptoms. Moreover, many abiotic stress responses manifest subtly or asymptomatically in the early stages, making early diagnosis difficult through conventional means. To address these limitations, the integration of artificial intelligence (AI) technologies in agriculture has emerged as a transformative solution. AI, particularly its subfields, machine learning (ML) and deep learning (DL), enables data-driven agricultural decision-making through the analysis of large and complex datasets (Ahmed, S.F. *et al.*, 2023; Houetohossou *et al.*, 2023). Among these, DL has shown particular promise due to its ability to process high-dimensional, heterogeneous, and unstructured data (e.g. images, sensor data, and time-series signals) with minimal manual feature engineering. DL employs multilayered artificial neural networks to autonomously learn hierarchical features from raw data, enabling accurate, scalable, and robust performance across a range of applications. In the context of abiotic stress detection, convolutional neural networks (CNNs) have been widely used for image-based classification of stress symptoms, such as leaf discoloration or wilting, using RGB, hyperspectral, multispectral, thermal, and chlorophyll fluorescence images. Similarly, recurrent neural networks (RNNs) and long short-term memory (LSTM) architectures have proven effective in modeling temporal stress responses, such as changes in canopy temperature or soil moisture over time. These models have demonstrated the ability to detect subtle stress cues often invisible to the naked eye, facilitating early diagnosis and targeted intervention (Gao *et al.*, 2023; Kumar *et al.*, 2023).

Recent advancements in cloud computing, edge computing, the Internet of Things (IoT), and automated phenotyping platforms have enabled the real-time deployment of DL models in field environments. For example, IoT-enabled sensors can continuously capture environmental variables (e.g. soil moisture, temperature, and humidity) and transmit data to AI systems for analysis. Edge devices, such as mobile phones and drones equipped with embedded AI, enable real-time stress detection without requiring constant internet connectivity (Fan *et al.*, 2021; Ahmed *et al.*, 2024). Automated phenotyping platforms, including ground robots and UAVs, allow high-throughput image and sensor data collection across large agricultural landscapes. This integration facilitates site-specific and crop-specific management decisions, optimizing input use (e.g. irrigation and fertilizers), reducing environmental impact, and enhancing overall crop resilience to abiotic stress conditions. These systems not only enhance stress detection accuracy but also enable site-specific management practices that optimize input use, reduce environmental impact, and improve crop resilience (Sharma and Shivandu, 2024; Pintus *et al.*, 2025) (Fig. 1.1).

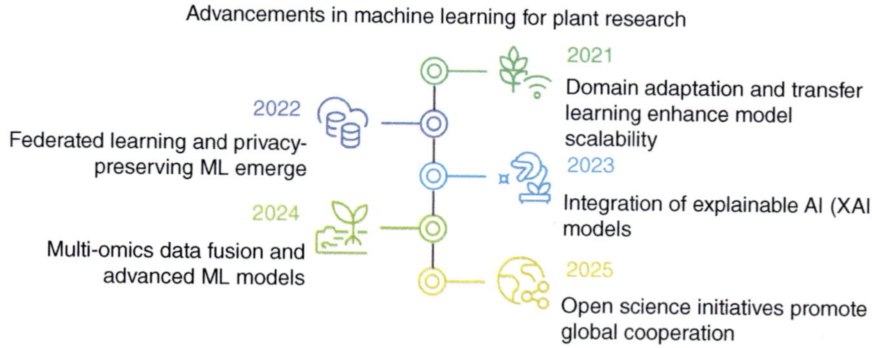

Fig. 1.1. Developmental stages of adapting machine learning. (Author's own image.)

1.2 Basics of Abiotic Stress in Plants

Abiotic stress refers to the negative impact of non-living environmental factors on plant growth, development, and productivity. Unlike biotic stress, which arises from pathogens and pests, abiotic stress is caused by physical and chemical conditions such as temperature extremes, drought, salinity, heavy metals, radiation, nutrient imbalances, and flooding. These stressors act either individually or in combination, often leading to significant physiological, morphological, and molecular disruptions in plants. The severity and duration of exposure to abiotic stresses largely determine the degree of damage and the plant's ability to recover or adapt (Farooq *et al.*, 2022; Kumari *et al.*, 2022; Saud *et al.*, 2022).

1.2.1 Types of abiotic stress

Several types of abiotic stress affect plants, each exerting specific effects on plant metabolism and survival. Drought stress is one of the most prevalent and damaging abiotic stressors, especially in arid and semiarid regions. It limits water availability, leading to reduced cell turgor, stomatal closure, impaired photosynthesis, and the accumulation of reactive oxygen species (ROS) (Teshome *et al.*, 2020; Bashir *et al.*, 2021; Raza *et al.*, 2024; Raza *et al.*, 2025). Salinity stress results from high concentrations of soluble salts in the soil, particularly sodium chloride. It affects plants by inducing osmotic stress, ionic toxicity, and nutrient imbalance, which together inhibit germination, root development, and overall biomass accumulation (Arif *et al.*, 2020; Dey *et al.*, 2021). Heat stress occurs when temperatures exceed the optimal range for plant metabolism, causing protein denaturation, membrane instability, and reduced pollen viability. Prolonged heat exposure can severely impair reproductive success and yield (Mazur *et al.*, 2020; Sadok *et al.*, 2021). Cold stress, including chilling and freezing, disrupts membrane fluidity, reduces enzymatic activity, and hampers water uptake, often leading to tissue damage and cell death. Waterlogging or submergence causes hypoxic or anoxic conditions in the root zone, affecting root respiration and nutrient uptake and triggering ethylene accumulation, which can lead to premature senescence (Xu *et al.*, 2022; Manghwar *et al.*, 2024). Nutrient stress arises from deficiencies or toxicities of essential macro- or micronutrients. For example, nitrogen deficiency leads to chlorosis and poor vegetative growth, while excess boron or aluminum can be toxic to roots. Each type of abiotic stress presents unique challenges, but many share common downstream effects, such as oxidative stress and metabolic imbalances, which form the basis for shared molecular and physiological response pathways (Abbas *et al.*, 2021; Dhaliwal *et al.*, 2022).

1.2.2 Physiological and biochemical effects

Abiotic stresses alter plant physiology at multiple levels. A common and early response is stomatal closure, which not only reduces transpiration and conserves water but also limits CO_2 uptake, thereby affecting photosynthesis. Chlorophyll degradation, often observed as yellowing or necrosis, is another typical symptom, indicating compromised photosynthetic capacity. Stress-induced accumulation of ROS, such as superoxide radicals and hydrogen peroxide, leads to oxidative damage of lipids, proteins, and DNA (Wahab *et al.*, 2022; Al-Khayri *et al.*, 2023; Shakespear *et al.*, 2025). To counteract this, plants activate antioxidant defense systems, including enzymatic components, such as superoxide dismutase, catalase, and peroxidases, and non-enzymatic antioxidants, such as ascorbate and glutathione. In addition, plants synthesize various osmoprotectants (e.g. proline, glycine betaine, and trehalose) and stress proteins (e.g. heat shock proteins and late embryogenesis abundant proteins) that stabilize cellular structures and improve stress tolerance. Hormonal regulation, particularly involving abscisic acid, ethylene, salicylic acid, and jasmonic acid, also plays a critical role in signaling stress perception and initiating protective responses (Wu *et al.*, 2022; Sarfraz *et al.*, 2025).

1.2.3 Molecular responses and signaling pathways

At the molecular level, plants respond to abiotic stress through a highly coordinated network of gene expression changes, signal transduction cascades, and transcriptional reprogramming. Stress perception typically begins at the cell surface via sensors and membrane receptors, which activate intracellular signaling molecules such as calcium ions (Ca^{2+}), mitogen-activated protein kinases (MAPKs), and phospholipids. These signals lead to the activation of transcription factors such as DREB (dehydration-responsive element binding), NAC, MYB, and bZIP families, which regulate downstream genes involved in stress adaptation (Saijo and Loo, 2020; Zhang *et al.*, 2023). The expression of stress-inducible genes enhances the synthesis of protective metabolites, transporters, detoxification enzymes, and structural proteins, thereby facilitating physiological adjustments and damage mitigation (Fig. 1.2). The emerging field of omics technologies, including transcriptomics, proteomics, metabolomics, and ionomics, has revealed the complexity and dynamic nature of plant responses to abiotic stress. Integration of these data layers has enhanced our understanding of stress resilience mechanisms and provided potential targets for crop improvement through genetic engineering or selective breeding (Raza *et al.*, 2021; Roychowdhury *et al.*, 2023).

1.2.4 Impact on agriculture and food security

The cumulative impact of abiotic stress on crops is profound, leading to significant reductions in yield, quality, and economic returns. Crops exposed to stress during critical stages, such as flowering or grain filling, may fail to develop viable seeds or fruits. In developing countries, where resources for irrigation and soil management are limited, the effects of abiotic stress are even more pronounced, exacerbating issues of poverty and malnutrition. With climate models projecting increased temperatures, erratic rainfall, and soil degradation in the coming decades, the vulnerability of agricultural systems to abiotic stress is expected to worsen. This necessitates the development of resilient crop varieties and adaptive management strategies, underpinned by advanced sensing, modeling, and decision support technologies (Dhaliwal *et al.*, 2022; Kumari *et al.*, 2022).

1.3 DL Architectures for Abiotic Stress Detection

DL architectures have revolutionized the way plant abiotic stress is detected and managed by enabling automated, accurate, and scalable analysis of complex agricultural data. Leveraging models such as CNNs for image-based classification and RNNs or LSTM networks for time-series analysis, these systems can detect subtle stress indicators from various data modalities—including RGB, thermal, multispectral, and environmental sensor data. Advanced architectures such as generative adversarial networks (GANs) and hybrid models further enhance stress detection capabilities by enabling data augmentation, improving model robustness, and integrating multisource information. As a result, DL has become a cornerstone of modern abiotic stress monitoring, offering real-time, non-invasive, and high-throughput solutions for precision agriculture (Gao *et al.*, 2020; Houetohossou *et al.*, 2023).

1.3.1 Overview of DL in agriculture

DL, a subfield of ML inspired by the structure and function of the human brain, has emerged

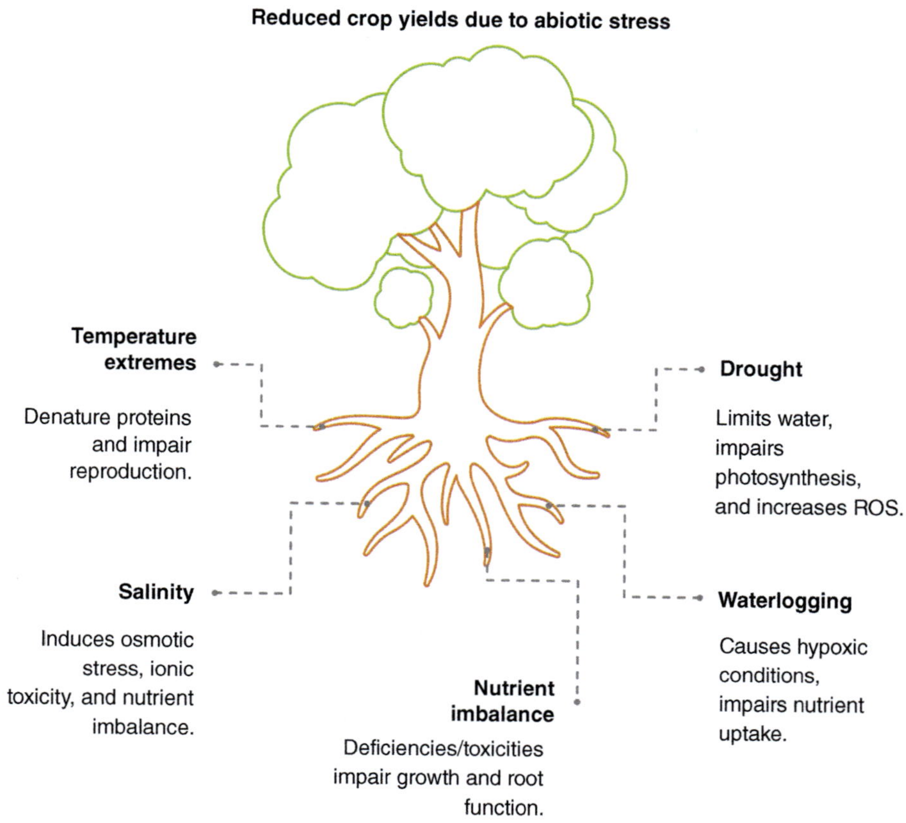

Fig. 1.2. Impact of abiotic stress on crop yield. (Author's own image.)

as a powerful tool in agricultural research and practice. It excels at learning hierarchical features from raw data and handling large, complex, and high-dimensional datasets. Unlike traditional ML approaches that require manual feature extraction and domain-specific expertise, DL models can autonomously learn relevant features from inputs such as images, time-series signals, or sensor data. This capability makes DL particularly suitable for detecting subtle and complex patterns associated with abiotic stress responses in plants. In agriculture, DL has been widely adopted for tasks such as crop disease detection, weed classification, yield prediction, and soil health monitoring (Fig. 1.3). When applied to abiotic stress detection, DL models help identify early stress symptoms, distinguish between different stress types, and quantify the severity of plant stress, often with higher precision and reliability than traditional methods (Santos *et al.*, 2020; Benos *et al.*, 2021; Saleem *et al.*, 2021).

1.3.2 Convolutional neural networks (CNNs)

CNNs are the most widely used DL architecture in visual recognition tasks, including plant stress detection. A CNN comprises multiple layers of interconnected nodes designed to extract spatial hierarchies of features—ranging from edges and textures in early layers to complex patterns, such as stress-specific leaf discoloration in deeper layers. CNNs have shown remarkable success in classifying plant stress from various imaging modalities, including RGB (color), multispectral, hyperspectral, thermal, and fluorescence images. For example, a CNN can be trained to distinguish between drought-stressed and healthy plants by learning from annotated images that capture changes in leaf color, wilting patterns, or thermal anomalies. These models can detect subtle stress symptoms earlier than human experts, enabling timely intervention. Variants such as

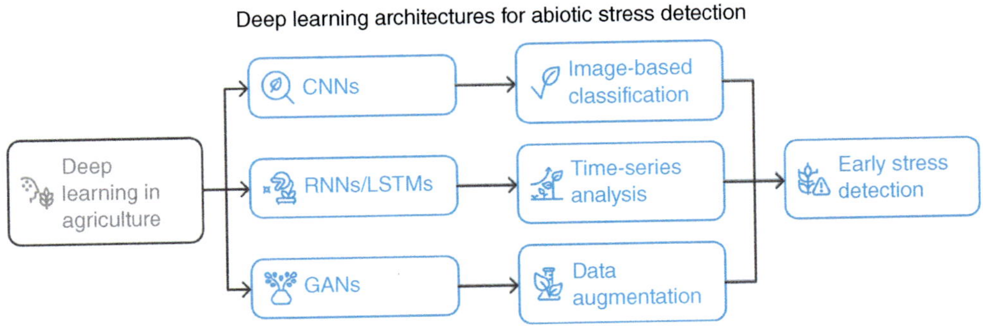

Fig. 1.3. Architecture of abiotic stress detection through deep learning. (Author's own image.)

ResNet, VGGNet, InceptionNet, and MobileNet have been adopted to improve classification performance, model depth, or computational efficiency. Transfer learning using pretrained CNNs fine-tuned on agricultural datasets has also been effective, especially when labeled stress-specific datasets are limited (Kujawa and Niedbała, 2021; Altalak *et al.*, 2022).

1.3.3 Recurrent neural networks (RNNs) and long short-term memory networks (LSTMs)

Abiotic stress is not always instantaneous; it often develops over time. This temporal dimension makes RNNs, particularly LSTM networks, valuable in stress modeling. These architectures are designed to capture sequential dependencies and time-series dynamics, making them suitable for analyzing environmental sensor data or time-lapse imaging. LSTMs are effective in modeling plant responses to time-varying stressors such as soil moisture decline, temperature fluctuations, or nutrient depletion. For instance, LSTM models can track the progression of drought stress by analyzing soil moisture levels, canopy temperature, and plant height over days or weeks. This time-aware prediction enables a more holistic understanding of stress development and enhances precision in stress forecasting and resource planning (Jaiswal *et al.*, 2022; Alzakari *et al.*, 2024).

1.3.4 Generative and hybrid architectures

Beyond conventional classification and regression tasks, generative models such as GANs have been explored in plant phenotyping and stress augmentation. GANs can generate synthetic images of stressed crops, which can be used to augment training datasets and improve model robustness in scenarios where real stress images are scarce. Hybrid DL architectures that combine CNNs with LSTMs or integrate DL with traditional ML techniques (e.g. random forests and support vector machines) have also been proposed. These models exploit the strengths of multiple learning paradigms and are particularly useful in multimodal stress analysis, where spatial, temporal, and contextual data need to be processed simultaneously (Dhanya *et al.*, 2022; Bal and Kayaalp,2024).

1.4 Datasets and Data Acquisition Methods for DL in Abiotic Stress Detection

The success of DL models in abiotic stress detection heavily depends on the quality, diversity, and scale of the datasets used for training and validation. These datasets encompass various data types—such as RGB, multispectral, hyperspectral, thermal images, and time-series sensor data—acquired through ground-based sensors, UAVs, satellites, and controlled environments. Effective annotation and preprocessing are critical for improving model accuracy and generalizability, especially given challenges such as data noise, imbalance, and variability across stress types and field conditions. While public datasets and phenotyping platforms have made significant contributions, there remains a need for more comprehensive, standardized, and open-access datasets tailored to abiotic stress. Future advances in real-time data acquisition, synthetic data generation, and collaborative data sharing will further accelerate the deployment of DL in stress-resilient agriculture (Al-Tamimi *et al.*, 2022; Zhang *et al.*, 2024).

1.4.1 Importance of high-quality datasets

The foundation of any effective DL model lies in the quality and quantity of the data used for training and validation. In the context of plant abiotic stress detection, datasets must be representative of diverse environmental conditions, crop types, developmental stages, and stress intensities. These datasets are crucial for enabling models to learn how to distinguish between various abiotic stressors, such as drought, salinity, or heat stress, and to generalize across field conditions. However, acquiring high-quality annotated data remains one of the major challenges in applying DL in agriculture, primarily due to the complexity and variability of plant stress responses and the need for expert interpretation during the labeling process (Azimi *et al.*, 2021; Gill *et al.*, 2022).

1.4.2 Types of data for abiotic stress detection

A variety of data types are used to detect abiotic stress in plants, each providing unique insights into plant health and physiology. RGB images are the most common and accessible form of data, capturing visible stress symptoms such as chlorosis, necrosis, and wilting. They are widely used due to the availability of standard cameras on smartphones, drones, and digital systems (Zubler and Yoon, 2020; Chandel *et al.*, 2022). Multispectral and hyperspectral images provide rich spectral information across narrow and broad wavelength bands, respectively. These images can detect subtle changes in pigment composition, water content, and nutrient levels before symptoms become visible to the naked eye (Gou *et al.*, 2024; Zhang *et al.*, 2024). Thermal imaging is used to monitor canopy temperature variations, which are indicative of stomatal conductance and transpiration rates. Elevated canopy temperature often suggests drought or heat stress (Wen *et al.*, 2024; Zhang *et al.*, 2024). Chlorophyll fluorescence imaging enables the assessment of photosynthetic efficiency under stress conditions and is particularly useful in controlled experimental setups (Long and Ma, 2022; Deng *et al.*, 2024). Time-series sensor data, such as soil moisture, temperature, humidity, and nutrient levels, provide valuable contextual information for tracking the onset and progression of stress over time. These data are typically acquired using IoT-enabled field sensors. Text and metadata, including cultivar information, phenological stages, management practices, and geographic coordinates, are often used as auxiliary inputs to enhance model precision and contextual relevance (Ali *et al.*, 2024; Wang *et al.*, 2024).

1.4.3 Data acquisition platforms

Modern data acquisition for abiotic stress detection relies on a range of platforms, which can be broadly classified into ground-based, aerial, and satellite systems. Ground-based platforms include handheld devices, stationary field sensors, and automated phenotyping rigs. These systems provide high-resolution data and precise control but are typically limited in spatial coverage (Gao *et al.*, 2020; Al-Tamimi *et al.*, 2022). UAVs, or drones, equipped with RGB, multispectral, or thermal cameras, have become essential tools for rapid, high-throughput phenotyping. UAVs offer the advantage of covering large areas efficiently while capturing fine-scale variability in plant stress indicators (Table 1.1). Satellite-based remote sensing provides large-scale and repeatable data acquisition but at lower spatial resolutions compared to UAVs (Fevgas *et al.*, 2023; Sharma *et al.*, 2025). Platforms such as Sentinel-2, Landsat, and MODIS offer multispectral data suitable for regional-level stress monitoring, especially when integrated with ground-truth data. Controlled environment systems, including growth chambers and greenhouse-based phenotyping platforms, are used to generate high-quality labeled datasets under precisely managed stress conditions. These datasets are critical for model training and benchmarking, especially when real-field annotation is difficult or impractical (Segarra *et al.*, 2020; Kganyago, 2022).

Table 1.1. Summary of key aspects of datasets and data acquisition methods for deep learning in abiotic stress detection. Author's own table.

Section	Summary	References
Importance of high-quality datasets	Diverse, annotated datasets are essential for training DL models to detect drought, salinity, heat, etc., but acquiring them is challenging due to complexity and the need for expert input.	Azimi *et al.* (2021); Gill *et al.* (2022)
Data acquisition platforms	Ground-based (sensors, phenotyping rigs), UAVs/drones (RGB, multispectral, thermal), satellites (Sentinel-2, Landsat, MODIS), and controlled environments (greenhouse, chambers).	Gao *et al.* (2020); Fevgas *et al.* (2023)
Annotation and labeling	Manual, semiautomated, or crowdsourced annotation with segmentation, and bounding boxes; need for standardization and expert validation to reduce variability.	Singh *et al.* (2021); Al-Tamimi *et al.* (2022)
Public datasets and repositories	PlantVillage, UAV-acquired datasets, PHENOROB, LemnaTec, TERRA-REF; but still lacking multi-stress, large-scale, field-labeled datasets.	Cándido-Mireles *et al.* (2023); Orka *et al.* (2023)

DL, deep learning; UAV, unmanned aerial vehicle.

1.4.4 Annotation and labeling techniques

A critical step in dataset preparation is annotation, accurately labeling data samples to reflect the type, stage, and severity of abiotic stress. Annotation can be manual, semiautomated, or crowdsourced. Manual annotation by experts ensures accuracy but is time-consuming and labor-intensive. Semiautomated methods leverage preexisting algorithms or AI-based tools to assist in the annotation process, often using segmentation masks, bounding boxes, or heat maps to localize stress symptoms. Crowdsourcing platforms and citizen science initiatives are also being explored to label large volumes of plant images. However, inter-annotator variability and misclassification risks must be carefully managed through consensus approaches and expert validation. Additionally, annotation consistency and metadata standardization are essential for building reproducible and interoperable datasets (Singh *et al.*, 2021; Al-Tamimi *et al.*, 2022).

1.4.5 Publicly available datasets and repositories

In recent years, several publicly available datasets and repositories have emerged, supporting the application of DL in plant stress detection. The PlantVillage Dataset contains over 50,000 annotated images of healthy and diseased leaves, including abiotic stress classes for some crops. Among UAV-based datasets, some research groups have released drone-acquired multispectral or thermal datasets targeting drought and nutrient stress in crops such as corn and wheat (Cándido-Mireles *et al.*, 2023; Orka *et al.*, 2023). High-throughput plant phenotyping platforms, such as PHENOROB, LemnaTec, and TERRA-REF, have contributed both image-based and environmental data for DL model development. Despite these resources, there remains a significant gap in the availability of large-scale, high-resolution datasets specifically labeled for multiple abiotic stresses under diverse field conditions. Therefore, continuous efforts in data sharing, collaborative labeling, and open-source dataset development are critical to advancing the field (Demidchik *et al.*, 2020; Watt *et al.*, 2020; Danilevicz *et al.*, 2021).

1.5 Applications of DL in Abiotic Stress Management

DL has found a wide range of applications in abiotic stress management (Table 1.2) by enabling

Table 1.2. Summary of deep learning applications in abiotic stress management. Author's own table.

Application area	Summary of role	References
Early detection and classification	DL models (e.g. CNNs) detect subtle early-stage stress symptoms (color, texture, shape changes) from RGB, thermal, or multispectral images, enabling timely intervention.	Gou *et al.* (2024); Zhang *et al.* (2024)
Stress severity assessment and quantification	DL frameworks (regression, multi-class classifiers) quantify stress intensity, for example, ranking drought severity or nutrient deficiency from hyperspectral data, supporting variable-rate input management.	Rico-Chávez *et al.* (2022); Esgario *et al.* (2024)
Stress-type differentiation and multi-stress classification	DL models distinguish between overlapping stress symptoms (e.g. drought vs nitrogen deficiency) and identify co-occurring stresses using multi-output or attention-based networks.	Islam *et al.* (2024); Varadharajan *et al.* (2025)
Integration with IoT and precision agriculture	DL integrated with IoT and edge devices (UAVs, field sensors) enables real-time, automated stress monitoring and management, improving sustainability and input efficiency.	Kumari and Nafchi (2024); Delfani *et al.* (2024)
Yield prediction and risk forecasting	RNNs, LSTMs, and other DL architectures use time-series and remote sensing data to forecast yield losses under stress, aiding farm management, supply chain planning, and policy decisions.	Kumar *et al.* (2023); Ali *et al.* (2024)
Breeding and genomic applications	High-throughput DL image analysis phenotypes stress-tolerant genotypes; integrated with genomic data, DL predicts genotype–phenotype associations, accelerating climate-resilient breeding programs.	Niazian and Niedbała (2020); Eftekhari *et al.* (2024)

CNNs, convolutional neural networks; DL, deep learning; IoT, Internet of Things; LSTMs, long short-term memory networks; RNNs, recurrent neural networks; UAVs, unmanned aerial vehicles.

early detection, quantification, and differentiation of stress symptoms in crops. From classifying stress types to assessing severity and predicting yield losses, DL models provide crucial insights that support proactive and site-specific management practices. The integration of DL with IoT systems and precision agriculture technologies further enhances real-time monitoring and decision-making, improving input efficiency and sustainability. Additionally, DL is transforming plant breeding efforts by accelerating the identification of stress-tolerant genotypes.

1.5.1 Early detection and classification of stress symptoms

One of the most impactful applications of DL in abiotic stress management is the early detection and accurate classification of stress symptoms in crops. Traditional approaches often rely on visual assessments or delayed physiological measurements, which may not capture the subtle, early-stage symptoms of stress. DL models, especially CNNs, can process large volumes of image data to detect slight variations in leaf color, texture, or shape that are indicative of drought, heat, or salinity stress, often well before they are visible to the human eye. These models have been successfully applied in greenhouse and field settings to identify stress-affected plants from RGB, thermal, or multispectral images, enabling timely interventions that minimize yield loss (Gou *et al.*, 2024; Zhang *et al.*, 2024).

1.5.2 Stress severity assessment and quantification

Beyond binary classification (stress vs no stress), DL models can be trained to assess the severity of abiotic stress. Regression-based deep networks and multi-class classification frameworks allow quantification of stress intensity, helping researchers and farmers understand how strongly a crop is affected. For example, using hyperspectral imaging data and DL algorithms, researchers can rank plants according to drought severity levels or quantify the extent of nutrient deficiency in a given field. This application is particularly useful in variable-rate management systems, where input applications (e.g. water and fertilizers) must be adjusted according to site-specific crop needs (Esgario *et al.*, 2020; Rico-Chávez *et al.*, 2022).

1.5.3 Stress-type differentiation and multi-stress classification

Abiotic stressors often manifest with overlapping visual and physiological symptoms, making it difficult to differentiate between them. DL models, when trained with properly labeled and diverse datasets, can learn to distinguish between various types of stress, such as differentiating between water stress and nitrogen deficiency, which may both cause leaf yellowing but with different patterns and progression rates. Furthermore, multi-output neural networks and attention-based DL models have been proposed to simultaneously identify co-occurring stresses, an essential capability for real-world farming conditions, where multiple abiotic stressors often act synergistically (Islam *et al.*, 2024; Varadharajan *et al.*, 2025).

1.5.4 Integration with IoT and precision agriculture systems

DL models are increasingly being integrated with IoT systems and smart farming technologies for real-time abiotic stress monitoring and decision-making. For instance, edge AI devices embedded in UAVs or field-deployed sensor nodes can run lightweight DL models to process images and environmental data on-site. These systems continuously monitor crop conditions and automatically flag signs of stress, sending alerts to farmers or triggering automated interventions such as localized irrigation or fertilizer application. This integration enables a shift from reactive to proactive management, significantly improving input efficiency, sustainability, and resilience (Delfani *et al.*, 2024; Kumari and Nafchi, 2024).

1.5.5 Yield prediction and risk forecasting

Abiotic stresses can drastically reduce crop yield if not managed appropriately. DL models are also being employed to predict potential yield losses under ongoing or forecasted stress conditions. Using time-series sensor data, remote sensing imagery, and historical weather patterns, DL architectures—particularly RNNs and LSTM networks—can model plant growth trajectories and simulate yield outcomes. These predictive insights are vital for planning harvest schedules, adjusting market supply chains, and making informed policy decisions in the face of climate-induced stress variability (Kumar *et al.*, 2023; Ali *et al.*, 2024).

1.5.6 Breeding and genomic applications

In plant breeding, DL models are being used to phenotype large populations for abiotic stress tolerance traits under both controlled and field conditions. Automated, high-throughput image analysis powered by DL helps breeders identify elite genotypes with superior resilience to drought, salinity, or heat. Moreover, when combined with genomic data, DL algorithms can predict genotype–phenotype associations and guide the selection of stress-tolerant cultivars. This accelerates the breeding cycle and supports the development of next-generation crops optimized for climate resilience (Niazian and Niedbała, 2020; Eftekhari *et al.*, 2024).

1.5.7 Decision support systems (DSS) for farmers

DL-based tools are increasingly being integrated into mobile apps, cloud platforms, and web-based DSS to assist farmers in stress identification and management. These systems provide actionable recommendations, such as when and where to irrigate, fertilize, or apply foliar treatments. By delivering real-time, site-specific guidance, such tools democratize access to cutting-edge technologies, especially in regions with limited extension services. Some applications are even capable of offline operation using embedded AI chips, making them accessible in remote or connectivity-limited areas (Van, 2022; Anand and Sandhu, 2024).

1.6 Challenges in DL for Abiotic Stress Management

While DL offers promising solutions for abiotic stress management in agriculture, several challenges remain, including data scarcity, model generalization, sensor quality issues, and the need for real-time processing. Furthermore, the interpretability of DL models and their integration with other agricultural technologies are critical for widespread adoption. However, advancements in synthetic data generation, federated learning, hybrid models, and AI hardware are paving the way for more efficient and scalable solutions. Addressing these challenges will enhance the potential of DL to revolutionize stress management, supporting more resilient and sustainable agricultural practices.

1.6.1 Data-related challenges

Despite the tremendous potential of DL in abiotic stress management, data-related challenges remain one of the primary hurdles. First and foremost, data scarcity in certain geographic regions and for specific stress types remains a significant issue. High-quality, annotated datasets for diverse crop types, stress conditions, and environmental factors are limited. This lack of data, especially for rare or extreme stress events, can hinder the generalization capabilities of DL models, making them less effective in real-world applications. Data imbalance is another challenge, where certain types of stress (e.g. drought) are often overrepresented in datasets, while others (e.g. nutrient deficiency) are underrepresented. This imbalance can cause DL models to perform poorly for underrepresented stress types, reducing the overall accuracy of the system. Additionally, variability in environmental conditions complicates data collection, as stress responses in plants can vary significantly based on soil types, weather conditions, and agronomic practices, adding another layer of complexity for model training and deployment (Williamson *et al.*, 2023; Murmu *et al.*, 2024).

1.6.2 Model generalization and overfitting

Another critical challenge in applying DL to plant stress management is ensuring model generalization. DL models are powerful, but they require large amounts of diverse data to avoid overfitting—a common problem in which models perform well on training data but fail to generalize to unseen data. This issue is particularly problematic in agricultural settings, where field conditions can differ widely, even within the same region. As a result, a model trained on a limited dataset from one geographic area or crop variety may not perform effectively in a different context. Moreover, overfitting can occur when models learn noise or irrelevant features in the data, making them overly sensitive to specific stress symptoms that do not generalize well to other plant varieties or environments. Transfer learning and domain adaptation methods are promising solutions to address this challenge, as they allow models to leverage pretrained knowledge from one context and apply it to new, unseen scenarios (Ali *et al.*, 2024; Zhang *et al.*, 2024).

1.6.3 Sensor and image quality issues

While the use of high-resolution sensors and imaging technologies (e.g. RGB, multispectral, hyperspectral, and thermal cameras) has advanced significantly, challenges still exist in sensor quality and data consistency. Variations in image quality due to environmental factors such as lighting conditions, sensor calibration, or motion blur can introduce noise and inaccuracies into model predictions. Inconsistent image quality, such as differences in resolution or sensor specifications, further complicates the integration of data from various sources, making it difficult to build unified, reliable models. Moreover, capturing the spatiotemporal variability of abiotic stress across large fields remains a challenge. Stress symptoms may not appear uniformly across a crop, and the effects of abiotic stressors such as drought or salinity often vary over time and space. This heterogeneity requires DL models to account for intricate spatial relationships, which can be computationally expensive and require significant data preprocessing (Azimi *et al*., 2021; Al-Tamimi *et al*., 2022).

1.6.4 Scalability and real-time processing

For DL models to be widely adopted in real-world agriculture, they must be scalable and capable of real-time processing. While advances in cloud computing, edge AI, and the IoT are enabling real-time data collection and processing, deploying DL models at scale remains challenging, especially in large-scale farming operations. Real-time analysis of large volumes of data, including high-resolution images and time-series sensor data, demands significant computational resources. Edge computing offers a potential solution by performing analysis directly on devices such as UAVs, drones, or smart sensors, thereby reducing dependency on cloud infrastructure and enabling faster decision-making. However, edge AI also brings its own set of challenges, including limited processing power, storage capacity, and the need for low-latency communication systems. Designing energy-efficient and cost-effective edge devices that can perform complex DL tasks in real time remains an area of active research (Choudhary *et al*., 2023; Angidi *et al*., 2025).

1.6.5 Interpretability and model transparency

While DL models have demonstrated impressive accuracy, their interpretability—the ability to understand how the model makes decisions—remains a significant barrier to trust and adoption, especially in high-stakes environments such as agriculture. Farmers and agricultural stakeholders need to understand why a model classifies a plant as stressed or predicts a certain yield loss to make informed decisions. Developing explainable AI (XAI) methods for DL models is essential for increasing user confidence in DL-based systems. Visualizations such as saliency maps or attention mechanisms can help illustrate which features or areas of an image contributed most to a model's decision. Additionally, integrating traditional agronomic knowledge into DL models may improve interpretability and bridge the gap between data-driven predictions and expert understanding (Mostafa *et al*., 2023; Patra and Sahoo, 2024).

1.6.6 Integration with other agricultural technologies

The full potential of DL for abiotic stress management can be realized only when integrated with other emerging agricultural technologies. This includes precision agriculture tools, such as automated irrigation systems, drones, and robots, which rely on real-time data and feedback to optimize crop management practices. Combining DL models with ML-based DSS, climate modeling tools, and crop simulation models can provide more accurate predictions of stress impacts under future climatic conditions. Farm management systems that combine DL with Geographic Information Systems and weather forecasting tools can further enhance decision-making by providing a comprehensive view of how abiotic stress affects crop growth and yield at both local and regional scales. However, the integration of these technologies often requires significant investment in infrastructure, data sharing, and interoperability, which can be a barrier for smallholders and those in resource-limited settings (Kamarudin *et al*., 2021; Wang *et al*., 2024).

1.7 Case Studies and Real-World Implementations

The real-world application of DL in abiotic stress management has shown significant promise across diverse agricultural systems. Case studies highlight the power of DL in detecting various stressors such as drought, salinity, heat, and nutrient deficiencies in crops, such as wheat, rice, tomato, corn, and sugarcane. These systems, often combined with UAVs, drones, thermal imaging, and hyperspectral sensors, enable early stress detection, real-time decision-making, and precise intervention strategies. From precision irrigation to climate change adaptation, DL is revolutionizing how farmers manage crop health and optimize resources, ultimately leading to improved productivity and sustainability in modern agriculture.

1.7.1 Early detection of drought stress in wheat using UAV imagery and DL

One notable case study involved the use of UAVs combined with DL for the early detection of drought stress in wheat crops. In this study, researchers utilized a UAV equipped with multispectral and thermal cameras to capture high-resolution images of wheat fields. These images were then processed using a CNN to identify early stress indicators, such as changes in leaf color, temperature, and water content, which are subtle in the early stages of drought. The model achieved high accuracy in identifying drought-affected areas even before visible symptoms were noticeable, allowing for targeted irrigation and other interventions. This case highlights the potential of DL in transforming precision agriculture, where rapid and accurate identification of stressors can reduce water usage and minimize crop loss (Su *et al.*, 2020; Dong *et al.*, 2024).

1.7.2 Salinity stress detection in rice using hyperspectral imaging and DL

Another example comes from a study that utilized hyperspectral imaging to detect salinity stress in rice fields. Salinity can significantly impact rice growth, and its early detection is critical for mitigating yield losses. The researchers applied hyperspectral sensors to capture detailed spectral information across various wavelengths, which was then fed into a DL model (a CNN variant). The model was able to distinguish between stressed and non-stressed plants based on subtle changes in spectral reflectance associated with salinity-induced physiological alterations. This technology was deployed in large rice paddies, where the model's ability to process hyperspectral data in real time allowed farmers to apply corrective measures, such as adjusting irrigation water salinity or selecting more resistant varieties for future planting cycles (Feng *et al.*, 2020; Gou *et al.*, 2024).

1.7.3 Monitoring heat stress in tomato plants using RGB and thermal imaging

A successful case study in greenhouse settings involved monitoring of heat stress in tomato plants using both RGB and thermal imaging cameras. Heat stress is a significant limiting factor for tomato production, especially in regions experiencing increasing temperatures due to climate change. Researchers integrated DL algorithms with RGB and thermal images to monitor changes in leaf temperature, leaf wilting, and chlorophyll content—signs indicative of heat stress. The model, trained using labeled datasets of heat-stressed and healthy plants, accurately predicted the onset and severity of heat stress before visible symptoms appeared. This early detection allowed for precise adjustments to greenhouse temperature control systems, optimizing growing conditions and minimizing crop loss. Such systems can be integrated into smart greenhouses for autonomous climate control, demonstrating the potential of DL in controlled environment agriculture (Fevgas *et al.*, 2023; Wen *et al.*, 2023).

1.7.4 Drought and nutrient deficiency detection in corn using drone-based RGB imaging and DL

In a large-scale corn farming operation, a drone-based system was developed to monitor drought and nutrient deficiencies simultaneously. This case study involved the use of drones equipped with RGB cameras to capture high-resolution aerial images of corn fields. These images were processed using DL models to classify areas experiencing drought stress and various nutrient deficiencies (e.g. nitrogen and phosphorus). By analyzing patterns such as leaf chlorosis, changes in canopy structure, and overall plant health, the model provided real-time, spatially resolved maps of stress distribution across the field. The system enabled farmers to target specific zones for irrigation or fertilizer application, leading to more efficient resource use and a reduction in environmental impact. This case demonstrates the power of combining drone technology and DL for precision agriculture (Sudhakar and Priya, 2023; Tejasri *et al.*, 2023).

1.7.5 Real-time stress detection in sugarcane fields using thermal imaging and edge computing

A groundbreaking implementation of DL in abiotic stress management was demonstrated in sugarcane fields, where researchers applied a thermal imaging-based DL model for real-time heat stress detection. Sugarcane is sensitive to both drought and heat stress, which can dramatically reduce its yield. To address this, a network of low-cost thermal sensors was installed in the field, transmitting data to edge devices that processed the thermal images using DL algorithms. The model detected subtle temperature variations across the plants and classified areas showing signs of heat stress. This real-time detection system allowed for immediate corrective actions, such as irrigation scheduling and shading, improving crop resilience under fluctuating weather conditions. The integration of edge computing enabled on-site data processing, reducing latency and dependence on cloud infrastructure (Kardile *et al.*, 2022; Dhaliwal *et al.*, 2024).

1.7.6 Precision irrigation using DL models in cotton fields

In a case study focused on precision irrigation, DL models were used to optimize water usage in cotton farming, particularly under drought conditions. The approach combined weather data, soil moisture readings, and historical crop performance data with imagery from multispectral satellites. A deep neural network was trained to predict water stress levels in different sections of the cotton field. The model provided irrigation recommendations, directing water precisely to the areas that needed it most, thereby preventing overwatering or underwatering. This precision irrigation system was integrated with an automated irrigation system, where sensors on the ground fed real-time data back into the model to make instantaneous adjustments. This case demonstrated the potential of DL to improve water-use efficiency and conserve resources in water-scarce regions (Chen *et al.*, 2023; Jin *et al.*, 2024).

1.7.7 Application of DL in climate change adaptation in coffee plantations

In coffee plantations, where both heat and water stress are growing concerns due to climate change, DL has been used to predict the vulnerability of plants to these stresses. Researchers employed CNNs to analyze images taken under different environmental conditions, including high temperatures and varied precipitation patterns. The model was trained to identify early stress indicators such as leaf curling, discoloration, and reduced chlorophyll content. By assessing the stress levels across large coffee farms, the DL model was able to forecast potential yield losses and recommend adaptive practices, such as altering planting schedules or introducing heat-tolerant varieties. This case study emphasizes the role of DL in supporting climate change adaptation strategies, particularly in agricultural systems that are highly sensitive to environmental changes (Esgario *et al.*, 2020; de Carvalho *et al.*, 2022).

1.7.8 Case study on multi-stress detection in soybeans using multispectral and thermal sensors

A unique multi-stress detection study was conducted in soybean crops to address the simultaneous occurrence of multiple abiotic stress factors, including drought, salinity, and nutrient deficiencies. The study used multispectral and thermal sensors mounted on UAVs to capture both spectral and temperature data, which were then processed through a DL model. The DL algorithm integrated multispectral reflectance indices with thermal temperature data to classify stress types and severity levels in real time. The ability of the model to detect multiple stressors simultaneously allowed farmers to manage complex stress interactions more effectively, tailoring their interventions for maximum crop resilience. This case highlights the value of multimodal data fusion and the versatility of DL in handling complex agricultural challenges (Chawade *et al.*, 2019; Lassalle, 2021).

1.7.9 Real-world deployment of AI-based stress detection systems in large-scale farming operations

Several large-scale farming operations worldwide have begun deploying AI-powered stress detection systems as part of their digital farming strategies. These systems, which incorporate a combination of satellite imagery, IoT sensors, and DL models, are used to continuously monitor fields for early signs of abiotic stress. For example, a large-scale rice farm in India integrated a DL-based system that uses drone-captured multispectral images to assess plant health under varying stress conditions. The system automatically identifies areas with potential water or nutrient stress and recommends corrective measures, such as targeted irrigation or fertilizer application. These AI systems have proven to significantly increase yield efficiency and reduce input costs, demonstrating the potential of DL in transforming traditional farming practices at scale (Adewusi *et al.*, 2024; Neog *et al.*, 2024; Upadhyay *et al.*, 2025).

1.8 Future Directions in DL for Abiotic Stress Management

The future of DL in abiotic stress management is filled with exciting possibilities, from integrating multimodal data and enhancing edge AI for real-time processing to developing climate-aware stress prediction models and leveraging synthetic data. Advancements in transfer learning, XAI, and precision agriculture tools will further improve the scalability, accuracy, and applicability of DL models across diverse agricultural systems. Collaboration across disciplines and the development of open-source initiatives will accelerate innovation, while long-term efforts in AI-driven, climate-resilient agriculture hold the potential to revolutionize the way we manage crop stress and ensure sustainable food production in the face of climate change (Fig. 1.4).

1.8.1 Integration of multimodal data for improved stress detection

One of the most promising future directions for DL in abiotic stress management is the integration

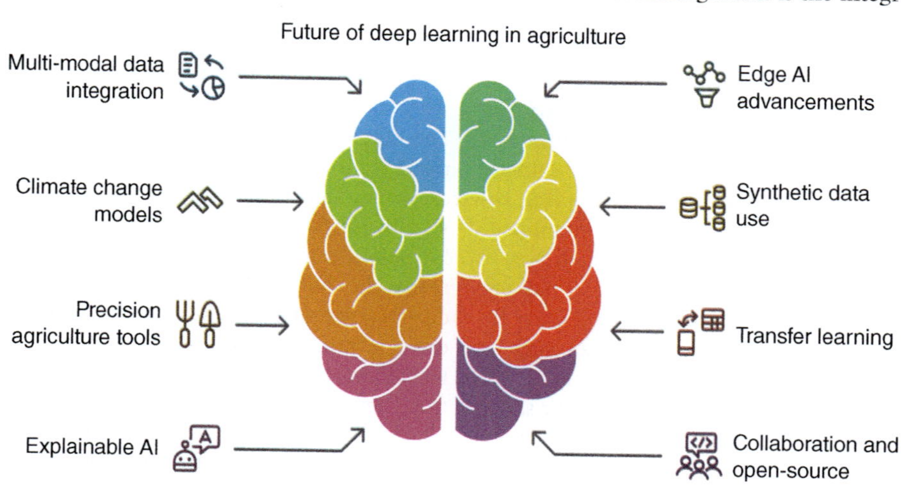

Fig. 1.4. Future aspects of deep learning in agriculture. (Author's own image.)

of multimodal data to improve stress detection and management strategies. While current models often rely on a single data source, such as images or sensor readings, combining data from various modalities—such as multispectral, hyperspectral, thermal, and RGB imaging—along with environmental parameters (e.g. soil moisture, temperature, and atmospheric conditions) could significantly enhance the accuracy of stress detection models. By integrating multimodal data, DL models will be able to leverage complementary information from different sensors, enabling more robust and comprehensive stress assessments. For instance, thermal imaging can capture temperature changes due to heat stress, while multispectral data can reveal changes in chlorophyll levels associated with drought or nutrient deficiencies, providing a more holistic understanding of crop health. This approach will not only improve early detection capabilities but also allow for better predictions of stress progression, enabling farmers to implement more effective, targeted, and timely interventions. Advances in data fusion techniques and ML algorithms designed to handle heterogeneous data will be crucial in realizing this potential (Kamarudin *et al.*, 2021; Angidi *et al.*, 2025).

1.8.2 Advancements in edge AI for real-time, on-site processing

The future of DL in abiotic stress management will likely involve continued development of edge AI, where data is processed directly at the source (e.g. on drones, sensors, or mobile devices) rather than being sent to centralized cloud servers. This shift will enable faster, real-time processing of agricultural data, essential for timely interventions in the field. Edge computing reduces the need for high-bandwidth communication and minimizes data latency, which is particularly important for real-time decision-making in dynamic environments such as crop fields. By embedding DL models into edge devices, such as UAVs, robots, and autonomous tractors, it will be possible to detect and respond to abiotic stress conditions instantly, making agriculture more adaptive and resilient. Moreover, edge AI can reduce the costs and complexities associated with cloud-based data storage and computation, democratizing access to advanced technologies, especially in remote or resource-constrained regions (Nahiyoon *et al.*, 2024; Hidayatullah and Shafik, 2025).

1.8.3 Incorporating climate change models into stress prediction systems

As climate change accelerates, the frequency and severity of abiotic stresses are expected to increase, making climate-informed stress prediction models a critical tool for future agriculture. DL can be integrated with climate models to predict how various abiotic stresses, such as drought or heat waves, will affect crops under future climate scenarios. This integration can provide predictive insights into stress risks months or even years in advance, allowing farmers to adapt their crop management strategies accordingly. For example, DL models could be trained to forecast regional climate patterns and assess their potential impact on crops, integrating this information with historical crop performance data to predict yield losses due to upcoming environmental stressors. By incorporating seasonal weather forecasts, long-term climate trends, and data from local weather stations, these models can offer valuable guidance on crop selection, planting times, irrigation scheduling, and pest management, thereby improving climate resilience in agriculture (Kumar *et al.*, 2023; Saxena *et al.*, 2023).

1.8.4 Use of synthetic data and generative models for data augmentation

The development of synthetic data generation using techniques such as GANs and other generative models is a rapidly evolving area with significant potential for improving DL models for abiotic stress management. In agricultural research, there is often a lack of sufficiently diverse, high-quality data, especially for rare or extreme stress events. GANs and other generative models can simulate realistic synthetic data that mimics real-world agricultural conditions, offering a solution to data scarcity. By augmenting existing datasets with synthetic images or sensor data, DL models can be trained on a broader range of scenarios, improving their robustness and generalization capabilities. This is particularly useful for underrepresented stress types or rare environmental conditions that may not be well represented in available datasets. Moreover, synthetic data can be tailored to simulate future environmental stress scenarios, helping farmers prepare for potential stressors associated with climate change. As generative models become more advanced, they could play a critical role in overcoming data limitations and expanding the range of predictive capabilities for DL models (Akkem *et al.*, 2024; Kar and Adinarayana, 2024).

1.8.5 Integration with precision agriculture tools and autonomous systems

The next frontier in DL for abiotic stress management lies in its integration with precision agriculture tools and autonomous systems. The combination of DL with technologies such as drones, automated tractors, and robotic harvesters will allow for real-time, automated monitoring and management of abiotic stress. For example, autonomous systems could continuously monitor crops for early signs of stress and adjust irrigation, fertilization, or pesticide application in response to detected conditions, reducing the need for manual intervention and improving overall resource efficiency. Furthermore, the integration of DL models with DSS and farm management platforms could streamline operations by providing real-time recommendations based on environmental conditions, crop status, and stress levels. These systems can optimize the use of inputs, reduce waste, and enhance productivity, ultimately leading to more sustainable and resilient farming practices. As the development of autonomous systems advances, we can expect DL to play an even more central role in automating field operations, from planting to harvesting (Mohamed, 2023; Taha *et al.*, 2025).

1.8.6 Advancements in transfer learning and domain adaptation

A promising future direction for DL in abiotic stress management is the development of transfer learning and domain adaptation techniques. Transfer learning allows models trained on one dataset to be adapted for use in different, but related, contexts. In agriculture, this means that a model trained on a specific crop type or geographic region can be fine-tuned to work with other crops or environments without requiring an entirely new dataset. For example, a DL model trained to detect drought stress in wheat might be adapted to detect the same stress in rice or corn with relatively few adjustments. Similarly, models developed in one region could be adapted for use in different climates or agricultural systems, even when the available data are limited. This adaptability is crucial for scaling DL applications across diverse farming systems and regions. Transfer learning and domain adaptation will help overcome the challenges posed by data scarcity and model generalization, enabling DL systems to be more widely applicable in global agriculture (Parul and Abhilasha, 2025; Wang *et al.*, 2025).

1.8.7 Development of XAI for agriculture

As DL models become more integral to agricultural decision-making, the need for XAI becomes more pronounced. The "black-box" nature of many DL algorithms, where the decision-making process is not easily understood, can be a barrier to adoption, especially in sectors such as agriculture, where stakeholders require transparency and trust in AI-driven decisions. Future advancements in XAI will allow farmers and agricultural experts to better understand how DL models reach conclusions, providing insights into which features of the data are most important for predictions. For example, saliency maps or attention mechanisms could be used to highlight which parts of an image or sensor reading were most indicative of stress symptoms, enhancing the credibility and interpretability of the model. Making DL models more transparent and interpretable will not only increase trust among farmers but also foster collaboration between AI researchers and agronomists, ensuring that DL tools align with traditional agricultural knowledge (Malashin *et al.*, 2024; Mohan *et al.*, 2025).

1.8.8 Collaboration across disciplines and open-source initiatives

The future of DL in abiotic stress management will likely involve increased collaboration across disciplines—combining expertise from agriculture, computer science, environmental science, and engineering. By working together, researchers can develop more effective, holistic solutions to the challenges posed by abiotic stress. Open-source initiatives will play a critical role in accelerating innovation by making datasets, models, and tools publicly available, allowing a wider community of developers and farmers to contribute to and benefit from these advancements. Collaborative efforts, such as shared research platforms and open-access agricultural databases, will enable the rapid development and dissemination of new DL techniques, helping to bridge the gap between research and practical field applications. This open-source approach will promote innovation, reduce barriers to entry, and accelerate the global adoption of DL in agriculture (Coleman and Salter, 2023; Zhang *et al.*, 2024).

1.8.9 Long-term vision: artificial intelligence-driven climate-resilient agriculture

Looking to the long-term future, the integration of DL with other advanced technologies has the potential to create AI-driven, climate-resilient agriculture systems. These systems will not only optimize resource use, minimize environmental impact, and increase yields but also enable farmers to make data-driven decisions based on real-time environmental conditions and predictive models of climate change. The combination of AI, precision agriculture, and climate modeling will help agriculture adapt to future environmental challenges, ensuring food security in an increasingly unpredictable climate (Gryshova *et al.*, 2024; Angidi *et al.*, 2025).

References

Abbas, S., Javed, M. T., Ali, Q., Azeem, M., & Ali, S. (2021). Nutrient deficiency stress and relation with plant growth and development. In *Engineering tolerance in crop plants against abiotic stress* (pp. 239-262). CRC Press.

Adewusi, A. O., Asuzu, O. F., Olorunsogo, T., Iwuanyanwu, C., Adaga, E., & Daraojimba, D. O. (2024). AI in precision agriculture: A review of technologies for sustainable farming practices. *World Journal of Advanced Research and Reviews*, 21(1), 2276-2285.

Ahmed, B., Haque, M. A., Iquebal, M. A., Jaiswal, S., Angadi, U. B., Kumar, D., & Rai, A. (2023). DeepAProt: Deep learning based abiotic stress protein sequence classification and identification tool in cereals. *Frontiers in plant science*, 13, 1008756.

Ahmed, N., & Shakoor, N. (2024). From Field to Cloud: IoT and Machine Learning Innovations in High-Throughput Phenotyping. In *International Conference on Emerging Applications of Information Technology* (pp. 125-137). Singapore: Springer Nature Singapore.

Ahmed, S.F., Alam, M.S.B., Hassan, M. *et al.* (2023). Deep learning modelling techniques: current progress, applications, advantages, and challenges. *Artificial Intelligence Review* 56, 13521–13617. https://doi.org/10.1007/s10462-023-10466-8.

Akkem, Y., Biswas, S. K., & Varanasi, A. (2024). A comprehensive review of synthetic data generation in smart farming by using variational autoencoder and generative adversarial network. *Engineering Applications of Artificial Intelligence*, 131, 107881.

Ali, T., Rehman, S. U., Ali, S., Mahmood, K., Obregon, S. A., Iglesias, R. C., & Ashraf, I. (2024). Smart agriculture: utilizing machine learning and deep learning for drought stress identification in crops. *Scientific Reports*, 14(1), 1-16.

Al-Khayri, J. M., Rashmi, R., Surya Ulhas, R., Sudheer, W. N., Banadka, A., Nagella, P., & Almaghasla, M. I. (2023). The role of nanoparticles in response of plants to abiotic stress at physiological, biochemical, and molecular levels. *Plants*, 12(2), 292.

Altalak, M., Ammad uddin, M., Alajmi, A., & Rizg, A. (2022). Smart agriculture applications using deep learning technologies: A survey. *Applied Sciences*, 12(12), 5919.

Al-Tamimi, N., Langan, P., Bernád, V., Walsh, J., Mangina, E., & Negrão, S. (2022). Capturing crop adaptation to abiotic stress using image-based technologies. *Open Biology*, 12(6), 210353.

Alzakari, S. A., Alhussan, A. A., Qenawy, A. S. T., Elshewey, A. M., & Eed, M. (2024). An enhanced long short-term memory recurrent neural network deep learning model for potato price prediction. *Potato Research*, 1-19.

Anand, S., & Sandhu, S. K. (2024). Biotic Stress Management in Field Crops Using Artificial Intelligence Technologies. In *Artificial Intelligence and Smart Agriculture: Technology and Applications* (pp. 315-335). Singapore: Springer Nature Singapore.

Angidi, S., Madankar, K., Tehseen, M. M., & Bhatla, A. (2025). Advanced High-Throughput Phenotyping Techniques for Managing Abiotic Stress in Agricultural Crops—A Comprehensive Review. *Crops*, 5(2), 8.

Arif, Y., Singh, P., Siddiqui, H., Bajguz, A., & Hayat, S. (2020). Salinity induced physiological and biochemical changes in plants: An omic approach towards salt stress tolerance. *Plant Physiology and Biochemistry*, 156, 64-77.

Azimi, S., Wadhawan, R., & Gandhi, T. K. (2021). Intelligent monitoring of stress induced by water deficiency in plants using deep learning. *IEEE Transactions on Instrumentation and Measurement*, 70, 1-13.

Bal, F., & Kayaalp, F. (2023). A novel deep learning-based hybrid method for the determination of productivity of agricultural products: Apple case study. *IEEE Access*, 11, 7808-7821.

Bashir, S. S., Hussain, A., Hussain, S. J., Wani, O. A., Zahid Nabi, S., Dar, N. A., & Mansoor, S. (2021). Plant drought stress tolerance: Understanding its physiological, biochemical and molecular mechanisms. *Biotechnology & Biotechnological Equipment*, 35(1), 1912-1925.

Benos, L., Tagarakis, A. C., Dolias, G., Berruto, R., Kateris, D., & Bochtis, D. (2021). Machine learning in agriculture: A comprehensive updated review. *Sensors*, 21(11), 3758.

Cándido-Mireles, M., Hernández-Gama, R., & Salas, J. (2023). Detecting vineyard plants stress in situ using deep learning. *Computers and Electronics in Agriculture*, 210, 107837.

Chandel, N. S., Rajwade, Y. A., Dubey, K., Chandel, A. K., Subeesh, A., & Tiwari, M. K. (2022). Water stress identification of winter wheat crop with state-of-the-art AI techniques and high-resolution thermal-RGB imagery. *Plants*, 11(23), 3344.

Chawade, A., Van Ham, J., Blomquist, H., Bagge, O., Alexandersson, E., & Ortiz, R. (2019). High-throughput field-phenotyping tools for plant breeding and precision agriculture. *Agronomy*, 9(5), 258.

Chen, Y., Yu, Z., Han, Z., Sun, W., & He, L. (2023). A decision-making system for cotton irrigation based on reinforcement learning strategy. *Agronomy*, 14(1), 11.

Choudhary, M., Sentil, S., Jones, J. B., & Paret, M. L. (2023). Non-coding deep learning models for tomato biotic and abiotic stress classification using microscopic images. *Frontiers in Plant Science*, 14, 1292643.

Coleman, G. R., & Salter, W. T. (2023). More eyes on the prize: open-source data, software and hardware for advancing plant science through collaboration. *AoB Plants*, 15(2), plad010.

Danilevicz, M. F., Bayer, P. E., Nestor, B. J., Bennamoun, M., & Edwards, D. (2021). Resources for image-based high-throughput phenotyping in crops and data sharing challenges. *Plant physiology*, 187(2), 699-715.

de Carvalho Alves, M., Sanches, L., Pozza, E. A., Pozza, A. A., & da Silva, F. M. (2022). The role of machine learning on Arabica coffee crop yield based on remote sensing and mineral nutrition monitoring. *Biosystems Engineering*, 221, 81-104.

Delfani, P., Thuraga, V., Banerjee, B., & Chawade, A. (2024). Integrative approaches in modern agriculture: IoT, ML and AI for disease forecasting amidst climate change. *Precision Agriculture*, 25(5), 2589-2613.

Demidchik, V. V., Shashko, A. Y., Bandarenka, U. Y., Smolikova, G. N., Przhevalskaya, D. A., Charnysh, M. A., & Medvedev, S. S. (2020). Plant phenomics: fundamental bases, software and hardware platforms, and machine learning. *Russian Journal of Plant Physiology*, 67, 397-412.

Deng, Y., Xin, N., Zhao, L., Shi, H., Deng, L., Han, Z., & Wu, G. (2024). Precision Detection of Salt Stress in Soybean Seedlings Based on Deep Learning and Chlorophyll Fluorescence Imaging. *Plants*, 13(15), 2089.

Dey, G., Banerjee, P., Sharma, R. K., Maity, J. P., Etesami, H., Shaw, A. K., & Chen, C. Y. (2021). Management of phosphorus in salinity-stressed agriculture for sustainable crop production by salt-tolerant phosphate-solubilizing bacteria—A review. *Agronomy*, 11(8), 1552.

Dhaliwal, S. S., Sharma, V., & Shukla, A. K. (2022). Impact of micronutrients in mitigation of abiotic stresses in soils and plants—A progressive step toward crop security and nutritional quality. *Advances in Agronomy*, 173, 1-78.

Dhaliwal, S. S., Sharma, V., Shivay, Y. S., Gupta, R. K., Verma, V., Kaur, M., & Hossain, A. (2024). Assessment and detection of biotic and abiotic stresses in field crops through remote and proximal sensing techniques—evidence from earlier findings. *Arabian Journal of Geosciences*, 17(6), 188.

Dhanya, V. G., Subeesh, A., Kushwaha, N. L., Vishwakarma, D. K., Kumar, T. N., Ritika, G., & Singh, A. N. (2022). Deep learning based computer vision approaches for smart agricultural applications. *Artificial Intelligence in Agriculture*, 6, 211-229.

Dong, H., Dong, J., Sun, S., Bai, T., Zhao, D., Yin, Y., & Wang, Y. (2024). Crop water stress detection based on UAV remote sensing systems. *Agricultural Water Management*, 303, 109059.

Eftekhari, M., Ma, C., & Orlov, Y. L. (2024). Applications of artificial intelligence, machine learning, and deep learning in plant breeding. *Frontiers in Plant Science*, 15, 1420938.

Esgario, J. G., Krohling, R. A., & Ventura, J. A. (2020). Deep learning for classification and severity estimation of coffee leaf biotic stress. *Computers and Electronics in Agriculture*, 169, 105162. https://doi.org/10.1016/j.compag.2019.105162.

Fan, J., Zhang, Y., Wen, W., Gu, S., Lu, X., & Guo, X. (2021). The future of Internet of Things in agriculture: Plant high-throughput phenotypic platform. *Journal of Cleaner Production*, 280, 123651.

Farooq, M. S., Uzair, M., Raza, A., Habib, M., Xu, Y., Yousuf, M., Yang, S. H., & Ramzan Khan, M. (2022). Uncovering the Research Gaps to Alleviate the Negative Impacts of Climate Change on Food Security: A Review. *Frontiers in Plant Science*, 13, 927535. https://doi.org/10.3389/fpls.2022.927535

Feng, X., Zhan, Y., Wang, Q., Yang, X., Yu, C., Wang, H., & He, Y. (2020). Hyperspectral imaging combined with machine learning as a tool to obtain high-throughput plant salt-stress phenotyping. *The Plant Journal*, 101(6), 1448-1461.

Fevgas, G., Lagkas, T., Argyriou, V., & Sarigiannidis, P. (2023). Detection of biotic or abiotic stress in vineyards using thermal and RGB images captured via IoT sensors. *IEEE Access*, 11, 105902-105915.

Gao, Z., Luo, Z., Zhang, W., Lv, Z., & Xu, Y. (2020). Deep learning application in plant stress imaging: a review. *AgriEngineering*, 2(3), 29.

Gill, T., Gill, S. K., Saini, D. K., Chopra, Y., de Koff, J. P., & Sandhu, K. S. (2022). A comprehensive review of high throughput phenotyping and machine learning for plant stress phenotyping. *Phenomics*, 2(3), 156-183.

Gou, C., Zafar, S., Fatima, N., Hasnain, Z., Aslam, N., Iqbal, N., & Abbas, M. (2024). Machine and deep learning: Artificial intelligence application in biotic and abiotic stress management in plants. *Frontiers in Bioscience-Landmark*, 29(1).

Gryshova, I., Balian, A., Antonik, I., Miniailo, V., Nehodenko, V., & Nyzhnychenko, Y. (2024). Artificial intelligence in climate smart in agricultural: toward a sustainable farming future. *Access J*, 5(1), 125-40.

Guo, Y., Lan, Y. and Chen, X (2022). CST: Convolutional swin transformer for detecting the degree and types of plant diseases. *Computers, Electronics in Agriculture,* 202, 107407. https://doi.org/10.1016/j.compag.2022.107407.

Habib-ur-Rahman, M., Ahmad, A., Raza, A., Hasnain, M. U., Alharby, H. F., Alzahrani, Y. M., & El Sabagh, A. (2022). Impact of climate change on agricultural production; Issues, challenges, and opportunities in Asia. *Frontiers in Plant Science*, 13, 925548.

Hidayatullah, A. F., & Shafik, W. (2025). Revolutionizing Agriculture With Automated Plant Disease Detection: Techniques, Applications, Challenges, Future Directions, and Sustainability Impacts. In *Artificial Intelligence and Data Science for Sustainability: Applications and Methods* (pp. 267-296). IGI Global Scientific Publishing.

Houetohossou, S. C. A., Houndji, V. R., Hounmenou, C. G., Sikirou, R., & Kakaï, R. L. G. (2023). Deep learning methods for biotic and abiotic stresses detection and classification in fruits and vegetables: State of the art and perspectives. *Artificial Intelligence in Agriculture*, 9, 46-60.

Islam, S., Reza, M. N., Samsuzzaman, S. A., Cho, Y. J., Noh, D. H., Chung, S. O., & Hong, S. J. (2024). Machine vision and artificial intelligence for plant growth stress detection and monitoring: A review. *Precision Agriculture*, 6(1), 34.

Jaiswal, R., Jha, G. K., Kumar, R. R., & Choudhary, K. (2022). Deep long short-term memory based model for agricultural price forecasting. *Neural Computing and Applications*, 34(6), 4661-4676.

Jalil, S. U., & Ansari, M. I. (2020). Stress implications and crop productivity. *Plant ecophysiology and adaptation under climate change: mechanisms and perspectives I: general consequences and plant responses*, 73-86.

Jin, K., Zhang, J., Wang, Z., Zhang, J., Liu, N., Li, M., & Ma, Z. (2024). Application of deep learning based on thermal images to identify the water stress in cotton under film-mulched drip irrigation. *Agricultural Water Management*, 299, 108901.

Kamarudin, M. H., Ismail, Z. H., & Saidi, N. B. (2021). Deep learning sensor fusion in plant water stress assessment: A comprehensive review. *Applied Sciences*, 11(4), 1403.

Kar, S., & Adinarayana, J. (2024). Deep Learning and Reinforcement Learning Methods for Advancing Sustainable Agricultural and Natural Resource Management. In *Harnessing Data Science for Sustainable Agriculture and Natural Resource Management* (pp. 201-223). Singapore: Springer Nature Singapore.

Kardile, P. B., Dalvi, D. G., & Kalyankar, S. V. (2022). Applications of Digital Sensors and Imaging Techniques in Agriculture and Abiotic Stress Management of Field Crops. *Advances in Agricultural and Horticultural Sciences*, 638.

Kganyago, M. L. (2022). *Remote Sensing of Crop Biophysical and Biochemical Parameters Using Sentinel-2 and Machine Learning Algorithms* (Doctoral dissertation, University of the Witwatersrand, Johannesburg (South Africa)).

Kopecká, R., Kameniarová, M., Černý, M., Brzobohatý, B., & Novák, J. (2023). Abiotic stress in crop production. *International Journal of Molecular Sciences*, 24(7), 6603.

Kujawa, S., & Niedbała, G. (2021). Artificial neural networks in agriculture. *Agriculture*, 11(6), 497.

Kumar, M., Saifi, Z., & Krishnananda, S. D. (2023). Decoding the physiological response of plants to stress using deep learning for forecasting crop loss due to abiotic, biotic, and climatic variables. *Scientific Reports*, 13(1), 8598.

Kumari, K., & Nafchi, A. M. (2024). Sustainable Agriculture with AI, Machine Learning, Deep Learning, and IoT for Future Farming. *In 2024 ASABE Annual International Meeting* (p. 1). American Society of Agricultural and Biological Engineers.

Kumari, V. V., Banerjee, P., Verma, V. C., Sukumaran, S., Chandran, M. A. S., Gopinath, K. A., & Awasthi, N. K. (2022). Plant nutrition: An effective way to alleviate abiotic stress in agricultural crops. *International Journal of Molecular Sciences*, 23(15), 8519.

Lassalle, G. (2021). Monitoring natural and anthropogenic plant stressors by hyperspectral remote sensing: Recommendations and guidelines based on a meta-review. *Science of the Total Environment*, 788, 147758.

Long, Y., & Ma, M. (2022). Recognition of drought stress state of tomato seedling based on chlorophyll fluorescence imaging. *Ieee Access*, 10, 48633-48642.

Malashin, I., Tynchenko, V., Gantimurov, A., Nelyub, V., Borodulin, A., & Tynchenko, Y. (2024). Predicting sustainable crop yields: Deep learning and explainable AI tools. *Sustainability*, 16(21), 9437.

Manghwar, H., Hussain, A., Alam, I., Khoso, M. A., Ali, Q., & Liu, F. (2024). Waterlogging stress in plants: Unraveling the mechanisms and impacts on growth, development, and productivity. *Environmental and Experimental Botany*, 105824.

Mazur, R., Gieczewska, K., Kowalewska, Ł., Kuta, A., Proboszcz, M., Gruszecki, W. I., & Garstka, M. (2020). Specific composition of lipid phases allows retaining an optimal thylakoid membrane fluidity in plant response to low-temperature treatment. *Frontiers in Plant Science*, 11, 723.

Mohamed, M. (2023). Agricultural sustainability in the age of deep learning: current trends, challenges, and future trajectories. *Sustainable Machine Intelligence Journal*, 4, 2-1.

Mohan, R. J., Rayanoothala, P. S., & Sree, R. P. (2025). Next-gen agriculture: integrating AI and XAI for precision crop yield predictions. *Frontiers in Plant Science*, 15, 1451607.

Mostafa, S., Mondal, D., Panjvani, K., Kochian, L., & Stavness, I. (2023). Explainable deep learning in plant phenotyping. *Frontiers in Artificial Intelligence*, 6, 1203546.

Murmu, S., Sinha, D., Chaurasia, H., Sharma, S., Das, R., Jha, G. K., & Archak, S. (2024). A review of artificial intelligence-assisted omics techniques in plant defense: current trends and future directions. *Frontiers in Plant Science*, 15, 1292054.

Nahiyoon, S. A., Ren, Z., Wei, P., Li, X., Li, X., Xu, J., & Yuan, H. (2024). Recent development trends in plant protection UAVs: A journey from conventional practices to cutting-edge technologies—A comprehensive Review. *Drones*, 8(9), 457.

Neog, D. R., Singha, G., Dev, S., & Prince, E. H. (2024). Artificial Intelligence and Its Application in Disaster Risk Reduction in the Agriculture Sector. In *Disaster Risk Reduction and Rural Resilience: With a Focus on Agriculture, Water, Gender and Technology* (pp. 279-305). Singapore: Springer Nature Singapore.

Niazian, M., & Niedbała, G. (2020). Machine learning for plant breeding and biotechnology. *Agriculture*, 10(10), 436.

Orka, N. A., Uddin, M. N., Toushique, F. M., & Hossain, M. S. (2023). OLID I: an open leaf image dataset for plant stress recognition. *Frontiers in Plant Science*, 14, 1251888.

Oshunsanya, S. O., Nwosu, N. J., & Li, Y. (2019). Abiotic stress in agricultural crops under climatic conditions. *Sustainable agriculture, forest and environmental management*, 71-100.

Parul, S., & Abhilasha, S. (2025). Deep learning to improve plant biosecurity and agriculture sustainability: A systematic literature review. *CABI Reviews*, 20(1), 0009.

Patra, A. K., & Sahoo, L. (2024). Explainable light-weight deep learning pipeline for improved drought stress identification. *Frontiers in Plant Science*, 15, 1476130.

Pintus, M., Colucci, F., & Maggio, F. (2025). Emerging Developments in Real-Time Edge AIoT for Agricultural Image Classification. *IoT*, 6(1), 13.

Prajapati, H. A., Yadav, K., Hanamasagar, Y., Kumar, M. B., Khan, T., Belagalla, N., & Malathi, G. (2024). Impact of climate change on global agriculture: Challenges and adaptation. *Int. J. Environ. Clim. Change*, 14(4), 372-379.

Raza, A., Gangurde, S. S., Singh Sandhu, K., & Lv, Y. (2024). Omics-assisted crop improvement under abiotic stress conditions. *Plant Stress*, 14, 100626. https://doi.org/10.1016/j.stress.2024.100626

Raza, A., Li, Y., Prakash, C. S., & Hu, Z. (2025). Panomics to manage combined abiotic stresses in plants. *Trends in Plant Science.* https://doi.org/10.1016/j.tplants.2025.03.001

Raza, A., Razzaq, A., Mehmood, S. S., Hussain, M. A., Wei, S., He, H., & Hasanuzzaman, M. (2021). Omics: The way forward to enhance abiotic stress tolerance in Brassica napus L. *GM crops & food*, 12(1), 251-281.

Rico-Chávez, A. K., Franco, J. A., Fernandez-Jaramillo, A. A., Contreras-Medina, L. M., Guevara-González, R. G., & Hernandez-Escobedo, Q. (2022). Machine learning for plant stress modeling: A perspective towards hormesis management. *Plants*, 11(7), 970.

Roychowdhury, R., Das, S. P., Gupta, A., Parihar, P., Chandrasekhar, K., Sarker, U., & Sudhakar, C. (2023). Multi-omics pipeline and omics-integration approach to decipher plant's abiotic stress tolerance responses. *Genes*, 14(6), 1281.

Sadok, W., Lopez, J. R., & Smith, K. P. (2021). Transpiration increases under high-temperature stress: Potential mechanisms, trade-offs and prospects for crop resilience in a warming world. *Plant, Cell & Environment*, 44(7), 2102-2116.

Saijo, Y., & Loo, E. P. I. (2020). Plant immunity in signal integration between biotic and abiotic stress responses. *New Phytologist*, 225(1), 87-104.

Saleem, M. H., Potgieter, J., & Arif, K. M. (2021). Automation in agriculture by machine and deep learning techniques: A review of recent developments. *Precision Agriculture*, 22(6), 2053-2091.

Santos, L., Santos, F. N., Oliveira, P. M., & Shinde, P. (2020). Deep learning applications in agriculture: A short review. In *Robot 2019: Fourth Iberian Robotics Conference: Advances in Robotics*, Volume 1 (pp. 139-151). Springer International Publishing.

Sarfraz, Z., Zarlashat, Y., Ambreen, A., Mujahid, M., Iqbal, M. S., Fatima, S. A., & Fiaz, S. (2025). *Plant Biochemistry in the Era of Omics: Integrated Omics Approaches to Unravel the Genetic Basis of Plant Stress Tolerance*. Plant Breeding.

Saud, S., Shi, Z., Xiong, L., Danish, S., Datta, R., Ahmad, I., & Banout, J. (2022). Recognizing the basics of phytochrome-interacting factors in plants for abiotic stress tolerance. *Plant Stress*, 3, 100050.

Saxena, R., Joshi, A., Joshi, S., Borkotoky, S., Singh, K., Rai, P. K., & Sharma, R. (2023). The role of artificial intelligence strategies to mitigate abiotic stress and climate change in crop production. In *Visualization Techniques for Climate Change with Machine Learning and Artificial Intelligence* (pp. 273-293). Elsevier.

Segarra, J., Buchaillot, M. L., Araus, J. L., & Kefauver, S. C. (2020). Remote sensing for precision agriculture: Sentinel-2 improved features and applications. *Agronomy*, 10(5), 641.

Shakespear, S., Sivaji, M., Kumar, V. *et al.* (2025). Navigating Through Harsh Conditions: Coordinated Networks of Plant Adaptation to Abiotic Stress. *J Plant Growth Regul* 44, 1396–1414. https://doi.org/10.1007/s00344-023-11224-4

Sharma, H., Sidhu, H., & Bhowmik, A. (2025). Remote Sensing Using Unmanned Aerial Vehicles for Water Stress Detection: A Review Focusing on Specialty Crops. *Drones*, 9(4), 241.

Sharma, K., & Shivandu, S. K. (2024). Integrating artificial intelligence and internet of things (IoT) for enhanced crop monitoring and management in precision agriculture. Sensors *International*, 100292.

Singh, A., Jones, S., Ganapathysubramanian, B., Sarkar, S., Mueller, D., Sandhu, K., & Nagasubramanian, K. (2021). Challenges and opportunities in machine-augmented plant stress phenotyping. *Trends in Plant Science*, 26(1), 53-69.

Su, J., Coombes, M., Liu, C., Zhu, Y., Song, X., Fang, S., & Chen, W. H. (2020). Machine learning-based crop drought mapping system by UAV remote sensing RGB imagery. *Unmanned systems*, 8(01), 71-83.

Sudhakar, M., & Priya, R. M. (2023). Computer Vision Based Machine Learning and Deep Learning Approaches for Identification of Nutrient Deficiency in Crops: A Survey. *Nature Environment & Pollution Technology*, 22(3).

Taha, M. F., Mao, H., Zhang, Z., Elmasry, G., Awad, M. A., Abdalla, A., & Elsherbiny, O. (2025). Emerging Technologies for Precision Crop Management Towards Agriculture 5.0: A Comprehensive Overview. *Agriculture*, 15(6), 582.

Tejasri, N., Pachamuthu, R., Naik, B., & Desai, U. B. (2023). *Intelligent drought stress monitoring on spatio-spectral-temporal drone based crop imagery using deep networks*. In 2nd AAAI Workshop on AI for Agriculture and Food Systems.

Teshome, D. T., Zharare, G. E., & Naidoo, S. (2020). The threat of the combined effect of biotic and abiotic stress factors in forestry under a changing climate. *Frontiers in plant science*, 11, 601009.

Upadhyay, A., Patel, A., Patel, A., Chandel, N. S., Chakraborty, S. K., & Bhalekar, D. G. (2025). *Leveraging AI and ML in Precision Farming for Pest and Disease Management: Benefits, Challenges, and Future Prospects*. Ecologically Mediated Development: Promoting Biodiversity Conservation and Food Security, 511-528.

Van Rooyen, J. C. (2022). *Concept demonstrator for a decision support tool for agricultural applications* (Doctoral dissertation, Stellenbosch University. Faculty of Engineering. Dept. of Industrial Engineering.).

Varadharajan, V., Rajendran, R., Muthuramalingam, P., Runthala, A., Madhesh, V., Swaminathan, G., & Ramesh, M. (2025). Multi-Omics Approaches Against Abiotic and Biotic Stress—A Review. *Plants*, 14(6), 865.

Wahab, A., Abdi, G., Saleem, M. H., Ali, B., Ullah, S., Shah, W., & Marc, R. A. (2022). Plants' physio-biochemical and phyto-hormonal responses to alleviate the adverse effects of drought stress: A comprehensive review. *Plants*, 11(13), 1620.Wang, J., Wang, Y., Li, G., & Qi, Z. (2024). Integration of remote sensing and machine learning for precision agriculture: a comprehensive perspective on applications. Agronomy, 14(9), 1975.

Wang, L., Zhang, H., Bian, L., Zhou, L., Wang, S., & Ge, Y. (2024). Poplar seedling varieties and drought stress classification based on multi-source, time-series data and deep learning. *Industrial Crops and Products*, 218, 118905.

Wang, S., Xu, D., Liang, H., Bai, Y., Li, X., Zhou, J., & Wei, W. (2025). Advances in Deep Learning Applications for Plant Disease and Pest Detection: A Review. *Remote Sensing*, 17(4), 698.

Watt, M., Fiorani, F., Usadel, B., Rascher, U., Muller, O., & Schurr, U. (2020). Phenotyping: new windows into the plant for breeders. *Annual review of plant biology*, 71(1), 689-712.

Wen, T., Li, J. H., Wang, Q., Gao, Y. Y., Hao, G. F., & Song, B. A. (2023). Thermal imaging: The digital eye facilitates high-throughput phenotyping traits of plant growth and stress responses. *Science of The Total Environment*, 899, 165626.

Wen, C., Margolis, M., Dai, R., Zhang, P., Przytycki, P. F., Vo, D. D. *et al*. (2024). Cross-ancestry atlas of gene, isoform, and splicing regulation in the developing human brain. *Science*, 384(6698), eadh0829. https://doi.org/10.1126/science.adh0829.

Williamson, H. F., Brettschneider, J., Caccamo, M., Davey, R. P., Goble, C., Kersey, P. J., & Leonelli, S. (2023). Data management challenges for artificial intelligence in plant and agricultural research. *F1000Research*, 10, 324.

Wu, J., Nadeem, M., Galagedara, L., Thomas, R., & Cheema, M. (2022). Recent insights into cell responses to cold stress in plants: Signaling, defence, and potential functions of phosphatidic acid. *Environmental and Experimental Botany*, 203, 105068.

Xu, H., Huang, C., Jiang, X., Zhu, J., Gao, X., & Yu, C. (2022). Impact of cold stress on leaf structure, photosynthesis, and metabolites in Camellia weiningensis and C. oleifera seedlings. *Horticulturae*, 8(6), 494.

Zhang, K., Yan, F., & Liu, P. (2024). The application of hyperspectral imaging for wheat biotic and abiotic stress analysis: A review. *Computers and Electronics in Agriculture*, 221, 109008.

Zhang, X., Ibrahim, Z., Khaskheli, M. B., Raza, H., Zhou, F., & Shamsi, I. H. (2024). Integrative approaches to abiotic stress management in crops: combining bioinformatics educational tools and artificial intelligence applications. *Sustainability*, 16(17), 7651.

Zhang, Y., Xu, J., Li, R., Ge, Y., Li, Y., & Li, R. (2023). Plants' response to abiotic stress: Mechanisms and strategies. *International Journal of Molecular Sciences*, 24(13), 10915.

Zubler, A. V., & Yoon, J. Y. (2020). Proximal methods for plant stress detection using optical sensors and machine learning. *Biosensors*, 10(12), 193.

2 Deep Learning in Plant Water Stress Assessment

Saniya[1], Sarfraz Ahmad[2*], and Mohammad Ghani Raghib[3]

[1]Department of Horticulture, ICAR-National Research Centre for Grapes, Pune, Maharashtra, India; [2]Department of Genetics and Plant Breeding, College of Agriculture Mandawa, Swami Keshwanand Rajasthan Agriculture University, Bikaner, Rajasthan, India; [3]Department of Horticulture, Faculty of Agriculture, Nangarhar University, Daronta, Nangarhar, Afghanistan

Abstract

Plant water stress is a critical factor limiting agricultural productivity and sustainability, particularly under changing climatic conditions. Traditional methods for assessing water stress are labor-intensive, localized, and often fail to capture the spatial and temporal variability across fields. Deep learning (DL), a rapidly evolving subfield of artificial intelligence (AI), offers transformative potential for automated, accurate, and scalable water stress detection. This chapter explores the application of DL techniques, including convolutional neural networks, recurrent neural networks, and transformers, using diverse data sources, such as RGB, hyperspectral, thermal imagery, and physiological sensors. We review the current state of research, practical applications in field and greenhouse environments, and the challenges related to data quality, model generalization, and interpretability. Furthermore, we discuss emerging trends such as explainable AI, domain adaptation, and edge computing for real-time stress monitoring. The integration of DL with remote sensing technologies is poised to revolutionize precision agriculture and water resource management.

Keywords: convolutional neural networks, deep learning, hyperspectral imaging, precision agriculture, remote sensing, smart irrigation, stress detection, water stress

2.1 Introduction

Water is indispensable for plant growth and survival, playing a vital role in physiological processes such as photosynthesis, nutrient transport, and cellular turgor maintenance. As agriculture faces increasing pressures from climate change, population growth, and resource scarcity, the accurate and timely assessment of plant water status has emerged as a critical necessity for ensuring food security and sustainable agricultural practices (Shang and Xie, 2024). Water stress, defined as the condition in which water availability is insufficient to meet the needs of a plant, can significantly impair crop development, reduce yields, and ultimately threaten agricultural productivity on a global scale (Ondrasek, 2013).

Traditional water stress assessment methods, such as pressure chamber techniques, stomatal conductance measurements, and gravimetric analysis, have been foundational in plant physiological research. However, these methods

*E-mail: sarfraz_firoz@rediffmail.com

 Deep Learning in Plant Science and Agricultural Biotechnology (Jen-Tsung Chen ed.)
DOI: 10.1079/9781836991120.0002

are often labor-intensive, destructive, and spatially limited, making them unsuitable for high-throughput agricultural monitoring or large-scale ecosystem management (Pagay, 2014). The advent of remote sensing technologies, such as multispectral and thermal imaging, offered new possibilities by enabling non-invasive, large-scale observation of plant physiological states (Galieni *et al.*, 2021). Nonetheless, extracting meaningful, stress-specific information from remote sensing data remains challenging due to the complexity, high dimensionality, and noise inherent in such datasets.

In this context, deep learning (DL), a subfield of artificial intelligence (AI) that focuses on neural networks with multiple processing layers, has emerged as a powerful tool for overcoming the limitations of conventional image analysis and machine learning approaches. DL models, particularly convolutional neural networks (CNNs), recurrent neural networks (RNNs), and transformer-based architectures, have demonstrated superior capabilities in automatically learning hierarchical feature representations from complex, high-dimensional data (Mienye and Swart, 2024). Unlike traditional machine learning models that require manual feature engineering, DL algorithms autonomously extract and optimize relevant features, thereby enhancing the robustness, generalizability, and accuracy of plant water stress assessments.

Several factors make DL particularly well suited for plant water stress detection. First, water stress manifests as subtle and dynamic physiological and morphological changes in plants, such as alterations in leaf temperature, canopy structure, spectral reflectance, and stomatal behavior, all of which can be captured using various imaging modalities. DL models excel at identifying such complex patterns, often imperceptible to human observers or traditional algorithms (Archana and Jeevaraj, 2024). Second, the integration of multimodal data sources, combining RGB, hyperspectral, thermal, and physiological sensor data, enhances the ability to accurately and reliably detect stress responses. DL frameworks can naturally handle such heterogeneous data inputs, learning rich feature embeddings that correlate with plant health status.

A growing body of research has demonstrated the effectiveness of DL for water stress detection. For instance, CNNs applied to hyperspectral imagery have enabled early detection of water stress in corn and soybean before visible symptoms appear, achieving classification accuracies exceeding 90% (Cho *et al.*, 2024). Similarly, thermal imagery processed with DL models has facilitated the identification of stressed plants by detecting slight increases in canopy temperature associated with reduced transpiration rates (Wen *et al.*, 2023). Moreover, advancements in unmanned aerial vehicle (UAV) technologies, equipped with multispectral and thermal sensors, coupled with onboard DL algorithms, have enabled real-time, high-resolution field monitoring, significantly enhancing precision irrigation management (Acharya *et al.*, 2021).

Despite these advancements, several challenges remain. One of the major obstacles is the scarcity of large, labeled datasets specific to plant water stress across diverse species, developmental stages, and environmental conditions. Data augmentation, transfer learning, and synthetic data generation using techniques such as generative adversarial networks (GANs) are increasingly employed to mitigate this limitation (Fayaz *et al.*, 2024). Another critical issue is the generalization ability of DL models: models trained on a specific crop type or environmental condition may not perform adequately when deployed in different settings. Consequently, domain adaptation and meta-learning approaches are gaining attention to improve model robustness across variable agricultural contexts (de Andrade Porto *et al.*, 2023).

Interpretability remains another pressing concern. DL models are often regarded as "black boxes," making it difficult to trace their decision-making processes or correlate predictions with biological phenomena. The field of explainable AI (XAI) aims to address this issue by developing methods that provide insights into model behavior, feature importance, and decision rationales, thus facilitating trust and broader adoption in agricultural practice (Martin *et al.*, 2024).

Furthermore, the computational and infrastructural demands of DL models, particularly during training phases, pose challenges for deployment in resource-constrained environments. Recent developments in edge computing and lightweight neural architectures, such as MobileNets and EfficientNets, offer promising solutions for implementing DL-based stress assessment systems directly on agricultural robots, UAVs, and Internet of Things (IoT) devices (Keskes, 2025).

In the future, the integration of DL with IoT platforms, real-time sensing networks, and cloud-based decision support systems holds the potential to transform plant water stress monitoring into a continuous, adaptive, and autonomous process. Such integration would not only enable early detection and mitigation of water stress but also optimize resource utilization, reduce water wastage, and support sustainable agricultural intensification.

This chapter provides a comprehensive review of the application of DL in plant water stress assessment. We begin by discussing the physiological basis of water stress and the traditional methods used for its evaluation. We then explore various DL architectures and their applications in processing diverse types of plant stress data. Case studies highlighting recent advancements are presented, followed by a critical discussion of the challenges, limitations, and future research directions. By synthesizing current knowledge and identifying emerging trends, this chapter aims to contribute to the advancement of precision agriculture and support the development of smart, resilient farming systems in an era of increasing environmental uncertainty.

2.2 Understanding Plant Water Stress

Water stress in plants occurs when water availability falls below optimal levels, leading to a physiological imbalance that affects growth, development, and survival. As water is a primary medium for nutrient transport, biochemical reactions, and cellular turgidity, even minor deficits can cascade into severe morphological and physiological consequences (Chada *et al.*, 2023). Understanding plant water stress from physiological, biochemical, and ecological perspectives is essential for developing effective monitoring and mitigation strategies, particularly as the frequency and intensity of drought events increase under climate change scenarios (Ali *et al.*, 2025).

2.2.1 Physiological basis of water stress

Water stress influences plants at multiple organizational levels. At the cellular level, reduced water availability leads to a loss of turgor pressure, impairing cell expansion and division, which in turn affects overall plant growth (Bhattacharya, 2021). Stomatal closure is one of the earliest responses to water deficit, aimed at minimizing transpiration losses but concurrently reducing carbon dioxide intake, thereby inhibiting photosynthesis (Kaur *et al.*, 2021). Prolonged stomatal closure results in decreased biomass accumulation and eventual yield loss.

On a systemic level, water stress disrupts vascular transport systems. Hydraulic conductivity is reduced, and embolisms, air bubbles in xylem vessels, may form under severe drought, permanently damaging the plant's water transport capacity (Zwieniecki and Secchi, 2015). Additionally, plants under water stress accumulate osmoprotectants, such as proline, glycine betaine, and soluble sugars, to maintain cell turgor and protect cellular structures (Ozturk *et al.*, 2021).

2.2.2 Types and causes of water stress

Plant water stress can be categorized based on cause and duration:

- *Drought stress*. It occurs due to a prolonged absence of rainfall or insufficient irrigation. It is the most common form of water stress and affects both rainfed and irrigated agriculture globally (Begna, 2022). Drought severely hampers plant growth and is a leading cause of yield loss. Key traits for drought resistance include improved root systems, leaf structure, osmotic balance, water content, and stomatal regulation (Ahmad *et al.*, 2023).
- *Soil salinity stress*. High salt concentrations in the soil reduce the osmotic potential, making water less available to plant roots even when moisture appears adequate (Sheldon *et al.*, 2017).
- *Flooding-induced stress*. Paradoxically, waterlogging can lead to oxygen deprivation in

roots, impairing water uptake and simulating water stress symptoms (Zhang *et al.*, 2025).

- *Atmospheric water stress.* High vapor pressure deficits due to elevated temperatures and low humidity can drive excessive transpiration, leading to desiccation even when soil moisture is sufficient (Grossiord *et al.*, 2020).

Each type presents distinct challenges for monitoring and requires different sensing modalities and modeling approaches for accurate detection and diagnosis.

2.2.3 Plant adaptation mechanisms to water stress

Drought resistance in plants involves three key mechanisms: escape through early maturation, avoidance via traits such as deep roots or specialized leaves, and tolerance through molecular and physiological adaptations at the tissue and cellular levels (Jain *et al.*, 2023). Plants have evolved diverse morphological, physiological, and biochemical adaptations to cope with water scarcity:

- *Morphological adaptations* include reduced leaf area, deeper root systems, and waxy leaf surfaces to limit water loss (Zahedi *et al.*, 2025).
- *Physiological adaptations* such as stomatal regulation, osmotic adjustment, and alterations in leaf hydraulic conductance help maintain water status during moderate stress (Sanders and Arndt, 2012).
- *Biochemical responses* involve the upregulation of stress-responsive genes, production of antioxidant enzymes, and synthesis of protective metabolites such as abscisic acid, a key hormonal signal during drought (Parwez *et al.*, 2022).

Understanding these adaptations is crucial when interpreting plant responses in stress detection systems. For instance, hyperspectral imaging can capture subtle biochemical changes associated with pigment degradation or osmolyte accumulation before morphological symptoms become apparent (Falcioni *et al.*, 2023).

2.2.4 Indicators of water stress

Effective plant water stress assessment relies on identifying reliable indicators across multiple scales:

- *Canopy temperature.* Elevated leaf or canopy temperatures relative to ambient conditions often signal reduced transpiration due to stomatal closure, measurable via thermal imaging (Still *et al.*, 2019).
- *Spectral reflectance.* Changes in reflectance patterns, particularly in near-infrared and shortwave infrared bands, correlate with plant water content and structural integrity (Holzman *et al.*, 2021).
- *Chlorophyll fluorescence.* Water stress can impair photosystem II efficiency, detectable through changes in chlorophyll fluorescence parameters (Zhuang *et al.*, 2020).
- *Leaf water potential.* Direct measurements using pressure chambers provide precise assessments but are invasive and impractical for large-scale monitoring (Carella, 2024).
- *Vegetation indices.* Metrics such as the Normalized Difference Vegetation Index (NDVI), Photochemical Reflectance Index, and Water Band Index offer indirect but scalable proxies for water stress detection (Skendžić *et al.*, 2023).

The integration of these indicators through advanced sensor platforms and analytical models, particularly DL frameworks, enables early, nondestructive, and spatially extensive monitoring of plant water stress.

2.2.5 Importance of early water stress detection

Early detection of plant water stress is critical for mitigating yield losses, optimizing water use, and informing irrigation scheduling (Ihuoma and Madramootoo, 2017). Conventional irrigation practices often rely on fixed schedules or visual assessments, leading to either over-irrigation or water deficits. In contrast, precision irrigation, guided by real-time stress monitoring, can significantly enhance water-use efficiency and crop productivity.

Moreover, early stress detection provides opportunities for adaptive management strategies, including the deployment of drought-tolerant varieties, dynamic fertigation plans, and targeted biostimulant applications. With the increasing unpredictability of weather patterns, proactive stress management based on accurate monitoring is poised to become a cornerstone of climate-resilient agriculture (Sarma *et al.*, 2024).

2.3 Role of DL in Water Stress Assessment

The increasing complexity of plant responses to environmental stressors necessitates advanced analytical approaches capable of extracting meaningful patterns from diverse, high-dimensional datasets. DL, a subset of machine learning based on artificial neural networks with multiple layers, has emerged as a powerful tool for modeling complex biological systems, including plant water stress assessment (Kamarudin *et al.*, 2021). By leveraging large volumes of sensor, imaging, and spectral data, DL enables the early, accurate, and scalable detection of water stress across different plant species and agricultural systems.

2.3.1 Advantages of DL over traditional methods

Traditional methods for plant water stress detection, such as threshold-based remote sensing indices (e.g. NDVI and CWSI) and statistical regression models, are often limited by their reliance on manually selected features and their inability to model non-linear relationships (Gommes *et al.*, 2010). DL addresses these limitations through the following:

- *Automatic feature extraction.* CNNs and other architectures can autonomously identify critical features related to stress without requiring domain-specific feature engineering (Dzieżyc *et al.*, 2020).
- *Non-linear modeling.* DL models can capture highly non-linear interactions between physiological traits, environmental variables, and spectral signatures that traditional linear models cannot (Lu *et al.*, 2024).
- *Scalability and transferability.* Once trained, DL models can be transferred across different crops and geographies with minimal retraining, enabling broader applicability (Wang *et al.*, 2022).
- *Integration of multimodal data.* DL architectures can simultaneously process heterogeneous data types, including RGB images, hyperspectral data, thermal imagery, and environmental sensor data, enhancing prediction accuracy (Wang *et al.*, 2022).

These advantages make DL particularly suited for operationalizing precision agriculture and stress management strategies under real-world conditions.

2.3.2 Key DL architectures for water stress assessment

Several DL architectures have demonstrated particular efficacy in assessing plant water stress, which are discussed in the following sections.

Convolutional neural networks (CNNs)

CNNs are particularly well suited for processing spatial data such as aerial images, satellite imagery, and proximal sensing datasets (Peng *et al.*, 2024). In water stress assessment, CNNs can classify stressed versus non-stressed plants based on subtle visual and spectral differences invisible to the naked eye (Islam *et al.*, 2024). They are also adept at handling various image scales and illumination conditions through their hierarchical feature extraction. For instance, Asaari *et al.* (2019) employed a deep CNN to analyze hyperspectral images for early drought stress detection in corn, achieving classification accuracies exceeding 90%.

Recurrent neural networks (RNNs) and long short-term memory (LSTM) networks

Temporal patterns of stress progression, as captured by time-series imaging or sensor data, are effectively modeled using RNNs and their improved variant, LSTM networks (Acikmese and Alptekin, 2019). These architectures maintain temporal dependencies, enabling the prediction of future stress states based on past observations.

For example, Sun *et al.* (2023) applied an LSTM model to a time series of vegetation indices and meteorological data, successfully predicting water stress events several days in advance.

Transformer models

Recently, transformer architectures, originally developed for natural language processing, have been adapted for image analysis and time-series forecasting in agricultural contexts (Ibañez and Monterola, 2023). Their self-attention mechanism allows for the modeling of long-range dependencies across spatial and temporal dimensions, offering superior performance compared to CNNs or LSTMs in some scenarios (Lin *et al.*, 2020). Transformers are particularly promising for combining satellite-based time-series imagery with ground-truth measurements to track regional drought dynamics (Danandeh Mehr *et al.*, 2025).

2.3.3 Data sources and preprocessing for DL models

Successful deployment of DL for water stress assessment requires high-quality data from various sources:

- *RGB and multispectral imagery*. Collected via UAVs (drones), satellites, or ground-based systems to capture canopy structure, color changes, and reflectance properties (Kulbacki Ouyang *et al.*, 2020).
- *Thermal imagery*. Provides information on canopy temperature, a direct proxy for stomatal conductance and transpiration rates (Struthers *et al.*, 2015).
- *Hyperspectral data*. Offers fine-grained spectral information, enabling the detection of subtle biochemical changes associated with stress (Borol *et al.*, 2024).
- *Environmental sensors*. Measure soil moisture, temperature, humidity, and radiation, providing context for interpreting plant responses (Rundel *et al.*, 2009).

Preprocessing steps such as noise filtering, data augmentation, normalization, and dimensionality reduction are critical to improving model generalization and robustness, especially when dealing with field conditions characterized by high variability.

2.3.4 Challenges and limitations

While DL offers transformative potential for water stress assessment, several challenges must be addressed:

- *Data scarcity and labeling*. Obtaining large, labeled datasets of stressed and non-stressed plants under diverse conditions is resource-intensive (Dwivedi *et al.*, 2023).
- *Overfitting*. Deep models are prone to overfitting, especially when trained on small datasets or when the stress symptoms are subtle and easily confounded by other factors such as disease or nutrient deficiency (Yi *et al.*, 2020).
- *Interpretability*. DL models often operate as "black boxes," complicating biological interpretation and limiting trust among agronomists and farmers (Keskes, 2025).
- *Generalization across crops and environments*. Models trained in one environment or for a specific crop variety may perform poorly elsewhere unless domain adaptation or transfer learning strategies are employed (Yi *et al.*, 2020).

Addressing these limitations through improved model architectures, XAI techniques, and collaborative data-sharing initiatives will be critical for scaling DL solutions in real-world agricultural settings.

2.3.5 Future perspectives

Emerging trends suggest promising avenues for enhancing DL-based water stress assessment:

- *Integration of multimodal and multiscale data*. Fusing UAV, ground, and satellite data streams could enable continuous, scalable monitoring (Igwenagu *et al.*, 2025).
- *Explainable DL*. Incorporating saliency maps, SHapley Additive exPlanations (SHAP) values, or attention heatmaps could make model predictions more transparent and actionable (Ahmad *et al.*, 2024).

- *Edge computing and IoT integration.* Deployment of lightweight DL models directly on farm-based devices (e.g. smart sensors and drones) could allow for real-time stress detection and response (O'Grady *et al.*, 2019).
- *Self-supervised learning and few-shot learning.* Reducing the dependence on large labeled datasets by leveraging unlabeled data or minimal supervision is a growing area of research (Su *et al.*, 2020).

By aligning DL innovations with agronomic insights and farmer needs, the next generation of stress monitoring systems could substantially improve agricultural resilience and sustainability under water-limited conditions.

2.4 DL Models and Applications for Water Stress Detection in Practice

The transition of DL models from controlled research settings to practical agricultural applications is a critical step toward sustainable crop production under increasing water scarcity. Recent advances have demonstrated the potential for real-world deployment of DL techniques in water stress detection across a range of crops, environments, and scales.

2.4.1 DL models applied in water stress detection

Various DL models have been tailored for detecting, classifying, and quantifying plant water stress. Each architecture brings unique strengths and is often selected based on the nature of the available data (e.g. imagery and sensor time series) and the specific agricultural use case.

Convolutional neural networks (CNNs)

CNNs remain the most widely used architecture for image-based stress detection due to their superior performance in visual pattern recognition.

- *Case study.* Al-Tamimi *et al.* (2022) analyzed 25,000 RGB images of soybean to capture crop adaptation to stress.
- *Key insight.* Deeper CNN architectures (e.g. ResNet and DenseNet) outperform shallower networks by capturing more abstract and complex stress indicators (Gao *et al.*, 2020).

Long short-term memory networks (LSTMs)

LSTMs are predominantly used when dealing with time-series datasets, particularly from environmental sensors and satellite observations.

- *Case study.* Alibabaei *et al.* (2023) applied LSTM models to soil moisture and weather station data, accurately predicting the onset and progression of water stress in vineyards.
- *Key insight.* LSTMs can capture lagged effects, such as cumulative heat stress or gradual soil moisture depletion, which are critical for predictive water management.

Attention-based models and Vision Transformers (ViTs)

Recent innovations such as attention mechanisms and ViTs offer a powerful alternative for modeling spatial and temporal patterns.

- *Case study.* Guo *et al.* (2024) implemented a transformer-based approach to predict yield in wheat, achieving over 95% precision.
- *Key insight.* Attention models improve interpretability by highlighting which regions of a plant canopy or landscape are most associated with stress signals.

Hybrid models

Hybrid models that combine CNNs, LSTMs, and attention mechanisms show promise for capturing both spatial and temporal dynamics simultaneously.

- *Example*: A CNN–LSTM pipeline was used to first extract spatial features from drone imagery, followed by temporal modeling of stress progression, providing a dynamic water stress risk map for cotton fields (Reddy *et al.*, 2024).

2.4.2 Sensor platforms and data collection strategies

The success of any DL application is deeply tied to the sensor technologies and data acquisition methods used. A brief description of common platforms and their applications is presented in Table 2.1.

Multisource integration has become a trend, where data fusion from UAVs, ground stations, and satellites enhances model robustness and prediction accuracy (Samadzadegan *et al.*, 2025).

2.4.3 Workflow of a typical DL-based water stress monitoring system

Figure 2.1 represents a systematic pipeline integrating multidisciplinary data and AI for precise, scalable crop monitoring. Academically, it captures the convergence of agronomy, remote sensing, and computational modeling in modern plant stress phenotyping. An end-to-end DL pipeline for practical water stress assessment generally includes the following stages:

- *Data acquisition.* Multimodal data collection via UAVs, ground sensors, or satellite platforms.
- *Preprocessing.* Image correction (radiometric and geometric), noise removal, and normalization.
- *Annotation/Labeling.* Manual or semiautomatic labeling of stress and non-stress classes.
- *Model development.* Selection of appropriate DL architectures (e.g. CNN, LSTM, and transformer).
- *Training and validation.* Model training with cross-validation to prevent overfitting.
- *Deployment.* Real-time integration into farm management systems or mobile apps.
- *Monitoring and updating.* Continual model refinement with new data to improve performance over time.

2.4.4 Real-world applications and case studies

Precision irrigation management

DL-based stress detection enables site-specific irrigation, conserving water without compromising yield.

- *Example*: A model applied in vineyards allowed growers to target irrigation only where water stress was detected (Peeters *et al.*, 2024).

Early warning systems for drought

When integrated with weather forecasts and crop growth models, DL systems provide early warnings of impending drought stress, giving farmers critical lead time to respond.

- *Example*: An LSTM-based drought early warning system for wheat improved preparedness and reduced crop losses (Parashar *et al.*, 2024).

Crop breeding and phenotyping

High-throughput phenotyping platforms using DL can screen large populations of crop varieties for drought resistance traits.

- *Example*: Deep CNNs classified drought-resilient versus susceptible soybean varieties from UAV imagery (Liang *et al.*, 2024).

2.4.5 Limitations and challenges in practical deployment

Despite promising applications, several barriers remain to fully operationalize DL for field-based water stress assessment:

Table 2.1. Overview of common sensor platforms, data types collected, and their applications.

Platform	Data type	Application
UAVs (drones)	RGB, multispectral, thermal imagery	Fine-resolution, flexible, real-time field monitoring
Ground sensors	Soil moisture, plant temperature, VWC sensors	Localized, high-frequency data for root zone analysis
Satellite imagery	Multispectral, SAR, hyperspectral	Large-scale drought and stress monitoring
IoT networks	Weather, soil, irrigation sensors	Automated and continuous environmental tracking

IoT, Internet of Things; SAR, synthetic aperture radar; UAVs, unmanned aerial vehicles; VWC, volumetric water content.

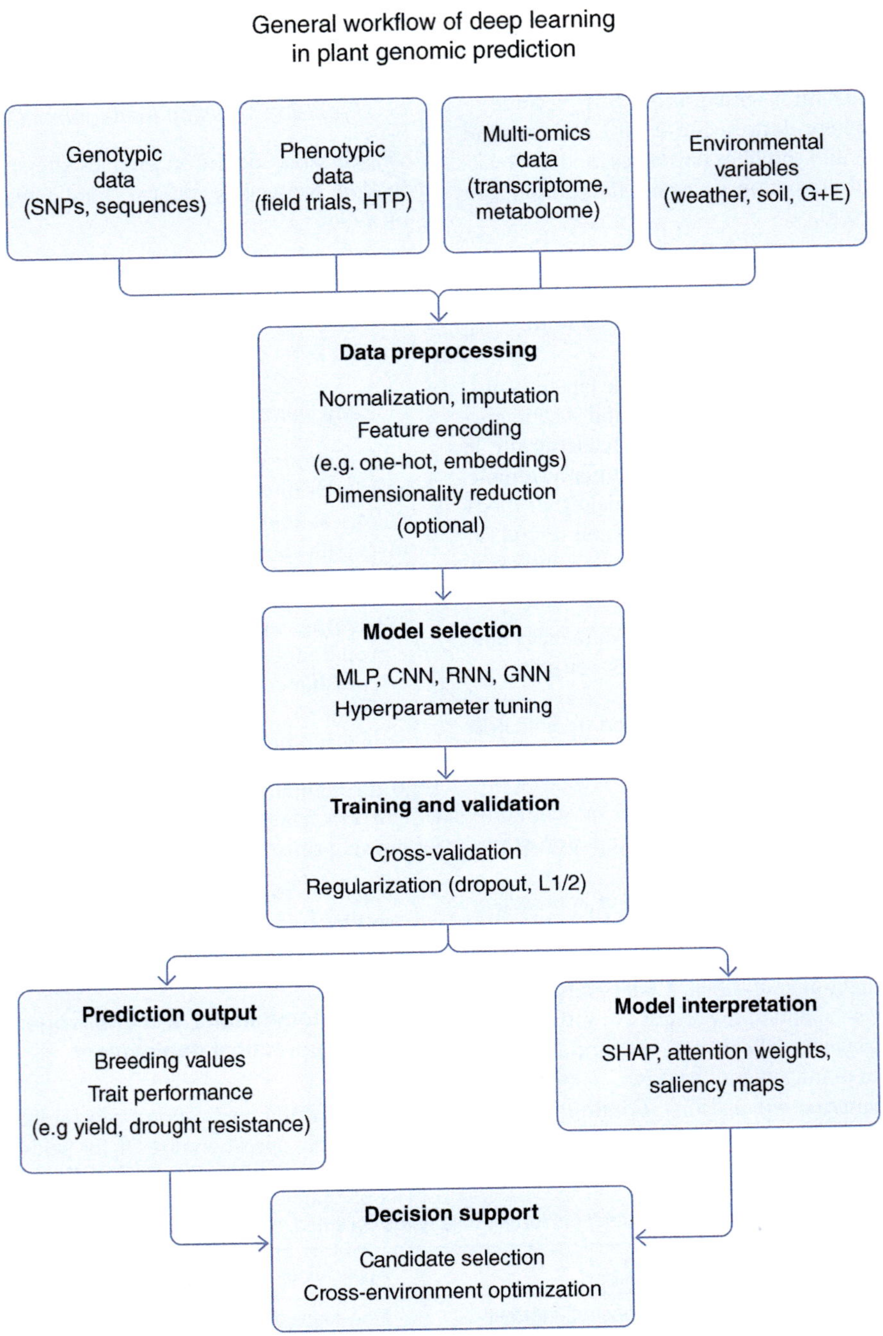

Fig. 2.1. General workflow of deep learning in water stress assessment. CNNs, convolutional neural networks; GNN, graph neural network; HTP, high-throughput phenotyping; MLP, multilayer perceptron; RNNs, recurrent neural networks; SHAP, SHapley Additive exPlanations; SNPs, single-nucleotide polymorphisms. (Author's own image.)

- *Data generalization.* Models trained in one region may fail in others due to differences in soil type, weather patterns, or plant variety.
- *Data annotation bottlenecks.* Creating high-quality labeled datasets is labor-intensive and expensive.
- *Computational resources.* Training large DL models requires significant computational power, which may not be available in rural farming communities.
- *Explainability and farmer trust.* Farmers need transparent, understandable AI systems to gain confidence in the technology.

Addressing these challenges through collaborative, multidisciplinary research and policy support is essential for sustainable integration of AI into agriculture.

2.4.6 Emerging trends

Several emerging technologies and strategies are poised to further enhance the application of DL in water stress management:

- *Few-shot and meta-learning.* Enabling models to learn stress patterns from very few labeled examples (Gharoun *et al.*, 2024).
- *Synthetic data generation.* Using GANs to create synthetic stressed plant images for model training (Benfenati *et al.*, 2022).
- *Federated learning.* Allowing model training across decentralized farms without requiring raw data sharing, thus protecting farmer privacy (Dembani *et al.*, 2025).
- *On-device AI (edge computing).* Running lightweight DL models directly on drones or mobile phones for real-time stress detection (Jouini *et al.*, 2024).

2.5 Challenges, Limitations, and Future Prospects

The integration of DL in plant water stress assessment has demonstrated considerable promise; however, significant technical, practical, and ethical challenges must be addressed to enable wider adoption in real-world agricultural systems. This section critically analyzes current limitations and explores emerging solutions and future opportunities.

2.5.1 Technical challenges

Data scarcity and quality

DL models are highly data-intensive, requiring large, diverse, and high-quality datasets for training and validation (Najafabadi *et al.*, 2015). However, in agricultural contexts:

- *Scarcity of labeled data.* Manually annotated datasets for plant water stress are rare, especially across multiple crops and climatic regions.
- *Sensor noise and variability.* Field-collected data often suffer from inconsistencies due to environmental factors such as lighting changes, wind, or equipment calibration issues (Aubard *et al.*, 2025).

These issues can limit model generalization and reduce prediction reliability.

Model generalization across regions and crops

Models trained under specific environmental or crop conditions often fail when applied to different regions, soil types, or plant varieties. This phenomenon, termed domain shift, challenges the universal applicability of a single model (Priyatikanto *et al.*, 2023).

Potential solutions:

- domain adaptation techniques, and
- meta-learning approaches enabling models to rapidly adapt to new contexts with minimal retraining.

Computational demands

Training and deploying sophisticated DL architectures, such as transformers or hybrid models, often require high-performance computing infrastructure, which may not be accessible in rural or resource-limited farming communities (Sapkota *et al.*, 2024).

Emerging solutions:

- model compression (pruning, quantization), and
- edge AI and lightweight model deployment for mobile platforms.

2.5.2 Practical and operational challenges

High implementation costs

The initial setup cost, including acquisition of drones, sensors, and computational infrastructure, is often prohibitive for small- and medium-sized farms, limiting widespread adoption.

Potential mitigation:

- cooperative ownership models, and
- government subsidies and public–private partnerships to democratize access.

Farmer trust and technology acceptance

Many farmers express skepticism toward “black-box” AI systems whose internal decision-making processes are opaque (Uslu, 2024).

Critical needs:

- XAI tools that provide transparent and interpretable outputs, and
- user-friendly interfaces integrating stress predictions with actionable agronomic advice.

Data privacy and ethical concerns

The deployment of networked sensors, UAVs, and remote sensing platforms raises concerns about data ownership, privacy, and potential misuse (Mekdad *et al.*, 2023).

Guidelines needed:

- clear data governance policies, and
- ethical frameworks protecting farmer rights and ensuring responsible AI deployment.

2.5.3 Future prospects

Despite current hurdles, the future of DL in plant water stress assessment is promising, fueled by rapid technological advancements and interdisciplinary collaborations.

Multimodal and multisensor fusion

Future systems will increasingly leverage multimodal data fusion, integrating visual (RGB, multispectral, thermal), biochemical (leaf chlorophyll fluorescence), and environmental (soil moisture, temperature) data streams for more holistic water stress diagnosis (Kamarudin *et al.*, 2021).

- *Example*: Combining hyperspectral imagery with *in situ* soil moisture data for early drought detection at both canopy and root levels.

Real-time decision support systems

Advances in 5G/6G networks, edge computing, and AIoT (Artificial Intelligence of Things) will enable real-time water stress monitoring, allowing immediate irrigation scheduling or remedial actions (Nguyen-Tan and Le-Trung, 2024).

Predictive and prescriptive analytics

Beyond detection, DL models will evolve toward predictive modeling (forecasting future stress based on weather predictions) and prescriptive analytics (suggesting optimal irrigation strategies based on predicted stress patterns).

- *Example*: Deep reinforcement learning models dynamically optimizing irrigation timing and quantity to maximize water-use efficiency.

Participatory machine learning and citizen science

Integrating farmer knowledge into the machine learning pipeline (e.g. through mobile apps where farmers annotate stressed areas) can

- improve dataset diversity
- build farmer trust in AI-driven decision support systems, and
- democratize AI development (Allen *et al.*, 2019).

Sustainable and climate-resilient agriculture

In the broader context of climate change adaptation, DL-driven water stress assessment tools will be critical for

- enhancing drought resilience
- improving crop productivity under variable rainfall conditions, and
- supporting smart irrigation in precision agriculture (Bwambale *et al.*, 2022).

2.5.4 Concluding remarks

DL holds transformative potential in improving the detection, understanding, and management of plant water stress. However, realizing this potential requires addressing key challenges related to data quality, model generalization, infrastructure needs, ethical deployment, and farmer engagement. Future innovations in multimodal sensing, edge AI, explainable systems, and participatory machine learning will likely drive the next wave of breakthroughs, leading to a more resilient and sustainable agricultural future.

2.6 Applications and Case Studies in DL for Water Stress Assessment

To better understand the practical applications and impact of DL in water stress assessment, this section highlights several case studies from diverse geographical regions and crop types. Additionally, expanded DL models and their applications for water stress detection are presented in Table 2.2. These real-world examples demonstrate the diverse methodologies, successful implementations, and transformative potential of AI-driven water management solutions.

Table 2.2. Expanded deep learning models and their applications for water stress detection. Author's own table.

Crop	Deep learning model	Application	References
Grapevine	CNN	Water stress detection from UAV imagery	Peeters *et al.* (2024)
Wheat	Hybrid (CNN + LSTM)	Drought monitoring and early stress prediction	Parashar *et al.* (2024)
Cotton	CNN + LSTM	Water stress classification and irrigation optimization	Reddy *et al.* (2024)
Soybean	CNN	Early drought stress detection from multispectral imagery	Liang *et al.* (2024)
Tomato	HSI	Drought stress identification	Weng *et al.* (2023)
Corn	CNN	Temporal drought stress prediction	Cho *et al.* (2024)
Rice	CNN	Water stress severity estimation from UAV data	Li *et al.* (2023)
Sunflower	Transfer learning (pretrained CNNs)	Drought stress detection	Lazić *et al.* (2024)
Corn	ResNet-50	Early detection of water stress	Goyal *et al.* (2024)
Blueberries	Transformer-based model	Hyperspectral image analysis for phenotyping drought tolerance	Rahman *et al.* (2025)
Lettuce	GAN	Synthetic stress data generation for improved model training	Cho *et al.* (2024)
Potato	Deep residual CNN	Visual symptom detection of water deficiency	Lapajne *et al.* (2024)
Potato	YOLO model	Real-time detection of drought-stressed plants	Takalloo (2024)
Cotton	VGG-16	Water stress identification	Jin *et al.* (2024)
Peanuts	LSTM	Rapid peanut phenotyping and water quality monitoring	Bagherian (2022)

CNN, convolutional neural network; GAN, generative adversarial network; HSI, hyperspectral imaging; LSTM, long short-term memory; UAV, unmanned aerial vehicles; YOLO, You Only Look Once.

2.6.1 Case study 1: precision irrigation in vineyards—California, USA

Objective

A study led by Peeters *et al.* (2024) aimed to utilize DL techniques for precision irrigation in California vineyards, one of the most water-intensive crops in the state. The goal was to improve water-use efficiency by accurately identifying and predicting water stress conditions in grapevines using UAV-based RGB and multispectral imagery.

Methodology

- *Data acquisition.* UAVs equipped with high-resolution cameras and multispectral sensors were used to capture aerial images of vineyard fields.
- *Model development.* A CNN was employed to classify the images based on water stress levels. This model was trained to detect subtle differences in canopy reflectance associated with water-deficit conditions.
- *Training data.* Annotated datasets of stress and non-stress images were created through manual ground truthing.
- *Validation.* Models were validated using ground-based measurements of soil moisture and vine physiological parameters, such as stomatal conductance.

Results

- The CNN achieved a classification accuracy of over 90% for water stress detection.
- Precision irrigation systems were integrated with the model, enabling automatic adjustment of irrigation schedules based on real-time stress detection.
- *Impact.* The precision irrigation system resulted in a 30% reduction in water use, maintaining high grape yield and quality during drought periods.

Key insights

- UAVs combined with DL can provide fine-resolution, real-time assessments of water stress across large agricultural areas.
- The success of the system was highly dependent on accurate data labeling and model calibration to minimize environmental variations.

2.6.2 Case study 2: drought monitoring in wheat—Australia

Objective

The Wheat Drought Early Warning System project, developed by the Australian Government and the University of Queensland, aimed to monitor and predict drought stress in wheat crops, using a combination of remote sensing data and DL models (Parashar *et al.*, 2024).

Methodology

- *Data acquisition.* Satellite-based multispectral imagery (e.g. Landsat and Sentinel-2) was used to capture field conditions over the growing season. Soil moisture data, temperature, and precipitation forecasts were also integrated into the model.
- *Model development.* A hybrid model combining LSTM networks for time-series analysis and CNNs for spatial feature extraction was employed. The model was trained to predict water stress levels and anticipate drought conditions.
- *Training data.* The dataset consisted of historical weather, soil moisture, and crop yield data, along with remote sensing imagery of wheat fields during drought events.

Results

- The model successfully predicted early drought conditions, enabling farmers to adjust irrigation schedules and implement water conservation practices.
- *Impact.* The system increased farmers' drought preparedness by providing actionable insights 3–4 weeks in advance, reducing crop losses by up to 15% during the 2018–2020 drought period.
- The tool was integrated into farm management platforms, offering real-time decision support.

Key insights

- Time-series models such as LSTMs are crucial for detecting temporal patterns in weather and soil data that influence water stress.
- Combining spatial data from remote sensing with temporal data from weather stations improves prediction accuracy and timeliness.
- Early warning systems based on DL can significantly reduce the economic impact of droughts by providing farmers with proactive, data-driven decision-making tools.

2.6.3 Case study 3: cotton water stress detection—India

Objective

In collaboration with the Indian Council of Agricultural Research (ICAR), a project focused on detecting water stress in cotton crops was initiated to optimize irrigation practices and minimize water wastage in water-scarce regions of India (Awais *et al.*, 2022).

Methodology

- *Data acquisition.* The study utilized a combination of UAV-based multispectral imagery and ground-based sensor networks for continuous monitoring of soil moisture and environmental variables.
- *Model development.* A DL framework combining CNNs for image-based stress detection and LSTMs for analyzing the temporal progression of stress was developed.
- *Training data.* The dataset consisted of labeled aerial images and soil moisture measurements collected during the growing season, from fields with varying water management practices.

Results

- The DL model was able to accurately classify different levels of water stress in cotton crops, achieving an overall accuracy of 92%.
- *Impact.* The model enabled farmers to optimize irrigation schedules, reducing water usage by 20–25% without compromising yield. It also provided valuable insights for the development of drought-resistant cotton varieties.
- The system was integrated into mobile apps, providing farmers with real-time notifications and recommendations for irrigation management.

Key insights

- Combining high-resolution imagery with ground-based sensor data yields a more comprehensive understanding of water stress conditions.
- The use of mobile platforms to deliver real-time feedback to farmers facilitates immediate actions that can mitigate the impacts of water stress.

2.6.4 Case study 4: early detection of drought stress in soybeans—USA

Objective

A study led by Liang *et al.* (2024) aimed to develop a DL-based tool for detecting early drought stress in soybeans, one of the key crops in the USA, using UAV-derived imagery and deep neural networks.

Methodology

- *Data acquisition.* High-resolution UAV-based multispectral imagery was collected across several soybean fields in the Midwest during periods of water scarcity. Various wavelengths, including infrared, were targeted to detect physiological changes in the crop under water stress.
- *Model development.* A CNN architecture was employed to classify stress levels based on leaf canopy characteristics. The model was trained on multispectral images to capture water stress-induced changes in leaf color, texture, and canopy structure.

- *Training data.* Labeled data was obtained from on-ground phenotypic measurements such as leaf water content, photosynthetic rate, and chlorophyll fluorescence.

Results

- The CNN model detected early signs of drought stress up to 2 weeks before visible symptoms appeared in the crops, with an accuracy of 94%.
- *Impact.* Early detection allowed farmers to make timely adjustments to irrigation, preventing significant yield loss and improving water-use efficiency by 15%.
- The tool was also integrated with agronomic models, providing detailed recommendations on irrigation and crop management strategies.

Key insights

- DL models, especially CNNs, are highly effective in detecting subtle early changes in plant physiology that indicate water stress before they become visible.
- Real-time monitoring of crops using UAVs can provide significant advantages for early intervention in large-scale agricultural systems.

2.6.5 Summary of key insights from case studies

These case studies highlight the effectiveness of DL in water stress assessment across different crops, regions, and agricultural contexts. The main takeaways include the following:

- UAVs and remote sensing are critical tools for obtaining high-resolution, real-time data to assess water stress.
- Hybrid DL models that combine CNNs and LSTMs, or other temporal models, show great promise in capturing both spatial and temporal patterns of water stress.
- Early detection of water stress provides actionable insights for water-saving measures, reducing irrigation needs without compromising crop yield.
- Mobile and cloud-based platforms for real-time feedback empower farmers, particularly in resource-limited regions, to make data-driven decisions.

2.7 Conclusion

The application of DL in plant water stress assessment represents a transformative advancement in precision agriculture, enabling accurate detection, prediction, and management of water deficits. By integrating remote sensing, UAVs, and AI-driven analytics, researchers can now identify subtle physiological indicators of stress, facilitating timely interventions to optimize irrigation practices. This chapter reviewed the use of DL models such as CNNs, LSTM networks, and hybrid architectures across various crops and agro-climatic zones, demonstrating improvements in drought resilience, water-use efficiency, and overall resource conservation. Despite these advancements, challenges such as limited annotated datasets, variability across environments, computational burdens, and data privacy concerns hinder broader adoption. Looking ahead, future developments such as multimodal data fusion, edge computing, explainable AI, and participatory machine learning promise to overcome these barriers and shift water stress management from reactive responses to predictive and prescriptive strategies. As climate change, resource scarcity, and population growth intensify, the role of DL in intelligent water management will be crucial to achieving global food security and sustainability. However, success will depend on inclusive approaches that ensure equitable access, farmer participation, and ethical data governance. Ultimately, DL is poised to revolutionize water stress monitoring, driving smarter and more resilient agricultural systems worldwide.

References

Acharya, B. S., Bhandari, M., Bandini, F., Pizarro, A., Perks, M., Joshi, D. R., & Sharma, S. (2021). Unmanned aerial vehicles in hydrology and water management: Applications, challenges, and perspectives. *Water Resources Research*, 57(11), e2021WR029925.

Acikmese, Y., & Alptekin, S. E. (2019). Prediction of stress levels with LSTM and passive mobile sensors. *Procedia Computer Science*, 159, 658-667.

Ahmad, S., Belwal, V., Punia, S. S., Ram, M., Dalip, Rajput, S. S., Kunwar, R. Meena, M. K., Gupta, D., Kumawat, G. L. Hussain, T. & Mohamed, H. I. (2023). Role of plant secondary metabolites and phytohormones in drought tolerance: a review. *Gesunde Pflanzen*, 75(4), 729-746.

Ahmad, T., Katari, P., Pamidi Venkata, A. K., Ravi, C., & Shaik, M. (2024). Explainable AI: Interpreting Deep Learning Models for Decision Support. *Advances in Deep Learning Techniques*, 4(1), 80-108.

Ali, S., Mir, R. A., Haque, M. A., Danishuddin, Almalki, M. A., Alfredan, M., & Mir, Z. A. (2025). Exploring physiological and molecular dynamics of drought stress responses in plants: challenges and future directions. *Frontiers in Plant Science*, 16, 1565635.

Alibabaei, K., Gaspar, P. D., Campos, R. M., Rodrigues, G. C., & Lopes, C. M. (2023). Evaluation of a deep learning approach for predicting the fraction of transpirable soil water in vineyards. *Applied Sciences*, 13(5), 2815.

Allen, B., Agarwal, S., Kalpathy-Cramer, J., & Dreyer, K. (2019). Democratizing ai. *Journal of the American College of Radiology*, 16(7), 961-963.

Al-Tamimi, N., Langan, P., Bernád, V., Walsh, J., Mangina, E., & Negrão, S. (2022). Capturing crop adaptation to abiotic stress using image-based technologies. *Open Biology*, 12(6), 210353.

Archana, R., & Jeevaraj, P. E. (2024). Deep learning models for digital image processing: a review. *Artificial Intelligence Review*, 57(1), 11.

Asaari, M. S. M., Mertens, S., Dhondt, S., Inzé, D., Wuyts, N., & Scheunders, P. (2019). Analysis of hyperspectral images for detection of drought stress and recovery in maize plants in a high-throughput phenotyping platform. *Computers and Electronics in Agriculture*, 162, 749-758.

Aubard, M., Madureira, A., Teixeira, L., & Pinto, J. (2025). Sonar-Based Deep Learning in Underwater Robotics: Overview, Robustness, and Challenges. *IEEE Journal of Oceanic Engineering*.

Awais, M., Li, W., Cheema, M. J. M., Zaman, Q. U., Shaheen, A., Aslam, B., & Liu, C. (2022). UAV-based remote sensing in plant stress imagine using high-resolution thermal sensor for digital agriculture practices: A meta-review. *International Journal of Environmental Science and Technology*, 1-18.

Bagherian, K. (2022). *Rapid peanut phenotyping and water quality monitoring using remote sensing and machine learning techniques* (Master's thesis, Auburn University).

Begna, T. (2022). Impact of drought stress on crop production and its management options. *Int. J. Res. Agron*, 8, 1-13.

Benfenati, A., Bolzi, D., Causin, P., & Oberti, R. (2022). A deep learning generative model approach for image synthesis of plant leaves. *Plos one*, 17(11), e0276972.

Bhattacharya, A. (2021). Effect of soil water deficit on growth and development of plants: a review. *Soil water deficit and physiological issues in plants*, 393-488. Springer Singapore.

Borol, Y. D., Thilagham, K. T., Nagpal, A., Harika, A., & Shnawa, A. H. (2024). Hyperspectral Information with Big Data and Machine Learning for Agriculture. In *2024 4th International Conference on Innovative Practices in Technology and Management (ICIPTM)* (pp. 1-6). IEEE.

Bwambale, E., Abagale, F. K., & Anornu, G. K. (2022). Smart irrigation monitoring and control strategies for improving water use efficiency in precision agriculture: A review. *Agricultural Water Management*, 260, 107324.

Carella, A. (2024). *Development of an effective and sustainable system to monitor fruit tree water status with precision devices*.

Chada, S., Asiedu, S., & Ofoe, R. (2023). An overview of plant morpho-physiology, biochemicals, and metabolic pathways under water stress. *Horticult Int J*, 7(4), 115-25.

Cho, S. B., Soleh, H. M., Choi, J. W., Hwang, W. H., Lee, H., Cho, Y. S., & Kim, G. (2024). Recent Methods for Evaluating Crop Water Stress Using AI Techniques: A Review. *Sensors*, 24(19), 6313.

Danandeh Mehr, A., Ghavifekr, A. A., Ghazaei, E., Safari, M. J. S., Ke, C. Q., & Nourani, V. (2025). S-Transformer: a new deep learning model enhanced by sequential transformer encoders for drought forecasting. *Earth Science Informatics*, 18(2), 341.

de Andrade Porto, J. V., Dorsa, A. C., de Moraes Weber, V. A., de Andrade Porto, K. R., & Pistori, H. (2023). Usage of few-shot learning and meta-learning in agriculture: A literature review. *Smart Agricultural Technology*, 5, 100307.

Dembani, R., Karvelas, I., Akbar, N. A., Rizou, S., Tegolo, D., & Fountas, S. (2025). Agricultural data privacy and federated learning: A review of challenges and opportunities. *Computers and Electronics in Agriculture*, 232, 110048.

Dwivedi, S. L., Chapman, M. A., Abberton, M. T., Akpojotor, U. L., & Ortiz, R. (2023). Exploiting genetic and genomic resources to enhance productivity and abiotic stress adaptation of underutilized pulses. *Frontiers in Genetics*, 14, 1193780.

Dzieżyc, M., Gjoreski, M., Kazienko, P., Saganowski, S., & Gams, M. (2020). Can we ditch feature engineering? end-to-end deep learning for affect recognition from physiological sensor data. *Sensors*, 20(22), 6535.

Falcioni, R., Antunes, W. C., Demattê, J. A., & Nanni, M. R. (2023). Biophysical, biochemical, and photochemical analyses using reflectance hyperspectroscopy and chlorophyll a fluorescence kinetics in variegated leaves. *Biology*, 12(5), 704.

Fayaz, S., Ahmad Shah, S. Z., ud din, N. M., Gul, N., & Assad, A. (2024). Advancements in Data Augmentation and Transfer Learning: A Comprehensive Survey to Address Data Scarcity Challenges. *Recent Advances in Computer Science and Communications (Formerly: Recent Patents on Computer Science)*, 17(8), 14-35.

Galieni, A., D'Ascenzo, N., Stagnari, F., Pagnani, G., Xie, Q., & Pisante, M. (2021). Past and future of plant stress detection: an overview from remote sensing to positron emission tomography. *Frontiers in Plant Science*, 11, 609155.

Gao, Z., Luo, Z., Zhang, W., Lv, Z., & Xu, Y. (2020). Deep learning application in plant stress imaging: a review. *AgriEngineering*, 2(3), 29.

Gharoun, H., Momenifar, F., Chen, F., & Gandomi, A. H. (2024). Meta-learning approaches for few-shot learning: A survey of recent advances. *ACM Computing Surveys*, 56(12), 1-41.

Gommes, R., Das, H., Mariani, L., Challinor, A., Tychon, B., Balaghi, R., & Dawod, M. A. (2010). Agrometeorological forecasting.

Goyal, P., Sharda, R., Saini, M., & Siag, M. (2024). A deep learning approach for early detection of drought stress in maize using proximal scale digital images. *Neural Computing and Applications*, 36(4), 1899-1913.

Grossiord, C., Buckley, T. N., Cernusak, L. A., Novick, K. A., Poulter, B., Siegwolf, R. T., & McDowell, N. G. (2020). Plant responses to rising vapor pressure deficit. *New phytologist*, 226(6), 1550-1566.

Guo, F., Wang, P., Tansey, K., Zhang, Y., Li, M., Liu, J., & Zhang, S. (2024). A novel transformer-based neural network under model interpretability for improving wheat yield estimation using remotely sensed multi-variables. *Computers and Electronics in Agriculture*, 223, 109111.

Holzman, M. E., Rivas, R. E., & Bayala, M. I. (2021). Relationship between TIR and NIR-SWIR as indicator of vegetation water availability. *Remote Sensing*, 13(17), 3371.

Ibañez, S. C., & Monterola, C. P. (2023). A global forecasting approach to large-scale crop production prediction with time series transformers. *Agriculture*, 13(9), 1855.

Igwenagu, U. T. I., Debnath, R., Ahmed, A. A., & Alam, M. J. B. (2025). An Integrated Approach for Earth Infrastructure Monitoring Using UAV and ERI: A Systematic Review. *Drones*, 9(3), 225.

Ihuoma, S. O., & Madramootoo, C. A. (2017). Recent advances in crop water stress detection. *Computers and Electronics in Agriculture*, 141, 267-275.

Islam, S., Reza, M. N., Samsuzzaman, S. A., Cho, Y. J., Noh, D. H., Chung, S. O., & Hong, S. J. (2024). Machine vision and artificial intelligence for plant growth stress detection and monitoring: A review. *Precision Agriculture*, 6(1), 34.

Jain, S. K., Wettberg, E. J. V., Punia, S. S., Parihar, A. K., Lamichaney, A., Kumar, J., Gupta, D. S., Ahmad, S., Pant, N. C., Dixit, G. P., Sar, H., Sari, D., Maruf, A., Toker, P., Toler, C. (2023). Genomic-mediated breeding strategies for global warming in chickpeas (Cicer arietinum L.). *Agriculture*, 13(9), 1721.

Jin, K., Zhang, J., Wang, Z., Zhang, J., Liu, N., Li, M., & Ma, Z. (2024). Application of deep learning based on thermal images to identify the water stress in cotton under film-mulched drip irrigation. *Agricultural Water Management*, 299, 108901.

Jouini, O., Sethom, K., Namoun, A., Aljohani, N., Alanazi, M. H., & Alanazi, M. N. (2024). A survey of machine learning in edge computing: Techniques, frameworks, applications, issues, and research directions. *Technologies*, 12(6), 81.

Kamarudin, M. H., Ismail, Z. H., & Saidi, N. B. (2021). Deep learning sensor fusion in plant water stress assessment: A comprehensive review. *Applied Sciences*, 11(4), 1403.

Kamarudin, M. H., Ismail, Z. H., & Saidi, N. B. (2021). Deep learning sensor fusion in plant water stress assessment: A comprehensive review. *Applied Sciences*, 11(4), 1403.

Kaur, H., Kohli, S. K., Khanna, K., & Bhardwaj, R. (2021). Scrutinizing the impact of water deficit in plants: Transcriptional regulation, signaling, photosynthetic efficacy, and management. *Physiologia Plantarum*, 172(2), 935-962.

Keskes, M. (2025). Review of The Current State of Deep Learning Applications in Agriculture.

Lapajne, J., Vojnović, A., Vončina, A., & Žibrat, U. (2024). Enhancing water-deficient potato plant identification: assessing realistic performance of attention-based deep neural networks and hyperspectral imaging for agricultural applications. *Plants*, 13(14), 1918.

Lazić, O., Cvejić, S., Dedić, B., Kupusinac, A., Jocić, S., & Miladinović, D. (2024). Transfer Learning in Multimodal Sunflower Drought Stress Detection. *Applied Sciences*, 14(14), 6034.

Li, M. W., Chan, Y. K., & Yu, S. S. (2023). Use of CNN for Water Stress Identification in Rice Fields Using Thermal Imagery. *Applied Sciences*, 13(9), 5423.

Liang, H., Zhou, Y., Lu, Y., Pei, S., Xu, D., Lu, Z., & Li, H. (2024). Evaluation of Soybean Drought Tolerance Using Multimodal Data from an Unmanned Aerial Vehicle and Machine Learning. *Remote Sensing*, 16(11), 2043.

Lin, Z., Li, M., Zheng, Z., Cheng, Y., & Yuan, C. (2020, April). Self-attention convlstm for spatiotemporal prediction. In *Proceedings of the AAAI conference on artificial intelligence* (Vol. 34, No. 07, pp. 11531-11538).

Lu, J., Li, J., Fu, H., Tang, X., Liu, Z., Chen, H., & Ning, X. (2024). Deep learning for multi-source data-driven crop yield prediction in northeast China. *Agriculture*, 14(6), 794.

Martin, R. J., Mittal, R., Malik, V., Jeribi, F., Siddiqui, S. T., Hossain, M. A., & Swapna, S. L. (2024). XAI-Powered Smart Agriculture Framework for Enhancing Food Productivity and Sustainability. *IEEE Access*.

Mekdad, Y., Aris, A., Babun, L., El Fergougui, A., Conti, M., Lazzeretti, R., & Uluagac, A. S. (2023). A survey on security and privacy issues of UAVs. *Computer networks*, 224, 109626.

Mienye, I. D., & Swart, T. G. (2024). A comprehensive review of deep learning: Architectures, recent advances, and applications. *Information*, 15(12), 755.

Najafabadi, M. M., Villanustre, F., Khoshgoftaar, T. M., Seliya, N., Wald, R., & Muharemagic, E. (2015). Deep learning applications and challenges in big data analytics. *Journal of big data*, 2, 1-21.

Nguyen-Tan, T., & Le-Trung, Q. (2024). A novel 5G PMN-driven approach for AI-powered irrigation and crop health monitoring. *IEEE access.*

O'Grady, M. J., Langton, D., & O'Hare, G. M. P. (2019). Edge computing: A tractable model for smart agriculture? *Artificial Intelligence in Agriculture*, 3, 42-51.

Ondrasek, G. (2013). Water scarcity and water stress in agriculture. In *Physiological Mechanisms and Adaptation Strategies in Plants Under Changing Environment: Volume 1* (pp. 75-96). New York, NY: Springer New York.

Ouyang, J., De Bei, R., Fuentes, S., & Collins, C. (2020). UAV and ground-based imagery analysis detects canopy structure changes after canopy management applications. *Oeno One*, 54(4), 1093-1103.

Ozturk, M., Turkyilmaz Unal, B., García-Caparrós, P., Khursheed, A., Gul, A., & Hasanuzzaman, M. (2021). Osmoregulation and its actions during the drought stress in plants. *Physiologia plantarum*, 172(2), 1321-1335.

Pagay, V. (2014). Physiological responses of grapevine shoots to water stress and The development of a microtensiometer to continuously measure water potential.

Parashar, N., Johri, P., Khan, A. A., Gaur, N., & Kadry, S. (2024). An Integrated Analysis of Yield Prediction Models: A Comprehensive Review of Advancements and Challenges. *Computers, Materials & Continua*, 80(1).

Parwez, R., Aftab, T., Gill, S. S., & Naeem, M. (2022). Abscisic acid signaling and crosstalk with phytohormones in regulation of environmental stress responses. *Environmental and Experimental Botany*, 199, 104885.

Peeters, A., Cohen, Y., Bahat, I., Ohana-Levi, N., Goldshtein, E., Netzer, Y., & Ben-Gal, A. (2024). A spatial machine-learning model for predicting crop water stress index for precision irrigation of vineyards. *Computers and Electronics in Agriculture*, 227, 109578.

Peng, M., Liu, Y., Khan, A., Ahmed, B., Sarker, S. K., Ghadi, Y. Y., & Ali, Y. A. (2024). Crop monitoring using remote sensing land use and land change data: Comparative analysis of deep learning methods using pre-trained CNN models. *Big Data Research*, 36, 100448.

Priyatikanto, R., Lu, Y., Dash, J., & Sheffield, J. (2023). Improving generalisability and transferability of machine-learning-based maize yield prediction model through domain adaptation. *Agricultural and Forest Meteorology*, 341, 109652.

Rahman, M. H., Busby, S., Ru, S., Hanif, S., Sanz-Saez, A., Zheng, J., & Rehman, T. U. (2025). Transformer-Based hyperspectral image analysis for phenotyping drought tolerance in blueberries. *Computers and Electronics in Agriculture*, 228, 109684.

Reddy, J., Niu, H., & Duffield, N. (2024). A CNN-LSTM model for cotton water stress classification. In *2024 ASABE Annual International Meeting* (p. 1). American Society of Agricultural and Biological Engineers.

Rundel, P. W., Graham, E. A., Allen, M. F., Fisher, J. C., & Harmon, T. C. (2009). Environmental sensor networks in ecological research. *New Phytologist*, 182(3), 589-607.

Samadzadegan, F., Toosi, A., & Dadrass Javan, F. (2025). A critical review on multi-sensor and multi-platform remote sensing data fusion approaches: current status and prospects. *International Journal of Remote Sensing*, 46(3), 1327-1402.

Sanders, G. J., & Arndt, S. K. (2012). Osmotic adjustment under drought conditions. In *Plant responses to drought stress: From morphological to molecular features* (pp. 199-229). Berlin, Heidelberg: Springer Berlin Heidelberg.

Sapkota, R., Qureshi, R., Hassan, S. Z., Shutske, J., Shoman, M., Sajjad, M., & Karkee, M. (2024). Multi-modal LLMs in agriculture: A comprehensive review. *Authorea Preprints*.

Sarma, H. H., Borah, S. K., Dutta, N., Sultana, N., Nath, H., & Das, B. C. (2024). Innovative approaches for climate-resilient farming: strategies against environmental shifts and climate change. *International Journal of Environment and Climate Change*, 14(9), 217-241.

Shang, M., & Xie, J. (2024). Agricultural sustainable development: Soil, water resources, biodiversity, climate change, and technological innovation. *Advances in Resources Research*, 4(2), 181-204.

Sheldon, A. R., Dalal, R. C., Kirchhof, G., Kopittke, P. M., & Menzies, N. W. (2017). The effect of salinity on plant-available water. *Plant and Soil*, 418, 477-491.

Skendžić, S., Zovko, M., Lešić, V., Pajač Živković, I., & Lemić, D. (2023). Detection and evaluation of environmental stress in winter wheat using remote and proximal sensing methods and vegetation indices—A review. *Diversity*, 15(4), 481.

Still, C., Powell, R., Aubrecht, D., Kim, Y., Helliker, B., Roberts, D., & Goulden, M. (2019). Thermal imaging in plant and ecosystem ecology: applications and challenges. *Ecosphere*, 10(6), e02768.

Struthers, R., Ivanova, A., Tits, L., Swennen, R., & Coppin, P. (2015). Thermal infrared imaging of the temporal variability in stomatal conductance for fruit trees. *International Journal of Applied Earth Observation and Geoinformation*, 39, 9-17.

Su, J. C., Maji, S., & Hariharan, B. (2020). When does self-supervision improve few-shot learning?. In *European conference on computer vision* (pp. 645-666). Cham: Springer International Publishing.

Sun, Y., Lao, D., Ruan, Y., Huang, C., & Xin, Q. (2023). A deep learning-based approach to predict large-scale dynamics of normalized difference vegetation index for the monitoring of vegetation activities and stresses using meteorological data. *Sustainability*, 15(8), 6632.

Takalloo, S. (2024). *Water Stress Detection in Potato Crops Using Multispectral Imaging and Advanced Object Detection Models* (Doctoral dissertation, Politecnico di Torino).

Uslu, S. (2024). *Trustworthy and Causal Artificial Intelligence in Environmental Decision Making*. Purdue University.

Wang, D., Cao, W., Zhang, F., Li, Z., Xu, S., & Wu, X. (2022). A review of deep learning in multiscale agricultural sensing. *Remote Sensing*, 14(3), 559.

Wen, T., Li, J. H., Wang, Q., Gao, Y. Y., Hao, G. F., & Song, B. A. (2023). Thermal imaging: The digital eye facilitates high-throughput phenotyping traits of plant growth and stress responses. *Science of The Total Environment*, 899, 165626.

Weng, S., Ma, J., Tao, W., Tan, Y., Pan, M., Zhang, Z., & Zhao, J. (2023). Drought stress identification of tomato plant using multi-features of hyperspectral imaging and subsample fusion. *Frontiers in Plant Science*, 14, 1073530.

Yi, J., Krusenbaum, L., Unger, P., Hüging, H., Seidel, S. J., Schaaf, G., & Gall, J. (2020). Deep learning for non-invasive diagnosis of nutrient deficiencies in sugar beet using RGB images. *Sensors*, 20(20), Priyatikanto, R., Lu, Y., Dash, J., & Sheffield, J. (2023). Improving generalisability and transferability of machine-learning-based maize yield prediction model through domain adaptation. *Agricultural and Forest Meteorology*, *341*, 109652. 5893.

Zahedi, S. M., Karimi, M., Venditti, A., Zahra, N., Siddique, K. H., & Farooq, M. (2025). Plant adaptation to drought stress: the role of anatomical and morphological characteristics in maintaining the water status. *Journal of Soil Science and Plant Nutrition*, 25(1), 409-427.

Zhang, Y., Chen, X., Geng, S., & Zhang, X. (2025). A review of soil waterlogging impacts, mechanisms, and adaptive strategies. *Frontiers in Plant Science*, 16, 1545912.

Zhuang, J., Wang, Y., Chi, Y., Zhou, L., Chen, J., Zhou, W., & Ding, J. (2020). Drought stress strengthens the link between chlorophyll fluorescence parameters and photosynthetic traits. *PeerJ*, 8, e10046.

Zwieniecki, M. A., & Secchi, F. (2015). Threats to xylem hydraulic function of trees under 'new climate normal' conditions. *Plant, Cell & Environment*, 38(9), 1713-1724.

3 Deep Learning for Plant Disease Detection and Identification

Siddhartha Das[1]* and Priyanka Paul[2]
[1]*Department of Plant Pathology, M.S. Swaminathan School of Agriculture, Centurion University of Technology and Management, Paralakhemundi, Odisha, India;*
[2]*Department of Agronomy and Agroforestry, M.S. Swaminathan School of Agriculture, Centurion University of Technology and Management, Paralakhemundi, Odisha, India*

Abstract

Plant diseases pose a significant threat to farming operations and food supply chains by exposing crops to harmful pathogens, thereby affecting the livelihoods of those involved. The swift progress in deep learning (DL) has revolutionized conventional methods for detecting plant diseases, providing unmatched levels of accuracy, speed, and scalability. This chapter explores the identification of various plant diseases such as blight, rust, leaf spot, and powdery mildew. Advanced image augmentation, data balancing, and feature extraction can contribute to the resilience of the models. The implementation of automated leaf disease diagnosis using the Internet of Things, DL, and artificial intelligence sensors is being explored for enhanced analysis and detection. Data visualization through recurrent neural networks and convolutional neural networks models established excellent milestones for plant diseases, alterations with gene regulation, and ion concentration spectrum differences in tissues of plants. This chapter provides a concise draft of plant pathogen detection through modern DL techniques.

Keywords: deep learning, pathogens, plant diseases, sensors

3.1 Introduction

Agriculture is a vital sector for improving a country's economy but faces several problems that impact grain yield and security of nutrition. Factors such as human development, global warming, scarce land for agriculture, plant diseases, and the coronavirus disease 2019 pandemic have exacerbated these issues, with plant diseases posing a significant risk to crops. Diseases, driven by climate change, water stress, and insect infestations, cause annual losses of about 40% of food crops, including the severe loss of wheat crops due to yellow rust (Beddow *et al.*, 2015). The ongoing Russian–Ukrainian war is expected to worsen wheat losses, as both countries are major producers. Reduced crop yields contribute to food insecurity and malnutrition, highlighting the need for early disease identification to enhance food productivity. Traditional disease detection methods, such as human inspection and ground-level machinery, are laborious and costly, keeping them inadequate for large-scale farming surveillance (Zhang *et al.*,

*E-mail: siddhartha.das10@gmail.com

DOI: 10.1079/9781836991120.0003

2021). These technologies enhance resource utilization and optimize crop quality (Maddikunta *et al.*, 2021). Machine vision and developed deep learning (DL) techniques have shown considerable promise in recognizing plant diseases and pests, shifting traditional methods of manual feature extraction. To address the problem, the agricultural industry is increasingly using automated technologies, including Internet of Things (IoT), machine learning, sensors, and robotics, as a part of shift toward Smart Farming or Precision Agriculture due to the lack of technological skills and infrastructure (Shaikh *et al.*, 2022). In countries like India, where agriculture plays an economic role and crop damage is significant, prospects for DL-powered identification of plant diseases are especially high (Agarwal *et al.*, 2019). Convolutional neural networks (CNNs), a form of DL, are especially profitable in analyzing crop images to identify disease-related visual patterns (Shaikh *et al.*, 2022). However, challenges remain, as environmental factors such as poor lighting, low contrast, and noise in images can impact the accuracy of these models (Zhang *et al.*, 2021). The World Health Organization forecasts that by 2030, over 600 million people will face hunger, with plant diseases being a significant component (Munjal *et al.*, 2023). Early disease detection is essential, as fungal diseases alone account for 83% of plant diseases globally (Hlophe-Ginindza and Mpandeli, 2021). With developments in technology, many farmers still depend on traditional visual inspections for disease detection.

3.2 Plant Disease Detection through Conventional Strategies and Cutting-Edge Deep Learning Algorithms

Conventional methods for identifying plant diseases depend on visual symptoms, field analysis, and laboratory testing (Fig. 3.1). Laboratory-based tests, such as microscopy, serological testing, and DNA analysis, offer accurate verification and early identification of pathogens. However, they tend to be more costly and time-consuming, requiring skilled personnel and proper sample handling (Ahmad *et al.*, 2021). While these developed methods provide precise data, delays and the need for learned individuals can prevent timely disease management (Bhagwat *et al.*, 2021). Visual inspection involves assessing symptoms such as discoloration, leaf stains, and abnormal growth, providing a simple and attainable diagnostic tool for farmers and field workers (Yakkundimath *et al.*, 2022). However, this method can be subjective, causing biases and delayed detection, especially since some diseases only show symptoms in the following stages (Singh *et al.*, 2017).

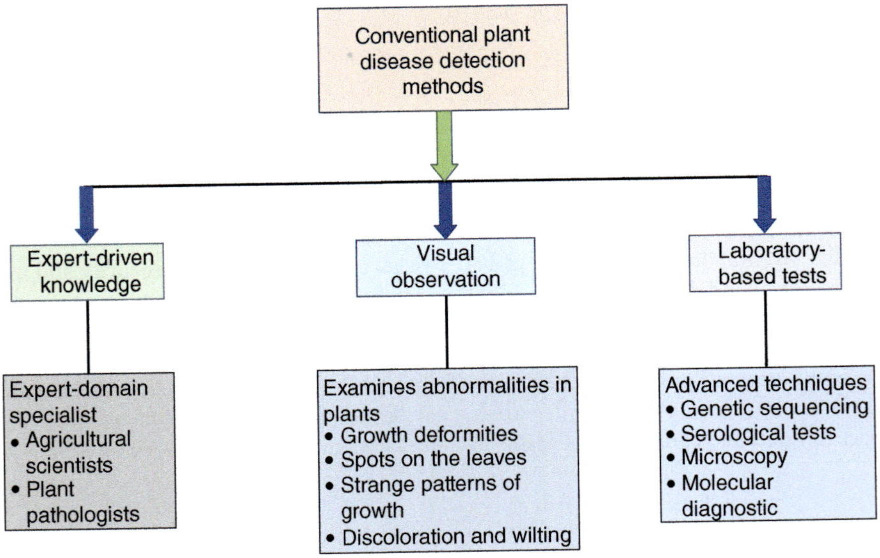

Fig. 3.1. Conventional methods of plant disease detection.

3.3 Deep Learning Algorithms are Currently Being Used to Detect Plant Diseases in Leaves

Deep learning trains neural networks, which are computer systems capable of collecting and interpreting information directly from data (Taye *et al.*, 2023). It has demonstrated effectiveness in a wide range of applications, such as artificial intelligence (AI), natural language processing, recognition of speech, and system of recommendations (Sarker *et al.*, 2021). The ability to learn classification algorithms from data in its entirety, paired with advances in computing and huge-scale information sets, has led to substantial advances in machine intelligence. DL approaches, including Transformers, CNNs, recurrent neural networks, and generative adversarial networks (GANs) (Fig. 3.2), are effective tools for identifying diseases in plants. The methods are discussed in the sections below.

3.4 Using Convolutional Neural Networks to Identify Diseases in Plants

Convolutional neural networks have been shown to be highly effective in identifying plant diseases, as they can continuously learn and collect crucial properties from images, including texture, color, and pattern. The stimulation maps emphasize areas crucial for decision-making. The system is formed of convex, grouping, and fully linked layers, with the capacity to capture spatial patterns and generate feature maps for accurate classification (Zhang *et al.*, 2019). CNNs can distinguish between healthy and diseased plants with high accuracy, determining complicated patterns linked to various diseases (Shamshirband *et al.*, 2021).

3.5 Techniques that Use Content-Based Filtering to Offer Treatment Ideas for Plant Diseases

A system based on a rule-based approach to plant disease management provides ideas based on known diseases, plant species, and environmental factors. The system compares plant symptoms against established rules and suggests treatments, such as isolating affected plants or using bactericides for specific infections. For example, proficient systems have been developed for diagnosing and controlling disorders in watermelon and onion plants, respectively (Alajrami and Abu-Naser *et al.*, 2018). These systems analyze disease characteristics and relevant factors to offer personalized treatment solutions.

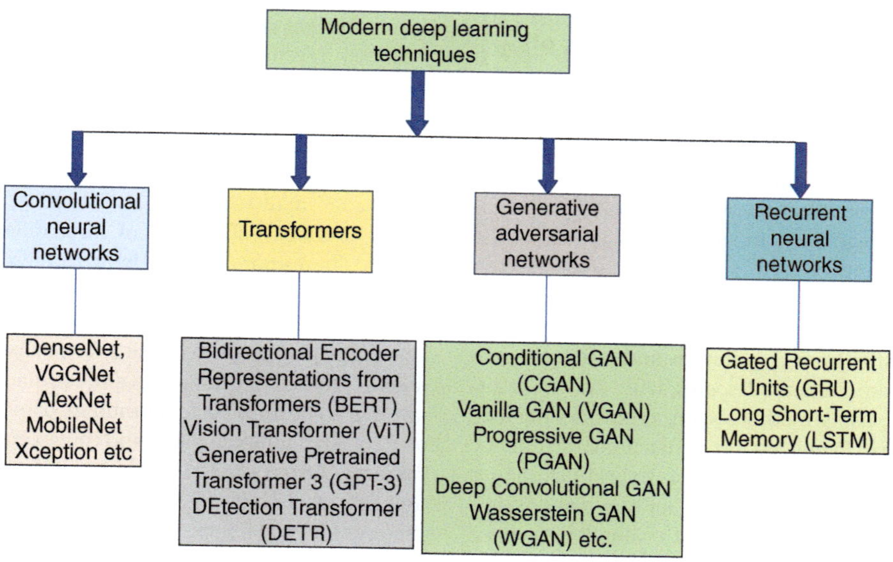

Fig. 3.2. Modern deep learning techniques for plant disease detection.

3.5.1 Constraints of content-based filtered methods in proposing treatments for plant diseases

Content-based filtering (CBF) is a technique for plant disease treatment that evaluates plant and disease characteristics to suggest appropriate solutions (Akulwar *et al.*, 2020). While CBF can be effective, it has limitations, including dependence on data quality and accessibility, which can contribute to insufficient suggestions (Wang *et al.*, 2019). Feature-based filtering identifies important features, such as crop features, disease symptoms, and regional conditions, to provide aimed treatment recommendations. It uses techniques like combination filtering, which integrates content-based and group screening to improve accuracy with additional information sources, such as treatment records and user feedback (Afoudi *et al.*, 2021). However, CBF does not encompass insights from other users or skills, reducing its scope. It may also omit factors like genetic differences and environmental changes that affect disease management (Son *et al.*, 2017). Combining CBF with other approaches, like dynamic filtering, can improve system efficiency by addressing these limitations.

3.6 Probable Ways to Integrate Content-Based Methods of Screening with Algorithms for Deep Learning to Supply Targeted and Effective Plant Disease Management Suggestions

Integrating machine learning and content-driven filtering offers an operational method for controlling plant leaf diseases. Multimodal information integration, including images, text, and environmental factors, improves disease identification and treatment suggestions (Zhao *et al.*, 2023). DL techniques such as CNNs detect characteristics from crop images and utilize them to calculate compared scores with known treatments, using methods such as cosine similarity and term frequency-inverse document frequency (TF-IDF) (Zhu *et al.*, 2020). Hybrid techniques combine neural networks for the diagnosis of disease with content-based evaluation to make suggestions for therapy (Subramanian *et al.*, 2022). Advanced learning models, such as Asian or triplet systems, can identify interactions across the foliage and medicinal recommendations (Smetanin *et al.*, 2021).

3.7 Collections of Data for the Training and Evaluation of Deep Neural Networks and Filtering Based on Content Methods to Detect and Manage Diseases in Plants

Plant infection records are essential for the advancement of plant pathology and for enhancing disease detection, management, and crop improvement (Shoaib *et al.*, 2023). Publicly obtainable datasets, containing tagged images of healthy and diseased plants, are widely used in training and evaluating AI models, thereby aiding the development of effective plant disease detection and classification methods. The plant pathology dataset consists of photographs developed for a competition called Kaggle to consider apple diseases by their foliage. The dataset contains 1200 images of apple scab, 1399 cedar apple rot, 187 mixed indications of disease, and 865 healthy leaves. The photographs were taken using a Canon Rebel T5i DSLR and various iPhones under different field conditions, offering diverse learning and evaluation opportunities. The PlantVillage collection contains 54,303 images of both healthy and diseased leaves, organized into 38 groups dependent on category and condition. It represents 14 varieties of crops and contains images of 17 fungal, four bacterial, two viral, two infectious, and one mite-borne diseases. It also includes healthy leaves from 12 species. Most of the images were captured in controlled lab environments. However, the dataset's class imbalance poses difficulties for model training (Ahmad *et al.*, 2023). The Rice Leaf dataset contains 120 images of three different diseases. The authors obtained images of disease-infected leaves at 2848 × 4288 pixels (Fenu *et al.*, 2021). These images were captured in the same coffee plantation field using a smartphone camera. The RoCoLe dataset evaluates machine learning algorithms for determining plant diseases through image division and categorization (Parraga-Alava *et al.*, 2019). The New Plant Diseases dataset contains 87,000 images of healthy and diseased crop foliage, divided into 38 types. The dataset is divided into 80% for instruction and 20% for confirmation, with 33 test images for forecasting. Offline augmentation methods were applied. Plant disease datasets are crucial for enhancing detection, management, and prediction models. Improving the quality and availability supports eco-friendly farming techniques.

3.8 Multiple-Observation Dataset

Generally speaking, human experts rely on multiple observations, rather than one, to identify plant diseases. Motivated by this circumstance, we argue that multiple observations can also enhance DL techniques. In simple terms, diverse knowledge is dispersed by various observations. Datasets containing multiple methods for similar diseases of plants are referred to as multimodal datasets. For example, different optical texts and image types can be generated from a leaf affected by a plant disease. Large datasets can improve model generalization across various computer vision tasks (Zhai *et al.*, 2022). Although collecting huge datasets for plant disease identification is time-consuming, challenging, and costly, but nevertheless worthwhile. Collaboration allows individuals in various locations to capture and submit images to a portal. These images can then be evaluated by the public. The obtained datasets exhibit significant variety, which helps generalize models (Xu *et al.*, 2023). Furthermore, datasets with time-series can be created by, for example, photographing diseases in plants at various points in time. Specifically, as diseases develop over time, visual indicators become more distinct and easier to identify. The difficulty of early plant disease detection may be lessened with periods of time records. Multiple spatial datasets, which include shooting pictures from various angles and levels, can also result in improved test outcomes for data from images. Certain plant diseases, for instance, have distinct designs on the front as well as the rear of their leaves.

3.9 Metrics for Evaluating the Accuracy and Potency of Content-Based Filtering and Deep Learning Techniques

Evaluating treatment options for plant diseases involves several key metrics to assess model effectiveness. Precision assesses the total accuracy of forecasts, giving how many correct treatment suggestions among all cases. The F1 score combines accuracy with recall to provide an in-depth assessment of disease management (Islam *et al.*, 2022). The mean absolute error (MAE) defines the precision of predicted treatment effectiveness, with lower MAE indicating more precise recommendations (Afify *et al.*, 2017). Precision measures the ability of the model to make correct positive forecasts, while recall (or sensitivity) assesses the ability to identify all actual positive cases and prescribe important procedures (Jiang *et al.*, 2021). These metrics offer an understanding of model abilities and drawbacks in plant disease treatment recommendations. Finally, execution time refers to the amount of time needed for the learning, indicating how quickly the model can make estimates on new data.

3.10 Determine the Gaps and Areas for Development in the Current Situation as well as Potential Future Directions

This section focuses on the constraints, prospects for development, and prospective future developments in plant disease diagnostics and treatment recommendations using pharmaceutical systems.

3.11 Deficiencies and Improvements

Despite innovations in the recognition of images and DL, identification of plant disease methods remains inefficient. These technologies' efficiency is dependent on both the accuracy and amount of information used for training, which may not represent all disease types or account for geographical differences. Current methods rely on static images or periodic data, failing to capture the exciting origin of root infections. Real-time monitoring is important to observe sickness transmission quickly. However, the combination of IoT devices and satellite imagery offers the possibility for real-time disease monitoring and early detection, resulting in faster interventions. Predictive models using historical data and weather patterns show assurance but still require further development. Furthermore, plant disease detection systems demand substantial analyzing power to manage large datasets and complex algorithms, this can be difficult to implement in rural regions with limited resources.

Specialized expertise can improve system efficiency by obtaining relevant features from incoming data.

3.12 Historical Perspective Deep Learning Techniques for the Identification of Plant Diseases

Crop-identifying diseases evaluate the occurrence, intensity, and impact of crop infections on growth, as diseases can impair plant function within a specific period (Abdu *et al.*, 2020). Machine vision methods are used to detect viruses by analyzing images (Liu and Wang, 2021). Disease classification categorizes crops into groups based on measures and color differences between healthy and infected leaves (Halder *et al.*, 2018). This classification can be supervised or unsupervised, using methods such as color data, image boundaries, and division. Crop disease severity is determined as amount of infected tissue related to total plant tissue, which is an ongoing variable demonstrating a pair of values (0 or 1). Crop severity, which reflects the force of disease symptoms, is crucial for forecasting and controlling production (Wu *et al.*, 2019). AI methods are promoting essential in precision agriculture for assessing quality and combining multiple image sources for better evaluation (Bock, 2022).

3.13 Categorization of Several Deep Learning Models

Crop diseases identifying and categorizing using AI are gaining popularity in agriculture due to their ability to diagnose plant diseases quickly and accurately. These strategies use data and algorithmic models as sources of information to assist scientists and workers in identifying, monitoring, and managing plant diseases. Below are some of the common techniques used in this field.

3.13.1 Plant disease identification methods include clustering through k-means, KNN, and GLCM

Plant disease recognition and categorization in agriculture can be achieved through clustering techniques such as support vector machine (SVM), k-means, and K-nearest neighbors (KNN). Diseased plant leaves were identified and extracted with the use of k-means clustering, followed by extracting features.

With the gray-level co-occurrence matrix (GLCM) approach, we utilized SVM and KNN for classification. The algorithm was verified on a private dataset of 62 images, categorized as either normal or unsafe, and was successfully classified with a perfect score of 100%. The approach aims to help farmers discover plant diseases, make informed decisions, and quickly select effective solutions (Anjna *et al.*, 2020).

3.13.2 Crop disease identification and categorization using SVM

An automated system was developed to send images of diseased plants to a central repository for analysis, categorization, and treatment recommendations for the author. Artificial neural network (ANN), SVM, and transfer learning (TL) were utilized to generate plants that improved agricultural disease identification efficiency.

ANN has been employed for recognition because it has interconnected layers that may acquire knowledge and interpret input. SVM was employed for hyperplane grouping, and TL was utilized again to increase model reliability and effectiveness, because the dataset of paddy stalks was relatively small. The collection included 2149 paddy leaf images, both healthy and diseased samples (Suresh, 2023).

3.14 Crop Disease Identification with Random Forest Methods

A crop disease diagnosis model was developed using the random forest (RF) and graph of an oriented gradient (HOG) method. HOG was employed to extract features from a labeled collection of 160 papaya leaf images. The collected features are used to generate an RF model. The algorithm had an accuracy in classification of 70.14%. The creators suggest enhancing the total amount of images utilized for conditioning the algorithm for better accuracy. To increase and improve the quality of the model, a variety of global and local image features, including SIFT, DENSE, and BOVW, are widely used.

3.14.1 Recognition of crop diseases based on image functions, KNN, and GLCM

A crop disease identification method combines KNN, Cetalatran Efficiency, and Relief algorithms. The alleviation technique selects significant features of hyperspectral UV images using significant associations. The KNN classifier is learned using these selected features, and Cetalatran optimization is used to fine-tune the classifier. This technique achieved 89.12% accuracy, 93.39% sensitivity, and 92.77% specificity (Gaikwad and Musande, 2023). Additionally, image processing techniques such as histogram equalization and K-means clustering, combined with feature extraction methods including discrete wavelet transform, principal component analysis, and GLCM, improve the forecasting of plant health, with KNN, CNN, and SVM used for classification (Harakannanavar *et al.*, 2022).

3.14.2 A technique based on deep learning for predicting crop disease intensity

The use of DL methods for estimating crop disease sensitivity can effectively handle complex and dynamic trends. Precision agriculture may optimize treatment procedures, decrease the use of insecticides, and enhance cultivation methods. These tools enable early detection and analysis of diseases, averting significant crop losses. The next sections give a quick summary of various DL-based techniques for assessing crop disease incidence.

3.14.3 Crop disease severity prediction with ResNet50

In this method, a computer visual system detects diseases, identifies new ones, screens for them, and estimates the foliar disease severity. The suggested method uses ResNet50 to detect earlier unreported cases of disorders, while the U-Net DL modeling architecture is used to measure disease intensity. They emphasized the need to assess the disease's severity and recurrence. Intensity is a more precise assessment of the affected area via leaf covering, whereas incidence relates to disease prevalence at the crop level (Wspanialy and Moussa, 2020).

3.14.4 Crop disease level estimation with VGG16

A DL model is used to calculate the intensity of plant diseases from images, drawing inspiration from advancements in image-based disease recognition. CNNs of varying depths were evaluated, ranging from beginnings to four fine-tuned advanced deep models: ResNet50, VGG19, Inception V3, and VGG16 (Das *et al.*, 2022). Compared to other networks, fine-tuned deep neural models significantly improve performance on small datasets (Wang *et al.*, 2017).

3.14.5 Crop disease level assessment with Inception V3

Six deep convolutional generative adversarial networks (DC-GANs) were employed to increase the test set, increasing its size from 5406 to 14,056 images. DC-GANs were then applied to see if the improved information could boost the learning effectiveness of the model that was selected (Zeng, 2020).

3.14.6 Crop disease identification with MobileNet

DL techniques are used to reduce disease-related agricultural losses. Researchers utilized Google's MobileNetV2, an ANN designed to improve mobile visual verification, enabling item discovery and categorization. The system in question was adopted for its great accuracy and ability to operate with minimal resources (Gandhi *et al.*, 2021).

3.14.7 Crop disease detection with AlexNet, VGG16, ResNet, and EfficientNet

To obtain disease features for machine learning, 52 texture features were manually extracted using local binary patterns, the GLCM technique, and character extraction (color). ResNet34, utilizing ML/DL techniques, obtained the highest recall, precision, accuracy, and F1-score (99.7%) (Tan *et al.*, 2021).

3.15 Using an Immediate Kinetic Global Positioning System to Estimate Crop Disease Intensity

A high-quantities assessment instrument for indicators of disease in testing peanut plots. A study has developed and replicated different sickness frequency and seriousness levels. An existing CNN was trained using TL to determine and evaluate previously unnoticed symptoms and background zones in site-level images. The trained model analyzed the photographs and correlated them with research plots using real-time kinematic GPS (RTK-GPS) data (Clohessy *et al.*, 2021).

3.15.1 Crop disease identification with VGG16, ResNet50, and GoogLeNet

A ResNet9 model was developed to identify blight disease in potato and tomato leaf images for farmers to use. ResNet9 makes predictions based on leaf morphology, diseased patches, and overall green areas. The suggested model utilized the PlantVillage dataset of 3990, which was later supplemented with training data and thorough parameter optimization. The model was trained with hyperparameters and evaluated on a batch of 1331 images (Anim-Ayeko *et al.*, 2023).

3.15.2 Crop identification of diseases with ResNet34

Convolutional neural networks are cutting-edge image recognition technologies that offer fast and accurate disease diagnoses. This study investigates the efficiency of a training ResNet34 simulation of diseases in crop identification using a web application capable of detecting seven plant maladies. A controlled collection of 8685 images was created to train and validate the model. Enhancement and sharing of data allowed CNN to generalize more precisely (Singh *et al.*, 2017).

3.15.3 Crop disease level estimation with CNN

This section outlines the challenges associated with CNN-based plant pathology extent assessment systems and presents proposed research solutions to address them. They reviewed the problems of using CNN-based plant disease level assessment methods in everyday situations. Based on the survey, CNN-based categorization theory is commonly used to estimate disease severity by labeling each pixel and estimating the proportion of infected areas (Shi *et al.*, 2023).

3.16 Customized Deep Learning Structure for Agricultural Identification of Diseases

The Rice PlantVillage (21,198) dataset was used to enhance a modified PDICNet system for crop disease identification. ResNet50, a well-known segmentation algorithm, was applied at the model level to extract deep features from the images. To predict diseases in rice crops, we utilized MRDOA to select specific characteristics and a CNN classifier to optimize them (Reddy *et al.*, 2023).

3.16.1 Crop identification of diseases with CNN

An end-to-end Android application was developed using TFLite and CNN algorithms to diagnose agricultural diseases. To efficiently support greenhouse producers, a DL detection method discovers crop diseases by capturing images of leaflets and juxtaposing them with other data collections. Users are directed to a website containing information on pesticides, including cost and usage guidelines. The process involved visual data collection and segmentation to enhance clarity and assess the ability of visual displays. Segmentation was accomplished by reducing the total squared proximity between elements associated with groups, using several algorithms. SVM was chosen due to its simplicity and reliability.

The proposed program diagnoses many diseases in a single system, resulting in improved precision than CNN (Rajendra *et al.*, 2020).

3.16.2 Assessment of open-source deep learning tools

TensorFlow, Torch/PyTorch, Caffe, and Theano are famous external free-to-use neural network technologies. Table 3.1 shows the various properties of every free-to-download tool.

The four most widely used neural network-free applications are on multiple platforms and can run on Linux, Windows, iOS, and Android devices.

3.17 A Network Intended for Categorization

Recognizing pests and diseases of plants in real-world environments is challenging due to variations in shape, size, color, texture, background, and lighting conditions. CNN-based classification networks are the most commonly utilized tools for plant diseases and pest identification because of their ability to gather characteristics effectively. Common CNN architectures for plant classification include GoogLeNet (Szegedy *et al.*, 2015), ResNet (Xie *et al.*, 2017), DenseNets (Huang *et al.*, 2017), and MobileNet (Howard *et al.*, 2017). Many studies have also designed networks tailored to specific problems. These networks can be categorized based on their tasks, such as feature extraction, direct classification, or defect localization.

3.18 Frequently Used Datasets for Detecting Pests and Plant Diseases

The most common approach to plant disease and pest identification using CNNs is the direct classification of lesions via classification networks. This can involve division of the original image or classification after detecting Region of Interest. In the original image categorization, complete images showing plant diseases and pest infestations are uploaded to the network for training. It employed ResNet50 for pest and disease identification, achieving 95.61% accuracy using the Adam optimization method and focal loss (Fang *et al.*, 2020). In ROI-based classification, a fixed area of the image is analyzed for lesions. A multidimensional deep convolutional neural network was used to detect soybean stem rot, achieving 95.73% accuracy. In multicategory classification, contextual data are integrated to train models that classify diseases across multiple crops, minimizing classifier errors by 71% and improving adjusted accuracy to 0.98 (Picon *et al.*, 2019).

3.18.1 Systems for visual remote sensing depending on UAVs are used to detect crop diseases

Unmanned aerial vehicles are important for agricultural tasks, filled with sensors to perform tasks like navigation, crop disease detection, and converting crop conditions. These capabilities significantly improve crop productivity, reduce resource use, and direct previously expensive tasks (Dammer *et al.*, 2022). These UAVs use

Table 3.1. Evaluation of software applications of machine learning.

Devices	Creator	Compatible devices	Software interface	Usage
TensorFlow	Google	Mobile, GPU, CPU	Python, C	Adjustable growth, mobility, and reliable efficiency, along with the assistance of application distribution
Caffe	BAIR	GPU, CPU	Matlab, Python	High clarity, easy to extend, rapid performance, vast number of individuals, and a diverse society
Theano	MILA	GPU, CPU	Python	Expandable and outstanding efficiency

algorithms to operate with minimal human action (Lytridis *et al.*, 2021). UAV-based aerial imagery has become a useful tool in agriculture, enabling farmers to identify crops efficiently.

3.18.2 By using RGB cameras

RGB cameras are commonly utilized in agriculture for tasks such as identifying crop diseases, offering benefits such as low cost, portability, and high resolution. UAVs filled with RGB cameras have been used during studies to identify plant diseases, such as soybean leaf diseases with a Sony EXMOR camera and pine wilt disease with a DJI Phantom 4 RTK UAV. These cameras are particularly useful for small-scale farmers with limited budgets, as they provide helpful visual data for identifying disease symptoms in the early vegetative phase (Heidarian Dehkordi *et al.*, 2020). For more detailed data, RGB-D cameras, which provide depth data, can assist in object identification (Li *et al.*, 2021), while multispectral sensors, though more expensive, are perfect for NDVI mapping and crop scenario assessment (Costa *et al.*, 2020). However, RGB cameras are limited in early disease identification due to their sensitivity to environmental factors such as sunlight and shadows, which can affect diagnostic accuracy (Ganchenko and Doudkin, 2019).

3.18.3 With cameras that are multispectral and hyperspectral

The electromagnetic spectrum ranging from the optical to the near-infrared region can be applied to calculate secure plant indicators such as the Normalized Difference Vegetation Index (NDVI). Spectral cameras are widely used sensors for identifying crop and plant diseases with UAV platforms. Hyperspectral cameras offer more operations than multispectral imaging cameras but are very cost prohibitive. It is difficult to differentiate between plant stress categories using multispectral data. Hyperspectral cameras give detailed measurements of vegetation indices (VIs), which can correctly differentiate stress types (The'au *et al.*, 2020).

Infrared thermal cameras

Other sensors, such as thermal infrared cameras, can identify crop diseases from overhead images. The infrared area is divided into several bands: mid-wave infrared, far infrared, shortwave infrared, long-wave infrared, and near infrared. These cameras are versatile and can be used at various times of the day, including early morning and the end of the day. Thermal imaging systems mounted on UAVs are now frequently employed in the farming industry to observe water stress and crop diseases, delivering significant information. Thermal imaging sensors measure infrared spectra, delivering more details about plant state than hyperspectral, visible, or multispectral cameras. When using thermal pictures, it is important to consider factors such as image resolution, ambient conditions, diversity of crops, development stage, flying height, and viewing perspective (Raj *et al.*, 2021).

Deep learning algorithms that use UAV-captured visual data to identify crop diseases

Deep learning-based machine perception techniques have shown promising results in various fields, which include agriculture. Diseased crops often exhibit common symptoms including plant shadow, leaf twisting, leaf marks, and fruit marks, making DL an ideal choice for disease identification. Algorithms utilizing DL have gained popularity for diagnosing crop diseases using UAV-captured imagery (Pan *et al.*, 2021).

3.19 Recognizing the Main Grain Crop Diseases

Recent studies have utilized UAV-based imagery and DL algorithms to identify diseases in crops such as wheat and corn. By using algorithms for computer vision to recognize yellow rust sickness, which is a major threat to wheat that causes significant yield loss, this method can easily predict the disease. Among the DL models evaluated, PSPNet (98%) and SVM (96%) achieved the best achievement in detecting yellow rust (Pan *et al.*, 2021). BPNN and hyperspectral visuals for fungal head blight detection, obtained 98% accuracy (Liu *et al.*, 2020).

3.20 Recognition of Diseases in Vineyards

The LeNet-5 framework was used to identify grapevine pollution from RGB images captured by UAVs. The model obtained 95.92% accuracy using a 64 × 64 patch size and a 5 × 5 filter. The model performed well throughout different color wavelength combinations, with YUV and RGI achieving the highest accuracy (95.73%) (Kerkech *et al.*, 2018).

3.21 Disease Identification in Pine Tree Crops

Several studies have used UAV-based imagery to identify diseases in pine trees. The Spatial-Context-Attention Network (SCANet) was applied to detect pine worm diseases using images with multiple spectral bands, attaining an F1-score over 88%, outperforming other models like DeepLabV3+ and HRNet (Hu *et al.*, 2020).

3.22 Detecting Diseases in Different Crops

Deep learning and UAV technology have been applied to determine diseases in several crops, such as soybeans and bananas. Among several deep CNN models tested, Inception-v3 outperformed the others, achieving 99.04% accuracy, surpassing Xception, ResNet-50, and VGG-19, while also reducing training time by 6, 31, and 42 h, respectively (Selvaraj *et al.*, 2020).

3.23 Automatic Classification Powered by Artificial Intelligence for Identifying Vegetable Diseases

Crop diseases are categorized into two categories: biotic and abiotic. Mold, microbes, and pathogens cause biotic diseases, while abiotic disorders result from environmental and non-living factors such as nutrient deficiencies, chemical imbalances, and physical trauma (Husin *et al.*, 2012). Abiotic diseases can also arise from unfavorable climatic conditions such as high temperatures, excess rainfall, inadequate light, or soil imbalances. Identifying plant diseases is challenging, especially since many have similar symptoms like distorted leaves, slow growth, and weak pods. Identifying the specific plant disease-causing harm might be challenging due to its similar appearance. Identifying uneven leaf development, distorted leaf coloration, slower growth, weakening pods, and other symptoms might be challenging (Manavalan, 2021). Effective disease detection is essential for maintaining healthy plants and ensuring agricultural viability (Mitra, 2021). Studies show that healthy leaves improve a plant's defense against diseases, emphasizing the importance of leaf health in preventing viral infections (Qiu *et al.*, 2022). Vegetable plant diseases, which often affect leaves, can spread quickly and cause significant damage (Qi *et al.*, 2021).

3.23.1 Automated disease detection for tomato plants

Tomatoes (*Solanum lycopersicum*) are widely grown for human consumption and are rich in nutrients such as vitamins C and E, beta-carotene, and plenty of the mineral potassium. They are vulnerable to a variety of diseases caused by yeast, microorganisms, and insects (Lal, 2021). The model, consisting of pooling layers, four convolutional layers, and dense layers with a sigmoid activation function, achieved an impressive 87% accuracy using 3663 images from the PlantVillage dataset. It used machine learning to detect diseases in tomatoes. The authors used 200 images with features like texture, color, and shape, resulting in high detection accuracy (Basavaiah and Anthony, 2020).

3.23.2 Automated identification of diseases in chilies

Chili peppers, a key agricultural commodity in India, are popular in regional and global cuisines for their flavor and medicinal properties. However, chili plants are vulnerable to approximately 45 different viruses, of which 24 occur naturally, while the rest potentially caused by vaccinations or other factors (Duranova *et al.*, 2022). Diseases like leaf curl, geminivirus, *Cercospora*, and leaf spot affect the plants due to bacterial, viral, or fungal infections. One study compared DL and machine learning methods for detecting chili leaves disease using a dataset with five major diseases. VGG19 achieved 83.54% accuracy without augmentation, while DarkNet53 performed better with augmentation (Naik *et al.*, 2022). Another study analyzed 974 self-collected images from Malaysia to detect pests and diseases in chili peppers. Using SVM, RF, and ANN classifiers along with features from six machine learning and DL methods, they found that DL strategies combined with SVM achieved the highest accuracy at 92.10%. These findings highlight the utility of automated image analysis for identifying chili plant diseases.

3.23.3 Automated disease identification for potatoes

Potatoes, the world's fourth-largest crop, are highly susceptible to various diseases, including common scab, black scurf, and black leg, making them susceptible in agriculture (Wani *et al.*, 2022). One technique, combining Maximum–Minimum Color Difference and three texture features, achieved 91.67% accuracy in determining healthy potato leaves from those affected by late blight and early blight, using 300 images from the PlantVillage database. Another study applied machine learning to classify potato leaf diseases based on color and texture (Suttapakti and Bunpeng, 2019). These studies suggest the utility of machine learning to enhance potato disease diagnosis.

3.23.4 Cucumber disease identification is automated

Cucumbers, called high water content, are a popular and healthy vegetable from the *Cucurbitaceae* family. However, they are subjected to diseases like anthrax, angular plant spots, and powdery fungal infections, particularly in the later stages of growth (Vishnoi *et al.*, 2021). In a related study, researchers developed a broad action on cucumber leaf disease detection, consisting of different stages like deep content extraction, image improvement, feature selection, disease classification, and segmentation (Khan *et al.*, 2020). This shows its potential as an efficient and secure technique for identifying and categorizing cucumber leaf diseases.

3.24 Applying Deep Learning Methods to Analyze Multispectral Images for Crop Diseases

Machine vision-based techniques for agricultural and disease management evaluation typically involve three key tasks: image categorization, object recognition, and segmentation. In addition, various evaluation metrics for assessing computer vision performance in this field are discussed to guide the utility of such models in the real-world scenarios. This section provides a detailed study of some of the usually employed DL-powered machine vision methods for determining diseases in crops, assisting farmers and scientists in selecting the best algorithms and supplies for their specific needs.

3.25 Scenario for Image Categorization

Classifying images is a common neural network approach for crop disease identification, where the classifier assigns labels to images based on predefined categories. CNNs are the primary neural network structures used for identifying plant and crop diseases due to their superior image classification performance. The evaluated publications identified VGGNet, Xception, Inception, ResNet, and AlexNet as common CNN architectures for classifying plant diseases (Das *et al.*, 2022). It used four advanced CNN structures for classifying soybean diseases, including ResNet-50, Inception-V3, Xception, and VGG-19, a bespoke CNN architecture for disease classification (Dang *et al.*, 2020).

3.25.1 Scenario for identifying objects

Identifying objects is a critical AI task for identifying agricultural and plant diseases in UAV imagery. DL-based object recognition algorithms can recognize and categorize multiple diseases in an image, drawing boundaries with the relevant class. There are two types of algorithms for detecting crop or agricultural diseases in aerial images like single-stage and two-stage (Wu *et al.*, 2021). Single-stage detectors are primarily based on YOLO versions.

Other studies have found that combining object recognition with image segmentation approaches can improve model performance.

3.25.2 An example of image segmentation

Image segmentation classifies pixels based on their class. DL-based image segmentation algorithms improve plant and crop disease detection by identifying each pixel depending on the disease kind. Image segmentation approaches are more suitable for detecting diseases in cluttered and/or obstructed situations, as object detection algorithms remain unreliable. Several studies have applied image segmentation technique to diagnose agricultural and plant diseases using UAV data. Semantic analysis is the most frequent image segmentation approach utilized in various research. However, their performance varies depending on parameters such as information category, aircraft elevation, plant kind, or disease type.

3.25.3 Assessment metrics

Models based on deep learning have been assessed based on a variety of metrics, including precision, recall, and F1-score. Table 3.2 summarizes the requirements frequently employed for assessing the results of DL-powered AI theories. Abbreviations such as TP, TN, FP, and FN are used for indicating True Positives (the quantity of properly identified diseases), True Negatives (the quantity of identified diseases), False Positives (the quantity of wrongly determined diseases), and False Negatives (the quantity of non-identified diseases). Additionally, Po and Pe denote the observed percentage concurrence and predicted agreement by coincidence, respectively.

3.25.4 Artificial intelligence's limitations in diagnosing diseases and possible future directions

Earlier, we examined how AI can be utilized to improve agriculture, particularly in identifying diseases in crops such as vegetables. We analyzed several automation models and systems provided by researchers worldwide and published in the literature. AI has significant potential in agriculture, especially for identifying plant diseases. To improve disease identification, it is important to address the limitations of these models. This section outlines the main obstacles that prevent the proper application of automatic identification of plant diseases and classification.

3.25.5 Examination of background and noise

Agricultural research photographs of plant diseases sometimes contain unnecessary noise, colorful backdrops, and other components such as roots, grass, and soil. It is vital to identify and separate these elements. Segmentation isolates polluted regions from collected images. To accurately identify diseased plants in real-time, the suggested automated system should remove unnecessary components from the image and isolate only the relevant segment.

3.25.6 Variables impacting image acquisition

Current databases generally include photographs collected in controlled environments, such as laboratories. Obtaining an equal image might be tough due to factors such as sunlight magnitude, levels of humidity, or climatic conditions. To achieve investigation objectives, photographs of the same plant object from varied viewpoints, at various times, and situations in nature are essential.

The choice of image acquisition tools has a significant impact on system performance. Forecast accuracy is affected by various elements such as sampling device, brightness of the light, date and time of each day, and humidity conditions.

Table 3.2. Some of the most commonly utilized metrics for comparing deep learning-powered vision programs. Author's own table.

Measured	Overview	Equation
Precision	The precision rate of incorrect diagnoses provided by the simulation for a certain ailment class. A higher accuracy rate indicates fewer false-positive expectations.	P = T P / (T P + F P)
Average precision	The AP value is a determinant that evaluates the effectiveness of an object identification system and is obtained for each individual class.	AP = P(R)d(R)
Accuracy	The most commonly used evaluation method is accuracy, which assesses the number of times an example correctly predicts a task involving classification.	ACC = (TP + TN) / (TP + TN + FP + FN)
F1-score	F1-score measurement takes the harmonic average of both recall and precision proportions.	F1 = 2*P*R/(P + R)
Recall (sensitivity)	Recall rate evaluates the simulation's effectiveness when identifying plant disease types. An increased recall rate indicates fewer false-negative expectations.	R = TP (TP + FN)
FPS	The FPS calculates detection velocity.	FPS = Number of frames / (current time-start time)
Kappa coefficient	Kappa coefficient measures the inter-annotator of a contract.	Kappa = (Po-Pe)/(1-Pe)

To effectively address these challenges, it is necessary to combine training with the immediate use of an automated sickness prediction model.

3.26 Identifying and Distinguishing a Disease's Symptoms

Digital processing of images is vital in agricultural research for detecting and isolating comparable symptoms of diseases. Identifying and categorizing symptoms with similar characteristics improves performance. However, this effort becomes challenging when multiple diseases share similar signs and factors in the environment. Alternative segmentation methods are needed to identify vegetable disorders with distinct symptoms.

3.27 Understanding the Amount of Data Affects the Efficiency of the Model

Research suggests that training systems for specific crop diseases require more data than the typical few thousand images used by researchers. A DL-based data augmentation strategy improves the overall volume of training images. In this case, a covariate shift occurs when training data differ from the data utilized for model implementation. Using large datasets can enhance the efficiency of models, but it can also increase computing loads.

3.28 Disparities in How Data are Represented for Various Diseases

Automated detection methods confront challenges due to unbalanced patterns in training data. Vegetable diseases have minimal data and are not homogeneous among classes, as previously discussed. To provide reliable analysis and prediction, diseases should be symbolized by sample vegetables of equal size, encompassing both healthy and diseased vegetables, in order to prevent bias.

Multiple diseases, dietary deficits, and pests can coexist in a single photograph, making the assumption of only one condition inaccurate. Identifying and tracking a disease can be tricky, as symptoms may differ by region. Therefore, it is important to consider these elements while assessing photographs.

3.28.1 Conditions with comparable symptoms

Detecting agricultural infections is challenging due to their identical signs and patterns. Scientists frequently explore across the visual range, but using light wavelengths to distinguish diseases can be complex, costly, and challenging. Innovative approaches have the potential to reduce the need for large datasets and minimize the risk of errors in agricultural activities.

3.29 Datasets in Public Resources for Identifying Plant Diseases

In recent years, techniques for DL have demonstrated the capacity to autonomously identify plant diseases (Salman *et al.*, 2023). The efficiency of DL models varies heavily on the accuracy of information used for training, as superior generalization and test results are expected from well-constructed datasets (Krishna *et al.*, 2017). However, datasets for plant disease detection have received relatively little attention, and the accuracy of current machine learning algorithms is often overstated. Research indicates that DL models struggle with expansion when the data used for training and test datasets exhibit distinct frequencies (Cui and Athey, 2022). This issue becomes obvious when models trained with controlled background images perform poorly on uncontrolled test data (Wu *et al.*, 2023). Therefore, creating credible test datasets is essential in assessing the model's performance in real-life situations. To construct a plant disease identification dataset, domain expertise in agriculture and plants is necessary, unlike generic machine vision benchmarks like ImageNet that focus on everyday items. Simultaneously, understanding deep acquiring included limitations associated with modern advanced DL methods and data labeling schemes. Hence, examining dataset parameters from both AI and disease of plants perspectives is an important issue.

3.30 Conclusion

Diseases of plants are a major hazard to world farming, reducing food production, performance, and financial security. Traditional methods for determining crop diseases are limited by the need for substantial investments in labor, resources, and skills. DL, ML, and AI provide satisfying opportunities to automate disease identification, developing efficiency and accuracy. Additionally, UAVs equipped with various sensors, such as multispectral, thermal cameras, hyperspectral, and RGB, offer an innovative method for identifying diseases from aerial imagery. Different UAV types, including rotary-wing, fixed-wing, and hybrid VTOL, have specific applications in crop production. Despite this potential, challenges remain in model selection, image processing, and environmental factors. The successful deployment of AI-based disease identification systems depends on the use of exact datasets, real-time assessment, or optimization of image-based methods for early disease identification. AI may evaluate plant leaves and estimate diseases, but challenges reside in improving their accuracy and efficacy. This requires large, diverse datasets, selecting different training images, and determining environmental conditions like lighting. DL has improved disease identification precision, enabling early operation to maintain plants and assure food security.

References

Abdu, A. M., Mokji, M. M. and Sheikh, U. U. (2020) Machine learning for plant disease detection: An investigative comparison between support vector machine and deep learning. IAES *International Journal of Artifcial Intelligence* 9(4), 670–683. DOI: 10.11591/ijai. v9.i4. pp670-683

Afify, Y. M., Moawad, I. F., Badr, N. L. and Tolba, M. F. (2017) An enhanced distance based similarity measure for user based recommendations. In *Proceedings of the International Conference on Advanced*

Intelligent Systems and Informatics 2016 2 (pp. 42-52). Springer International Publishing. DOI: 10.1007/978-3-319-48308-5_5

Afoudi, Y., Lazaar, M. and Al Achhab, M. (2021) Hybrid recommendation system combined content-based filtering and collaborative prediction using artificial neural network. *Simulation Modelling Practice and Theory* 113, 102375. DOI: 10.1016/j.simpat.2021.102375

Agarwal, A., Sarkar, A. and Dubey, A. K. (2019) Computer vision-based fruit disease detection and classification. In *Smart Innovations in Communication and Computational Sciences: Proceedings of ICSICCS-2018* (pp. 105-115). Springer Singapore. DOI: 10.1007/978-981-13-2414-7_11

Ahmad, A., Saraswat, D. and El Gamal, A. (2023) A survey on using deep learning techniques for plant disease diagnosis and recommendations for development of appropriate tools. *Smart Agricultural Technology* 3, 100083. DOI: 10.1016/j.atech.2022.100083

Ahmad, M., Abdullah, M., Moon, H. and Han, D. (2021) Plant disease detection in imbalanced datasets using efficient convolutional neural networks with stepwise transfer learning. *IEEE Access* 9, 140565-140580. DOI: 10.1109/access.2021.3119655

Akulwar, P. (2020) A recommended system for crop disease detection and yield prediction using machine learning approach. *Recommender System with Machine Learning and Artificial Intelligence: Practical Tools and Applications in Medical, Agricultural and Other Industries* 141-163. DOI: 10.1002/ 9781119711582.ch8

Alajrami, M. A. and Abu-Naser, S. S. (2018). Onion Rule Based System for Disorders Diagnosis and Treatment. *Int. J. Acad. Pedagog. Res* 2 (8), 1–9. DOI: http://dstore. alazhar.edu.ps/xmlui/handle/123456789/360

Anim-Ayeko, A. O., Schillaci, C. and Lipani, A. (2023) Automatic blight disease detection in potato (Solanum tuberosum L.) and tomato (Solanum lycopersicum, L. 1753) plants using deep learning. *Smart Agricultural Technology* 4, 100178. DOI: 10.1016/j.atech.2023.100178

Anjna, Sood, M. and Singh, P. K. (2020) Hybrid system for detection and classifcation of plant disease using qualitative texture features analysis. *Procedia Computer Science* 167(2019), 1056–1065. DOI: 10.1016/j.procs. 2020.03.404

Basavaiah, J. and Anthony, A. A. (2020) Tomato leaf disease classification using multiple feature extraction techniques. *Wireless Pers. Commun* 115, 633–515. DOI: 10.1007/s11277-020-07590-x

Beddow, J. M., Pardey, P. G., Chai, Y., Hurley, T. M., Kriticos, D. J., Braun, H. J. and Yonow, T. (2015) Research investment implications of shifts in the global geography of wheat stripe rust. *Nature Plants* 1(10), 1-5. DOI: https://doi.org/10.1038/nplants.2015.132

Bhagwat, R. and Dandawate, Y. (2021) A review on advances in automated plant disease detection. *International Journal of Engineering and Technology Innovation* 11(4), 251. DOI: 10.46604/ijeti.2021.8244

Bock, C. H. (2022) Plant disease severity estimated visually: A century of research, best practices, and opportunities for improving methods and practices to maximize accuracy (pp. 25–42).

Clohessy, J. W., Sanjel, S., Brien, G. K. O., Barocco, R., Kumar, S., Adkins, S., Tillman, B., Wright, D. L. and Small, I. M. (2021) Development of a high-throughput plant disease symptom severity assessment tool using machine learning image analysis and integrated geolocation. *Computers and Electronics in Agriculture* 184, 106089. DOI: 10.1016/j.compag.2021.106089

Costa, L., Nunes, L. and Ampatzidis, Y. (2020) A new visible band index (vNDVI) for estimating NDVI values on RGB images utilizing genetic algorithms. *Computers and Electronics in Agriculture* 172, 105334. DOI: 10.1016/j.compag.2020.105334

Cui, P. and Athey, S. (2022) Stable learning establishes some common ground between causal inference and machine learning. *Nat. Mach. Intell* 4, 110–115. DOI: 10.1038/s42256-022-00445-z

Dammer, K. H., Garz, A., Hobart, M. and Schirrmann, M. (2022) Combined UAV-and tractor-based stripe rust monitoring in winter wheat under field conditions. *Agronomy Journal* 114(1), 651-661.DOI: 10.1002/ agj2.20916

Dang, L. M., Wang, H., Li, Y., Min, K., Kwak, J. T., Lee, O. N. and Moon, H. (2020) Fusarium wilt of radish detection using RGB and near infrared images from Unmanned Aerial Vehicles. *Remote Sensing* 12(17), 2863.DOI: 10.3390/ rs12172863

Das, S., Pattanayak, S., Behera, P.R. (2022) Application of macine learning: a recent advancement in plant diseases detection. *Journal of Plant Protection Research* 62(2), 122-135. DOI: 10.24425/jppr.2022.141360

Duranova, H., Valkova, V. and Gabriny, I. (2022) Chili peppers (*Capsicum spp.*): the spice not only for cuisine purposes: an update on current knowledge. *Phytochem. Rev* 21, 1379–14135. DOI: 10.1007/s11101-021-09789-7

Fang, T., Chen, P., Zhang, J. and Wang, B. (2020) Crop leaf disease grade identification based on an improved convolutional neural network. *Journal of Electronic Imaging* 29(1), 013004-013004. DOI: 10.1117/1.JEI.29.1.013004

Fenu, G. and Malloci, F. M. (2021) DiaMOS plant: A dataset for diagnosis and monitoring plant disease. *Agronomy* 11(11), 2107.DOI: 10.3390/ agronomy11112107

Gaikwad, V. P. and Musande, V. (2023) Advanced prediction of crop diseases using cetalatran-optimized deep KNN in multispectral imaging. *Traitement Du Signal* 40(3), 1093–1106. DOI: 10.18280/ts.400325

Ganchenko, V. and Doudkin, A. (2019) Agricultural vegetation monitoring based on aerial data using convolutional neural networks. *Optical Memory and Neural Networks* 28, 129-134. DOI: 10. 3103/ S1060992X1902005X

Gandhi, V., Bhide, A., Dharmawat, S. and Aware, M. (2021) Detection of crop diseases using deep learning via android application 3307, 305–311.

Halder, M., Sarkar, A. and Bahar, H. (2018) Plant disease detection by image processing: A literature review. *SDRP Journal of Food Science & Technology* 3(6), 534–538. DOI:10.25177/jfst.3.6.6

Harakannanavar, S. S., Rudagi, J. M., Puranikmath, V. I., Siddiqua, A. and Pramodhini, R. (2022) Plant leaf disease detection using computer vision and machine learning algorithms. *Global Transitions Proceedings* 3(1), 305–310. DOI: 10.1016/j.gltp.2022.03.016

Heidarian Dehkordi, R., El Jarroudi, M., Kouadio, L., Meersmans, J. and Beyer, M. (2020) Monitoring wheat leaf rust and stripe rust in winter wheat using high-resolution UAV-based red-green-blue imagery. *Remote Sensing* 12(22), 3696. DOI: 10.3390/rs12223696

Hlophe-Ginindza, S.N. and Mpandeli, N.S. (2021) Ensuring Food Security in Africa. *Food security in Africa* 63. DOI: 10.5772/intechopen.91694

Howard, A. G., Zhu, M., Chen, B., Kalenichenko, D., Wang, W., Weyand, T. and Adam, H. (2017) Mobilenets: Efficient convolutional neural networks for mobile vision applications. *arXiv preprint arXiv:*1704.04861. DOI: 10.48550/arXiv.1704.04861

Hu, G., Yin, C., Wan, M., Zhang, Y. and Fang, Y. (2020) Recognition of diseased Pinus trees in UAV images using deep learning and AdaBoost classifier. *Biosystems Engineering* 194, 138-151. DOI: 10.1016/j. biosystemseng.2020.03.021

Huang, G., Liu, Z., Van Der Maaten, L. and Weinberger, K. Q. (2017) Densely connected convolutional networks. In *Proceedings of the IEEE conference on computer vision and pattern recognition* (pp. 4700-4708).

Husin, Z. B., Shakaff, A. Y. B. M., Aziz, A. H. B. A. and Farook, R. B. S. M. (2012, February) Feasibility study on plant chili disease detection using image processing techniques. In *2012 Third International Conference on Intelligent Systems Modelling and Simulation* (pp. 291-296). IEEE. DOI: 10.1109/ISMS.2012.33

Islam, S. S., Haque, M. S., Miah, M. S. U., Sarwar, T. B. and Nugraha, R. (2022) Application of machine learning algorithms to predict the thyroid disease risk: an experimental comparative study. *PeerJ Computer Science* 8, e898. DOI: 10.7717/peerj-cs.898

Jiang, S. and Song, J. (2021, May) Evaluation metrics for personalized recommendation systems. In *Journal of Physics: Conference Series* 1920 (1), p. 012109. IOP Publishing. DOI: 10.1088/1742-6596/ 1920/1/012109

Kerkech, M., Hafiane, A. and Canals, R. (2018) Deep leaning approach with colorimetric spaces and vegetation indices for vine diseases detection in UAV images. *Computers and electronics in agriculture* 155, 237-243.DOI: 10.1016/j.compag.2018.10.006

Khan, M. A., Akram, T., Sharif, M., Javed, K., Raza, M. and Saba., T. (2020) An automated system for cucumber leaf diseased spot detection and classification using improved saliency method and deep features selection. *Multimedia Tools Appl* 79, 18627–18565. DOI: 10.1007/s11042-020-08726-8

Krishna, R., Zhu, Y., Groth, O., Johnson, J., Hata, K.and Kravitz, J. (2017) Visual genome: Connecting language and vision using crowdsourced dense image annotations. *Int. J. Comput Vision* 123, 32–73. DOI: 10.1007/s11263-016-0981-7

Lal, B. (2021) A review study on tomato and its health benefits. *Asian J. Res. Soc. Sci. Humanities* 11, 197–202. DOI: 10.5958/2249-7315.2021.00337.3

Li, D., Sun, X., Elkhouchlaa, H., Jia, Y., Yao, Z., Lin, P. and Lu, H. (2021) Fast detection and location of longan fruits using UAV images. *Computers and Electronics in Agriculture* 190, 106465.DOI: 10.1016/j. compag. 2021.106465

Li, L., Zhang, S. and Wang, B. (2021b) Plant disease detection and classifcation by deep learning—A review. *IEEE Access*, 9, 56683–56698

Liu, J. and Wang, X. (2021) Plant diseases and pests detection based on deep learning: A review. *Plant Methods* 17, 225. DOI: 10.1186/s13007-021-00722-9

Liu, L., Dong, Y., Huang, W., Du, X. and Ma, H. (2020) Monitoring wheat fusarium head blight using unmanned aerial vehicle hyperspectral imagery. *Remote Sensing* 12(22), 3811.DOI:10.3390/rs12223811

Lytridis, C., Kaburlasos, V. G., Pachidis, T., Manios, M., Vrochidou, E., Kalampokas, T. and Chatzistamatis, S. (2021) An overview of cooperative robotics in agriculture. *Agronomy* 11(9), 1818.DOI: 10.3390/agronomy11091818

Maddikunta, P. K. R., Hakak, S., Alazab, M., Bhattacharya, S., Gadekallu, T. R., Khan, W. Z. and Pham, Q. V. (2021) Unmanned aerial vehicles in smart agriculture: Applications, requirements, and challenges. *IEEE sensors journal* 21(16), 17608-17619. DOI: 10.1109/JSEN.2021.3049471

Makandar, A. and Bhagirathi, H. (2015) Image enhancement techniques using highpass and lowpass filters. *Int. J. Comput. Appl* 109, 21–275. DOI: 10.5120/19256-0999

Manavalan, R. (2021) Cucumber leaves diseases detection through computational approaches: A review. *Asian J. Res. Biosciences October* 143–53.

Mitra, D. (2021). Emerging plant diseases: research status and challenges. *Emerging trends in plant pathology*, 1-17. DOI: 10.1007/978-981-15-6275-4_1

Munjal, D., Singh, L., Pandey, M. and Lakra, S. (2023) A systematic review on the detection and classification of plant diseases using machine learning. *International Journal of Software Innovation (IJSI)* 11(1), 1-25. DOI: 10.4018/IJSI.315657

Naik, Nageswararao, B., Malmathanraj, R. and Palanisamy, P. (2022) Detection and classification of chilli leaf disease using a squeeze-and-excitation-based CNN model. *Ecol. Inf* 69, 101663. DOI: 10.1016/j.ecoinf.2022.101663

Pan, Q., Gao, M., Wu, P., Yan, J. and Li, S. (2021) A deep-learning-based approach for wheat yellow rust disease recognition from unmanned aerial vehicle images. *Sensors* 21(19), 6540.DOI: 10.3390/s21196540

Parraga-Alava, J., Cusme, K., Loor, A. and Santander, E. (2019) RoCoLe: A robusta coffee leaf images dataset for evaluation of machine learning based methods in plant diseases recognition. *Data in brief* 25, 104414. DOI: 10.1016/j.atech.2022.100083

Picon, A., Seitz, M., Alvarez-Gila, A., Mohnke, P., Ortiz-Barredo, A. and Echazarra, J. (2019) Crop conditional Convolutional Neural Networks for massive multi-crop plant disease classification over cell phone acquired images taken on real field conditions. *Computers and Electronics in Agriculture* 167, 105093.DOI: 10.1016/j.compag.2019.105093

Qi, S., Zhang, S., Islam, Md M., El-Sappah, A. H., Zhang, F. and Liang., Y. (2021) Natural resources resistance to tomato spotted wilt virus (TSWV) in tomato (*Solanum lycopersicum*). *Int. J. Mol. Sci* 22, 109785. DOI: 10.3390/ijms222010978

Qiu, Z., Verma, J. P., Liu, H., Wang, J., Batista, B. D., Kaur, S. and Singh, B. K. (2022) Response of the plant core microbiome to Fusarium oxysporum infection and identification of the pathobiome. *Environmental Microbiology* 24(10), 4652-4669. DOI: 10.1111/ 1462-2920.16194

Raj, M., Gupta, S., Chamola, V., Elhence, A., Garg, T., Atiquzzaman, M. and Niyato, D. (2021) A survey on the role of Internet of Things for adopting and promoting Agriculture 4.0. *Journal of Network and Computer Applications* 187, 103107.DOI: 10. 1016/j.jnca.2021.103107

Rajendra, A. B., Rajkumar, N. and Shetty, P. D. (2020) Areca nut disease detection using Image processing. *Advances in Intelligent Systems and Computing* 1154(03), 925–931. DOI: /10.1007/978-981-15-4032-5_83

Ramesh, S., Hebbar, R., Niveditha, M., Pooja, R., Shashank, N. and Vinod, P. V. (2018, April) Plant disease detection using machine learning. In *2018 International conference on design innovations for 3Cs compute communicate control (ICDI3C)* (pp. 41-45). IEEE. DOI: 10.1109/ICDI3C.2018.00017

Reddy, S. R. G., Varma, G. P. S. and Davuluri, R. L. (2023) Resnet-based modifed red deer optimization with DLCNN classifer for plant disease identifcation and classifcation. *Computers and Electrical Engineering* 105, 108492. DOI: 10.1016/j.compeleceng.2022.108492

Salman, Z., Muhammad, A., Piran, M. J. and Han, D. (2023) Crop-saving with AI: latest trends in deep learning techniques for plant pathology. *Front. Plant Sci* 14.DOI: 10.3389/fpls.2023.1224709

Sarker, I. H. (2021) Deep learning: a comprehensive overview on techniques, taxonomy, applications and research directions. *SN computer science* 2(6), 1-20. DOI: 10.1007/s42979-021-00815-1

Selvaraj, M. G., Vergara, A., Montenegro, F., Ruiz, H. A., Safari, N., Raymaekers, D. and Blomme, G. (2020) Detection of banana plants and their major diseases through aerial images and machine learning methods: A case study in DR Congo and Republic of Benin. *ISPRS Journal of Photogrammetry and Remote Sensing* 169, 110-124. DOI: 10.1016/j.isprsjprs.2020.08.025

Shaikh, T. A., Rasool, T. and Lone, F. R. (2022) Towards leveraging the role of machine learning and artificial intelligence in precision agriculture and smart farming. *Computers and Electronics in Agriculture* 198, 107119. DOI: 10.1016/j.compag.2022.107119

Shamshirband, S., Fathi, M., Dehzangi, A., Chronopoulos, A. T. and Alinejad-Rokny, H. (2021) A review on deep learning approaches in healthcare systems: Taxonomies, challenges, and open issues. *Journal of Biomedical Informatics* 113, 103627. DOI: 10.1016/j.jbi.2020.103627

Shi, T., Liu, Y., Zheng, X., Hu, K., Huang, H., Liu, H. and Huang, H. (2023) Recent advances in plant disease severity assessment using convolutional neural networks. *Scientifc Reports* 13(1), 1–13. DOI: 10.1038/ s41598-023-29230-7

Shoaib, M., Shah, B., Ei-Sappagh, S., Ali, A., Ullah, A., Alenezi, F. and Ali, F. (2023) An advanced deep learning models-based plant disease detection: A review of recent research. *Frontiers in Plant Science* 14, 1158933. DOI: 10.3389/fpls.2023.1158933

Singh, V. and Misra, A. K. (2017) Detection of plant leaf diseases using image segmentation and soft computing techniques. *Information processing in Agriculture* 4(1), 41-49. DOI: 10.1016/j.inpa.2016.10.005

Smetanin, A., Uzhinskiy, A., Ososkov, G., Goncharov, P. and Nechaevskiy, A. (2021, September) Deep learning methods for the plant disease detection platform. In *AIP Conference Proceedings 2377(1)*. AIP Publishing. DOI:10.1063/5.0068797

Son, J. and Kim, S. B. (2017) Content-based filtering for recommendation systems using multiattribute networks. *Expert Systems with Applications* 89, 404-412. DOI: 10.1016/j.eswa.2017.08.008

Subramanian, M., Shanmugavadivel, K. and Nandhini, P. S. (2022) On fine-tuning deep learning models using transfer learning and hyper-parameters optimization for disease identification in maize leaves. *Neural Computing and Applications* 34(16), 13951-13968. DOI: 10.1007/s00521-022-07246-w

Suresh, A. (2023) Crop disease prevention and detection using a hybrid of CNN and SVM ISSN no: 1869-9391 crop disease prevention and detection using a hybrid of CNN and SVM.

Suttapakti, U. and Bunpeng, A. (2019, September) Potato leaf disease classification based on distinct color and texture feature extraction. In *2019 19th international symposium on communications and information technologies (ISCIT)* (pp. 82-85). IEEE. DOI: 10.1109/ISCIT.2019.8905128

Szegedy, C., Liu, W., Jia, Y., Sermanet, P., Reed, S., Anguelov, D. and Rabinovich, A. (2015) Going deeper with convolutions. In *Proceedings of the IEEE conference on computer vision and pattern recognition* (pp. 1-9).

Tan, L., Lu, J. and Jiang, H. (2021) Tomato leaf diseases classifcation based on leaf images: A comparison be classical machine learning and deep learning methods. *Agri Engineering* 3(3), 542–558. DOI: 10.3390/ agriengineering3030035

Taye, M. M. (2023) Theoretical understanding of convolutional neural network: Concepts, architectures, applications, future directions. *Computation* 11(3), 52. DOI: 10.3390/computation11030052

Théau, J., Gavelle, E. and Ménard, P. (2020) Crop scouting using UAV imagery: a case study for potatoes. *Journal of Unmanned Vehicle Systems* 8(2), 99-118.DOI: 10.1139/juvs-2019-0009

Thenmozhi, K. and Reddy, U. S. (2019) Crop pest classification based on deep convolutional neural network and transfer learning. *Computers and Electronics in Agriculture* 164, 104906.DOI: 10.1016/j.compag.2019.104906

Vishnoi, V. K., Kumar, K. and brajesh, k. (2021) Plant disease detection using computational intelligence and image processing. *J. Plant Dis. Prot* 128, 19–535. DOI: 10.1007/s41348-020-00368-0

Wang, G., Sun, Y. and Wang, J. (2017) Automatic image-based plant disease severity estimation using deep learning. *Computational Intelligence and Neuroscience* 2017. DOI: 10.1155/2017/2917536

Wang, H., Zhao, Y., Wang, Q. and Zhou, B. (2019, December) Latent factor models fusing user & item attributes. In *2019 IEEE Symposium Series on Computational Intelligence (SSCI)* (pp. 3201-3206). IEEE. DOI: 10.1109/ssci44817.2019.9002724

Wani, J. A., Sharma, S., Muzamil, M., Ahmed, S., Sharma, S. and Singh., S. (2022) Machine learning and deep learning based computational techniques in automatic agricultural diseases detection: methodologies, applications, and challenges. *Arch. Comput. Methods Eng* 29, 641–775. DOI: 10.1007/s11831-021-09588-5

Wspanialy, P. and Moussa, M. (2020) A detection and severity estimation system for generic diseases of tomato greenhouse plants. *Computers and Electronics in Agriculture* 178, 105701. DOI: 10.1016/j.compag.2020.105701

Wu, B., Liang, A., Zhang, H., Zhu, T., Zou, Z., Yang, D. and Su, J. (2021) Application of conventional UAV-based high-throughput object detection to the early diagnosis of pine wilt disease by deep learning. *Forest Ecology and Management* 486, 118986. DOI: 10.1016/j.foreco.2021. 118986

Wu, Q., Zhang, K. and Meng, J. (2019) Identifcation of soy- bean leaf diseases via deep learning. DOI: 10. 1007/s40030-019-00390-y

Wu, X., Fan, X., Luo, P., Choudhury, S. D., Tjahjadi, T. and Hu, C. (2023) From laboratory to field: unsupervised domain adaptation for plant disease recognition in the wild. *Plant Phenom* 5, 0038. DOI: 10.34133/plantphenomics.0038

Xie, S., Girshick, R., Dollár, P., Tu, Z. and He, K. (2017) Aggregated residual transformations for deep neural networks. In *Proceedings of the IEEE conference on computer vision and pattern recognition* (pp. 1492-1500).

Xu, M., Kim, H., Yang, J., Fuentes, A., Meng, Y. and Yoon, S. (2023) Embracing limited and imperfect training datasets: opportunities and challenges in plant disease recognition using deep learning. *Front. Plant Sci.* 14. DOI: 10.3389/fpls.2023. 1225409

Yakkundimath, R., Saunshi, G. and Palaiah, S. (2022) Automatic methods for classification of visual based viral and bacterial disease symptoms in plants. *International Journal of Information Technology* 14(1), 287-299. DOI: 10.1007/s41870-021-00701-2

Zeng, Q., Ma, X., Cheng, B., Zhou, E. and Pang, W. (2020) Gans-based data augmentation for citrus disease severity detection using deep learning. *IEEE Access* 8, 172882-172891.

Zhai, X., Kolesnikov, A., Houlsby, N. and Beyer, L. (2022) "Scaling vision transformers," in Proceedings of the IEEE/CVF *Conference on Computer Vision and Pattern Recognition.* (New Orleans Louisiana) 12104–12113.

Zhang, S., Zhang, S., Zhang, C., Wang, X. and Shi, Y. (2019) Cucumber leaf disease identification with global pooling dilated convolutional neural network. *Computers and Electronics in Agriculture* 162, 422-430. DOI:10.1016/j.compag.2019.03.012

Zhang, T., Xu, Z., Su, J., Yang, Z., Liu, C., Chen, W. H. and Li, J. (2021) Ir-unet: Irregular segmentation u-shape network for wheat yellow rust detection by UAV multispectral imagery. *Remote Sensing* 13(19), 3892.DOI: 10.3390/ rs13193892

Zhao, Z., Zhao, Z., Li, Q., Zhuang, J. and Ju, X. (2023, March) Multi-modal cross-attention learning on detecting anomalies of electrical equipment. In *2023 International Conference on Pattern Recognition, Machine Vision and Intelligent Algorithms (PRMVIA)* (pp. 137-142). IEEE. DOI: 10.1109/ prmvia58252.2023.00030

Zhu, R., Tu, X. and Huang, J. X. (2020) Deep learning on information retrieval and its applications. In *Deep learning for data analytics* (pp. 125-153). Academic Press. DOI: 10.1016/B978-0-12-819764-6.0000

4 Deep Learning for Plant Leaf Disease and Stress Detection

Mohammed Radi[1,2], Ilham Dehbi[1,3], Chayma Ikan[4], Fatima-Ezzahra Soussani[4,5], Hakima Achetoui[1], Abdelilah Meddich[4,6], and Rachid Lahlali[1*]

[1]Department of Plant Protection, Phytopathology Unit, Ecole Nationale d'Agriculture de Meknès, Km10, Rte Haj Kaddour, BP S/40, Meknès, Morocco; [2]Environment and Valorization of Microbial and Plant Resources Unit, Faculty of Sciences, Moulay Ismail University, Meknes, Morocco; [3]Laboratory of Plant Biotechnology and Molecular Biology, Faculty of Sciences, Moulay Ismail University, Zitoune, Meknes, Morocco; [4]Centre d'Agrobiotechnologie et Bioingénierie, Unité de Recherche labellisée CNRST (Centre AgroBiotech-URL-CNRST-05), Université Cadi Ayyad, Marrakech, Morocco; [5]Faculty of Sciences, Laboratory of Plant, Animal, and Agro-Industry Productions, University Ibn Toufail, Kenitra, Morocco; [6]African Sustainable Agriculture Research Institute (ASARI), Mohammed VI Polytechnic University (UM6P), Laâyoune, Morocco

Abstract

Early identification and diagnosis of plant leaf diseases and stress play a vital role in enhancing crop yield and maintaining food security. Current methods for identifying these diseases are time-consuming and error-prone, which has prompted the use of deep learning (DL) approaches to reduce the workload and improve diagnostic accuracy. This chapter introduces how DL algorithms capable of mathematically analyzing images, particularly convolutional neural networks, can take a step toward the automatic identification of different plant stresses from the plant leaf images. We have performed a review of the current state-of-the-art architectures and methodologies, ranging from data acquisition to preprocessing methods specifically designed for agricultural needs. Finally, we summarize the state of the art, outline the challenges of deploying DL here, and address possible future steps toward improvement through advanced technology for more accessible and efficient precision agriculture.

Keywords: convolutional neural network, diagnosis, diagnostic accuracy, image analysis, precision agriculture

4.1 Introduction

Early and high-precision identification of plant diseases and abiotic stress factors is the foundation of sustainable agriculture, strongly affecting food security, ecosystem resilience, and economic productivity. Globally, plant pathogens and pest-related diseases are estimated to cause losses of up to 40% of crop yields every year (Savary *et al.*, 2019). Apart from biotic threats, fungi, bacteria, and viruses, crops are affected by abiotic stressors, including drought, heat waves,

*E-mail: rlahlali@enameknes.ac.ma

DOI: 10.1079/9781836991120.0004

salinity, and nutrient deficiencies, that interfere with plant physiology and yield quality (Gou *et al.*, 2024).

Climate change, soil degradation, and increasingly intensified systems of agriculture further compound such challenges and necessitate effective management of disease and stress. Traditional detection approaches are based on visual inspection, symptom scoring, and further laboratory testing, which are time-intensive, subjective, and lack scalability (Pujari *et al.*, 2015; Mahlein, 2016). Moreover, the accuracy of traditional methods may diminish in the field when symptoms overlap or multiple concurrent stressors are present. Consequently, the need for accessible, cost-effective, rapid, scalable diagnostic tools has become a priority in precision agriculture (LeCun *et al.*, 2015; Pantazi *et al.*, 2019).

Artificial intelligence (AI), particularly deep learning (DL), has arisen as a powerful alternative to these limitations in plant disease and stress detection within a relatively short amount of time. This allows DL models to learn complex representations from raw input data, especially images, without handcrafting features, which is especially useful for visual tasks and leaf-based diagnoses (Kamilaris and Prenafeta-Boldú, 2018; Barbedo, 2019). A widely adopted approach in plant phenotyping is to employ convolutional neural networks (CNNs), which extract spatial features through a series of convolution and pooling layers followed by non-linear activation functions. A good performance of CNNs in classification tasks such as plant disease classification, object detection, and semantic segmentation has led to their significant success in outperforming classical machine learning models in generalization and robustness (Brahimi *et al.*, 2017; Ferentinos, 2018).

They have been validated in several studies for distinguishing healthy from infected leaves across multiple crop types and disease classes under various lighting, background conditions, and growth stages (Mohanty *et al.*, 2016; Sladojevic *et al.*, 2016; Too *et al.*, 2019). The widening use of large-scale annotated datasets, such as PlantVillage, and the affordability of imaging instruments (smartphones, unmanned aerial vehicles (UAVs), multispectral sensors) have supplemented the effectiveness of CNNs in agricultural diagnostics (Hughes and Salathé, 2015; Lu and Young, 2020). Furthermore, advancements in transfer learning, data augmentation, and model compression techniques have also allowed DL models to run on edge devices, allowing real-time plant health monitoring within high-tech and low-resource environments (Dyrmann *et al.*, 2016).

The chapter gives a comprehensive overview of the applications of DL techniques for plant leaf disease and stress identification. It explains the theoretical principles behind neural networks in image analysis, introduces important methods for collecting and preprocessing data, and covers popular CNN architectures including Visual Geometry Group (VGG), ResNet, Inception, and U-Net. The chapter uses selected case studies to demonstrate the application of DL to practical agricultural issues and concludes by discussing current limitations, future directions, and trends of AI-enabled plant disease monitoring.

4.2 Fundamentals of DL in Image Analysis

4.2.1 Explanation of core DL concepts (neural networks, convolutional layers)

DL is one of the most successful techniques in machine learning (ML), accomplishing complex tasks sometimes even better than humans themselves (Taye, 2023). One key strength of DL is its capacity to learn from massive amounts of data, which has pioneered development in cybersecurity, natural language processing, bioinformatics, robotics, and even the processing of medical information (Rane *et al.*, 2024). While specialization in certain applications can be beneficial, having a broader knowledge of DL is equally important to maximize its capabilities (Sarker, 2021).

The building blocks of advanced DL applications indeed include CNNs and convolutional filters, two sets of core concepts of DL (Yoshua Bengio, 2019). The key constituents of neural networks are interconnected layers of nodes (neurons) (Thakur, 2021; Montesinos López *et al.*, 2022). The output of a neuron is ultimately defined by the weighted inputs coming from the neurons in the previous layer of the network. In CNNs, filters are used in convolutional layers and allow the model to identify characteristics such as borders, textures, and other patterns contained in images (Kattenborn *et al.*, 2021). This allows the model to understand complex spatial hierarchies, which is particularly useful in tasks such as image classification and object recognition (Krichen, 2023). Yet CNNs, which excel in image and spatial data processing, are made up of convolutional layers that use filters (kernels) to detect features such as edges, textures, and patterns, producing feature maps (Nawrocka *et al.*, 2023). By piling up these layers, CNNs learn hierarchical features, enabling them to recognize both simple and complex patterns. Pooling layers, frequently max-pooling, lower spatial dimensions for computational efficiency, while fully linked layers collect learned characteristics, with the final SoftMax layer transforming outputs into classification probabilities (Aziz, 2020). CNNs have transformed applications such as biometric identification, which recognizes individual face traits. Prominent CNN designs, including LeNet, AlexNet, VGG Net, and Network-in-Network (NiN), have revolutionized image analysis by incorporating innovations such as deeper layers, smaller filters, and 1 × 1 convolutions, broadening the reach of DL in a variety of areas (Swapna *et al.*, 2020).

4.2.2 Overview of CNNs and their relevance to plant disease and stress detection

Deep neural networks (DNNs), particularly CNNs, have demonstrated great performance in a variety of computer vision tasks such as object identification, recognition, classification, and biometrics (Jin, 2022). CNNs' convolutional layers function as filters learned directly from data, resulting in a succession of visual representations that are optimal for the task at hand (Hao, 2021). CNNs create a model comprising weights and biases that are suited to execute the job for which they were trained. One of the main benefits of CNNs is their capacity to generalize, which allows them to handle previously unknown data (Salehi *et al.*, 2023; Liu *et al.*, 2024). This capacity ensures resilience against fluctuations in the backdrop, image capture settings, and intra-class disparities. CNNs are widely used in agriculture for plant disease detection because they excel at distinguishing patterns in complex data like leaf textures and spotting disease symptoms (Masibo, 2024). For example, CNNs can be trained to differentiate between healthy and diseased leaves by analyzing features such as color, shape, and texture. This has applications in detecting specific plant diseases, nutrient deficiencies, and stress caused by environmental factors.

Crop production can be greatly reduced due to various diseases, which seriously endanger food security. Thus, detecting plant diseases accurately is necessary and urgent. CNN has gained widespread application in plant disease classification (Iftikhar *et al.*, 2024). They have solved or partially solved problems of traditional classification methods and represent state-of-the-art technology in this field. However, it is observed that CNN scans a dataset containing leaf images and identifies and extracts the essential features before classifying the diseases. CNN's decision-making approach is comparable to that of humans (Shi *et al.*, 2023). It captures the textures and color lesions on plant leaves to aid in disease detection. CNN's feature-learning capacity automatically recognizes significant elements from a leaf image using a variety of non-linear filters. As a result, the overall performance of a model improves over models with hand-engineered characteristics (Latif *et al.*, 2022).

4.2.3 Popular architectures in leaf image analysis

Multiple renowned architectures, including ResNet, VGG, and Inception, have been widely employed in leaf image analysis because of their excellent skills in both feature extraction and classification applications.

ResNet, or Residual Network, is an innovative approach that uses recursive connections to solve the gradient disparity problem in deep networks (Sengupta *et al.*, 2020). ResNet allows for successful creation and stability by allowing gradients to propagate directly across these connections, even in designs with hundreds of layers. This idea overcomes the degradation problem in deep networks while retaining crucial properties and allows for the capture of complex models (Hasanah *et al.*, 2023). Furthermore, ResNet-50, a commonly used variation, has established new standards in DL by exploiting residual blocks to improve gradient flow and optimize feature extraction (Duan *et al.*, 2024). Its shape makes it especially useful for difficult tasks such as plant disease detection, where minute distinctions between healthy and sick leaves are crucial. ResNet has performed very well in the early identification of disease leaf blasts, a key image classification task in precision agriculture (Li *et al.*, 2021). Its capacity to distinguish between healthy and diseased leaf patterns makes it extremely useful for disease control. ResNet, a foundation of neural network topologies, provides a strong and adaptable framework that facilitates learning across layers, driving improvements in multiple machine vision applications spanning from medical imaging to autonomous systems (Crespi, 2023).

The VGG model is widely renowned for its ability to increase performance metrics across a variety of image categorization tasks. The VGG model excels in feature extraction and pattern recognition because of its deep architecture, which uses small 3 × 3 filters and is made up of successive convolutional layers with small receptive fields (Zakaria and Hassim, 2023; Zakaria *et al.*, 2024). The tool's simplicity and flexibility make it ideal for both benchmarking and practical applications. However, the VGG architecture, including popular variations such as VGG-16 and VGG-19, performs particularly well in tasks that require high-resolution image processing. These models use a consistent design approach of stacking numerous convolutional layers followed by max-pooling, which gradually captures the spatial structure of inputs (Shafapourtehrany *et al.*, 2023; Zhang *et al.*, 2023). This approach allows VGG networks to learn sophisticated visual representations, improving their performance across a variety of datasets and problem areas. Despite its processing complexity when compared to more recent design principles, the VGG model remains a fundamental tool in computer vision. It has been effectively utilized in domains such as medical imaging, object identification, and style transfer, proving its flexibility and resilience in producing enhanced performance metrics (Zhang *et al.*, 2023).

The Inception network, notably Inception-v3, transformed DL with its multiscale design, which combines convolutions of varying sizes inside the same layer to collect characteristics at several scales (Horasan and Güneş, 2024). This approach allows for the simultaneous recording of fine-grained details and wider contextual data, making it excellent for applications such as plant disease identification, where symptoms vary in size and appearance (Shoaib *et al.*, 2023). Inception-v3 improves efficiency further with approaches such as factorized convolutions, filter-based dimensionality reduction, and batch normalization, which reduce computing complexity while retaining excellent accuracy. Inception's adaptability and efficiency have made it a standard choice for challenging image processing tasks (Ahmmed *et al.*, 2023).

4.3 Data Acquisition and Preprocessing

4.3.1 Types of data used for leaf disease and stress detection (red, green, blue (RGB) images, multispectral, and hyperspectral)

Plant diseases are one of the main challenges that cause substantial economic losses to producers and farmers. Conventional approaches for diagnosing stressed plants are labor-intensive and traditional, which makes their use very limited. Rapid techniques are essential to address this issue (Vurro *et al.*, 2010). Advancements in ML and enhanced detection technologies have triggered precision agriculture revolutions based on DL and big data. Different forms of data are used to identify diseases, providing comprehensive information on plant health that allows for faster action and more effective management (Ramesh *et al.*, 2018). Currently, intense research is focused on creating optical imaging technologies for plant disease detection. Optical imaging approaches for plant disease detection are of great interest because they provide non-invasive, fast, and cost-effective tools for monitoring plant health (Gudkov *et al.*, 2023). Optical imaging stands out from traditional approaches that rely on visual assessment by offering rapid and non-contact measurement of changes produced by biotic stresses in plant physiology (Mahlein, 2016). These techniques use several parts of the electromagnetic spectrum to gather precise information about plant properties that may indicate the presence of disease or stress.

In general, common imaging methods used to assess crop stress include the following optical sensors:

- *RGB images*. The most basic and widely used technique of optical imaging is RGB imagery, which captures the three major colors of visible light—red, green, and blue. The wavelengths collected are around 400–499 nm for blue light (maximum at 475 nm), 500–549 nm for green light (maximum at 520 nm), and 550–750 nm for red light (maximum at 650 nm) (Skala, 2023). In this regard, RGB imaging may be regarded as a subset of multispectral imaging. This approach is frequently used to identify visual indicators of plant disease, such as lesions, discoloration, or wilting (Padmavathi and Thangadurai, 2016). These RGB images also give a simple and straightforward depiction of plant health, making them perfect for quick visual evaluation and identification of observable indicators such as leaf discoloration, pathogen-caused lesions, or spots (Jha *et al.*, 2024). However, while RGB imaging data are typically collected using a digital camera or smartphone, multispectral imaging necessitates more specialized equipment or instruments. Despite this, they are normally considered independently (Kumar *et al.*, 2012). The key advantages of RGB imaging are the inexpensive price and the small dimensions of the portable sensor. Smartphones already feature RGB image sensors, which have been effectively used to assess plant pressures such as iron deficiency chlorosis in soybean (Kinnikar *et al.*, 2015), different nutrient deficiencies in black beans (Arboleda *et al.*, 2018), late blight and mildew in potato plants (Sugiura *et al.*, 2016), and fusarium head blight in wheat (Qiu *et al.*, 2019). However, RGB images are restricted in identifying stress or disease-related symptoms that are not evident to the human eye, particularly in the initial phases.
- *Multispectral imaging* is a revolutionary method that collects comprehensive data from a variety of wavelengths in the electromagnetic spectrum, including visible light, infrared, and ultraviolet bands. Multispectral imaging uses specialist sensors, cameras, or satellites to collect data from many spectral bands at the same time, resulting in a rich, multidimensional image of the environment (Kerkech *et al.*, 2020). Multispectral imaging sensors, which typically include three to ten spectral bands, are intended to capture images at various wavelengths (Pu *et al.*, 2015). These sensors typically gather one or more images that encompass the visible spectrum and the near-infrared range (Ma *et al.*, 2023). When plants are stressed, their visible light reflectance increases due to decreased chlorophyll and light absorption (*Veys et al.*, 2019). At the same time, changes in leaf tissue composition cause a reduction in near-infrared (NIR) reflectance. Therefore, the most widely employed spectral bands for these sensors are green, red, red edge, and NIR (Li *et al.*, 2016). When paired with UAVs, multispectral sensors are commonly employed in remote sensing for plant disease detection (Deng *et al.*, 2018). However, one shortcoming of these sensors is that they typically function with only a few spectral bands, making it difficult to precisely detect the severity of plant diseases. It helps to improve precision farming in agriculture by identifying plant stress, insect infestations, and nutritional deficits. Changes in agricultural practices, deforestation, and pollution levels may all be tracked in environmental monitoring. Multispectral imaging can be utilized in medicine for non-invasive diagnostics, delivering information on tissue health and disease detection early on. This technology is transforming remote sensing, environmental research, defense, and healthcare by delivering enriched, actionable data that enhances decision-making, sustainability, and resource management (Ilişanu *et al.*, 2023).
- *Hyperspectral imaging*. Hyperspectral analysis is a novel field that combines optical spectroscopy and image analysis techniques to analyze physiological and morphological characteristics at the same time. It is a cutting-edge remote sensing technology that captures data over many small, contiguous spectral bands, allowing for significantly more detailed analysis than typical multispectral images (Cheshkova, 2022).

Hyperspectral sensors operate on a similar concept as RGB and multispectral cameras. All of these devices detect the quantity of light that reaches the sensor and store the data. Hyperspectral sensors, in contrast to RGB cameras, which record light in three spectral bands, and multispectral cameras, which generally employ less than 20 bands, may measure up to several hundred bands over the electromagnetic spectrum. Each of these bands has a tiny range of wavelengths, generally only a few nanometers, resulting in great spectral resolution for hyperspectral sensors (Nagasubramanian *et al.*, 2018). The distinction between multispectral and hyperspectral imaging is that the latter captures information in hundreds of unique bands over the electromagnetic spectrum, from ultraviolet to infrared, whereas multispectral is confined to a few broad spectral bands (Lowe *et al.*, 2017). This high spectral resolution enables precise material identification, detection of small compositional fluctuations, and surface property characterization. Environmental monitoring, agriculture, mining exploration, defense, and health all make use of hyperspectral imaging. It may be used in agriculture to evaluate plant health and soil conditions, as well as to identify pests and diseases. It contributes to the monitoring of water quality, pollution, and vegetation changes in the environment (Zhang *et al.*, 2020). It is critical in defense for observation and identification of targets. Its capacity to collect specific spectral fingerprints makes it a useful tool for applications that need great accuracy as well as the ability to discern between identical materials or environments. The majority of hyperspectral scanning cameras used in laboratories are placed on specialized moving platforms that allow for linear movement and camera stabilization (Mahlein *et al.*, 2017). Fixed rail systems are commonly employed in greenhouses for this purpose (Rayhana *et al.*, 2023). For field applications (Thomas *et al.*, 2018) or drones (Abdulridha *et al.*, 2023), hyperspectral imagery is widely utilized. One significant disadvantage of scanning sensors is their relatively long image acquisition time, which varies depending on the size of the area being scanned, making it difficult to record moving objects. This issue is addressed by the Specim IQ portable camera, which has an integrated scanner for quicker image collection (Behmann *et al.*, 2018; Nguyen *et al.*, 2021).

4.3.2 Techniques for data collection, including drones, field cameras, and controlled environments

Methods of collecting data to identify plant diseases have advanced considerably thanks to the use of technologies such as drones, field cameras, and controlled conditions, enabling greater precision and efficiency in the monitoring, diagnosis, and management of plants (Buja *et al.*, 2021; Abbas *et al.*, 2023). Drones equipped with hyperspectral or multispectral sensors are particularly useful for monitoring vast areas (Heincke *et al.*, 2019). They collect high-resolution images and spectral data, making it possible to detect the first symptoms of stress, disease, or damage caused by pests. The idea of using drones to obtain high-resolution images of healthy and damaged plants has recently emerged (Chin *et al.*, 2023). Drones are equipped with sensors that monitor reflectance, temperature, and fluorescence, making it possible to move, detect, and locate potential crop diseases from visual data, while providing a map of crop conditions that farmers or other machines collaborating with drones can use to perform various tasks autonomously with little or no human intervention (León-Rueda *et al.*, 2021). According to Stavrakoudis *et al.* (2019), drones equipped with digital and multispectral cameras took photos of research plots containing 67 cultivars and a few elite lines. Additionally, field cameras mounted on permanent or mobile supports, together with AI and ML algorithms, enable real-time analysis of visual data, making it easier to diagnose and monitor signs of disease (Mahendran *et al.*, 2021; Shaikh *et al.*, 2022). The ground-based normalized difference vegetation index and the image-based normalized difference vegetation index were calculated, and the two Normalized Difference Vegetation Index (NDVI) data indices had a significant connection (Bhandari *et al.*, 2020). Multispectral images were then used to quantify the different degrees of rice sheath blast disease in the plots, with an accuracy of over 60%. The results showed that drones equipped with digital and multispectral cameras are the most effective tools for identifying rice sheath blast disease in the field (Zhang DongYan *et al.*, 2018). These techniques can detect early disease symptoms and track their progression, providing farmers and managers with useful tools for targeted crop management (Zhou *et al.*, 2021). In addition, controlled environments such as greenhouses and growth chambers can mimic the exact conditions under which diseases occur. The incidence of late blight (*Phytophthora infestans*) in potatoes was assessed by describing the spectrum changes associated with late blight development at low levels of disease severity (Franceschini *et al.*, 2019; León-Rueda *et al.*, 2021; Abbas *et al.*, 2023). They used optical imagery from subdecimetric drones to collect data on early changes in the potato crop associated with disease incidence. Spectral variability was summarized using simplex volume maximization (Rümer *et al.*, 2012; Kersting *et al.*, 2016). UAV photographs were taken at four different times during the growing season, before and after late blight disease was discovered in the field (Franceschini *et al.*, 2019). The ability of these techniques to cover large agricultural areas rapidly makes them essential tools for precision agriculture.

4.3.3 Key preprocessing steps: annotation, image augmentation, normalization, and segmentation to enhance model performance

Several preprocessing procedures are required in plant disease detection to improve model performance and assure accurate symptom identification. These procedures include annotation, image enhancement, normalization, and segmentation, which prepare the data for the model to properly analyze diseases under various settings (Mamat *et al.*, 2022). The combination of these preprocessing stages guarantees that plant disease detection models are fed high-quality, diverse, and well-structured data. This provides early disease identification and more effective crop management, allowing farmers to respond swiftly and maximize harvests (Ngugi *et al.*, 2021; Xu *et al.*, 2022). Annotation is the first essential stage, which involves recognizing and labeling parts of an image that indicate signs of disease, such as spots, discoloration, or lesions (Dong *et al.*, 2022, 2023). This stage is critical for supervised learning because it provides the model with explicit instances of what it needs to learn to identify, namely the distinctions between healthy and affected plants. The more exact the annotation, the more accurately the model can detect disease. After the annotation has been completed, images are enriched. This method entails creating more data by applying different changes, such as rotation, scaling, or brightness adjustment (Singh *et al.*, 2020). It is especially effective when the number of images accessible is restricted, since it enhances the variety of instances and strengthens the model. With augmentation, the model can more accurately generalize, accounting for variations in illumination and plant appearance (Zeng *et al.*, 2020). Image normalization standardizes pixel values by readjusting them to a consistent scale, such as 0 to 1, or centering them around the mean and standard deviation. This step prevents external factors, such as variations in light or image quality, from interfering with model learning. By making it easier for the model to focus on the essential features of the disease, normalization improves detection accuracy and accelerates the learning process (Fuentes *et al.*, 2017; Chowdhury *et al.*, 2021; Li *et al.*, 2021). Finally, segmentation divides the image into different regions of interest, such as infected areas of plants (Singh and Misra, 2017). According to Kaur *et al.* (2024), this allows the model to focus specifically on the parts that are relevant for identifying the disease, such as affected leaves or stems of tomatoes. By isolating these regions, segmentation simplifies image processing and allows the model to focus on the characteristic signs of the disease, improving detection efficiency and accuracy (Dhingra *et al.*, 2018; Mzoughi and Yahiaoui, 2023; Shoaib *et al.*, 2025).

4.4 DL Techniques for Disease and Stress Detection

4.4.1 Supervised classification approaches for identifying healthy vs stressed/diseased plants

Supervised classification methods, particularly those using DL, have proven very effective in distinguishing between healthy and stressed plants. Traditional ML techniques, such as support vector machines (SVM) and random forest inference (RF), are widely used to classify plant diseases using spectral and imaging data (Sujatha *et al.*, 2021). However, recent advances in CNNs, recurrent neural networks (RNNs), and hybrid DL frameworks have significantly improved the accuracy and reliability of classification in practical farming (Kamilaris and Prenafeta-Boldú, 2018; Peng *et al.*, 2024). DL models, especially CNNs, are capable of learning autonomously from plant stress indicators such as leaf color, texture, and vein patterns (Wang *et al.*, 2022; Islam *et al.*, 2024). Research has shown that when hyperspectral and multispectral imaging data are processed by DL models, they are more accurate in detecting diseases and facilitate early intervention strategies (Signoroni *et al.*, 2019; Khan *et al.*, 2022) (Table 4.1). Furthermore, incorporating attentional and transfer learning has improved classification efficiency while minimizing computational costs.

Prior to hyperspectral imaging becoming widely accessible, scientists used point-source tools such as spectroradiometers or multispectral imaging to measure plant color information. Hyperspectral imaging, in contrast to conventional imaging, necessitates specialized acquisition and produces sizable numerical datasets, both of which require advanced analytical techniques (Manley, 2014). One common technique for managing hyperspectral data is to concentrate on significant wavelengths and compute reflectance indices at different wavelengths to account for variations in lighting. Hardware limitations or biological traits are used to create these indices (Lowe *et al.*, 2017). One popular vegetation index is the NDVI, which uses NIR and visible light absorption to measure plant health (Stamford *et al.*, 2023). In order to detect stress, such as wheat damage brought on by *Eurygaster integriceps* (Sunn pest), NDVI has been used (Motie *et al.*, 2023). Nevertheless, a large number of indices are dataset-specific and not generally relevant. Disease-specific indices, such as the Leaf Rust Disease Severity Index, have demonstrated 87–91% accuracy in identifying wheat leaf rust (*Puccinia triticina*), despite the fact that their generalizability has not been tested (Ashourloo *et al.*, 2014; Khosrokhani and Nasr, 2022). Another important hyperspectral characteristic is the red edge position (690–740 nm), where NIR reflectance sharply rises and chlorophyll absorption falls. This spectral region is used to detect stress and plant diseases (Lowe *et al.*, 2017). In a pertinent field study, hyperspectral data collected from a wheat experiment were processed by first utilizing the NDVI to eliminate non-leaf elements and focus solely on leaf-specific data. Following this, analysis of covariance was used to pinpoint wavelength bands that were statistically linked to disease symptoms. These identified bands were then employed as inputs for quadratic discriminant analysis (QDA) to categorize spectra into healthy and yellow rust-infected leaf groups. This methodology exemplifies a typical approach in hyperspectral analysis, consisting of three primary steps: (i) segmenting the region of interest, (ii) selecting

Table 4.1. Different CNN approaches for detecting leaf diseases in certain plants. Author's own table.

Detection approaches	Plant	Plant stress detection	Accuracy	References
Autonomous Mobile Robot (2D LiDARs and RTK GNSS)	Apple	Early-stage disease	99.34%	Karpyshev *et al.* (2021)
RGB images and HSI	Corn	Early disease detection	Increases by 6.14%	Fu *et al.* (2022)
Two-dimensional CNN model	Wheat	Early-stage disease detection	98.84%	Hossen *et al.* (2022)
CNN and MSC (CNN-MSC), CNN and WT (CNN-WT), CNN and D1 (CNN-D1), and CNN and D2 (CNN-D2)	Potato	PLB	CNN (98%) CNN-WT and CNN-D2 (100%)	Zhang *et al.* (2021)
CNN	Tomato	Early blight, late blight, and leaf mold	98%	Karthik *et al.* (2020)

CNN, convolutional neural network; D1, first-order difference; D2, second-order difference; HSI, hyperspectral images; MSC, multiple scattering correction; PLT, potato late blight; RGB, red, green, blue; WT, wavelet transform.

predictive spectral features, and (iii) classifying based on spatial and spectral data. The QDA model, utilizing just four optimized wavebands, achieved an overall classification accuracy of 92% (Lowe *et al.*, 2017; Kurouski, 2021).

Convolutional neural networks have been extensively applied in the detection of plant diseases (Fig. 4.1) (Karpyshev *et al.*, 2021). In one research, CNNs were used to identify 26 diseases across 14 different crops using a database containing 54,306 color images (Lu *et al.*, 2021). The models GoogLeNet and AlexNet, when trained from scratch, achieved classification rates of 97.82% and 98.36%, respectively. These rates improved to 99.27% and 99.34% with the application of transfer learning. However, performance significantly decreased to 31.4% when the models were tested on images with conditions differing from the training set (Brahimi *et al.*, 2018). Another investigation utilized CNNs to detect 13 diseases in apple, pear, and grapevine crops, achieving 96.3% accuracy with the CaffeNet model (Sladojevic *et al.*, 2016). Although DL is effective for image classification, its application in hyperspectral data analysis remains limited (Pan *et al.*, 2018). Hyperspectral imaging faces challenges such as high data dimensionality, computational demands, and redundant wavelength noise (Jaiswal *et al.*, 2023). Successful DL implementation depends on having ample labeled datasets and robust computing resources (Nguyen *et al.*, 2019). Other non-DL methods have also been explored for plant disease detection. For instance, Fisher's linear discriminant analysis was used to classify yellow rust and powdery mildew in wheat with 93% accuracy using optimal wavelengths (531 nm, 570–654 nm, and 685–717 nm) (Yao *et al.*, 2019). Image processing techniques, such as morphological operators, have been employed to segment disease features from binary images (Wani *et al.*, 2022). In a study on cucumber leaves infected with downy mildew, principal component analysis (PCA) was used to reduce data size, followed by erosion and dilation operations to emphasize disease characteristics, achieving 90% accuracy with a small dataset of 20 samples (Lin *et al.*, 2021; Omer, 2023). Hyperspectral imaging has also been combined with microscopy for detailed analysis. A study on barley genotypes examined spectral differences between healthy and *Blumeria graminis*-infected plants. Spectral analysis showed significant variations over time, except in genotypes with the *mlo* gene, which provides resistance (Brugger *et al.*, 2021; Aroge *et al.*, 2024). These studies highlight the potential of both DL and traditional statistical methods in plant disease classification. While DL offers high accuracy, its application to hyperspectral data is still a developing research area that needs further exploration.

4.4.2 Object detection and segmentation techniques (e.g. YOLO, mask R-CNN, and U-NET) used in precise localization and identification

In recent times, DL has emerged as a leading method for processing and analyzing digital images, with significant applications in various industries, including agriculture (Wang *et al.*,

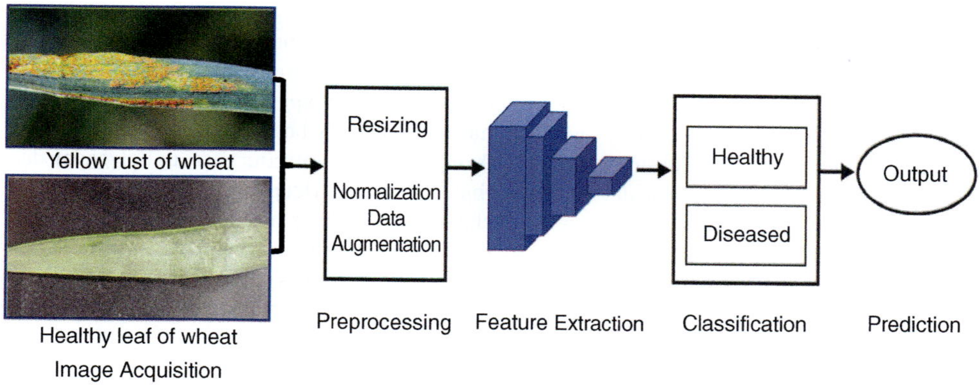

Fig. 4.1. A typical DL framework using CNNs for stripe rust detection in wheat. CNNs, convolutional neural networks; DL, deep learning. (Author's own image.)

2022). While DL is a subset of ML, its models are more intricate, offering a layered representation of data through multiple convolutions (Alzubaidi *et al.*, 2021). Deep CNNs, the most common type of DL neural network, have demonstrated remarkable success in tasks such as object detection and semantic image segmentation (Dhillon and Verma, 2020). Instance segmentation models, which heavily rely on CNNs, can autonomously learn intricate patterns and features in images, extracting detailed information with high precision (Zhao *et al.*, 2022; Shoaib *et al.*, 2023). Consequently, these models achieve more accurate feature extraction (Zhao *et al.*, 2022) compared to traditional ML methods such as linear/logistic regression, SVM, k-nearest neighbors (KNN), decision trees (DT), and RF. There is a burgeoning interest in research focused on developing DL-based machine vision solutions for precise detection and pattern recognition challenges in agricultural image datasets (Charisis and Argyropoulos, 2024). Dias *et al.* (2018) explored the use of a CNN combined with an SVM model for detecting apple flowers in an orchard. The work compared the performance of some well-known DL-based object detection frameworks, that is R-CNN (Thoke and Rai, 2024), Fast R-CNN (Zhao and Su, 2024), and Faster R-CNN (Hu *et al.*, 2022), for detecting strawberry flowers. Parvathi and Tamil Selvi (2021) proposed a Faster R-CNN model for coconut detection and maturity estimation with higher performance compared to other DL object detection models. Similarly, Sa *et al.* (2016) experimented with a Faster R-CNN model for sweet pepper detection by fusing information from RGB images and NIR images.

Notably, the majority of the research on DL-based object detection in the literature has concentrated on the two leading object detection families (i) Faster R-CNN and YOLO (You Only Look Once) and (ii) U-Net (Redmon *et al.*, 2016; Wang *et al.*, 2021). These models have significantly aided the advancement of object detection and have served as foundations for developing other DL-based object detection algorithms by modifying and improving existing architectures. However, following the introduction of Mask R-CNN (He *et al.*, 2017), a DL architecture that extended object detection to include pixel-level instance segmentation, numerous studies have been published examining both standard and customized Mask R-CNN-based architectures to innovate instance segmentation tasks in agriculture (Charisis and Argyropoulos, 2024). For example, Mask R-CNN was applied to detect leaf diseases of apple using the PlantVillage dataset in the experimental process, achieving a top accuracy of 96.6% with the Ensemble Subspace Discriminant Analysis classifier (Rehman *et al.*, 2021). In contrast, the R-CNN series operates at a slower pace but offers the benefit of potentially reducing both commission and omission recognition errors (Adarsh *et al.*, 2020). Therefore, the advantage of the YOLO series is its fast recognition speed, which can meet immediate requirements. Examples of applications include employing the YOLOv3 model, which features four scale detection layers, or utilizing YOLOv7 to identify and categorize road surface damage using Google Street View images (Shiu *et al.*, 2023). Several previous studies have proposed related applications in detection of leaf diseases on plants. Alhwaiti *et al.* (2025) customized YOLOv3 and YOLOv4 for the detection of healthy peach and strawberry leaves, as well as peach leaves affected by bacterial spots and strawberry leaves with scorch disease. YOLOv5 was used to detect rice leaf disease (Kumar *et al.*, 2023). The YOLOv7 method's detection and identification outcomes were confirmed by key statistical measures, including detection accuracy and precision, which are 97.3% and 96.7%, respectively (Soeb *et al.*, 2023), while YOLOv8 was used to detect yellow rust in wheat with 95.8% accuracy (Hassan *et al.*, 2024). Furthermore, U-Net, a semantic segmentation model developed earlier, has well-established uses in the medical sector. Yang and Qiu (2022) introduced a computed tomography image segmentation algorithm leveraging DL, which enhanced segmentation accuracy by altering the U-Net model and incorporating an attention mechanism (Dice coefficient 0.9594). In recent years, applications based on U-Net have increasingly been adopted in agriculture. A research study introduced a dual-phase model that integrates DeepLabV3+ with U-Net to classify the severity of cucumber leaf diseases (DUNet) in intricate backgrounds (Wang *et al.*, 2021). Other studies using Ir-U-Net, MFBP-U-Net, and U-Net with InceptionV3 transfer learning models to improve accuracy for yellow rust, pear leaf, and pepper bell leaf diseases, respectively (Zhang *et al.*, 2021; AlArfaj *et al.*, 2023; Wang *et al.*, 2023) (Table 4.2).

Table 4.2. The different segmentation techniques used to improve detection accuracy in certain plants. Author's own table.

Segmentation techniques	Plant leaves	Stress detection	Segmentation accuracy	References
YOLOv3 YOLOv4	Peach and strawberry leaves	Healthy peach and strawberry leaves, as well as peach leaves affected by bacterial spots and strawberry leaves with scorch disease	YOLOv3 (97%) YOLOv4 (98%)	Alhwaiti *et al.* (2025)
Algorithm YOLOX, and ResNet101 network	Kiwi fruit leaf	Two-stage disease detection algorithm	96.6%	Yao *et al.* (2022)
YOLOv7	Tea leaf	Tea leaf disease	97.3%	Soeb *et al.* (2023)
YOLO v5 (RLD dataset)	Rice leaf	Rice leaf disease	94.87%	Kumar *et al.* (2023)
YOLOv8	Wheat leaf	Yellow rust disease	95.8%	Hassan *et al.* (2024)
U-Net, with inceptionV3 transfer learning model	Pepper bell leaf	Pepper bell leaf disease	99.48%	AlArfaj *et al.* (2023)
Ir-U-Net	Wheat leaf	Yellow rust disease	97.13%	Zhang *et al.* (2021)
DeepLabV3+ and U-Net	Cucumber leaf	Leaf diseases spots	93.27%	Wang *et al.* (2021)
MFBP-U-Net	Pear leaf	Leaf diseases	90.89%	Wang *et al.* (2023)
Mask R-CNN	Apple leaf	Leaf diseases	96.6%	Rehman *et al.* (2021)

4.4.3 Review of training processes and common evaluation metrics

Deep learning has significantly advanced the training processes and evaluation metrics for plant leaf disease and stress detection (Shoaib *et al.*, 2023). Various models, including CNNs, VGG architectures, and advanced techniques such as LSTM and RNN, have been employed to enhance detection accuracy (Alzubaidi *et al.*, 2021; Jabir *et al.*, 2021). The training processes often involve extensive datasets, data augmentation, and transfer learning to improve model performance (Shin *et al.*, 2016). The evaluation of these models relies on several common metrics, which are crucial for assessing their effectiveness (Kamilaris and Prenafeta-Boldú, 2018).

Models are trained on large, annotated datasets, such as the 9127 images used in one study (Shivaprasad and Wadhawan, 2023). Techniques such as rotation, scaling, and flipping are employed to enhance dataset diversity and model robustness (Balamurugan *et al.*, 2024). Pretrained models are fine-tuned on specific datasets to leverage existing knowledge and improve training efficiency (Balamurugan *et al.*, 2024). Accuracy measures the proportion of correct predictions. For instance, CNNs achieved an accuracy of 97% in one study (Shivaprasad and Wadhawan, 2023). Precision indicates the accuracy of positive predictions, while recall measures the model's ability to identify all relevant instances. Models have reported precision rates around 94% and recall rates of 96% (Balamurugan *et al.*, 2024). This harmonic mean (F1 score) of precision and recall provides a balance between the two metrics, with some models achieving scores of 0.95 (Shivaprasad and Wadhawan, 2023). While DL models show promise in plant disease detection, challenges such as overfitting and the need for high-quality data remain significant hurdles that require ongoing research and innovation (Li *et al.*, 2021; Shivaprasad and Wadhawan, 2023).

4.5 Case Studies and Practical Applications

4.5.1 Plant disease detection using DL

The advent of accurate and automatic diagnostic technologies has changed the methods used in plant disease identification (Fig. 4.2), particularly

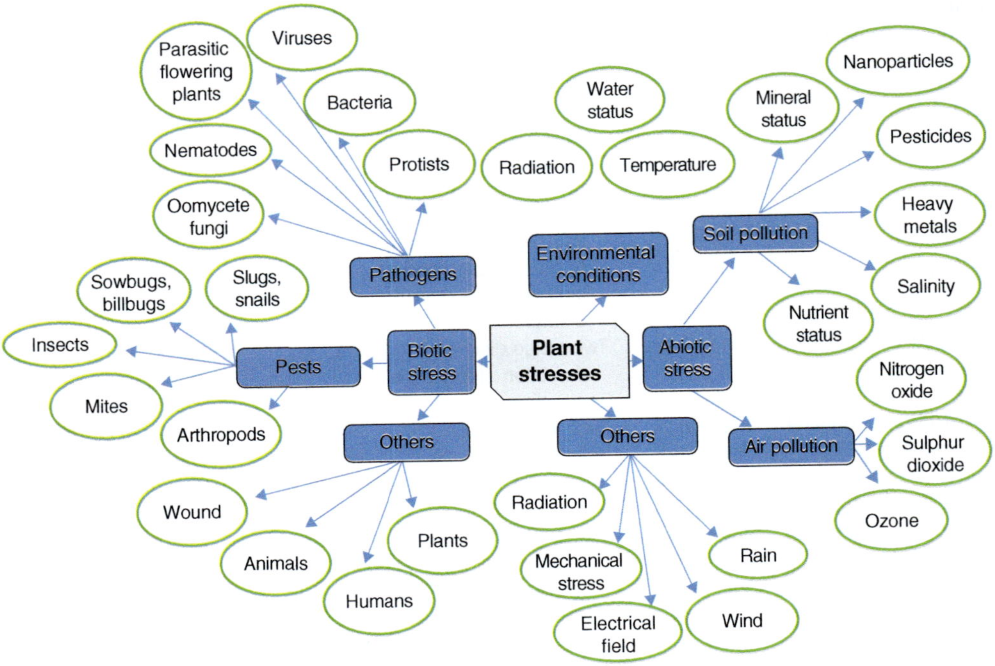

Fig. 4.2. Typology of biotic and abiotic stresses and their environmental origins in plant.

using DL tools. Many researchers during the past few years have shown that DL techniques are much better than conventional visual-based inspections or remote sensing indices in detecting specific agricultural diseases (Mohanty *et al.*, 2016; Nigam *et al.*, 2023). One well-known example is the identification of wheat rust (*Puccinia* spp.), a highly destructive fungal disease of wheat cultivated globally, which has been detected with the use of DL systems based on CNNs (Mohanty *et al.*, 2016). Their study claimed that more than 95% of the wheat rust symptoms captured by drone and mobile phone cameras are correctly diagnosed from the images. They also emphasized that models pretrained on ImageNet delivered better results when retrained with wheat disease data than models built with standard techniques. Other than that, DL approaches have also been used to identify the late potato blight pathogen, *Phytophthora infestans* (Rakotonindraina, 2012). In this regard, researchers have implemented hybrid CNN and RNN architectures for the analysis of multispectral images, which makes it possible to diagnose infections at a very early stage when no signs of the disease are visible. This technology has been very important for the reduction of agricultural damage by allowing the early intervention in controlling plant disease management (Chaigneau and Etude, 2014).

4.5.2 Detection of abiotic stress factors

Drought stress

Drought is one of the most insidious of all abiotic stresses, creeping in unnoticed until the damage is already reflected in the potential of a crop (Lesk *et al.*, 2016). Delays in perceiving the early indications of drought often translate into delayed responses, which, in turn, translate to considerable yield losses and pose a real menace to international food security (Ray *et al.*, 2015). To address this challenge, Talaei Khoei *et al.* (2023) introduced a novel methodology that integrates hyperspectral imaging with the U-Net DL framework, widely recognized for its outstanding performance in image segmentation tasks (Talaei Khoei *et al.*, 2023). It monitors fine physiological variations within wheat leaves, changes that the human eye normally cannot detect, and spots the very initial symptoms of water stress. In fact, their model turned out to be incredibly accurate around 98%, truly an ultra-sensitive approach with highly reliable detection of a problem well before it shows in the field (Talaei Khoei *et al.*, 2023). In perspective, such technology brings us closer to a future where farmers receive real-time warnings of drought stress affecting their fields (Sishodia *et al.*, 2020). Hyperspectral sensors mounted on drones or satellites can monitor vast areas of agriculture, and thus help make the right decision at the right time, about when to irrigate and fine-tune maximum water use—an increasingly critical capability in today's climate (Sadler *et al.*, 2005).

Detection of nutrient deficiencies

Although generally less sudden or dramatic than drought, nutrient deficiencies represent a persistent and extensive challenge in global agriculture. When crops are deprived of essential macronutrients such as nitrogen (N), phosphorus (P), or potassium (K), they exhibit restricted growth, chlorosis, necrosis, and poor root development, which collectively result in lower yields and diminished food quality. The problem is exacerbated by the fact that visual symptoms of nutrient deficiencies can be subtle, non-specific, and easily confused with signs of disease or abiotic stress (Musinguzi *et al.*, 2016).

Traditional detection methods are time-consuming, labor-intensive, and often impractical for large-scale monitoring like soil analysis or plant tissue analysis (Jones Jr, 2012). To address these limitations, researchers have turned to DL and other computer-based approaches as innovative tools to automate and facilitate the process of identifying nutrient deficiencies in crops (Xie *et al.*, 2021). One promising approach, proposed as an efficient solution, uses a CNN model trained on simple RGB images of corn leaves (Syarief and Setiawan, 2020).

Their strategy focused particularly on diagnosing nitrogen deficiencies, since they are among the most common nutrient problems and usually limit cereal crops yields (Sainju *et al.*, 2002). The model by CNN achieved quite a remarkable value of accuracy of 96.88%, in the diagnostics for nitrogen deficiency of target plants, offering fast, reliable, and non-invasive diagnostics without the need for costly hyperspectral or multispectral imaging devices (Xie *et al.*, 2021). That makes the technology more available to more people, not only for large-scale mechanized farms but also to smallholder farmers in developing regions, where resources and infrastructure may be limited (Balafoutis *et al.*, 2017).

The incorporation of DL-enabled diagnostic tools into UAVs or autonomous ground vehicles gives farmers and agronomists the ability to conduct real-time monitoring of nutrients on plants-by-plants basis throughout entire fields (Sankaran *et al.*, 2015). This fine deep knowledge enables site-specific nutrient management, where fertilizer applications can be applied precisely to the actual needs of different zones or even individual plants (Mulla, 2013).

This level of precision helps to minimize the risk of over-fertilization, which not only lowers input costs but also mitigates negative environmental effects, such as nutrient leaching, groundwater contamination, and greenhouse gas emissions from excessive nitrogen use (Zhang *et al.*, 2015). Another benefit is that when a proper nutrient balance is achieved, crop health—and consequently yields and the quality of harvested produce—will improve (Gan *et al.*, 2025).

In summary, the dawn of DL models for the detection of nutrient deficiencies culminates in a new age for precision agriculture based on data; something that places power in the hands of farmers to make more informed decisions and contributes to sustainable food production systems around the world (Wolfert *et al.*, 2017).

4.6 Moving Toward Sustainable and Smart Agriculture

In conclusion, DL technologies play a pivotal role in modern agriculture by enabling smarter and more sustainable techniques for farming practices. Their transformative impact extends beyond crop monitoring. In addition to the ability to carry out accurate data-driven decisions and effectively optimize resource management, it also supports early detection and responsive systems to biotic and abiotic stresses. These advances contribute directly to the concept of sustainable intensification, which seeks to increase agricultural productivity while simultaneously reducing environmental footprints and conserving natural resources (Pretty *et al.*, 2018). With the evolution and gradual incorporation of DL technologies with multimodal datasets, ranging from satellite imagery and ground-based sensor data to real-time weather forecasts, there will be also be a substantial improvement in predictive analytics as well as decision support systems in agriculture (Kamilaris *et al.*, 2017).

4.7 Challenges and Limitations

DL has made great progress in the diagnosis of plant leaf diseases, but several challenges still hinder its plague certification. These issues range from data availability, environmental conditions, and computational needs, which could negatively impact the operation of DL models in agriculture. A lack of labeled datasets for different plant diseases limits the training of reliable DL models (Li *et al.*, 2021; Mandal, 2023). Many existing databases may not reflect the diversity of plant species or the manifestations of the disease, which can lead to biased modeling (Wadnere and Ramteke, 2022). Accuracy rates may vary considerably depending on the environmental factors such as lighting, weather, and soil conditions, which complicate disease detection (Bhise *et al.*, 2022). Differentiated models trained in specific environments may not be well adapted to different agricultural environments, which may limit their practical applicability (Bhise *et al.*, 2022). While DL offers promising advances in plant disease detection, the challenges identified require ongoing research and innovation to improve model robustness and accessibility, which may be a barrier to widespread adoption in resource-constrained environments (Mandal, 2023). Time needed to train complex models may be lengthy, delaying the implementation of efficient solutions in real-world scenarios (Li *et al.*, 2021).

DL models are struggling with changes in lighting, weather, and environmental noise, all of which can affect image quality and model accuracy (Arsenovic *et al.*, 2019). Most existing datasets do not capture the variability of environmental conditions, and therefore produce models that are efficient in controlled environments but not in real life (Li *et al.*, 2021). Current models are excellent at identifying diseases in a particular plant species but not between different species, which limits their usefulness (Revathy and Meenakshi, 2024). The lack of large datasets covering a wide range of plant species and diseases makes it difficult to develop useful models for the precise identification of diseases between species (Hu *et al.*, 2024). Although DL has promising potential for plant disease detection, the challenges of environmental sensitivity and interspecies variability require further research and data diversity to improve the robustness of models and their applicability to agriculture.

DL is now an innovative technology used to detect leaf diseases, but its application is subject to ethical and pragmatic considerations, notably regarding data availability and privacy in low-resource environments. These should be addressed in order to ensure fair access to agricultural innovations and the protection of sensitive information. For example, the collection of agricultural data, such as images of plant diseases, may include sensitive information on farmers and agricultural practices. Ensuring that such information is anonymized and securely stored is crucial for protecting the privacy of farmers (Khadeer *et al.*, 2025). ML models can end up with biases if trained on non-representative datasets, resulting in inconclusive disease determinations in a variety of agricultural contexts (Das *et al.*, 2024). Farmers in resource-limited environments may be deprived of technology that can be used for DL applications. It is crucial to develop applications that are mobile-friendly and can run on low-end devices in order to enable widespread use (Mustofa *et al.*, 2023). Farmers need to be trained to make the most of these technologies and to understand the results presented to them effectively (Li *et al.*, 2021). While DL offers significant advancements in plant disease diagnosis, it is important to address these practical and ethical issues in order to make these advances available to farmers everywhere.

4.8 Future Perspectives and Technological Advancements

With the continuous evolution of AI, there are many new technologies that can be beneficial for plant leaf disease and stress detection. Such approaches are difficult to be scaled, and privacy-preserving model training methods such as federated learning (FL) have recently drawn attention. FL is a new training mechanism for DL models that allows multiple users or institutions to collaboratively train models without sharing raw data, in contrast to centralized learning. In agriculture, localized datasets and their wide variability often include sensitive or proprietary information, making this especially relevant (Caminha and Oliveira, 2024). While improved model generalization under data privacy constraints through the use of FL in crop disease detection has already shown promising results, FL remains immensely valuable in the context of multisite agricultural deployment in the real world.

A second interesting line of research involves leveraging generative adversarial networks (GANs) and other generative models as data augmentation. As labeled training data are voluminous for most DL models, GANs are becoming popular for synthetically generating realistic images of the diseased leaves. This is useful for handling the data scarcity problem; in particular, when there are few or no plant conditions represented in the data. Studies have shown that datasets augmented with GAN improve model robustness, lower the aliased classes, and improve classification performance (Barbedo, 2018; Shorten and Khoshgoftaar, 2019; Khan *et al.*, 2020). This would make DL models better suited for use in diverse agricultural environments with heterogeneous disease incidence and limited expert annotation capacity.

In parallel, the combination of DL with Internet of Things (IoT) systems and agricultural robotics is revolutionizing plant monitoring into a real-time, ongoing, and automated process. This is why IoT devices with smart sensors can retrieve high-resolution data on plant physiology and environmental conditions, which can then be processed by DL models to analyze the data on-site in edge computing units. This decreases latency as well as reliance on cloud infrastructure, enabling disease and stress condition diagnosis directly in the field (Wolfert *et al.*, 2017). Agricultural robots and autonomous drones (typically outfitted with cameras) can also be equipped with trained CNNs to monitor crops, highlight affected areas, and perform site-specific interventions (i.e. spraying and pruning) in a timely manner. Such closed-loop systems combine the entire process of observation, analysis, and intervention to create new efficiencies and higher degrees of sustainability in precision farming (Shamshiri *et al.*, 2018).

Yet there are limitations, and one of the major challenges is determining how to scale these technologies, particularly for smallholder farmers in developing parts of the world. Lightweight DL models that are deployable on smartphone hardware, along with low-cost imaging tools and open-source interface platforms, are being developed in increasing numbers to help scale adoption. Following the principles of human-centered AI, solutions have to be developed considering local cultural, linguistic, and technological realities (Kamilaris and Prenafeta-Boldú, 2018). Developing shared datasets, standardized evaluation systems, and viable business models depends on cooperative frameworks including institutions, governments, and industry players. Over time, such initiatives will help to create strong, data-driven agricultural ecosystems, thereby enabling farmers all around to make wise, timely, sustainable crop management decisions.

4.9 Conclusion

DL is one of the recent avenues that has emerged for detecting plant leaf diseases and stress, which can be more accurate, faster, and scalable when compared to traditional diagnostic systems. In this chapter, the fundamental principles of DL based on CNNs have been introduced, and their application on the image of leaves, detection of disease signs, and classification of healthy plants and stressed plants has been illustrated. We also discussed important aspects of data acquisition and preprocessing, along with important model architectures and evaluation strategies. Although there have been many case studies that demonstrate the successful application of these models within real-world agricultural settings, several challenges continue to persist, including limited access to annotated datasets, changes in the surrounding environment, and the demand for lightweight, easy-to-use solutions. Further developments in FL, synthetic data generation, and integration with IoT and robotics, for example, are expected to overcome these constraints and enable AI-powered tools in both commercial and smallholder farming systems.

Acknowledgments

This research was jointly funded by the Phytopathology Unit of the Ecole Nationale d'Agriculture de Meknès and the University of Cadi Ayyad.

References

Abbas, A., Zhang, Z., Zheng, H., Alami, M. M., Alrefaei, A. F., Abbas, Q., Naqvi, S. A. H., Rao, M. J., Mosa, W. F. A., & Abbas, Q. (2023). Drones in plant disease assessment, efficient monitoring, and detection: A way forward to smart agriculture. *Agronomy*, 13(6), 1524.

Abdulridha, J., Min, A., Rouse, M. N., Kianian, S., Isler, V., & Yang, C. (2023). Evaluation of stem rust disease in wheat fields by drone hyperspectral imaging. *Sensors*, 23(8), 4154.

Adarsh, P., Rathi, P., & Kumar, M. (2020). YOLO v3-Tiny: Object Detection and Recognition using one stage improved model. *2020 6th International Conference on Advanced Computing and Communication Systems (ICACCS)*, 687–694. https://doi.org/10.1109/ICACCS48705.2020.9074315

Ahmmed, S., Podder, P., Mondal, M. R. H., Rahman, S. M. A., Kannan, S., Hasan, M. J., & Rohan, A. (2023). Enhancing Brain Tumor Classification with Transfer Learning across Multiple Classes: An In-Depth Analysis. *BioMedInformatics*, 3(4), 1124–1144. https://doi.org/10.3390/biomedinformatics3040068

AlArfaj, A. A., Altamimi, A., Aljrees, T., Basheer, S., Umer, M., Samad, M. A., Alsubai, S., & Ashraf, I. (2023). Multi-Step Preprocessing With UNet Segmentation and Transfer Learning Model for Pepper Bell Leaf Disease Detection. *IEEE Access*, 11, 132254–132267. https://doi.org/10.1109/ACCESS.2023.3334428

Alhwaiti, Y., Khan, M., Asim, M., Siddiqi, M. H., Ishaq, M., & Alruwaili, M. (2025). Leveraging YOLO deep learning models to enhance plant disease identification. *Scientific Reports*, 15(1), 1–16. https://doi.org/10.1038/s41598-025-92143-0

Alzubaidi, L., Zhang, J., Humaidi, A. J., Al-Dujaili, A., Duan, Y., Al-Shamma, O., Santamaría, J., Fadhel, M. A., Al-Amidie, M., & Farhan, L. (2021). Review of deep learning: concepts, CNN architectures, challenges, applications, future directions. *Journal of Big Data*, 8(1), 53. https://doi.org/10.1186/s40537-021-00444-8

Arboleda, E. R., Fajardo, A. C., & Medina, R. P. (2018). An image processing technique for coffee black beans identification. *2018 IEEE International Conference on Innovative Research and Development (ICIRD)*, 1–5.

Aroge, T., Zhu, Y., Jin, D.-N., Dara, M. Z. N., Feng, J., Olajuyin, A. M., Abbas, A., & Liu, S.-Y. (2024). Omics and CRISPR-Cas9 molecular perception: A progressive review approach for powdery mildew disease management. *Physiological and Molecular Plant Pathology*, 130, 102217. https://doi.org/10.1016/j.pmpp.2024.102217

Arsenovic, M., Karanovic, M., Sladojevic, S., Anderla, A., & Stefanovic, D. (2019). Solving Current Limitations of Deep Learning Based Approaches for Plant Disease Detection. *Symmetry*, 11(7), 939. https://doi.org/10.3390/sym11070939

Ashourloo, D., Mobasheri, M., & Huete, A. (2014). Developing Two Spectral Disease Indices for Detection of Wheat Leaf Rust (Pucciniatriticina). *Remote Sensing*, 6(6), 4723–4740. https://doi.org/10.3390/rs6064723

Aziz, I. (2020). *Deep Learning : An Overview of Convolutional Neural Network (CNN). April*, 1–54.

Balafoutis, A., Beck, B., Fountas, S., Vangeyte, J., Van Der Wal, T., Soto, I., Gómez-Barbero, M., Barnes, A., & Eory, V. (2017). Precision agriculture technologies positively contributing to ghg emissions mitigation, farm productivity and economics. *Sustainability (Switzerland)*, 9(8), 1–28. https://doi.org/10.3390/su9081339

Balamurugan, M., Srividhya, N., Indhumathi, G., Arul, V., Rahul S, S. G., & Kalaiarasi, K. (2024). Deep Learning Innovations for Improved Plant Leaf Disease Detection in Smart Agriculture. *2024 Second International Conference on Data Science and Information System (ICDSIS)*, 1–6. https://doi.org/10.1109/ICDSIS61070.2024.10594188

Barbedo, J. G. A. (2018). Impact of dataset size and variety on the effectiveness of deep learning and transfer learning for plant disease classification. *Computers and Electronics in Agriculture*, 153, 46–53.

Barbedo, J. G. A. (2019). Plant disease identification from individual lesions and spots using deep learning. *Biosystems Engineering*, 180, 96–107.

Behmann, J., Acebron, K., Emin, D., Bennertz, S., Matsubara, S., Thomas, S., Bohnenkamp, D., Kuska, M. T., Jussila, J., & Salo, H. (2018). Specim IQ: evaluation of a new, miniaturized handheld hyperspectral camera and its application for plant phenotyping and disease detection. *Sensors*, 18(2), 441.

Bhandari, M., Ibrahim, A. M. H., Xue, Q., Jung, J., Chang, A., Rudd, J. C., Maeda, M., Rajan, N., Neely, H., & Landivar, J. (2020). Assessing winter wheat foliage disease severity using aerial imagery acquired from small Unmanned Aerial Vehicle (UAV). *Computers and Electronics in Agriculture*, 176, 105665.

Bhise, D., Kumar, S., & Mohapatra, H. (2022). Review on deep learning-based plant disease detection. *2022 6th International Conference on Electronics, Communication and Aerospace Technology*, 1106–1111. https://doi.org/10.1109/ICECA55336.2022.10009290

Brahimi, M., Arsenovic, M., Laraba, S., Sladojevic, S., Boukhalfa, K., & Moussaoui, A. (2018). *Deep Learning for Plant Diseases: Detection and Saliency Map Visualisation*. 93–117. https://doi.org/10.1007/978-3-319-90403-0_6

Brahimi, M., Boukhalfa, K., & Moussaoui, A. (2017). Deep learning for tomato diseases: classification and symptoms visualization. *Applied Artificial Intelligence*, 31(4), 299–315.

Brugger, A., Schramowski, P., Paulus, S., Steiner, U., Kersting, K., & Mahlein, A. (2021). Spectral signatures in the UV range can be combined with secondary plant metabolites by deep learning to characterize barley–powdery mildew interaction. *Plant Pathology*, 70(7), 1572–1582. https://doi.org/10.1111/ppa.13411

Buja, I., Sabella, E., Monteduro, A. G., Chiriacò, M. S., De Bellis, L., Luvisi, A., & Maruccio, G. (2021). Advances in plant disease detection and monitoring: From traditional assays to in-field diagnostics. *Sensors*, 21(6), 2129.

Caminha, P. V, & Oliveira, H. M. N. da S. (2024). Cloud-Driven Federated Learning for Plant Disease Detection in Agriculture. *2024 IEEE 13th International Conference on Cloud Networking (CloudNet)*, 1–4.

Chaigneau, A., & Etude, A. C. (2014). *a la d ´ecision Mileos R contre le mildiou de la pomme de terre : analyses de sensibilit ´e et ajustement Étude sur le paramétrage de l ' outil d ' aide à la décision Mileos ® contre le mildiou de la pomme de terre : analyses de sensibilité et ajustemen*.

Charisis, C., & Argyropoulos, D. (2024). Deep learning-based instance segmentation architectures in agriculture: A review of the scopes and challenges. *Smart Agricultural Technology*, 8(December 2023), 100448. https://doi.org/10.1016/j.atech.2024.100448

Cheshkova, A. F. (2022). A review of hyperspectral image analysis techniques for plant disease detection and identification. *Vavilov Journal of Genetics and Breeding*, 26(2), 202.

Chin, R., Catal, C., & Kassahun, A. (2023). Plant disease detection using drones in precision agriculture. *Precision Agriculture*, 24(5), 1663–1682.

Chowdhury, M. E. H., Rahman, T., Khandakar, A., Ayari, M. A., Khan, A. U., Khan, M. S., Al-Emadi, N., Reaz, M. B. I., Islam, M. T., & Ali, S. H. M. (2021). Automatic and reliable leaf disease detection using deep learning techniques. *AgriEngineering*, 3(2), 294–312.

Crespi, L. (2023). *Towards a Comprehensive Application of Deep Learning Methods in Medical Imaging: Segmentation, Feature Extraction, Synthetic Image Generation*. Doctoral Dissertation, Polytechnic of Milan.

Das, S., Banerjee, S., & Mondal, Prof. Dr. A. C. (2024). Plant Leaf Disease Detection by Deep Learning: A Literature Review. *International Journal of Scientific and Research Publications*, 14(4), 88–93. https://doi.org/10.29322/IJSRP.14.04.2023.p14811

Deng, L., Mao, Z., Li, X., Hu, Z., Duan, F., & Yan, Y. (2018). UAV-based multispectral remote sensing for precision agriculture: A comparison between different cameras. *ISPRS Journal of Photogrammetry and Remote Sensing*, 146, 124–136.

Dhillon, A., & Verma, G. K. (2020). Convolutional neural network: a review of models, methodologies and applications to object detection. *Progress in Artificial Intelligence*, 9(2), 85–112. https://doi.org/10.1007/s13748-019-00203-0

Dhingra, G., Kumar, V., & Joshi, H. D. (2018). Study of digital image processing techniques for leaf disease detection and classification. *Multimedia Tools and Applications*, 77, 19951–20000.

Dias, P. A., Tabb, A., & Medeiros, H. (2018). Apple flower detection using deep convolutional networks. *Computers in Industry*, 99, 17–28. https://doi.org/10.1016/j.compind.2018.03.010

Dong, J., Fuentes, A., Yoon, S., Kim, H., & Park, D. S. (2023). An iterative noisy annotation correction model for robust plant disease detection. *Frontiers in Plant Science*, 14, 1238722.

Dong, J., Lee, J., Fuentes, A., Xu, M., Yoon, S., Lee, M. H., & Park, D. S. (2022). Data-centric annotation analysis for plant disease detection: Strategy, consistency, and performance. *Frontiers in Plant Science*, 13, 1037655.

Duan, Z., Wang, F., Wang, B., Luo, G., & Jiang, Z. (2024). An Adapted ResNet-50 Architecture for Predicting Flow Fields of an Underwater Vehicle. *IEEE Access*, 12(May), 66398–66407. https://doi.org/10.1109/ACCESS.2024.3399077

Dyrmann, M., Karstoft, H., & Midtiby, H. S. (2016). Plant species classification using deep convolutional neural network. *Biosystems Engineering*, 151, 72–80.

Ferentinos, K. P. (2018). Deep learning models for plant disease detection and diagnosis. *Computers and Electronics in Agriculture*, 145, 311–318.

Franceschini, M. H. D., Bartholomeus, H., Van Apeldoorn, D. F., Suomalainen, J., & Kooistra, L. (2019). Feasibility of unmanned aerial vehicle optical imagery for early detection and severity assessment of late blight in potato. *Remote Sensing*, 11(3), 224.

Fu, J., Liu, J., Zhao, R., Chen, Z., Qiao, Y., & Li, D. (2022). Maize disease detection based on spectral recovery from RGB images. *Frontiers in Plant Science*, 13, 1056842. https://doi.org/10.3389/FPLS.2022.1056842/BIBTEX

Fuentes, A., Yoon, S., Kim, S. C., & Park, D. S. (2017). A robust deep-learning-based detector for real-time tomato plant diseases and pests recognition. *Sensors*, 17(9), 2022.

Gan, H., Li, L., Qiu, P., Cui, M., Sun, J., Xia, Y., & Zhu, W. (2025). High voltage Li-rich Mn-based cathode modified by silica-coated silver nanowires for next-generation high energy density lithium-ion batteries. *Electrochimica Acta*, 516(November 2024), 145728. https://doi.org/10.1016/j.electacta.2025.145728

Gou, C., Zafar, S., Fatima, N., Hasnain, Z., Aslam, N., Iqbal, N., Abbas, S., Li, H., Li, J., & Chen, B. (2024). Machine and deep learning: Artificial intelligence application in biotic and abiotic stress management in plants. *Frontiers in Bioscience-Landmark*, 29(1).

Gudkov, S. V, Matveeva, T. A., Sarimov, R. M., Simakin, A. V, Stepanova, E. V, Moskovskiy, M. N., Dorokhov, A. S., & Izmailov, A. Y. (2023). Optical methods for the detection of plant pathogens and diseases. *AgriEngineering*, 5(4), 1789–1812.

Hao, Y. (2021). Convolutional Neural Networks for Image Classification. *Proceedings - 2021 2nd International Conference on Artificial Intelligence and Computer Engineering, ICAICE 2021, May*, 342–345. https://doi.org/10.1109/ICAICE54393.2021.00073

Hasanah, S. A., Pravitasari, A. A., Abdullah, A. S., Yulita, I. N., & Asnawi, M. H. (2023). A Deep Learning Review of ResNet Architecture for Lung Disease Identification in CXR Image. *Applied Sciences (Switzerland)*, 13(24). https://doi.org/10.3390/app132413111

Hassan, A., Mumtaz, R., Mahmood, Z., Fayyaz, M., & Naeem, M. K. (2024). Wheat leaf localization and segmentation for yellow rust disease detection in complex natural backgrounds. *Alexandria Engineering Journal*, 107, 786–798. https://doi.org/10.1016/J.AEJ.2024.09.018

He, K., Gkioxari, G., Dollar, P., & Girshick, R. (2017). Mask R-CNN. *2017 IEEE International Conference on Computer Vision (ICCV)*, 2980–2988. https://doi.org/10.1109/ICCV.2017.322

Heincke, B., Jackisch, R., Saartenoja, A., Salmirinne, H., Rapp, S., Zimmermann, R., Pirttijärvi, M., Sörensen, E. V., Gloaguen, R., & Ek, L. (2019). Developing multi-sensor drones for geological mapping and mineral exploration: Setup and first results from the MULSEDRO project. *GEUS Bulletin*, 43.

Horasan, A., & Güneş, A. (2024). Advancing Prostate Cancer Diagnosis: A Deep Learning Approach for Enhanced Detection in MRI Images. *Diagnostics*, 14(17). https://doi.org/10.3390/diagnostics14171871

Hossen, M. H., Mohibullah, M., Muzammel, C. S., Ahmed, T., Acharjee, S., & Panna, M. B. (2022). Wheat Diseases Detection and Classification using Convolutional Neural Network (CNN). *International Journal of Advanced Computer Science and Applications*, 13(11), 719–726. https://doi.org/10.14569/IJACSA.2022.0131183

Hu, M., Long, S., Wang, C., & Wang, Z. (2024). Leaf disease detection using deep Convolutional Neural Networks. *Journal of Physics: Conference Series*, 2711(1), 012020. https://doi.org/10.1088/1742-6596/2711/1/012020

Hu, Q., Peng, X., Wang, C., Chen, Q., Qu, Y., & Yan, Z. (2022). Object Detection and Recognition based on Improved Faster R-CNN. *2022 International Conference on Image Processing and Media Computing (ICIPMC)*, 1–6. https://doi.org/10.1109/ICIPMC55686.2022.00009

Hughes, D., & Salathé, M. (2015). An open access repository of images on plant health to enable the development of mobile disease diagnostics. *ArXiv Preprint ArXiv:1511.08060*.

Iftikhar, M., Kandhro, I. A., Kausar, N., Kehar, A., Uddin, M., & Dandoush, A. (2024). Plant disease management: a fine-tuned enhanced CNN approach with mobile app integration for early detection and classification. *Artificial Intelligence Review*, 57(7), 1–29. https://doi.org/10.1007/s10462-024-10809-z

Ilişanu, M.-A., Moldoveanu, F., & Moldoveanu, A. (2023). Multispectral imaging for skin diseases assessment—state of the art and perspectives. *Sensors*, 23(8), 3888.

Islam, S., Reza, M. N., Samsuzzaman, S., Ahmed, S., Cho, Y. J., Noh, D. H., Chung, S.-O., & Hong, S. J. (2024). Machine vision and artificial intelligence for plant growth stress detection and monitoring: A review. *Precision Agriculture Science and Technology*, 6(1), 33–57. https://doi.org/10.12972/pastj.20240003

Jabir, B., Rabhi, L., & Falih, N. (2021). RNN- and CNN-based weed detection for crop improvement: An overview. *Foods and Raw Materials*, 9(2), 387–396. https://doi.org/10.21603/2308-4057-2021-2-387-396

Jaiswal, G., Rani, R., Mangotra, H., & Sharma, A. (2023). Integration of hyperspectral imaging and autoencoders: Benefits, applications, hyperparameter tunning and challenges. *Computer Science Review*, 50, 100584. https://doi.org/10.1016/j.cosrev.2023.100584

Jha, P., Dembla, D., & Dubey, W. (2024). Implementation of Machine Learning Classification Algorithm Based on Ensemble Learning for Detection of Vegetable Crops Disease. *International Journal of Advanced Computer Science & Applications*, 15(1).

Jin, J. (2022). Convolutional Neural Networks for Biometrics Applications. *SHS Web of Conferences*, 144, 03013. https://doi.org/10.1051/shsconf/202214403013

Jones Jr., J. B. (2012). Plant Analysis and Tissue Testing. *Plant Nutrition and Soil Fertility Manual*, 150–169. https://doi.org/10.1201/b11577-20

Kamilaris, A., Kartakoullis, A., & Prenafeta-Boldú, F. X. (2017). A review on the practice of big data analysis in agriculture. *Computers and Electronics in Agriculture*, 143(January), 23–37. https://doi.org/10.1016/j.compag.2017.09.037

Kamilaris, A., & Prenafeta-Boldú, F. X. (2018). A review of the use of convolutional neural networks in agriculture. *The Journal of Agricultural Science*, 156(3), 312–322. https://doi.org/10.1017/S0021859618000436

Karpyshev, P., Ilin, V., Kalinov, I., Petrovsky, A., & Tsetserukou, D. (2021). Autonomous Mobile Robot for Apple Plant Disease Detection based on CNN and Multi-Spectral Vision System. *2021 IEEE/SICE International Symposium on System Integration, SII 2021, March*, 157–162. https://doi.org/10.1109/IEEECONF49454.2021.9382649

Karthik, R., Hariharan, M., Anand, S., Mathikshara, P., Johnson, A., & Menaka, R. (2020). Attention embedded residual CNN for disease detection in tomato leaves. *Applied Soft Computing*, 86, 105933. https://doi.org/10.1016/J.ASOC.2019.105933

Kattenborn, T., Leitloff, J., Schiefer, F., & Hinz, S. (2021). Review on Convolutional Neural Networks (CNN) in vegetation remote sensing. *ISPRS Journal of Photogrammetry and Remote Sensing*, 173(December 2020), 24–49. https://doi.org/10.1016/j.isprsjprs.2020.12.010

Kaur, P., Harnal, S., Gautam, V., Singh, M. P., & Singh, S. P. (2024). Performance analysis of segmentation models to detect leaf diseases in tomato plant. *Multimedia Tools and Applications*, 83(6), 16019–16043.

Kerkech, M., Hafiane, A., & Canals, R. (2020). Vine disease detection in UAV multispectral images using optimized image registration and deep learning segmentation approach. *Computers and Electronics in Agriculture*, 174, 105446.

Kersting, K., Bauckhage, C., Wahabzada, M., Mahlein, A.-K., Steiner, U., Oerke, E.-C., Römer, C., & Plümer, L. (2016). Feeding the world with big data: Uncovering spectral characteristics and dynamics of stressed plants. *Computational Sustainability*, 99–120.

Khadeer, K., Khan, K. S., Bharathi, M., & Srinivas, T. A. S. (2025). From Pixels to Protection: Deep Learning Approaches for Plant Leaf Disease Detection. *Journal of Computer Networks and Virtualization*, 3(1), 8–14. https://doi.org/10.48001/jocnv.2025.318-14

Khan, A., Sohail, A., Zahoora, U., & Qureshi, A. S. (2020). A survey of the recent architectures of deep convolutional neural networks. *Artificial Intelligence Review*, 53, 5455–5516.

Khan, A., Vibhute, A. D., Mali, S., & Patil, C. H. (2022). A systematic review on hyperspectral imaging technology with a machine and deep learning methodology for agricultural applications. *Ecological Informatics*, 69, 101678. https://doi.org/10.1016/J.ECOINF.2022.101678

Khosrokhani, M., & Nasr, A. H. (2022). Applications of the remote sensing technology to detect and monitor the rust disease in the wheat – a literature review. *Geocarto International*, 37(26), 13268–13290. https://doi.org/10.1080/10106049.2022.2076922

Kinnikar, A., Desai, P., & Jahagirdar, S. (2015). Identification and detection of seed borne diseases of soybean using image processing-a survey. *Int. J. Emerg. Technol. Comput. Sci. Electron*, 14, 363–368.

Krichen, M. (2023). Convolutional Neural Networks: A Survey. *Computers*, 12(8), 1–41. https://doi.org/10.3390/computers12080151

Kumar, A., Lee, W. S., Ehsani, R. J., Albrigo, L. G., Yang, C., & Mangan, R. L. (2012). Citrus greening disease detection using aerial hyperspectral and multispectral imaging techniques. *Journal of Applied Remote Sensing*, 6(1), 63542.

Kumar, V. S., Jaganathan, M., Viswanathan, A., Umamaheswari, M., & Vignesh, J. (2023). Rice leaf disease detection based on bidirectional feature attention pyramid network with YOLO v5 model. *Environmental Research Communications*, 5(6). https://doi.org/10.1088/2515-7620/acdece

Kurouski, Dmitry. (2021). *Diagnostics of plant diseases*.

Latif, G., Abdelhamid, S. E., Mallouhy, R. E., Alghazo, J., & Kazimi, Z. A. (2022). Deep Learning Utilization in Agriculture: Detection of Rice Plant Diseases Using an Improved CNN Model. *Plants*, 11(17). https://doi.org/10.3390/plants11172230

LeCun, Y., Bengio, Y., & Hinton, G. (2015). Deep learning. *Nature*, 521(7553), 436–444.

León-Rueda, W. A., León, C., Caro, S. G.-, & Ramírez-Gil, J. G. (2021). Identification of diseases and physiological disorders in potato via multispectral drone imagery using machine learning tools. *Tropical Plant Pathology*, 1–16.

Lesk, C., Rowhani, P., & Ramankutty, N. (2016). Influence of extreme weather disasters on global crop production. *Nature*, 529(7584), 84–87. https://doi.org/10.1038/nature16467

Li, J., Chen, L., Huang, W., Wang, Q., Zhang, B., Tian, X., Fan, S., & Li, B. (2016). Multispectral detection of skin defects of bi-colored peaches based on vis–NIR hyperspectral imaging. *Postharvest Biology and Technology*, 112, 121–133.

Li, L., Zhang, S., & Wang, B. (2021). Plant Disease Detection and Classification by Deep Learning - A Review. *IEEE Access*, 9(Ccv), 56683–56698. https://doi.org/10.1109/ACCESS.2021.3069646

Lin, H., Sheng, H., Sun, G., Li, Y., Xiao, M., & Wang, X. (2021). Identification of pumpkin powdery mildew based on image processing PCA and machine learning. *Multimedia Tools and Applications*, 80(14), 21085–21099. https://doi.org/10.1007/S11042-020-10419-1

Liu, X., Yu, Z., & Tan, L. (2024). *Deep Learning for Lung Disease Classification Using Transfer Learning and a Customized CNN Architecture with Attention. Dl.*

Lowe, A., Harrison, N., & French, A. P. (2017). Hyperspectral image analysis techniques for the detection and classification of the early onset of plant disease and stress. *Plant Methods*, 13(1), 80.

Lu, J., Tan, L., & Jiang, H. (2021). Review on Convolutional Neural Network (CNN) Applied to Plant Leaf Disease Classification. *Agriculture*, 11(8), 707. https://doi.org/10.3390/agriculture11080707

Lu, Y., & Young, S. (2020). A survey of public datasets for computer vision tasks in precision agriculture. *Computers and Electronics in Agriculture*, 178, 105760.

Ma, F., Yuan, M., & Kozak, I. (2023). Multispectral imaging: Review of current applications. *Survey of Ophthalmology*, 68(5), 889–904.

Mahendran, J. K., Barry, D. T., Nivedha, A. K., & Bhandarkar, S. M. (2021). Computer vision-based assistance system for the visually impaired using mobile edge artificial intelligence. *Proceedings of the IEEE/CVF Conference on Computer Vision and Pattern Recognition*, 2418–2427.

Mahlein, A. K., Kuska, M. T., Thomas, S., Bohnenkamp, D., Alisaac, E., Behmann, J., Wahabzada, M., & Kersting, K. (2017). Plant disease detection by hyperspectral imaging: from the lab to the field. *Advances in Animal Biosciences*, 8(2), 238–243.

Mahlein, A.-K. (2016). Plant disease detection by imaging sensors–parallels and specific demands for precision agriculture and plant phenotyping. *Plant Disease*, 100(2), 241–251.

Mamat, N., Othman, M. F., Abdoulghafor, R., Belhaouari, S. B., Mamat, N., & Mohd Hussein, S. F. (2022). Advanced technology in agriculture industry by implementing image annotation technique and deep learning approach: A review. *Agriculture*, 12(7), 1033.

Mandal, S. (2023). Survey on Plant Disease Detection using Deep Learning based Frameworks. *International Journal of Medical, Pharmacy and Drug Research*, 7(2), 27–36. https://doi.org/10.22161/ijmpd.7.2.2

Manley, M. (2014). Near-infrared spectroscopy and hyperspectral imaging: non-destructive analysis of biological materials. *Chem. Soc. Rev.*, 43(24), 8200–8214. https://doi.org/10.1039/C4CS00062E

Masibo, B. (2024). *a Deep Learning Model for Maize Leaf Disease Identification: a Case of Trans-Nzoia County. April.*

Mohanty, S. P., Hughes, D. P., & Salathé, M. (2016). Using deep learning for image-based plant disease detection. *Frontiers in Plant Science*, 7, 215232.

Montesinos López, O. A., Montesinos López, A., & Crossa, J. (2022). Fundamentals of Artificial Neural Networks and Deep Learning. In *Multivariate Statistical Machine Learning Methods for Genomic Prediction*. https://doi.org/10.1007/978-3-030-89010-0_10

Motie, J. B., Saeidirad, M. H., & Jafarian, M. (2023). Identification of Sunn-pest affected (Eurygaster Integriceps put.) wheat plants and their distribution in wheat fields using aerial imaging. *Ecological Informatics*, 76, 102146. https://doi.org/10.1016/J.ECOINF.2023.102146

Mulla, D. J. (2013). Twenty five years of remote sensing in precision agriculture: Key advances and remaining knowledge gaps. *Biosystems Engineering*, 114(4), 358–371. https://doi.org/10.1016/j.biosystemseng.2012.08.009

Musinguzi, P., Ebanyat, P., Tenywa, J. S., Basamba, T. A., Tenywa, M. M., & Mubiru, D. N. (2016). Critical soil organic carbon range for optimal crop response to mineral Fertiliser nitrogen on a Ferralsol. *Experimental Agriculture*, 52(4), 635–653. https://doi.org/10.1017/S0014479715000307

Mustofa, S., Mehedi Hasan Munna, M., Rayhan Emon, Y., Rabbany, G., & Taimur Ahad Associate Professor, M. (2023). *A comprehensive review on Plant Leaf Disease detection using Deep learning.*

Mzoughi, O., & Yahiaoui, I. (2023). Deep learning-based segmentation for disease identification. *Ecological Informatics*, 75, 102000.

Nagasubramanian, K., Jones, S., Sarkar, S., Singh, A. K., Singh, A., & Ganapathysubramanian, B. (2018). Hyperspectral band selection using genetic algorithm and support vector machines for early identification of charcoal rot disease in soybean stems. *Plant Methods*, 14, 1–13.

Nawrocka, A., Nawrocki, M., & Kot, A. (2023). Research study of image classification algorithms based on Convolutional Neural Networks. *Proceedings of the 2023 24th International Carpathian Control Conference, ICCC 2023*, 299–302. https://doi.org/10.1109/ICCC57093.2023.10178933

Ngugi, L. C., Abelwahab, M., & Abo-Zahhad, M. (2021). Recent advances in image processing techniques for automated leaf pest and disease recognition–A review. *Information Processing in Agriculture*, 8(1), 27–51.

Nguyen, C., Sagan, V., Maimaitiyiming, M., Maimaitijiang, M., Bhadra, S., & Kwasniewski, M. T. (2021). Early detection of plant viral disease using hyperspectral imaging and deep learning. *Sensors*, 21(3), 742.

Nguyen, G., Dlugolinsky, S., Bobák, M., Tran, V., López García, Á., Heredia, I., Malík, P., & Hluchý, L. (2019). Machine Learning and Deep Learning frameworks and libraries for large-scale data mining: a survey. *Artificial Intelligence Review*, 52(1), 77–124. https://doi.org/10.1007/s10462-018-09679-z

Nigam, S., Jain, R., Marwaha, S., Arora, A., Haque, M. A., Dheeraj, A., & Singh, V. K. (2023). Deep transfer learning model for disease identification in wheat crop. *Ecological Informatics*, 75(March), 102068. https://doi.org/10.1016/j.ecoinf.2023.102068

Omer, S. (2023). *An Efficient Cucumber Leaf Disease Diagnosis using Deep Learning Algorithms*.

Padmavathi, K., & Thangadurai, K. (2016). Implementation of RGB and grayscale images in plant leaves disease detection–comparative study. *Indian Journal of Science and Technology*, 9(6), 1–6.

Pan, B., Shi, Z., & Xu, X. (2018). MugNet: Deep learning for hyperspectral image classification using limited samples. *ISPRS Journal of Photogrammetry and Remote Sensing*, 145, 108–119. https://doi.org/10.1016/j.isprsjprs.2017.11.003

Pantazi, X. E., Moshou, D., & Tamouridou, A. A. (2019). Automated leaf disease detection in different crop species through image features analysis and One Class Classifiers. *Computers and Electronics in Agriculture*, 156, 96–104.

Parvathi, S., & Tamil Selvi, S. (2021). Detection of maturity stages of coconuts in complex background using Faster R-CNN model. *Biosystems Engineering*, 202, 119–132. https://doi.org/10.1016/j.biosystemseng.2020.12.002

Peng, M., Liu, Y., Khan, A., Ahmed, B., Sarker, S. K., Ghadi, Y. Y., Bhatti, U. A., Al-Razgan, M., & Ali, Y. A. (2024). Crop monitoring using remote sensing land use and land change data: Comparative analysis of deep learning methods using pre-trained CNN models. *Big Data Research*, 36, 100448. https://doi.org/10.1016/J.BDR.2024.100448

Pretty, J., Benton, T. G., Bharucha, Z. P., Dicks, L. V., Flora, C. B., Godfray, H. C. J., Goulson, D., Hartley, S., Lampkin, N., Morris, C., Pierzynski, G., Prasad, P. V. V., Reganold, J., Rockström, J., Smith, P., Thorne, P., & Wratten, S. (2018). Global assessment of agricultural system redesign for sustainable intensification. *Nature Sustainability*, 1(8), 441–446. https://doi.org/10.1038/s41893-018-0114-0

Pu, H., Kamruzzaman, M., & Sun, D.-W. (2015). Selection of feature wavelengths for developing multispectral imaging systems for quality, safety and authenticity of muscle foods-a review. *Trends in Food Science & Technology*, 45(1), 86–104.

Pujari, J. D., Yakkundimath, R., & Byadgi, A. S. (2015). Image processing based detection of fungal diseases in plants. *Procedia Computer Science*, 46, 1802–1808.

Qiu, R., Yang, C., Moghimi, A., Zhang, M., Steffenson, B. J., & Hirsch, C. D. (2019). Detection of fusarium head blight in wheat using a deep neural network and color imaging. *Remote Sensing*, 11(22), 2658.

Rakotonindraina, T. (2012). *Analyse et modélisation des effets des pratiques culturales sur les épidémies de mildiou de la pomme de terre. Adaptation du modèle SIPPOM (Simulator for Integrated Pathogen POpulation Management) au pathosystème*. 160.

Ramesh, S., Hebbar, R., Niveditha, M., Pooja, R., Shashank, N., & Vinod, P. V. (2018). Plant disease detection using machine learning. *2018 International Conference on Design Innovations for 3Cs Compute Communicate Control (ICDI3C)*, 41–45.

Rane, N. L., Mallick, S. K., Kaya, Ö., & Rane, J. (2024). *Machine learning and deep learning architectures and trends : A review*.

Ray, D. K., Gerber, J. S., Macdonald, G. K., & West, P. C. (2015). Climate variation explains a third of global crop yield variability. *Nature Communications*, 6, 1–9. https://doi.org/10.1038/ncomms6989

Rayhana, R., Ma, Z., Liu, Z., Xiao, G., Ruan, Y., & Sangha, J. S. (2023). A review on plant disease detection using hyperspectral imaging. *IEEE Transactions on AgriFood Electronics*, 1(2), 108–134.

Redmon, J., Divvala, S., Girshick, R., & Farhadi, A. (2016). You Only Look Once: Unified, Real-Time Object Detection. *2016 IEEE Conference on Computer Vision and Pattern Recognition (CVPR)*, 779–788. https://doi.org/10.1109/CVPR.2016.91

Rehman, Z. ur, Khan, M. A., Ahmed, F., Damaševičius, R., Naqvi, S. R., Nisar, W., & Javed, K. (2021). Recognizing apple leaf diseases using a novel parallel real-time processing framework based on MASK RCNN and transfer learning: An application for smart agriculture. *IET Image Processing*, 15(10), 2157–2168. https://doi.org/10.1049/IPR2.12183

Revathy, N., & Meenakshi, A. (2024). Classification of Plant Leaf Diseases Identification Using Deep Learning. *2024 2nd International Conference on Networking and Communications (ICNWC)*, 1–9. https://doi.org/10.1109/ICNWC60771.2024.10537326

Rümer, C., Wahabzada, M., Ballvora, A., Pinto, F., Rossini, M., Panigada, C., Behmann, J., Léon, J., Thurau, C., & Bauckhage, C. (2012). Early drought stress detection in cereals: simplex volume maximisation for hyperspectral image analysis. *Functional Plant Biology*, 39(11).

Sa, I., Ge, Z., Dayoub, F., Upcroft, B., Perez, T., & McCool, C. (2016). DeepFruits: A Fruit Detection System Using Deep Neural Networks. *Sensors*, 16(8), 1222. https://doi.org/10.3390/s16081222

Sadler, E. J., Evans, R. G., Stone, K. C., & Camp, C. R. (2005). *Opportunities for conservation with.* 60(6), 371–379.

Sainju, U. M., Singh, B. P., & Whitehead, W. F. (2002). Long-term effects of tillage, cover crops, and nitrogen fertilization on organic carbon and nitrogen concentrations in sandy loam soils in Georgia, USA. *Soil and Tillage Research*, 63(3–4), 167–179. https://doi.org/10.1016/S0167-1987(01)00244-6

Salehi, A. W., Khan, S., Gupta, G., Alabduallah, B. I., Almjally, A., Alsolai, H., Siddiqui, T., & Mellit, A. (2023). A Study of CNN and Transfer Learning in Medical Imaging: Advantages, Challenges, Future Scope. *Sustainability (Switzerland)*, 15(7). https://doi.org/10.3390/su15075930

Sankaran, S., Khot, L. R., Espinoza, C. Z., Jarolmasjed, S., Sathuvalli, V. R., Vandemark, G. J., Miklas, P. N., Carter, A. H., Pumphrey, M. O., Knowles, R. R. N., & Pavek, M. J. (2015). Low-altitude, high-resolution aerial imaging systems for row and field crop phenotyping: A review. *European Journal of Agronomy*, 70, 112–123. https://doi.org/10.1016/j.eja.2015.07.004

Sarker, I. H. (2021). Deep Learning: A Comprehensive Overview on Techniques, Taxonomy, Applications and Research Directions. *SN Computer Science*, 2(6), 1–20. https://doi.org/10.1007/s42979-021-00815-1

Savary, S., Willocquet, L., Pethybridge, S. J., Esker, P., McRoberts, N., & Nelson, A. (2019). The global burden of pathogens and pests on major food crops. *Nature Ecology & Evolution*, 3(3), 430–439.

Sengupta, S., Basak, S., Saikia, P., Paul, S., Tsalavoutis, V., Atiah, F., Ravi, V., & Peters, A. (2020). A review of deep learning with special emphasis on architectures, applications and recent trends. *Knowledge-Based Systems*, 194, 105596. https://doi.org/10.1016/j.knosys.2020.105596

Shafapourtehrany, M., Rezaie, F., Jun, C., Heggy, E., Bateni, S. M., Panahi, M., Özener, H., Shabani, F., & Moeini, H. (2023). Mapping Post-Earthquake Landslide Susceptibility Using U-Net, VGG-16, VGG-19, and Metaheuristic Algorithms. *Remote Sensing*, 15(18). https://doi.org/10.3390/rs15184501

Shaikh, T. A., Rasool, T., & Lone, F. R. (2022). Towards leveraging the role of machine learning and artificial intelligence in precision agriculture and smart farming. *Computers and Electronics in Agriculture*, 198, 107119.

Shamshiri, R., Kalantari, F., Ting, K. C., Thorp, K. R., Hameed, I. A., Weltzien, C., Ahmad, D., & Shad, Z. M. (2018). Advances in greenhouse automation and controlled environment agriculture: A transition to plant factories and urban agriculture. *International Journal of Agricultural and Biological Engineering*, 11(1), 1–22.

Shi, T., Liu, Y., Zheng, X., Hu, K., Huang, H., Liu, H., & Huang, H. (2023). Recent advances in plant disease severity assessment using convolutional neural networks. *Scientific Reports*, 13(1), 1–13. https://doi.org/10.1038/s41598-023-29230-7

Shin, H.-C., Roth, H. R., Gao, M., Lu, L., Xu, Z., Nogues, I., Yao, J., Mollura, D., & Summers, R. M. (2016). Deep Convolutional Neural Networks for Computer-Aided Detection: CNN Architectures, Dataset Characteristics and Transfer Learning. *IEEE Transactions on Medical Imaging*, 35(5), 1285–1298. https://doi.org/10.1109/TMI.2016.2528162

Shiu, Y. S., Lee, R. Y., & Chang, Y. C. (2023). Pineapples' Detection and Segmentation Based on Faster and Mask R-CNN in UAV Imagery. *Remote Sensing*, 15(3). https://doi.org/10.3390/rs15030814

Shivaprasad, K., & Wadhawan, A. (2023). Deep Learning-based Plant Leaf Disease Detection. *2023 7th International Conference on Intelligent Computing and Control Systems (ICICCS)*, 360–365. https://doi.org/10.1109/ICICCS56967.2023.10142857

Shoaib, M., Sadeghi-Niaraki, A., Ali, F., Hussain, I., & Khalid, S. (2025). Leveraging deep learning for plant disease and pest detection: a comprehensive review and future directions. *Frontiers in Plant Science*, 16, 1538163.

Shoaib, M., Shah, B., El-Sappagh, S., Ali, A., Ullah, A., Alenezi, F., Gechev, T., Hussain, T., & Ali, F. (2023). An advanced deep learning models-based plant disease detection: A review of recent research. *Frontiers in Plant Science*, 14(March), 1–22. https://doi.org/10.3389/fpls.2023.1158933

Shorten, C., & Khoshgoftaar, T. M. (2019). A survey on image data augmentation for deep learning. *Journal of Big Data*, 6(1), 1–48.

Signoroni, A., Savardi, M., Baronio, A., & Benini, S. (2019). Deep Learning Meets Hyperspectral Image Analysis: A Multidisciplinary Review. *Journal of Imaging 2019, Vol. 5, Page 52*, 5(5), 52. https://doi.org/10.3390/JIMAGING5050052

Singh, D., Jain, N., Jain, P., Kayal, P., Kumawat, S., & Batra, N. (2020). PlantDoc: A dataset for visual plant disease detection. In *Proceedings of the 7th ACM IKDD CoDS and 25th COMAD* (pp. 249–253).

Singh, V., & Misra, A. K. (2017). Detection of plant leaf diseases using image segmentation and soft computing techniques. *Information Processing in Agriculture*, 4(1), 41–49.

Sishodia, R. P., Ray, R. L., & Singh, S. K. (2020). Applications of remote sensing in precision agriculture: A review. *Remote Sensing*, 12(19), 1–31. https://doi.org/10.3390/rs12193136

Skala, V. (2023). Multispectral image generation from RGB based on WSL color representation: Wavelength, saturation, and lightness. *Computers*, 12(9), 182.

Sladojevic, S., Arsenovic, M., Anderla, A., Culibrk, D., & Stefanovic, D. (2016). Deep neural networks based recognition of plant diseases by leaf image classification. *Computational Intelligence and Neuroscience*, 2016(1), 3289801.

Soeb, M. J. A., Jubayer, M. F., Tarin, T. A., Al Mamun, M. R., Ruhad, F. M., Parven, A., Mubarak, N. M., Karri, S. L., & Meftaul, I. M. (2023). Tea leaf disease detection and identification based on YOLOv7 (YOLO-T). *Scientific Reports*, 13(1), 1–16. https://doi.org/10.1038/s41598-023-33270-4

Stamford, J. D., Vialet-Chabrand, S., Cameron, I., & Lawson, T. (2023). Development of an accurate low cost NDVI imaging system for assessing plant health. *Plant Methods*, 19(1). https://doi.org/10.1186/S13007-023-00981-8

Stavrakoudis, D., Katsantonis, D., Kadoglidou, K., Kalaitzidis, A., & Gitas, I. Z. (2019). Estimating rice agronomic traits using drone-collected multispectral imagery. *Remote Sensing*, 11(5), 545.

Sugiura, R., Tsuda, S., Tamiya, S., Itoh, A., Nishiwaki, K., Murakami, N., Shibuya, Y., Hirafuji, M., & Nuske, S. (2016). Field phenotyping system for the assessment of potato late blight resistance using RGB imagery from an unmanned aerial vehicle. *Biosystems Engineering*, 148, 1–10.

Sujatha, R., Chatterjee, J. M., Jhanjhi, N. Z., & Brohi, S. N. (2021). Performance of deep learning vs machine learning in plant leaf disease detection. *Microprocessors and Microsystems*, 80(November 2020), 103615. https://doi.org/10.1016/j.micpro.2020.103615

Swapna, Munira., Sharma, Dr. Y. K., & Prasad, Dr. B. (2020). CNN Architectures: Alex Net, Le Net, VGG, Google Net, Res Net. *International Journal of Recent Technology and Engineering (IJRTE)*, 8(6), 953–959. https://doi.org/10.35940/ijrte.f9532.038620

Syarief, M., & Setiawan, W. (2020). Convolutional neural network for maize leaf disease image classification. *Telkomnika (Telecommunication Computing Electronics and Control)*, 18(3), 1376–1381. https://doi.org/10.12928/TELKOMNIKA.v18i3.14840

Talaei Khoei, T., Ould Slimane, H., & Kaabouch, N. (2023). Deep learning: systematic review, models, challenges, and research directions. *Neural Computing and Applications*, 35(31), 23103–23124. https://doi.org/10.1007/s00521-023-08957-4

Taye, M. M. (2023). Understanding of Machine Learning with Deep Learning : *Computers MDPI*, 12(91), 1–26.

Thakur, A. (2021). Fundamentals of Neural Networks. *International Journal for Research in Applied Science and Engineering Technology*, 9(VIII), 407–426. https://doi.org/10.22214/ijraset.2021.37362

Thoke, A., & Rai, S. (2024). Exploring Faster R-CNN Algorithms for Object Detection. *2024 First International Conference on Software, Systems and Information Technology (SSITCON)*, 1–5. https://doi.org/10.1109/SSITCON62437.2024.10796389

Thomas, S., Kuska, M. T., Bohnenkamp, D., Brugger, A., Alisaac, E., Wahabzada, M., Behmann, J., & Mahlein, A.-K. (2018). Benefits of hyperspectral imaging for plant disease detection and plant protection: a technical perspective. *Journal of Plant Diseases and Protection*, 125, 5–20.

Too, E. C., Yujian, L., Njuki, S., & Yingchun, L. (2019). A comparative study of fine-tuning deep learning models for plant disease identification. *Computers and Electronics in Agriculture*, 161, 272–279.

Veys, C., Chatziavgerinos, F., AlSuwaidi, A., Hibbert, J., Hansen, M., Bernotas, G., Smith, M., Yin, H., Rolfe, S., & Grieve, B. (2019). Multispectral imaging for presymptomatic analysis of light leaf spot in oilseed rape. *Plant Methods*, 15, 1–12.

Vurro, M., Bonciani, B., & Vannacci, G. (2010). Emerging infectious diseases of crop plants in developing countries: impact on agriculture and socio-economic consequences. *Food Security*, 2, 113–132.

Wadnere, P., & Ramteke, Dr. P. L. (2022). Detection and Classification of Plant Disease with Deep Learning. *International Journal for Research in Applied Science and Engineering Technology*, 10(4), 612–614. https://doi.org/10.22214/ijraset.2022.41291

Wang, C., Du, P., Wu, H., Li, J., Zhao, C., & Zhu, H. (2021). A cucumber leaf disease severity classification method based on the fusion of DeepLabV3+ and U-Net. *Computers and Electronics in Agriculture*, 189, 106373. https://doi.org/10.1016/J.COMPAG.2021.106373

Wang, D., Cao, W., Zhang, F., Li, Z., Xu, S., & Wu, X. (2022). A Review of Deep Learning in Multiscale Agricultural Sensing. *Remote Sensing*, 14(3). https://doi.org/10.3390/rs14030559

Wang, H., Ding, J., He, S., Feng, C., Zhang, C., Fan, G., Wu, Y., & Zhang, Y. (2023). MFBP-UNet: A Network for Pear Leaf Disease Segmentation in Natural Agricultural Environments. *Plants 2023, Vol. 12, Page 3209*, *12*(18), 3209. https://doi.org/10.3390/PLANTS12183209

Wani, J. A., Sharma, S., Muzamil, M., Ahmed, S., Sharma, S., & Singh, S. (2022). Machine Learning and Deep Learning Based Computational Techniques in Automatic Agricultural Diseases Detection: Methodologies, Applications, and Challenges. In *Archives of Computational Methods in Engineering* (Vol. 29, Issue 1). Springer Netherlands. https://doi.org/10.1007/s11831-021-09588-5

Wolfert, S., Ge, L., Verdouw, C., & Bogaardt, M. J. (2017a). Big Data in Smart Farming – A review. *Agricultural Systems*, 153, 69–80. https://doi.org/10.1016/j.agsy.2017.01.023

Xie, B., Wang, L., Li, H., Huo, H., Cui, C., Sun, B., Ma, Y., Wang, J., Yin, G., & Zuo, P. (2021). An interface-reinforced rhombohedral Prussian blue analogue in semi-solid state electrolyte for sodium-ion battery. *Energy Storage Materials*, 36(December 2020), 99–107. https://doi.org/10.1016/j.ensm.2020.12.008

Xu, M., Yoon, S., Fuentes, A., Yang, J., & Park, D. S. (2022). Style-consistent image translation: A novel data augmentation paradigm to improve plant disease recognition. *Frontiers in Plant Science*, 12, 773142.

Yang, J., & Qiu, K. (2022). An improved segmentation algorithm of CT image based on U-Net network and attention mechanism. *Multimedia Tools and Applications*, 81(25), 35983–36006. https://doi.org/10.1007/s11042-021-10841-z

Yao, H., Huang, Y., Tang, L., Tian, L., Bhatnagar, D., & Cleveland, T. E. (2019). Using Hyperspectral Data in Precision Farming Applications. *Advanced Applications in Remote Sensing of Agricultural Crops and Natural Vegetation*, 3–35. https://doi.org/10.1201/9780429431166-1/USING-HYPERSPECTRAL-DATA-PRECISION-FARMING-APPLICATIONS-HAIBO-YAO-YANBO-HUANG-LIE-TANG-LEI-TIAN-DEEPAK-BHATNAGAR-THOMAS-CLEVELAND

Yao, J., Wang, Y., Xiang, Y., Yang, J., Zhu, Y., Li, X., Li, S., Zhang, J., & Gong, G. (2022). Two-Stage Detection Algorithm for Kiwifruit Leaf Diseases Based on Deep Learning. *Plants*, 11(6), 1–12. https://doi.org/10.3390/plants11060768

yoshua bengio. (2019). Deep learning foundations and concepts. *Nature*, 29(7553), 1–73.

Zakaria, N., & Hassim, Y. M. M. (2023). Improved Image Classification Task Using Enhanced Visual Geometry Group of Convolution Neural Networks. *International Journal on Informatics Visualization*, 7(4), 2498–2505. https://doi.org/10.30630/joiv.7.4.1752

Zakaria, N., Mazwin, Y., & Hassim, M. (2024). A Review Study of the Visual Geometry Group Approaches for Image Classification. *Journal of Applied Sciences Technology and Computing*, 1(1), 14–28.

Zeng, Q., Ma, X., Cheng, B., Zhou, E., & Pang, W. (2020). Gans-based data augmentation for citrus disease severity detection using deep learning. *IEEE Access*, 8, 172882–172891.

Zhang DongYan, Z. D., Zhou XinGen, Z. X., Zhang Jian, Z. J., Lan YuBin, L. Y., Xu Chao, X. C., & Liang Dong, L. D. (2018). *Detection of rice sheath blight using an unmanned aerial system with high-resolution color and multispectral imaging.*

Zhang, F., Li, X., Qiu, S., Feng, J., Wang, D., Wu, X., & Cheng, Q. (2021). Hyperspectral imaging combined with convolutional neural network for outdoor detection of potato diseases. *Proceedings - 2021 6th International Symposium on Computer and Information Processing Technology, ISCIPT 2021*, 846–850. https://doi.org/10.1109/ISCIPT53667.2021.00177

Zhang, J., Li, C., Yin, Y., Zhang, J., & Grzegorzek, M. (2023). Applications of artificial neural networks in microorganism image analysis: a comprehensive review from conventional multilayer perceptron to popular convolutional neural network and potential visual transformer. In *Artificial Intelligence Review* (Vol. 56, Issue 2). Springer Netherlands. https://doi.org/10.1007/s10462-022-10192-7

Zhang, N., Yang, G., Pan, Y., Yang, X., Chen, L., & Zhao, C. (2020). A review of advanced technologies and development for hyperspectral-based plant disease detection in the past three decades. *Remote Sensing*, 12(19), 3188.

Zhang, T., Xu, Z., Su, J., Yang, Z., Liu, C., Chen, W. H., & Li, J. (2021). Ir-UNet: Irregular Segmentation U-Shape Network for Wheat Yellow Rust Detection by UAV Multispectral Imagery. *Remote Sensing 2021, Vol. 13, Page 3892*, *13*(19), 3892. https://doi.org/10.3390/RS13193892

Zhang, X., Davidson, E. A., Mauzerall, D. L., Searchinger, T. D., Dumas, P., & Shen, Y. (2015). Managing nitrogen for sustainable development. *Nature*, 528(7580), 51–59. https://doi.org/10.1038/nature15743

Zhao, J., & Su, M. (2024). Comparative Performance Analysis of Single-Shot Detector and Faster R-CNN for Object Detection. *Computer Science, Engineering and Information Technology*, 209–222. https://doi.org/10.5121/csit.2024.141919

Zhao, X., Zhang, J., Tang, A., Yu, Y., Yan, L., Chen, D., & Yuan, L. (2022). The Stress Detection and Segmentation Strategy in Tea Plant at Canopy Level. *Frontiers in Plant Science*, 13. https://doi.org/10.3389/fpls.2022.949054

Zhou, X.-G., Zhang, D., & Lin, F. (2021). UAV remote sensing: An innovative tool for detection and management of rice diseases. In *Diagnostics of plant diseases*. IntechOpen.

5 Deep Learning for Studying Plant–Microbe Interactions

Rahul Gogoi[1], G. Thirumala Devi[2], Madhumita Barooah[1], and Sudipta Sankar Bora[3]*

[1]Department of Agricultural Biotechnology, Assam Agricultural University, Jorhat, Assam, India; [2]Rythu Sadhikara Samstha, IGGAARL, Guntur, India; [3]Silchar Regional Forensic Science Laboratory, Ghungur, Silchar, Assam under the Directorate of Forensic Science, Kahilipara, Guwahati, Assam, India

Abstract

Plant–microbe interactions are highly complex biological processes, which encompass a significant impact on plant health and yield, and also on decently high physiognomy to environmental stressors. Thus, understanding these interactions is of paramount importance in furthering sustainable agriculture, improving crop resistance, and enhancing soil health. In recent years, deep learning (DL) methodologies have witnessed a tremendous impact on analyzing complex biological datasets, offering powerful tools to decipher the interconnected dynamics of plant–microbe interactions. This chapter reviews the applicability of DL techniques in the study of plant–microbe interactions, with a focus on various applications, including genotyping analysis, phenotyping prediction, microbial community profiling, and imaging-based diagnostics, with examples from recent studies and common reporting patterns. It also explores methodologies employed, such as convolutional neural networks, recurrent neural networks, graph neural networks, and transformer architectures, along with relevant case studies from current literature. The authors identify several other challenges of employing DL models to biological data, including data heterogeneity, interpretability, and large labeled datasets. Furthermore, the chapter discusses how multiomics integration and transfer-learning-based techniques can help address these challenges, along with model accuracy and robustness. This chapter aims to serve as a one-stop resource for allied researchers and practitioners who may be interested in using DL for unraveling plant–microbe interactions, which will open up avenues of innovations in agriculture and environmental science.

Keywords: agricultural biotechnology, data heterogeneity, genomic analysis, microbial community profiling, model interpretability, multiomics integration, phenotype prediction

5.1 Introduction

One of nature's biggest gifts, communication between organisms, is essential to evolution and the complexity of life on Earth (Divekar *et al.*, 2022). Plants have developed a variety of biochemical and structural defensive mechanisms against both biotic and abiotic threats since they are unable to flee in adverse environmental circumstances (Li *et al.*, 2020). Plants and microbes have long coexisted in nature, and their interactions, whether beneficial (plants interacting with soil

*E-mail: sudip.asm@gmail.com

DOI: 10.1079/9781836991120.0005

microbiota to increase plant growth and provide biotic and abiotic stress tolerance) or detrimental (host–pathogen interactions causing disease development), may have a big influence on sustainable agriculture (Tiwari *et al.*, 2024). Plants produce a variety of signals or indications that facilitate effective interactions between microorganisms and plants in order to transmit crucial data needed for development and survival (Jamil *et al.*, 2022). Effective communication between microbes and plants can result in the development of a number of beneficial associations, including commensalism, mutualism, and symbiosis, which can increase plant immunity (Khan *et al.*, 2024). Similarly, microorganisms such as bacteria, fungi, and others release a variety of chemicals that can alter the composition of root exudates, thereby impairing plant physiology (Chen and Liu, 2024). It has long been believed that the root system serves only as an anchor and a conduit for water and nutrients. Through the use of substances known as aromatic secondary metabolites, the shoot system also acts as a platform for these interactions.

Shoot-associated endophytic microbes have the ability to synthesize secondary metabolite products from their host plants. Furthermore, interactions between microbes and plants may present a new bioremediation possibility. Some of the main reasons for the decline in soil fertility and the damage done to plant tissue include soil pollutants, such as heavy metals, surfactants, emulsifiers, and other complex organic and inorganic contaminants (Anand *et al.*, 2023). In soil bioremediation and heavy metal sequestration, phytosiderophores have demonstrated exceptional performance. To address the issues caused by soil pollutants, such as heavy metal toxicity, pesticide residues, surfactants, and emulsifiers, in complex environments, several associations based on plant–microbial interactions are currently being utilized (Alengebawy *et al.*, 2021). Furthermore, diseases and new pathogenic species may emerge as a result of an infection's ability to resist or destroy plant defenses. The development of disease management methods, a challenging task for humans, would be made possible by a comprehensive understanding of pathogen action and plant responses (Pandit *et al.*, 2022). The plant–microbe relationship is greatly impacted by climatic variables such as temperature, light, and carbon flux (Cheng *et al.*, 2019). In nature, complex biological activities such as biofertilization, bioremediation, and biocontrol are governed by the association of plant microbes, which are regulated by physical, chemical, biological, and environmental signals (Lahlali *et al.*, 2022). Understanding the process of association between plants and their microbial partners in a constantly changing environment can help clarify both beneficial and destructive relationships, which have significant implications for sustainable agriculture. In this chapter, we provide an overview of the significance of plant–microbe interactions, with an emphasis on deep learning (DL) techniques, potential applications as methodologies and tools, and the challenges, future directions, and opportunities in sustainable agriculture.

5.2 Importance of Plant–Microbe Interactions

5.2.1 Role of plant–microbe interaction in plant health, crop yield, and environmental stress resilience

Finding practical uses that capitalize on the advantageous interactions between plants and microbes is essential for achieving sustainable agriculture. Mycorrhizae and rhizobia are examples of microbial inoculants that have been employed to enhance crop nutrient uptake and soil health. Beneficial bacteria and fungi are examples of biocontrol agents that can safeguard crops from diseases and pests while lowering the need for chemical pesticides. Since they are essential for encouraging plant growth and increasing agricultural productivity, the mechanisms of plant–microbe interactions have recently attracted significant attention (Suman *et al.*, 2022). Nutrient exchange is one of the main processes underlying interactions between microbes and plants. Microbes in the soil play a crucial role in improving nutrient uptake by plants through the production of enzymes and other substances that break down organic matter in the soil (Wang *et al.*, 2024). These enzymes help release vital nutrients in soil, such as phosphorus and nitrogen, increasing their availability for plant absorption. In return, the microorganisms use the carbon molecules that plants give them as a source of energy for their own growth and development

(Wahab *et al.*, 2023). Hormone signaling is another key mechanism underlying interactions between microbes and plants. Auxins, cytokinins, and gibberellins are among the hormones that microbes in the soil can create to promote plant growth and development (Koza *et al.*, 2022). Increased root and shoot growth results from these hormones' support in promoting cell proliferation, elongation, and differentiation in plant tissues. Microbes can occasionally alter the expression of certain plant genes, resulting in the synthesis of particular proteins that support defense and plant growth (Pandita *et al.*, 2023). Another significant mechanism underlying interactions between microbes and plants is induced systemic resistance. In this process, a plant's innate immune system is triggered in reaction to signals from microorganisms. Plants can activate a variety of defense mechanisms that help to guard them from harmful microorganisms when they are exposed to helpful microbes (Yu *et al.*, 2022). For example, beneficial microbes have the ability to trigger the synthesis of phytohormones, such as salicylic acid and jasmonic acid, which trigger particular defensive mechanisms in plants. Additionally, they have the ability to generate molecules that initiate the synthesis of antibiotic chemicals, such as phytoalexins, which help protect the plant from pathogens (Zehra *et al.*, 2021). Recent studies have also revealed the role that epigenetic changes play in plant–microbe interactions, in addition to these pathways. Changes in gene expression that do not affect the underlying DNA sequence are known as epigenetic modifications (Abdulraheem *et al.*, 2024). According to recent research, microbes can cause epigenetic modifications in plant tissues, which can result in different patterns of gene expression that support plant development and pathogen resistance (Tao *et al.*, 2023).

5.2.2 Need for advanced analytical tools

Limitations of traditional methods in studying these interactions

Studying plant–microbe interactions presents several limitations, including the complexity of microbial communities, the difficulty of cultivating many microbes in laboratory conditions, and the obstacles in interpreting complex datasets from "omics" technologies. Additionally, environmental factors and the dynamic nature of these interactions in the field create conditions that hinder a comprehensive understanding of these relationships.

1. *Microbial diversity and culturability.* The rhizosphere (soil around plant roots) harbors a vast and diverse microbial community, with many members being difficult or impossible to cultivate in the laboratory. This makes it challenging to study the full spectrum of microbes involved in these interactions. Advances in molecular biology, such as metagenomics, are helping to overcome this challenge by allowing researchers to explore microbial diversity without the need for cultivation; however, these methods also have limitations (Kumar and Dubey, 2020).

2. *Complex interactions and environmental influences.* Plant–microbe interactions are highly dynamic and influenced by various environmental factors, such as temperature, light, soil pH, and nutrient availability. Studying these interactions in the field, where these factors are constantly changing, is more challenging than in controlled laboratory settings. The complex interplay of multiple environmental parameters often makes it difficult to isolate the effects of specific factors on plant–microbe interactions (Sharma *et al.*, 2024).

3. *"Omics" data interpretation.* Advancements in "omics" technologies (genomics, transcriptomics, proteomics, etc.) provide valuable insights into plant–microbe interactions. However, interpreting the vast amounts of data generated by these technologies requires advanced computational tools and expertise. Integrating data from different "omics" platforms and translating this information into a comprehensive understanding of the interactions remains a significant challenge (Dai and Shen, 2022).

4. *Other challenges.* The study of plant–microbe interactions offers several hurdles, including technical difficulties, data interpretation, interdisciplinary collaboration, lack of defined techniques, and ethical considerations. The complex nature of testing and fieldwork needs specific equipment and methodologies. The interaction of elements makes it difficult to understand outcomes. Standardization of both experimental and computational techniques is required for data comparison and reproducibility (Cheng *et al.*, 2019). Ethical considerations come from the potential consequences of introducing genetically modified organisms (GMOs) into the environment.

Emergence of DL as a transformative approach

DL is emerging as a transformative approach for studying plant–microbe interactions, offering ways to analyze complex data and identify hidden relationships. This approach leverages powerful algorithms, such as convolutional neural networks (CNNs), to analyze data from various omics datasets, including plant stress detection and disease prediction. DL can also be used to understand the intricate connections between plant roots, soil microbes, and their impact on plant health and productivity.

1. *Transforming plant disease detection.* DL models, particularly CNNs, are being used for the early and precise identification of plant diseases through imagery data. This offers a significant advancement over traditional methods, potentially leading to more effective and timely intervention (Upadhyay *et al.*, 2025).
2. *Understanding plant–microbe interactions.* DL can help analyze the vast and complex data generated by omics studies, revealing hidden patterns and relationships between plant root exudates, soil microbes, and plant health. This can lead to a deeper understanding of how plant–microbe interactions impact plant growth, stress resistance, and overall productivity (Deng *et al.*, 2021).
3. *Improving plant stress detection.* DL algorithms can analyze image-based data to detect various forms of plant stress, including disease, pest infestation, and abiotic stress such as drought. This can increase crop productivity and assist farmers in making useful decisions regarding plant management (Shoaib *et al.*, 2023).
4. *Leveraging deep learning in microbiome research.* DL is particularly well suited for analyzing microbiome data due to its ability to handle complex, high-dimensional datasets. This can lead to a better understanding of microbial communities in different environments, including their roles in plant health, disease, and the environment (Przymus *et al.*, 2025).

5.3 Overview of DL Techniques

5.3.1 Basics of DL

The basics of DL are shown in Fig. 5.1.

Key concepts and architectures (e.g. neural networks, layers, and training)

Key concepts in neural networks include neurons, weights, biases, activation functions, layers, and training. Neural networks are structured into layers (input, hidden, and output), where neurons, the basic units, process data, learn through adjusted weights and biases, and utilize activation functions to introduce non-linearity. Training involves feeding the network data, adjusting weights and biases to minimize errors, and repeating this process to improve accuracy.

1. *Neurons.* Neurons are the fundamental computational units within a neural network. They receive input, perform a weighted sum of these inputs, add a bias, and then apply an activation function.
2. *Weights and biases.* Weights and biases represent the strength of connections between neurons at different levels, while biases permit predictions even when all inputs are zero. Weights and biases are changed throughout training in order to reduce the difference between expected and actual results.
3. *Activation functions.* By introducing non-linearity into the network, activation functions allow it to represent complex relationships. Sigmoid and rectified linear unit (ReLU) are a pair of examples.
4. *Layers.* Neural networks consist of input, hidden, and output layers. The input layer receives the initial data, hidden layers perform computations on it, and the output layer delivers the end predictions or classifications.
5. *Training.* Training involves feeding the network data (labeled or unlabeled) and adjusting weights and biases to minimize errors. This process is iterative, and the network learns to make more accurate predictions over time.
6. *Forward and backward propagation.* Forward and backward propagation are methods used to convey information from the input layer to the output layer, with errors estimated to alter weights and biases.

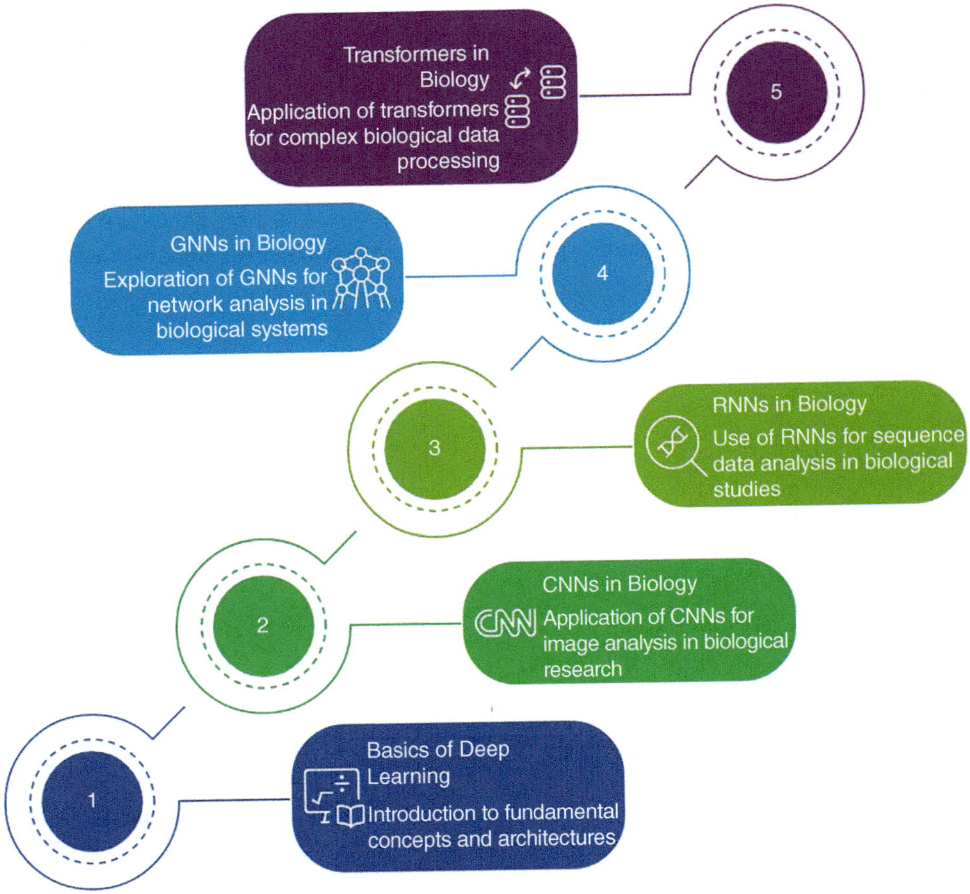

Fig. 5.1. Overview of deep learning. (Author's own image.)

7. *Loss function.* The loss function measures the difference between the network's predictions and the actual values. It guides the training process by indicating how much the network is "off."

5.3.2 DL models used in biological research

DL models are mostly used in biological research to analyze large and complex datasets, such as those generated from genomics, proteomics, and bioimaging. These models, particularly CNNs, recurrent neural networks (RNNs), and graph neural networks (GNNs), are adept at uncovering patterns and making predictions in areas such as gene expression prediction, protein structure prediction, and disease diagnosis (Duan *et al.*, 2024).

1. *Convolutional neural networks.* CNNs are particularly well suited for processing image and grid-like data, making them valuable for bioimage analysis and sequence analysis. CNNs are particularly effective in predicting gene expression from DNA sequences. CNNs are commonly used to process image and video tasks. Key components of a CNN include convolution layers, activation functions, pooling layers, and fully connected layers.

2. *Recurrent neural networks and long short-term memories.* RNNs and their variants, such as long short-term memory (LSTM), are used for analyzing sequential data, including DNA and RNA sequences. It is a type of neural network designed to process sequential data, such as time series or natural language. Unlike traditional feedforward neural networks, RNNs have feedback connections, allowing them to maintain an internal memory of previous inputs. This memory allows RNNs to learn patterns and relationships within sequences, making them suitable for tasks such as language translation, speech recognition, and time-series predictions.

Advanced RNNs include LSTM networks and gated recurrent units, which address the vanishing/exploding gradient problem and have better memory capacity, enabling them to learn from longer sequences and making them simpler to train.

1. *Graph neural networks*. This class of DL techniques is designed to infer information from graphs. Neural networks that can be directly applied to graphs are called GNNs. They offer a simple method for learning from graph data and for making predictions at the node, edge, and graph levels.
2. *Transformer-based architectures*. Transformer-based architectures are neural network models that leverage the attention mechanism to process sequential data, such as text or images. Unlike traditional RNNs, they can parallelize the processing of input sequences, enabling faster training and inference. This architecture is a core component in many modern large language models and is also used in computer vision tasks.
3. *Generative adversarial networks*. GANs can generate synthetic data, which can be used to train models or explore biological possibilities.

5.4 Applications of DL in Plant–Microbe Interactions

Research on plants and microbes has been transformed by DL, enabling new techniques to handle complicated datasets, including genomic, phenotypic, transcriptomic, and imaging data (Fig. 5.2). DL algorithms' ability to find subtle patterns and make accurate predictions based on enormous multidimensional datasets has greatly increased our knowledge of plant–microbe interactions, such as microbial colonization, disease resistance, plant development, and stress tolerance.

5.4.1 Image analysis in plant–microbe studies

Image analysis plays a crucial role in research on plant–microbe interactions, wherein researchers may visually detect plant health, microbial colonization, and disease development. In previous methodologies, investigations of interactions between plants and bacteria were primarily based on manual observation, which is highly laborious and prone to mistakes. However, new DL advances and computer vision algorithms have made picture analysis more precise, effective, and scalable. These developments have opened new possibilities for exploring the intricate connections between plants and microbes, ranging from severity of infection to root colonization and changes in plant morphology (Irigoyen *et al.*, 2020).

Microscopy image classification

The high-definition, exact data analysis provided by microscopy holds its status as a core investigation tool for studying plants and microorganisms. This is because it reveals pathogen entry patterns, microbial growth variability, cellular reactions, and tissue modifications within host plants. The process of using traditional microscopy for

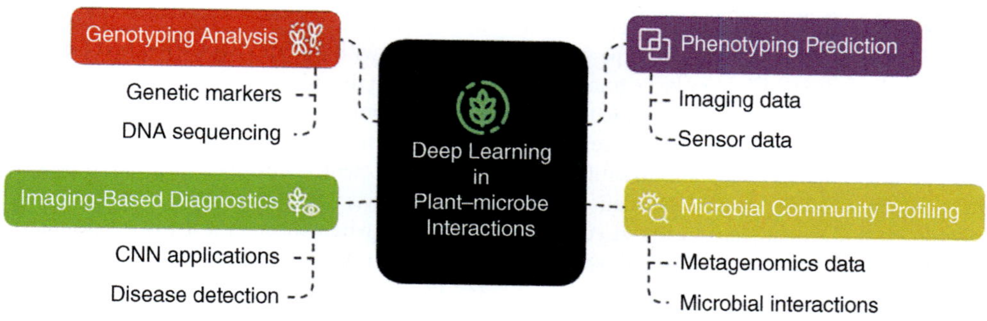

Fig. 5.2. Applications of deep learning. (Author's own image.)

identifying microbes and disease severity levels led to both immense complexity and substantial variations between different research teams when performing evaluations. The combination of the DL domain with CNNs enables automatic repeatable quantitative and qualitative methods for analyzing modified microscopic images. Categorizing images with CNNs demonstrates the suitability of this algorithm for automated image categorization, as it can identify hierarchical structures within images. The networks use spatial organization datasets to distinguish delicate differences between healthy and diseased tissue samples, as well as to identify multiple stages of microbial colonization and interaction types. The detection of plant histology and microscopy images became feasible through three updated design modifications—VGGNet, ResNet, and Vision Transformers (Yamashita *et al.*, 2018).

Algorithm models created for analyzing disease-infected plant images can automatically classify photographs into three phases of infection severity (mild, moderate, and severe), while also recognizing microbiological features, including hyphae, spores, and biofilms, through labeled training datasets. The leaf microscopy application of YOLOv4 and similar models leads to successful object recognition and segmentation techniques that analyze diseases in real time (Xu *et al.*, 2024). Explainable artificial intelligence (XAI) methods, such as Grad-CAM, provide explainability features that enable researchers to determine prediction influences while maintaining both interpretive properties and biological validity. CNN algorithms support research in plant–pathogen interactions by detecting different types of plant pathological structures, including fungal penetration structures (such as appressoria), bacterial colonization snapshots, and virus-induced damage effects within microscopic images (Shi *et al.*, 2023). The models allow researchers to perform vast phenotype screenings of multiple tissue slices in large quantities, and this process proves especially beneficial for resistance breeding and pathogen surveillance. DL technology enabled automatic quantification of powdery mildew disease spread across Arabidopsis and wheat plants, thus speeding up the disease assessment process (Caldwell and Iyer-Pascuzzi, 2019).

Considerably, in the case of mutually beneficial relationships such as mycorrhizal symbiosis, DL models have the capacity to determine vesicles and arbuscules in root cortex cells and, consequently, assess the colonization efficiency of the system under various environmental or therapeutic settings. These properties enable the high-throughput screening of host genotypes and microbial inoculants (Slimani *et al.*, 2024).

Root system architecture under microbial influence

Root system architecture (RSA) specifies the spatial arrangement and morphological aspects of a plant's root system, including root length, branching pattern, angle, and density. RSA plays a critical role in water and nutrient uptake, stress adaptation, and interactions with soil bacteria. Microbial communities—both beneficial and pathogenic—can drastically change RSA by modifying root growth dynamics, hormone signaling, and structural development. These findings illustrate how DL represents a potent new technique for non-invasive, high-throughput study of RSA under varied microbiological circumstances, exposing the hitherto unimagined complexity of plant–microbe interactions (Ranjan *et al.*, 2022). Beneficial microbes (e.g. arbuscular mycorrhizal fungi, plant growth-promoting rhizobacteria, and endophytes) can improve aspects of root structure (e.g. root branching, lateral roots, and root hairs) that are believed to mediate (i.e. modifications to phytohormones such as auxin, cytokinin, and ethylene) their influence on RSAs. In contrast, pathogenic microorganisms are associated with a decrease in root growth (i.e. delays, necrosis, or pathological architecture) from stress response(s) and altered phytohormone signaling pathways. Understanding how bacteria influence RSAs is important not only for agricultural sustainability (e.g. microbial inoculants, breeding qualities connected with microbial proliferation) but also for research techniques (Sagar *et al.*, 2021).

Traditional efforts to estimate RSA have depended on manual mapping or semiautomated procedures, which have proven to be slow and unproductive, with a possibility of human error (especially when working with huge datasets). Advances in DL approaches, particularly through the use of CNN and U-Net architectures, have provided automated approaches to extract traits (segment, classify, and extract data) from root images collected through scanners, rhizotrons, and in-soil imaging (e.g. X-ray computed tomography). For example, DL models built on annotated root images can differentiate primary roots from lateral and fine roots, quantify root length, density, and branching angles, and identify specific developmental changes as time elapses. RootNav 2.0 and DeepRoot are examples that leverage DL-type pipelines to assess RSA in plant species and have demonstrated superior performance compared to comparable approaches that utilized fibrous taxa to compare performance (Yasrab *et al.*, 2019).

Recently, DL research indicated dependency on assessing RSA changes subjected to therapies or organisms that are part of the possible microbiome. For example, with some PGPR strains in the plant *Arabidopsis thaliana*, there has been a deployment of CNN-based models to measure variables linked with lateral root quantity and lateral root length. Similarly, in maize, wheat, and other crops, the influence of mycorrhizal colonization on root system complexity has been deployed using rooting systems assessed with root phenotyping tools that have a DL-based function (Alonazi *et al.*, 2025). DL gives an opportunity for researchers to evaluate modest differences in growth connected with RSA, which are likely due to varying degrees of microbial colonization rates, nutritional capacity, and/or environmental conditions.

Segmentation of colonization zones

Results pertaining to plant–microbe research need proper microorganism identification together with precise internal tissue analysis to deliver sustainable study outcomes (Iqbal *et al.*, 2023). Research requires successful identification strategies to track the entry points of microbial colonization into host tissue for microbial species that require suitable living environments for integration (Yang *et al.*, 2024). The manual method of using microscopy to identify colonization sites is slow and inconsistent because researchers influence the observations.

Semantic segmentation is the pixel-level classification of images, where each pixel is allocated a category name (e.g. colonized vs uncolonized tissue), generally employing DL models or architectures (e.g. U-Net, DeepLab, and Mask R-CNN). DL models with architectures are crucial because they can provide both spatial information and contextual information from various scales in images, both of which can be useful for biological items such as plants and animals. The U-Net model has become a prominent modeling tool in the study of plant–microbe interactions due to its encoder–decoder design that offers perfect localization and classification of microbial features within complicated plant tissues (Paul *et al.*, 2025). U-Net-based models have demonstrated the capability in assessing images from fluorescence and bright-field microscopy to discriminate fungal hyphae, bacterial biofilms, and viral infection lesions.

The pathogenic segmentation process enables DL-based detection of pathological cells, including necrotic effects and fungal appressoria found within healthy tissue structures. This capability facilitates accurate quantification of infection intensity, monitoring of infection progression rates over time, and analysis of the spatial patterns of pathogen invasion. A Mask R-CNN model analyzed confocal images of *Arabidopsis* roots containing *Pseudomonas* bacterial invasions to produce essential data on microbial spreading patterns under different environmental and genetic conditions. Detection of nodules, vesicles, and arbuscules in root tissues through DL-based segmentation has proven effective for studying symbiotic and mutualistic relationships, such as rhizobia or mycorrhizas. The outstanding progress in key parameters related to colonization rate assessments enables high-throughput screening of different host genotypes and microbial strains for agricultural applications (Wang *et al.*, 2021).

Microscopic images utilized for segmentation can be gathered from several platforms, including fluorescence, confocal microscopy, and histochemical staining techniques. Nevertheless, the impact of DL models heavily depends on high-quality annotated training data (Kreiss *et al.*, 2023). Therefore, recently begun efforts built shared image repositories while applying semisupervised or weakly supervised learning techniques that do not require comprehensive annotation with little speed penalty. Data augmentation approaches, such as rotation, contrast change, and synthetic image synthesis, also help strengthen the resilience and generalizability of segmentation algorithms across species, tissues, and imaging technologies (Melanthota *et al.*, 2022).

5.4.2 Genomics and transcriptomics in plant–microbe studies

Genomics and transcriptomics are unquestionably critical areas of exploration in plant–microbe studies that assist researchers in discovering the genomic and regulatory processes underlying the molecular basis of plant responses to microbial interactors. Genomics involves sequencing and analysis of the plant genome, while transcriptomics focuses on gene activity and expression through analysis of RNA (e.g. RNA sequencing or RNA-seq). Together, genomics and transcriptomics help provide insights into the molecular basis of plant immune responses, symbiotic relationships, and stress tolerance in relation to microbial interactions in the plant host (Tyagi *et al.*, 2022). Recently, DL algorithms have been applied throughout genomes and transcriptomes to help us comprehend plant–microbe interactions at a scale and depth that was previously unfathomable.

DL in genomic data analysis

The vast amount of sequencing data generated from whole-genome sequencing and exome sequencing methods needs complex analytical methods. The fundamental technique in genomics research relies on sequence alignment that uses variant calling approaches through multiple database annotations. However, two fundamental obstacles exist for standard techniques because they need to choose a database, and they need increased processing times for handling enormous and complex datasets (Marshall *et al.*, 2020). Science research in genomics obtained exceptional data processing capabilities through DL when scientists gained the ability to find genetic variants alongside functional features in genomic data. Neural networks of current innovation perform effective pattern detection on DNA sequence information to display microbial symbiosis as well as disease resistance phenomena. CNN analysis of DNA sequences allows laboratory professionals to identify regulatory domains, including enhancers and promoters, without requiring prior data processing of unprocessed data. Through pattern detection, these algorithms discover novel plant–microbe genes together with improved genomic annotation by finding patterns that previous analytical techniques missed (Yue *et al.*, 2023). By finding significant genomic variants, DL algorithms' predictions help plants defend against stressors and disease resistance manifestations. Artificial intelligence (AI)-based analysis of GMOs combines genetic codes with environmental elements to determine modifications in plant–microbial sensitivity resulting from genetic feature enhancements (Kimotho and Maina, 2024).

DL in transcriptomics

Transcriptomics, particularly RNA-seq, gives a comprehensive snapshot of gene expression when organisms are subjected to a diversity of stimuli, including microbial pathogens and beneficial microbes. Understanding gene expression patterns in response to microbial pathogens and beneficial microorganisms is critical to deciphering the plant's immune response, microbial symbionts, and the genomic basis of plant–microbe interactions. Traditionally, transcriptome analysis typically focused on differential gene expression as the mean to discover genes that were upregulated or downregulated in response to a microbial interaction (Lowe *et al.*, 2017). However, standard transcriptome studies do not include the complex, non-linear relationships that exist between gene expression, environmental variables of the study, and the microbiological background that the plants are encountering (Zheng *et al.*, 2020).

The ability of DL technologies improves transcriptome analysis through their creation of exact gene control networks and microbial network reactions (Li *et al.*, 2023). RNA-seq data patterns that described the regulatory processes between microbes and plants were analyzed using autoencoders and deep neural networks (DNNs). DNNs provide researchers with the ability to predict plant gene expression by analyzing transcriptome data to identify genes that respond to stress and immune challenges after microbe–pathogen interactions (Wang *et al.*, 2020). DL technologies provide essential advantages for transcriptomic science and enable large-scale omics-level data analysis that includes genomes and transcriptomes together with proteomes and other omics levels. The investigation of biomarkers using DL models enhances both microbial inoculant development and plant breeding approaches, while also providing information about microbial–plant connections (Chen *et al.*, 2023).

Metagenomics and microbiome analysis

Metagenomics includes examining genetic material directly from environmental samples, allowing the study of microbial communities in their native settings without ever cultivating any of the organisms. Metagenomics has been essential in changing our understanding of how plant-associated microbiomes function similarly to soil microbiomes, including those in the rhizosphere (soil), phyllosphere (atmospheric), and other plant-associated microbiomes. Plant microbiomes are crucial to plant health and contribute to disease resistance, nutrient absorption, and stress tolerance. The difficulty with microbiome data is the sheer multitude of unique microbiome data (for hundreds of species, there are millions of genomes). Inputs linked with DL algorithms have been a powerful technique to evaluate metagenomic or microbiome data, enabling unique ways to monitor and estimate microbial community composition, function, and dynamics (Pérez-Cobas *et al.*, 2020).

DL in metagenomics

Research scientists obtain comprehensive data on microbial taxonomy and functionality of communities through the combination of shotgun metagenomic sequencing and 16S ribosomal RNA (rRNA) gene sequencing technologies. The investigation of classical metagenomic data requires both sequence-matching algorithms and grouping procedures; however, these big data approaches face significant processing challenges during database operations and the analysis of large, complex contemporary sequencing datasets (Hempel *et al.*, 2022).

Three DL models, specifically BERT transformers, CNNs, and RNNs, represent exceptional potential to solve difficulties. Sequence patterns that generate extractable information remain elusive to standard evaluation methods. The identification of 16S rRNA sequences required CNNs, but RNNs performed the evaluation of microorganism functional potential from shotgun metagenomic data (Przymus *et al.*, 2025).

Predicting microbial community function

The basics of metagenomic research heavily depend on DL technology to determine microbial community functioning. Active functional predictions help researchers identify which biosynthetic pathways are operational within microbiome communities but fail to maintain the functionality of the microbiome. DL algorithms generate functional assessments of metagenomic specimens by processing DNA sequences through KEGG and MetaCyc functional gene repositories (Roy *et al.*, 2024). As part of its operation, DeepMicrobe uses shotgun sequencing data to identify microbial community roles by applying DL methods. Sequence composition functions provide the algorithm with connected features that enable it to map sequences to various functional categories. Research on microbiome benefits for plant health needs a clear understanding of particular functional contributions, including nutrient fixation, disease prevention, and stress reduction capabilities (Deng *et al.*, 2021).

Microbial community profiling

DL is enabling scientists to characterize microbial communities from environmental sequencing data. These algorithms facilitate the detection of microbial species from metagenomic collections, even when organisms are present at very low concentrations in samples. Analyzing such minor species in plant–microbe studies is important because, within large microbial populations, the vast majority in terms of number are low-abundance organisms that might influence the environment (Mathieu *et al.*, 2022). Dimensional reduction serves as a miniaturization technique for some metagenomic data, while autoencoders attempt to infer latent community patterns without supervision. This approach is very much useful in the detection of rare organisms that are of environmental importance in plant–microbial communities (Roy *et al.*, 2024).

5.4.3 Prediction of molecular interactions

Biological functions arising from molecular interactions are particularly relevant to the interactions between microorganisms and plants. Molecular interactions that consist of complex signaling pathways, receptor–ligand binding, protein–protein interactions (PPIs), and other molecular mechanisms will ultimately decide the outcome of their partnerships (You-Yu Wang *et al.*, 2023). Understanding and predicting these linkages is especially important for microbial pathogenicity and symbiosis, plant responses to immunity, and certain ways of inducing resistance against diseases. However, the dynamic and highly diverse nature of molecular interactions has traditionally made them difficult to predict and describe. Using DL, which has become one of the best tools to predict molecular interaction, will enhance our understanding of plant–microbe interaction at the molecular level (Grassmann *et al.*, 2024).

DL for predicting protein–protein interactions

Protein–protein interactions are one of the most researched molecular interactions linked with plants and microorganisms with respect to how plant and pathogen proteins interact with one another to induce immune responses or to enable infections to infect plants. Knowledge and precise prediction regarding PPIs are vital for identifying the primary participants in plant immunology processes (including R proteins, plant immune modulators, and effector proteins derived from pathogens). Methodologies regularly utilized, including yeast two-hybrid screening and co-immunoprecipitation, may be informative but are time-consuming and often result in insufficient data (Lei *et al.*, 2024).

Protein–protein interactions prediction has greatly profited from DL algorithms, as these make it possible to leverage large protein datasets to learn laws. PPI has been predicted by CNN or RNN based on information relating to the sequence, structure, and phylogenetic information of proteins. DeepPPI and PIPR can be used, for instance, to predict probable sites of interaction between plant immune receptors and microbial effectors using features of amino acid sequence motifs and domain topologies (Bernett *et al.*, 2024). Users may design new strategies of virulence and defense by using these models to predict the interactions between newly discovered proteins.

Receptor–ligand interaction prediction

The receptor–ligand-based interaction is essential for recognizing and responding to pathogens. Plants possess receptor molecules, namely RLKs and NLRs, which recognize PAMPs and pathogen-specific effectors, respectively. Typically, a receptor–ligand interaction initiates the induction of some of the earliest innate defense responses in the plant downstream (Dodds *et al.*, 2024).

Emerging DL methods provide insights into receptor–ligand binding affinities and specificities involved in the self- or non-self-recognition of a plant. RNNs and GCNs are particularly well suited for analyzing the transient nature of receptor–effector interactions (Yuxiao Wang *et al.*, 2023). These models assess not only whether a receptor–ligand interaction can occur but also whether it should trigger an immune response. Such empirical models describe plant resistance mechanisms, how a plant survives or tolerates a particular pathogen, and thus, the breeding to enhance resistance can be aided through modifications of receptor–ligand interactions for general or specific resistance against that pathogen (Zhao *et al.*, 2024).

Functional annotation of unknown molecular interactions

In addition to predicting known linkages, DL can contribute to the functional annotation of undiscovered chemical interactions. Using large omics data (e.g. transcriptomics and proteomics), DL models can individually evaluate the likely functional role of uncharacterized proteins or genes affecting the plant–microbe system. Autoencoders and GNN approaches, for example, have been applied to designate functional connections within plant immune signaling pathways, calculating how a specific protein or metabolite might affect plant defense or microbe virulence (Trapotsi *et al.*, 2022). DL models demonstrate potential in instances where experimental characterization data are currently unavailable or where interactions are assumed to be context-dependent, for example, under different environmental circumstances or phases of plant development. With large-scale sequence data and functional annotations, DL models can potentially identify undetected chemical interactions and provide a basis for creating sustainable plant protection techniques.

5.4.4 Phenotype prediction and trait mapping

Plant breeding, agriculture, and crop development all depend on phenotype prediction and trait mapping. These approaches require linking genetic variation with phenotypes, or distinct qualities and attributes, such as yield and stress or disease resistance. Plant–microbe interactions make phenotype prediction much more challenging than in other cases, since the plant is reacting to a wide spectrum of microorganisms, including pathogens and helpful bacteria, all of which can impact the ultimate phenotype (Mu *et al.*, 2022). Until recently, phenotype prediction and trait mapping were guided by statistical techniques and simplistic models. DL breakthroughs and other statistical advancements have made these predictions much more possible, reproducible, and scalable, as they now permit predictions of complex features in plant–microbe systems (Crossa *et al.*, 2025).

DL in phenotype prediction

High-throughput phenotyping, genome sequencing, and other omics technologies have already shown that DL can correctly predict phenotype from huge volumes of data. To more precisely forecast or confirm basic plant phenotypes, DL models integrate and process multiple layers of data, including genomic, transcriptomic, proteomic, and/or metabolomic data features. This is in contrast to traditional techniques, which often concentrate on a few genes without the possibility to integrate with another set of traits (Sehrawat *et al.*, 2023). Specifically, there are CNNs, which can evaluate images of plants or plant organs and learn relationships of space with respect to qualities such as leaf form, disease symptoms, or root architecture. RNNs, on the other hand, can learn dynamic attributes such as growth or stress responses over time. This is particularly useful when phenotypic data are complex or highly dynamic, such as changes in plant development in response to microbial stress or environmental challenges (Gill *et al.*, 2022).

DL methods can also leverage high-dimensional genetic data alongside phenotypic data to predict complex traits. These models can be used to predict traits such as disease resistance, drought tolerance, or nutrient uptake efficiency by learning the relationships between genotype and phenotype across large populations. Additionally, they can identify potential key genetic markers for particular traits (Vourlaki *et al.*, 2024).

Trait mapping and quantitative trait loci mapping

The very first step that is critical for intricately determining the biology of phenotypic traits is trait mapping. This involves locating quantitative trait loci (QTLs) associated with features pertinent to microbial interactions, such as symbiotic efficacy or resistance to diseases in plant–microbe interactions. Traditional QTL mapping associates markers with traits by crossing two different kinds of plants and obtaining genotypic and phenotypic information on them. This ties badly with the number of traits one tries to study, hence being quite an expensive technique (Capilla-Pérez *et al.*, 2024).

A recent major advancement in QTL mapping is DL, especially with random forests and DNNs. Basically, these methods analyze huge, multidimensional datasets to pinpoint more accurately the genetic loci that underlie complex phenotypes (Farooq *et al.*, 2024). Combining DL with genomic data from hundreds of genotyped individuals and phenotypic data from plant–microbe interaction assays permits DL models to detect genetic loci responsible for microbial traits, such as resistance to particular pathogens or susceptibility to beneficial microbial symbiosis. Most importantly, environmental factors and other modifiers of plant traits may be taken into consideration in DL models; hence, the prediction can be made more precise (Sigala *et al.*, 2023). This application is particularly relevant in plant–microbe studies, where environmental factors, microbial diversity, and interactions will tremendously modify phenotypic responses.

Applications in plant breeding and crop improvement

DL offers new approaches to phenotype prediction and trait mapping as agricultural crops are bred or enhanced. It potentially creates a new opportunity for predicting how a plant will respond to microbial stresses or symbionts, enabling breeders to deliberately select plants with higher resistance to disease and performance under unfavorable conditions. DL systems can predict how plants will respond to inoculation with a pathogen or a beneficial bacterium (Chao *et al.*, 2025).

Deep-learning-based trait mapping may also accelerate efforts toward increased crop productivity, nutritional quality, and environmental compatibility by locating critical genetic loci related to these traits. If breeders know the critical genetic loci for these traits, they can expedite the design of crops with specific environmental conditions and market requirements in mind (Gill *et al.*, 2022).

5.5 Methodologies and Tools

Bioinformatics undergoes a revolution through DL, which enables automated data evaluation through process modeling, feature selection, prediction, and outcome estimation. This section addresses different DL methodologies while addressing issues during model training and evaluation phases, and presents the tools that facilitate these studies.

5.5.1 Key DL methodologies

Convolutional neural networks

CNNs are a type of DL algorithm that, using convolutional filters, are intended to automatically learn spatial hierarchies of features. These networks are commonly used for image analysis tasks, which makes them ideal for work in plant phenotyping, medical imaging, and protein structure studies. CNNs are typically used in plant disease identification, where they assess images of either plant leaves or their associated crops and assign disease statuses or stress that may lead to changes (Yamashita *et al.*, 2018). One example of this type of research is using CNNs to classify diseases of leaves based on photographs, which significantly increased the speed and accuracy of disease diagnosis in agricultural settings (Castillo-Girones *et al.*, 2025).

Recurrent neural networks

RNNs are designed to take sequential input for applications where context and temporal relationships are important. RNNs are also used in biology to study gene expression, where gene activity changes over time or under different conditions. In this application, RNNs are used to make predictions based on time-series data for plant development trends or to simulate dynamic gene regulatory networks (Monti *et al.*, 2022). LSTM, a subtype of RNN, has been known to be quite successful in capturing long-range dependencies in biological sequences, such as predicting future levels of gene expression based on earlier data (Erfanian *et al.*, 2023).

Graph neural networks

Graph neural networks offer a powerful DL approach for handling graph-structured data, which is often found in biological networks. In these systems, complex interactions occur among proteins, genes, and metabolites, justifying the utility of GNNs for simulating such interactions. Using GNNs, one could predict protein–protein interactions, propose new therapeutic targets, or simulate entire signaling cascades (Zhang *et al.*, 2021). GNNs have been harnessed in systems biology to model metabolic networks and also anticipate how perturbation of one component of a network could lead to downstream effects on cellular activities, hence enabling the discovery of possible therapeutic targets of interest in drug development (Evans *et al.*, 2024).

Transformers

The transformer comprises a family of DL architectures that deploy attention mechanisms to capture long-range dependencies in sequences, thereby freeing from the sequential processing required by RNNs. They promise usability in biological data analysis, while their superlative successes have been realized in natural language processing. The other burning application of transformer architectures is genomics in deoxyribonucleic acid (DNA) sequence analysis, where they have been well-versed in predicting gene function and finding regulatory regions (Choi and Lee, 2023). Architecturally modeled on BERT (Bidirectional Encoder Representations from

Transformers), BioBERT, which is further fine-tuned for biomedical text mining, has indeed proven its mettle in the task of extracting biological insights from scientific literature (Lee *et al.*, 2020).

5.5.2 Training and evaluation of models

Handling large datasets and annotation challenges

Protein structures, genetic sequences, and multiomics data are examples of biological datasets that are typically large and high-dimensional, requiring specific preparation and analysis techniques. To reduce noise and improve model performance, data preparation techniques such as dimensionality reduction (e.g. principal component analysis), normalization, and imputation are essential (Eicher *et al.*, 2020). Furthermore, a fundamental challenge for supervised learning is the normal deficiency of sufficient labeled data in biological datasets. To tackle this problem, transfer learning, active learning, and semisupervised learning have been devised. For example, researchers may employ transfer learning to improve model performance with few annotations by fine-tuning pretrained models, such as BioBERT, using fewer datasets (Lee *et al.*, 2020).

Metrics for evaluating model performance

Evaluating the effectiveness of DL models in biological environments requires measures that are suited to the nature of the data. In biological applications, traditional measures of performance, such as accuracy, are not always trustworthy, particularly when the dataset is unbalanced—where at least one class (e.g. rare diseases or some uncertain phenotypic attributes) is underrepresented (Ching *et al.*, 2018). Alternative measurements such as precision, recall, and F1-score provide further information on false-positive and false-negative trade-offs in these cases. Classification challenges tend to make use of area under the curve as a performance measure, which gives collective outcomes across all decision threshold settings (Ahmed *et al.*, 2023). The testing for generalization to the model needs k-fold cross-validation together with additional procedures because these processes confirm that the model performs effectively on various sections of data while avoiding overfitting.

5.5.3 Tools and frameworks

TensorFlow and PyTorch

The DL frameworks TensorFlow and PyTorch comprise the most prominent technologies that provide distinct functionality for multiple biological applications (Ahmed *et al.*, 2023). As a framework that scales well, TensorFlow works efficiently in production scenarios as well as massive datasets within biological research. Distributed computing support is added to this platform, which becomes crucial when working with massive bioinformatics datasets (Nguyen *et al.*, 2019). Academic researchers tend to select PyTorch because of its user-friendly interface combined with strong flexibility features. Researchers have access to PyTorch's dynamic computing network for simple investigation of both novel model designs and debugging approaches, making it highly relevant in the expanding field of computational biology (Jurj *et al.*, 2024).

Specialized bioinformatics tools

Different tools and libraries used in bioinformatics combine DL mechanisms with different biological data analysis techniques. DeepChem represents one such library that combines DL with chemical informatics to achieve tasks such as molecular property prediction and drug discovery (Askr *et al.*, 2023). DeepChem stands as a major computational tool that applies across drug research for virtual screening, drug optimization, and predictive toxicology investigations. BioBERT provides a specific tool that tailors the transformer model for biomedical text mining objectives to extract important data from substantial scientific literature (Lee *et al.*, 2020). BioBERT acts as a tool for extracting associations from scientific literature to identify links between genes and diseases, as well as between medications and genes, thereby supporting biological study development.

5.6 Challenges in Applying DL to Biological Data

Amid the promising advances, some challenges hinder the application of DL to plant–microbe interaction studies, arising due to the complexity of biological data itself. One major issue is data heterogeneity, as biological datasets are quite heterogeneous, covering a vast range of data types from genomic sequences to transcriptomic profiles, phenotypic traits, and metagenomic records (Murmu *et al.*, 2024). Thus, integrating all these data of varying sizes, formats, and noise types into one DL framework is rather difficult. Second, the limited number of high-quality labeled datasets restricts the training of DNNs, requiring large amounts of annotated data for best performance (Kumar *et al.*, 2024). In the case of biological samples, annotation is time-consuming and labor-intensive and often lacks consistency, particularly in plant–microbe studies where experimental validation adds to the inherent complexity (Ahmed *et al.*, 2025).

Interpretability of models is another pressing issue. DL models are considered "black boxes" because it is difficult to garner a biological insight from a prediction. This, therefore, impedes trust and application in biological research, where the reasoning behind the prediction has to be understood so as to be able to generate hypotheses and do experimental validations (Talaei Khoei *et al.*, 2023). Computational complexity also remains a challenge in resource-limited research settings. Training state-of-the-art transformers or a very large CNN requires highly computational resources, memory, and time, which places restrictions on many researchers for access (Wang *et al.*, 2023).

Together, these challenges necessitate interdisciplinary strategies that merge biological knowledge with computational innovation to render DL approaches more robust, interpretable, and usable for studies of plant–microbe interactions.

5.7 Strategies to Address Challenges

Potential solutions and approaches to the major challenges associated with the use of DL in investigations of plant–microbe interactions are discussed below. A promising approach to address these issues is multiomics integration, which merges data from various biological fields of study, including genomics, transcriptomics, proteomics, and metabolomics, into unified frameworks (Jain *et al.*, 2024). It also overcomes issues with data heterogeneity and adds biological insight by extracting information from a system-level perspective on plant–microbe interaction. The benefits of combining omics layers for improved predictive accuracy, as well as uncovering multilayer complex interdependencies, have been recently exemplified (Picard *et al.*, 2021).

Another effective technique is transfer learning, which consists of fine-tuning models previously trained on large datasets using smaller datasets focused on a particular domain. This method lessens the requirement for large, labeled datasets and is especially useful in agriculture, where collected labeled data are scarce (Singh *et al.*, 2024). Transfer learning has been successfully applied in crop disease detection, microbiome classification, and other areas, achieving a dramatic increase in performance with a very limited number of training examples. In order to make their models interpretable, many researchers have turned to using XAI methodologies such as SHAP (SHapley Additive exPlanations), LIME (Local Interpretable Model-agnostic Explanations), and attention mechanisms (Mathew *et al.*, 2025). These tools identify the most important features in the predictions of the model, helping extract biologically relevant information. In the case of plant–microbe interactions, such tools help in dissecting the genetic or microbial signatures behind a phenotype (Ali *et al.*, 2023).

Finally, stronger model robustness is fundamental for generalization to other biological scenarios. To prevent overfitting to the noise in the training set and failure to generalize well to new, out-of-sample data, techniques such as data augmentation, regularization methods such as dropout and batch normalization, or ensemble methods can be used to fit the model more robustly (Ahmed *et al.*, 2023). In agricultural biotechnology, these techniques are especially relevant in the confidence of extending models from a given crop or environmental scenario to another one (Sultan *et al.*, 2025). Collectively, these strategies represent a path forward for improving the efficacy, interpretability, and applicability of DL approaches in the complicated field of plant–microbe interactions.

5.8 Conclusion

Plants build beneficial microorganism connections that enhance their health status while improving yields and minimizing environmental risks. Existing examination systems often fail to deliver satisfactory outcomes in multidimensional biological interaction studies. DL provides specific answer sets to address current analytical challenges through its ability to develop novel approaches for investigating large biological data that lead to predictive models of complicated patterns. Research on plant–microbe interactions has greatly benefited from DL techniques, enabling visual disease diagnosis, microbiome analysis, and assessments of genetic variants as well as species attributes. DL models consisting of CNNs, RNNs, GNNs, and transformers enable biological researchers to process genomic sequences and metagenomic profiles together with high-resolution images. However, widespread DL applications are limited by many obstacles arising from diverse data sources, minimal labeled datasets, challenging interpretability, and demanding computational needs.

Study groups present different solutions by combining multiomics information with transfer learning and explainable AI coding approaches that improve investigation robustness. Research on robotics and sustainable farming using IoT technologies should create specific ecological management and agricultural application systems that can be implemented. DL technology provides us with exceptional capabilities to better understand plant microbiological functions. A complete implementation of these solutions requires close collaboration between plant biologists and data scientists, along with agricultural engineers to develop universal solutions that boost food security and environmental sustainability.

References

Abdulraheem, M.I. *et al.* (2024) 'Mechanisms of Plant Epigenetic Regulation in Response to Plant Stress: Recent Discoveries and Implications', *Plants*, 13(2), p. 163. Available at: https://doi.org/10.3390/plants13020163.

Ahmed, S.F. *et al.* (2023) 'Deep learning modelling techniques: current progress, applications, advantages, and challenges', *Artificial Intelligence Review*, 56(11), pp. 13521–13617. Available at: https://doi.org/10.1007/s10462-023-10466-8.

Ahmed, S.F. *et al.* (2025) 'Unveiling the frontiers of deep learning: Innovations shaping diverse domains', *Applied Intelligence*, 55(7), p. 573. Available at: https://doi.org/10.1007/s10489-025-06259-x.

Alengebawy, A. *et al.* (2021) 'Heavy Metals and Pesticides Toxicity in Agricultural Soil and Plants: Ecological Risks and Human Health Implications', *Toxics*, 9(3), p. 42. Available at: https://doi.org/10.3390/toxics9030042.

Ali, S. *et al.* (2023) 'The enlightening role of explainable artificial intelligence in medical; healthcare domains: A systematic literature review', *Computers in Biology and Medicine*, 166, p. 107555. Available at: https://doi.org/10.1016/j.compbiomed.2023.107555.

Alonazi, M.A. *et al.* (2025) 'Native Plant Growth-Promoting Rhizobacteria Containing ACC Deaminase Promote Plant Growth and Alleviate Salinity and Heat Stress in Maize (Zea mays L.) *Plants in Saudi Arabia*', *Plants*, 14(7), p. 1107. Available at: https://doi.org/10.3390/plants14071107.

Anand, U. *et al.* (2023) 'Current Scenario and Future Prospects of Endophytic Microbes: Promising Candidates for Abiotic and Biotic Stress Management for Agricultural and Environmental Sustainability', *Microbial Ecology*, 86(3), pp. 1455–1486. Available at: https://doi.org/10.1007/s00248-023-02190-1.

Askr, H. *et al.* (2023) 'Deep learning in drug discovery: an integrative review and future challenges', *Artificial Intelligence Review*, 56(7), pp. 5975–6037. Available at: https://doi.org/10.1007/s10462-022-10306-1.

Bernett, J., Blumenthal, D.B. and List, M. (2024) 'Cracking the black box of deep sequence-based protein–protein interaction prediction', *Briefings in Bioinformatics*, 25(2). Available at: https://doi.org/10.1093/bib/bbae076.

Caldwell, D. and Iyer-Pascuzzi, A.S. (2019) 'A Scanning Electron Microscopy Technique for Viewing Plant–Microbe Interactions at Tissue and Cell-Type Resolution', *Phytopathology®*, 109(7), pp. 1302–1311. Available at: https://doi.org/10.1094/PHYTO-07-18-0216-R.

Capilla-Pérez, L. *et al.* (2024) 'Enhanced recombination empowers the detection and mapping of Quantitative Trait Loci', *Communications Biology*, 7(1), p. 829. Available at: https://doi.org/10.1038/s42003-024-06530-w.

Castillo-Girones, S. *et al.* (2025) 'Artificial Neural Networks in Agriculture, the core of artificial intelligence: What, When, and Why', *Computers and Electronics in Agriculture*, 230, p. 109938. Available at: https://doi.org/10.1016/j.compag.2025.109938.

Chao, D. *et al.* (2025) 'MtCro: multi-task deep learning framework improves multi-trait genomic prediction of crops', *Plant Methods*, 21(1), p. 12. Available at: https://doi.org/10.1186/s13007-024-01321-0.

Chen, C. *et al.* (2023) 'Applications of multi-omics analysis in human diseases', *MedComm*, 4(4). Available at: https://doi.org/10.1002/mco2.315.

Chen, L. and Liu, Y. (2024) 'The Function of Root Exudates in the Root Colonization by Beneficial Soil Rhizobacteria', *Biology*, 13(2), p. 95. Available at: https://doi.org/10.3390/biology13020095.

Cheng, Y.T., Zhang, L. and He, S.Y. (2019) 'Plant-Microbe Interactions Facing Environmental Challenge', *Cell Host & Microbe*, 26(2), pp. 183–192. Available at: https://doi.org/10.1016/j.chom.2019.07.009.

Ching, T. *et al.* (2018) 'Opportunities and obstacles for deep learning in biology and medicine', *Journal of The Royal Society Interface*, 15(141). Available at: https://doi.org/10.1098/rsif.2017.0387.

Choi, S.R. and Lee, M. (2023) 'Transformer Architecture and Attention Mechanisms in Genome Data Analysis: A Comprehensive Review', *Biology*, 12(7), p. 1033. Available at: https://doi.org/10.3390/biology12071033.

Crossa, J. *et al.* (2025) 'Expanding genomic prediction in plant breeding: harnessing big data, machine learning, and advanced software', *Trends in Plant Science* [Preprint]. Available at: https://doi.org/10.1016/j.tplants.2024.12.009.

Dai, X. and Shen, L. (2022) 'Advances and Trends in Omics Technology Development.', *Frontiers in medicine*, 9, p. 911861. Available at: https://doi.org/10.3389/fmed.2022.911861.

Deng, Z. *et al.* (2021) 'Application of Deep Learning in Plant–Microbiota Association Analysis', *Frontiers in Genetics*, 12. Available at: https://doi.org/10.3389/fgene.2021.697090.

Divekar, P.A. *et al.* (2022) 'Plant Secondary Metabolites as Defense Tools against Herbivores for Sustainable Crop Protection', *International Journal of Molecular Sciences*, 23(5), p. 2690. Available at: https://doi.org/10.3390/ijms23052690.

Dodds, P.N., Chen, J. and Outram, M.A. (2024) 'Pathogen perception and signaling in plant immunity', *The Plant Cell*, 36(5), pp. 1465–1481. Available at: https://doi.org/10.1093/plcell/koae020.

Duan, J. *et al.* (2024) 'Deep learning based multimodal biomedical data fusion: An overview and comparative review', *Information Fusion*, 112, p. 102536. Available at: https://doi.org/10.1016/j.inffus.2024.102536.

Eicher, T. *et al.* (2020) 'Metabolomics and Multi-Omics Integration: A Survey of Computational Methods and Resources', *Metabolites*, 10(5), p. 202. Available at: https://doi.org/10.3390/metabo10050202.

Erfanian, N. *et al.* (2023) 'Deep learning applications in single-cell genomics and transcriptomics data analysis', *Biomedicine & Pharmacotherapy*, 165, p. 115077. Available at: https://doi.org/10.1016/j.biopha.2023.115077.

Evans, N.J. *et al.* (2024) 'Graph Structured Neural Networks for Perturbation Biology'. Available at: https://doi.org/10.1101/2024.02.28.582164.

Farooq, M.A. *et al.* (2024) 'Artificial intelligence in plant breeding', *Trends in Genetics*, 40(10), pp. 891–908. Available at: https://doi.org/10.1016/j.tig.2024.07.001.

Gill, T., Gill, Simranveer K, *et al.* (2022) 'A Comprehensive Review of High Throughput Phenotyping and Machine Learning for Plant Stress Phenotyping', *Phenomics*, 2(3), pp. 156–183. Available at: https://doi.org/10.1007/s43657-022-00048-z.

Gill, T., Gill, Simranveer K., *et al.* (2022) 'A Comprehensive Review of High Throughput Phenotyping and Machine Learning for Plant Stress Phenotyping', *Phenomics*, 2(3), pp. 156–183. Available at: https://doi.org/10.1007/s43657-022-00048-z.

Grassmann, G. *et al.* (2024) 'Computational Approaches to Predict Protein–Protein Interactions in Crowded Cellular Environments', *Chemical Reviews*, 124(7), pp. 3932–3977. Available at: https://doi.org/10.1021/acs.chemrev.3c00550.

Hempel, C.A. *et al.* (2022) 'Metagenomics versus total RNA sequencing: most accurate data-processing tools, microbial identification accuracy and perspectives for ecological assessments', *Nucleic Acids Research*, 50(16), pp. 9279–9293. Available at: https://doi.org/10.1093/nar/gkac689.

Iqbal, B. *et al.* (2023) 'Advancing environmental sustainability through microbial reprogramming in growth improvement, stress alleviation, and phytoremediation', *Plant Stress*, 10, p. 100283. Available at: https://doi.org/10.1016/j.stress.2023.100283.

Irigoyen, S. *et al.* (2020) 'Plant hairy roots enable high throughput identification of antimicrobials against Candidatus Liberibacter spp.', *Nature Communications*, 11(1), p. 5802. Available at: https://doi.org/10.1038/s41467-020-19631-x.

Jain, A. *et al.* (2024) 'Omics approaches in understanding the benefits of plant-microbe interactions', *Frontiers in Microbiology*, 15. Available at: https://doi.org/10.3389/fmicb.2024.1391059.

Jamil, F. *et al.* (2022) 'Rhizosphere Signaling: Insights into Plant–Rhizomicrobiome Interactions for Sustainable Agronomy', *Microorganisms*, 10(5), p. 899. Available at: https://doi.org/10.3390/microorganisms10050899.

Jurj, S.L., Nouri, S.B. and Strutwolf, J. (2024) 'SNNtrainer3D: Training Spiking Neural Networks Using a User-Friendly Application with 3D Architecture Visualization Capabilities', *Applied Sciences*, 14(13), p. 5752. Available at: https://doi.org/10.3390/app14135752.

Khan, Q. *et al.* (2024) 'An insight into conflict and collaboration between plants and microorganisms', *Chemical and Biological Technologies in Agriculture*, 11(1), p. 161. Available at: https://doi.org/10.1186/s40538-024-00684-9.

Kimotho, R.N. and Maina, S. (2024) 'Unraveling plant–microbe interactions: can integrated omics approaches offer concrete answers?', *Journal of Experimental Botany*. Edited by M. Gifford, 75(5), pp. 1289–1313. Available at: https://doi.org/10.1093/jxb/erad448.

Koza, N. *et al.* (2022) 'Microorganisms in Plant Growth and Development: Roles in Abiotic Stress Tolerance and Secondary Metabolites Secretion', *Microorganisms*, 10(8), p. 1528. Available at: https://doi.org/10.3390/microorganisms10081528.

Kreiss, L. *et al.* (2023) 'Digital staining in optical microscopy using deep learning - a review', *PhotoniX*, 4(1), p. 34. Available at: https://doi.org/10.1186/s43074-023-00113-4.

Kumar, A. and Dubey, A. (2020) 'Rhizosphere microbiome: Engineering bacterial competitiveness for enhancing crop production', *Journal of Advanced Research*, 24, pp. 337–352. Available at: https://doi.org/10.1016/j.jare.2020.04.014.

Kumar, B. *et al.* (2024) 'A comprehensive overview of microbiome data in the light of machine learning applications: categorization, accessibility, and future directions', *Frontiers in Microbiology*, 15. Available at: https://doi.org/10.3389/fmicb.2024.1343572.

Lahlali, R. *et al.* (2022) 'Biological Control of Plant Pathogens: A Global Perspective', *Microorganisms*, 10(3), p. 596. Available at: https://doi.org/10.3390/microorganisms10030596.

Lee, J. *et al.* (2020) 'BioBERT: a pre-trained biomedical language representation model for biomedical text mining', *Bioinformatics*. Edited by J. Wren, 36(4), pp. 1234–1240. Available at: https://doi.org/10.1093/bioinformatics/btz682.

Lei, C. *et al.* (2024) 'AraPathogen2.0: An Improved Prediction of Plant–Pathogen Protein–Protein Interactions Empowered by the Natural Language Processing Technique', *Journal of Proteome Research*, 23(1), pp. 494–499. Available at: https://doi.org/10.1021/acs.jproteome.3c00364.

Li, P. *et al.* (2020) 'The Lifecycle of the Plant Immune System', *Critical Reviews in Plant Sciences*, 39(1), pp. 72–100. Available at: https://doi.org/10.1080/07352689.2020.1757829.

Li, Z. *et al.* (2023) 'Applications of deep learning in understanding gene regulation', *Cell Reports Methods*, 3(1), p. 100384. Available at: https://doi.org/10.1016/j.crmeth.2022.100384.

Lowe, R. *et al.* (2017) 'Transcriptomics technologies', *PLOS Computational Biology*, 13(5), p. e1005457. Available at: https://doi.org/10.1371/journal.pcbi.1005457.

Marshall, C.R. *et al.* (2020) 'Best practices for the analytical validation of clinical whole-genome sequencing intended for the diagnosis of germline disease', *npj Genomic Medicine*, 5(1), p. 47. Available at: https://doi.org/10.1038/s41525-020-00154-9.

Mathew, D.E. *et al.* (2025) 'Recent Emerging Techniques in Explainable Artificial Intelligence to Enhance the Interpretable and Understanding of AI Models for Human', *Neural Processing Letters*, 57(1), p. 16. Available at: https://doi.org/10.1007/s11063-025-11732-2.

Mathieu, A. *et al.* (2022) 'Machine Learning and Deep Learning Applications in Metagenomic Taxonomy and Functional Annotation', *Frontiers in Microbiology*, 13. Available at: https://doi.org/10.3389/fmicb.2022.811495.

Melanthota, S.K. *et al.* (2022) 'Deep learning-based image processing in optical microscopy', *Biophysical Reviews*, 14(2), pp. 463–481. Available at: https://doi.org/10.1007/s12551-022-00949-3.

Monti, M. *et al.* (2022) 'Prediction of Time Series Gene Expression and Structural Analysis of Gene Regulatory Networks Using Recurrent Neural Networks', *Entropy*, 24(2), p. 141. Available at: https://doi.org/10.3390/e24020141.

Mu, H., Wang, B. and Yuan, F. (2022) 'Bioinformatics in Plant Breeding and Research on Disease Resistance', *Plants*, 11(22), p. 3118. Available at: https://doi.org/10.3390/plants11223118.

Murmu, S. *et al.* (2024) 'A review of artificial intelligence-assisted omics techniques in plant defense: current trends and future directions', *Frontiers in Plant Science*, 15. Available at: https://doi.org/10.3389/fpls.2024.1292054.

Nguyen, G. *et al.* (2019) 'Machine Learning and Deep Learning frameworks and libraries for large-scale data mining: a survey', *Artificial Intelligence Review*, 52(1), pp. 77–124. Available at: https://doi.org/10.1007/s10462-018-09679-z.

Pandit, M.A. *et al.* (2022) 'Major Biological Control Strategies for Plant Pathogens', *Pathogens*, 11(2), p. 273. Available at: https://doi.org/10.3390/pathogens11020273.

Pandita, D. *et al.* (2023) 'Mobile Signaling Peptides: Secret Molecular Messengers with a Mighty Role in Plant Life', *Journal of Plant Growth Regulation*, 42(10), pp. 6801–6834. Available at: https://doi.org/10.1007/s00344-023-11069-x.

Paul, N. *et al.* (2025) 'Deep learning for plant stress detection: A comprehensive review of technologies, challenges, and future directions', *Computers and Electronics in Agriculture*, 229, p. 109734. Available at: https://doi.org/10.1016/j.compag.2024.109734.

Pérez-Cobas, A.E., Gomez-Valero, L. and Buchrieser, C. (2020) 'Metagenomic approaches in microbial ecology: an update on whole-genome and marker gene sequencing analyses', *Microbial Genomics*, 6(8). Available at: https://doi.org/10.1099/mgen.0.000409.

Picard, M. *et al.* (2021) 'Integration strategies of multi-omics data for machine learning analysis', *Computational and Structural Biotechnology Journal*, 19, pp. 3735–3746. Available at: https://doi.org/10.1016/j.csbj.2021.06.030.

Przymus, P. *et al.* (2025) 'Deep learning in microbiome analysis: a comprehensive review of neural network models', *Frontiers in Microbiology*, 15. Available at: https://doi.org/10.3389/fmicb.2024.1516667.

Ranjan, A. *et al.* (2022) 'Shaping the root system architecture in plants for adaptation to drought stress', *Physiologia Plantarum*, 174(2). Available at: https://doi.org/10.1111/ppl.13651.

Roy, G. *et al.* (2024) 'Deep learning methods in metagenomics: a review', *Microbial Genomics*, 10(4). Available at: https://doi.org/10.1099/mgen.0.001231.

Sagar, A. *et al.* (2021) 'Plant Growth Promoting Rhizobacteria, Arbuscular Mycorrhizal Fungi and Their Synergistic Interactions to Counteract the Negative Effects of Saline Soil on Agriculture: Key Macromolecules and Mechanisms', *Microorganisms*, 9(7), p. 1491. Available at: https://doi.org/10.3390/microorganisms9071491.

Sehrawat, S., Najafian, K. and Jin, L. (2023) 'Predicting phenotypes from novel genomic markers using deep learning', *Bioinformatics Advances*. Edited by T. Lengauer, 3(1). Available at: https://doi.org/10.1093/bioadv/vbad028.

Sharma, V. *et al.* (2024) 'Unveiling the dynamic relationship of viruses and/or symbiotic bacteria with plant resilience in abiotic stress', *Stress Biology*, 4(1), p. 10. Available at: https://doi.org/10.1007/s44154-023-00126-w.

Shi, T.-T., Li, G.-H. and Zhao, P.-J. (2023) 'Appressoria—Small but Incredibly Powerful Structures in Plant–Pathogen Interactions', *International Journal of Molecular Sciences*, 24(3), p. 2141. Available at: https://doi.org/10.3390/ijms24032141.

Shoaib, M. *et al.* (2023) 'An advanced deep learning models-based plant disease detection: A review of recent research', *Frontiers in Plant Science*, 14. Available at: https://doi.org/10.3389/fpls.2023.1158933.

Sigala, R.E. *et al.* (2023) 'Machine Learning to Advance Human Genome-Wide Association Studies', *Genes*, 15(1), p. 34. Available at: https://doi.org/10.3390/genes15010034.

Singh, A. *et al.* (2024) 'Ensembled transfer learning approach for error reduction in landslide susceptibility mapping of the data scare region', *Scientific Reports*, 14(1), p. 29060. Available at: https://doi.org/10.1038/s41598-024-76541-4.

Slimani, A. *et al.* (2024) 'Molecular and Systems Biology Approaches for Harnessing the Symbiotic Interaction in Mycorrhizal Symbiosis for Grain and Oil Crop Cultivation', *International Journal of Molecular Sciences*, 25(2), p. 912. Available at: https://doi.org/10.3390/ijms25020912.

Sultan, T. *et al.* (2025) 'LeafDNet: Transforming Leaf Disease Diagnosis Through Deep Transfer Learning', *Plant Direct*, 9(2). Available at: https://doi.org/10.1002/pld3.70047.

Suman, A. *et al.* (2022) 'Microbial Community and Function-Based Synthetic Bioinoculants: A Perspective for Sustainable Agriculture', *Frontiers in Microbiology*, 12. Available at: https://doi.org/10.3389/fmicb.2021.805498.

Talaei Khoei, T., Ould Slimane, H. and Kaabouch, N. (2023) 'Deep learning: systematic review, models, challenges, and research directions', *Neural Computing and Applications*, 35(31), pp. 23103–23124. Available at: https://doi.org/10.1007/s00521-023-08957-4.

Tao, Z. *et al.* (2023) 'Regulatory roles of epigenetic modifications in plant-phytopathogen interactions', *Crop Health*, 1(1), p. 6. Available at: https://doi.org/10.1007/s44297-023-00003-y.

Tiwari, P. *et al.* (2024) 'Plant-Microbe Interactions under the Extreme Habitats and Their Potential Applications', *Microorganisms*, 12(3), p. 448. Available at: https://doi.org/10.3390/microorganisms12030448.

Trapotsi, M.-A., Hosseini-Gerami, L. and Bender, A. (2022) 'Computational analyses of mechanism of action (MoA): data, methods and integration', *RSC Chemical Biology*, 3(2), pp. 170–200. Available at: https://doi.org/10.1039/D1CB00069A.

Tyagi, P. *et al.* (2022) 'Upcoming progress of transcriptomics studies on plants: An overview', *Frontiers in Plant Science*, 13. Available at: https://doi.org/10.3389/fpls.2022.1030890.

Upadhyay, A. *et al.* (2025) 'Deep learning and computer vision in plant disease detection: a comprehensive review of techniques, models, and trends in precision agriculture', *Artificial Intelligence Review*, 58(3), p. 92. Available at: https://doi.org/10.1007/s10462-024-11100-x.

Vourlaki, I.-T. *et al.* (2024) 'Evaluation of deep learning for predicting rice traits using structural and single-nucleotide genomic variants', *Plant Methods*, 20(1), p. 121. Available at: https://doi.org/10.1186/s13007-024-01250-y.

Wahab, A. *et al.* (2023) 'Role of Arbuscular Mycorrhizal Fungi in Regulating Growth, Enhancing Productivity, and Potentially Influencing Ecosystems under Abiotic and Biotic Stresses', *Plants*, 12(17), p. 3102. Available at: https://doi.org/10.3390/plants12173102.

Wang, H. *et al.* (2020) 'Deep learning for plant genomics and crop improvement', *Current Opinion in Plant Biology*, 54, pp. 34–41. Available at: https://doi.org/10.1016/j.pbi.2019.12.010.

Wang, X. *et al.* (2021) 'Mycorrhizal symbiosis modulates the rhizosphere microbiota to promote rhizobia–legume symbiosis', *Molecular Plant*, 14(3), pp. 503–516. Available at: https://doi.org/10.1016/j.molp.2020.12.002.

Wang, X., Chi, Y. and Song, S. (2024) 'Important soil microbiota's effects on plants and soils: a comprehensive 30-year systematic literature review', *Frontiers in Microbiology*, 15. Available at: https://doi.org/10.3389/fmicb.2024.1347745.

Wang, You-Yu *et al.* (2023) 'Chemical and Biological Strategies for Profiling Protein-Protein Interactions in Living Cells', *Chemistry – An Asian Journal*, 18(14). Available at: https://doi.org/10.1002/asia.202300226.

Wang, Yuxiao *et al.* (2023) 'Prediction of protein-ligand binding affinity with deep learning', *Computational and Structural Biotechnology Journal*, 21, pp. 5796–5806. Available at: https://doi.org/10.1016/j.csbj.2023.11.009.

Wang, Y., Liu, L. and Wang, C. (2023) 'Trends in using deep learning algorithms in biomedical prediction systems', *Frontiers in Neuroscience*, 17. Available at: https://doi.org/10.3389/fnins.2023.1256351.

Xu, M. *et al.* (2024) 'Plant disease recognition datasets in the age of deep learning: challenges and opportunities', *Frontiers in Plant Science*, 15. Available at: https://doi.org/10.3389/fpls.2024.1452551.

Yamashita, R. *et al.* (2018) 'Convolutional neural networks: an overview and application in radiology', *Insights into Imaging*, 9(4), pp. 611–629. Available at: https://doi.org/10.1007/s13244-018-0639-9.

Yang, L. *et al.* (2024) 'Mechanisms of rhizosphere plant-microbe interactions: molecular insights into microbial colonization', *Frontiers in Plant Science*, 15. Available at: https://doi.org/10.3389/fpls.2024.1491495.

Yasrab, R. *et al.* (2019) 'RootNav 2.0: Deep learning for automatic navigation of complex plant root architectures', *GigaScience*, 8(11). Available at: https://doi.org/10.1093/gigascience/giz123.

Yu, Y. *et al.* (2022) 'Induced Systemic Resistance for Improving Plant Immunity by Beneficial Microbes', *Plants*, 11(3), p. 386. Available at: https://doi.org/10.3390/plants11030386.

Yue, T. *et al.* (2023) 'Deep Learning for Genomics: From Early Neural Nets to Modern Large Language Models', *International Journal of Molecular Sciences*, 24(21), p. 15858. Available at: https://doi.org/10.3390/ijms242115858.

Zehra, A. *et al.* (2021) 'Efficiency of microbial bio-agents as elicitors in plant defense mechanism under biotic stress: A review', *Current Research in Microbial Sciences*, 2, p. 100054. Available at: https://doi.org/10.1016/j.crmicr.2021.100054.

Zhang, X.-M. *et al.* (2021) 'Graph Neural Networks and Their Current Applications in Bioinformatics', *Frontiers in Genetics*, 12. Available at: https://doi.org/10.3389/fgene.2021.690049.

Zhao, C., Onyino, J. and Gao, X. (2024) 'Current Advances in the Functional Diversity and Mechanisms Underlying Endophyte–Plant Interactions', *Microorganisms*, 12(4), p. 779. Available at: https://doi.org/10.3390/microorganisms12040779.

Zheng, D., Liwinski, T. and Elinav, E. (2020) 'Interaction between microbiota and immunity in health and disease', *Cell Research*, 30(6), pp. 492–506. Available at: https://doi.org/10.1038/s41422-020-0332-7.

6 Deep Learning-Enhanced Unmanned Aerial Vehicles: Detection and Classification in Crop Production

Charan Babu Ankela[1], Suthari Vijay Kumar[2], Shreedevi Jinnur[3], K. Aruna[4], K. Ramakrishna[4], A. Jaya Kishore Reddy[1], K. Ramesh[5], Santanu Kundu[6]*, Maksud Hasan Shah[7]*, and Arafat Abdel Hamed Abdel Latef[8]

[1]IARI Mega University, Hyderabad hub, ICAR—Central Research Institute for Dryland Agriculture, Hyderabad, Telangana, India; [2]Uttar Banga Krishi Vishwavidyalaya, Pundibari, Cooch Behar, West Bengal, India; [3]University of Agricultural Sciences, Bengaluru, Karnataka, India; [4]Professor Jayashankar Telangana Agricultural University, Hyderabad, Telangana, India; [5]ICAR—Indian Institute of Oilseeds Research, Hyderabad, Telangana, India; [6]ICAR-Indian Agricultural Research Institute, New Delhi, India; [7]Bidhan Chandra Krishi Viswavidyalaya, Mohanpur, Nadia, West Bengal, India; [8]Department of Plant Physiology, South Valley University, Qena, Egypt

Abstract

Over the past few years, unmanned aerial vehicles (UAVs) have been revolutionizing various sectors. In the context of agriculture, UAVs have been increasingly employed by researchers for diverse applications, especially as remote sensing platforms. These advanced platforms enable efficient scientific data collection, offering significant advantages over traditional methods and satellite imagery. UAVs capture spectral imagery at exceptionally high resolutions on demand, and bypass cloud cover, providing near-real-time data acquisition. As the volume of high-quality data continues to grow, the demand for effective analysis, classification, and meaningful interpretation of agricultural parameters is becoming increasingly critical. Deep learning (DL), an advanced form of artificial intelligence (AI), plays a vital role in this endeavor through its neural networks, which enable the automatic feature extraction and the efficient handling of large datasets. This chapter delves into the DL methodologies, focusing on convolutional neural networks (CNNs), object detection, and image classification techniques applied to UAV imagery in agricultural landscapes.

Keywords: Convolutional neural networks, crop classification, deep learning, unmanned aerial vehicles

6.1 Introduction

Technological advancements over the past decade have revolutionized the agricultural sector (Sharma *et al.*, 2022). At present, artificial intelligence (AI), big data analytics, remote sensing technology, unmanned aerial vehicles (UAVs), and the Internet of Things (IoT) represent some of the most prominent technological trends. These significantly aid in effective and precise monitoring,

*E-mail: santanukundu120@gmail.com or maksudhasanshah@gmail.com

DOI: 10.1079/9781836991120.0006

management, and timely decision-making in agricultural landscapes.

In the context of digital agriculture, data plays a crucial role; therefore, the timely collection and acquisition of high-resolution scientific data are essential. This is where UAVs emerge as effective platforms for scientific data collection in agriculture. UAVs offer several advantages over traditional ground-based methods and satellites: they can fly at low altitudes as needed, follow flexible flight paths, capture very high-resolution imagery in both space and time, bypass cloud cover, and enable cost-effective, rapid data acquisition (Zhang and Zhu, 2023).

These capabilities make UAVs particularly valuable for the precise and timely detection of pests and diseases, crop classification, and crop area estimation applications that are highly beneficial not only for farmers but also for agro-based industries, government agencies, and policymakers. Moreover, these capabilities are critical for policymaking, market regulation, resource optimization, productivity improvement, and ensuring food security at a larger scale (Karmakar *et al.*, 2023). Although conventional methods, such as manual data collection and crop-cutting experiments, can be effective to some extent, they are labor-intensive, time-consuming, expensive, and often unable to represent large areas accurately (Liu *et al.*, 2020; Feng *et al.*, 2023). Furthermore, geographic information system (GIS) software tools (e.g. ERDAS, ArcGIS, and QGIS) are powerful, but they require specialized expertise, are time-consuming, and struggle to handle large volumes of data efficiently. While traditional machine learning (ML) techniques, such as random forests (RFs) and support vector machines (SVMs), have been popularly used for image analysis, a more advanced subset of ML, deep learning (DL), has opened new possibilities with exceptional capabilities in image processing and analysis (Machichi *et al.*, 2023). Among these, convolutional neural networks (CNNs) have displayed superior performance in recognizing complex visual patterns, enabling scalable, automated, and highly accurate detection and classification tasks in agriculture (Shen *et al.*, 2020).

This chapter explores the application of DL techniques to UAV-based imagery for agricultural crop monitoring, detection, and classification. It emphasizes the role of CNNs and reviews prominent object detection and image classification frameworks relevant to agricultural practices.

6.2 Unmanned Aerial Vehicles (UAVs)

UAVs are one of the most rapidly evolving technologies in modern agriculture. UAVs have become widely discussed and are now a popular subject of research, with significant market growth trends observed day by day. Commonly known as drones, UAVs are airborne vehicles that operate independently without a pilot onboard. Although initially developed for military and surveillance purposes (Makam *et al.*, 2024), UAVs have emerged as essential tools across various sectors due to their wide range of applications. Often referred to as “mechanical birds,” they use rotors to hover, ascend, and navigate (Laghari *et al.*, 2022).

Technological advancements—particularly in robotics, AI, and electronics—have tremendously enhanced the design and functionality of UAVs, transforming them into more sophisticated tools for various applications. Nowadays, UAVs are increasingly being adopted for critical tasks in the fields of agriculture, geoscience, disaster management, and several other sectors (Nex *et al.*, 2022). Being versatile instruments, UAVs are designed to carry a range of remote sensing sensors—and even multiple instruments at once without compromising flight autonomy (Katsigiannis *et al.*, 2016; Lambertini *et al.*, 2022). Their ability to operate under unfavorable weather conditions, fly on demand, follow flexible flight paths, capture very high-resolution imagery, and acquire data quickly—combined with their relatively cost-effective nature—makes UAVs a better choice over ground-based instruments and satellites for data collection.

In agriculture, UAVs have been widely employed for a range of applications, such as crop monitoring, weed mapping, soil mapping, stress detection, yield prediction, drought assessment, and pest and disease monitoring (Kim *et al.*, 2019; Toscano *et al.*, 2024). Additionally, they are used for various field operations such as pesticide spraying, fertilizer application, and seed planting (Hafeez *et al.*, 2022; Adithya *et al.*, 2024).

Various UAV platforms and sensor systems are discussed in the following sections.

6.2.1 Types of UAV platforms

Based on the design and flight mechanics, UAV platforms are broadly classified into four main categories, as discussed in the following section.

Fixed-wing UAVs

Fixed-wing UAVs are similar to conventional airplanes in both structure and function. They are characterized by stationary wings that generate aerodynamic lift, which enables takeoff from the ground and support for longer flight durations at higher speeds. These platforms are typically suitable for covering large areas, making them ideal for wide-area aerial mapping and spraying (Kim *et al.*, 2019). However, fixed-wing UAVs require a runway or launch system for takeoff and landing, and they are unable to hover or maneuver in tight spaces—these are the trade-offs for their extended range and endurance.

Rotary-wing UAVs

These are the most commonly used drone platforms and are further divided into helicopter-type and multirotor UAVs.

1. Helicopter-type UAVs utilize a single large propeller for lift and a smaller tail rotor for directional control. These UAVs offer higher payload capacity and better power efficiency, making them suitable for agricultural spraying and aerial photography. However, their mechanical complexity often results in higher vibrations and requires regular maintenance.
2. Multirotor UAVs (quad-, hexa-, and octocopters) have gained more popularity due to their ease of use and precise control. Their vertical takeoff and landing (VTOL) capabilities eliminate the need for runways and allow operation in uneven terrain or remote areas. Multirotor UAVs can hover, adjust altitude quickly, and perform intricate maneuvers, making them ideal for close-range sensing, monitoring, and mapping. These UAVs control lift and movement through brushless motors that independently regulate the rotation speed of each rotor. This enables rapid stabilization in response to environmental disturbances. They are often equipped with additional features such as obstacle detection systems. However, multirotor UAVs generally have lower power efficiency and shorter flight durations (Zhang and Zhu, 2023). The payload capacity of these platforms varies with the number of rotors—octocopters and helicopters can carry up to 9.5 kg, while quadcopters and hexacopters typically support between 1.25 and 2.6 kg. Therefore, larger models are preferred for spraying operations, while smaller variants are ideal for reconnaissance and precision mapping.

Hybrid UAVs (VTOL)

Hybrid UAVs, or VTOL drones, offer the combined advantages of fixed-wing and rotary-wing systems. These UAVs take off and land using a multirotor mode and switch to fixed-wing mode for forward flight. This dual-mode operation provides the best of both worlds: precise, runway-free takeoffs and landings, coupled with extended flight durations and improved energy efficiency during long-range missions.

The key features of the different types of UAV platforms are provided in Table 6.1. The choice of UAV platform depends on the operational requirements—whether high-endurance coverage, precise hovering for data collection, or adaptability in a challenging terrain.

6.2.2 Types of sensors

The effectiveness of UAVs in agricultural data collection operations is largely determined by the types of sensors they carry onboard. A variety of imaging and non-imaging sensors are available to capture spatial information from the surface across the electromagnetic spectrum. Generally, imaging sensors collect visual and spectral data, while 3D sensors capture spatial depth and the elevation of surface objects. The following section discusses the common types of sensors.

Red, green, and blue (RGB) cameras

These cameras operate in the visible range (blue, red, and green) of the light spectrum. They are primarily designed to mimic the human eye in the digital world. Due to the high energy levels

Table 6.1. Key features of different types of drones. Makam *et al.* (2024) CC BY 4.0.

Parameters	Fixed-wing	Single-rotor	Multirotor
Weight	Heavy	Heavy	Light
Payload capacity	2–15 kg	2–25 kg	0.2–2 kg
Flight time	1 h or more	1 h or more	20–45 min
Hovering capabilities	Cannot hover	Can hover	Can hover
Expertise required	Hard to fly	Very hard to fly	Easy to use
Key applications	Command area survey and mapping	Sowing, spraying, and harvesting	Surveillance, photography, and inspection, spraying operations in small areas

in the visible region, these cameras offer higher resolution compared to other sensor types. RGB cameras are widely used in applications such as vegetation mapping, land-use classification, and environmental monitoring. They are especially useful in situations where the visual appearance of the target is crucial, for instance, monitoring changes in crop coloration due to biotic or abiotic stress and detecting visible symptoms caused by pests and diseases.

UAVs may carry either of the following:

1. *Integrated RGB cameras.* These are lightweight, compact, and optimized for endurance.
2. *Digital single-lens reflex cameras (DSLR/SLR).* These cameras allow for interchangeable lenses with different focal lengths and can capture both close-up and wide-angle images but require larger UAVs with sufficient payload capacity.

Multispectral cameras

Multispectral cameras capture reflectance data across a limited number of discrete, predefined spectral bands commonly including red edge (RE), near-infrared (NIR), and shortwave infrared (SWIR). These bands are highly sensitive to vegetation characteristics, such as the content of the chlorophyll, water status, and leaf anatomy. These sensors are widely employed in agricultural monitoring due to their sensitivity and cost-effectiveness. Multispectral imagery is broadly applied in the following:

1. crop classification;
2. biomass and yield estimation;
3. mapping Leaf Area Index (LAI);
4. drought and stress detection;
5. vegetation indices (e.g. NDVI and NDRE); and
6. land use and land cover (LULC) assessment.

Hyperspectral cameras

Hyperspectral cameras operate across hundreds of narrow, contiguous spectral bands, capturing detailed spectral signatures of the vegetation. This fine spectral resolution enables the detection of subtle variations in plant physiology and composition that may be overlooked by multispectral sensors. While hyperspectral sensors offer rich and comprehensive information, they are typically more complex, data-intensive, and expensive compared to multispectral cameras.

Hyperspectral imagery is widely used in the following:

1. precise mapping of biophysical and biochemical parameters (e.g. chlorophyll, nitrogen, and moisture);
2. early detection of plant stress and diseases;
3. species-level vegetation classification;
4. crop quality assessment;
5. plant phenotyping—non-destructive, high-throughput analysis of growth stages, nutrient uptake, stress tolerance, and other characteristics; and
6. trait characterization.

Thermal cameras

Thermal sensors cover the long-wave infrared (LWIR) spectrum, typically within the 8–14 μm range. Unlike optical sensors, thermal cameras capture the emittance from the surface, enabling them to measure temperature anomalies with high sensitivity. In agricultural monitoring, canopy temperature and land surface temperature (LST) serve as indirect indicators of plant health, water stress, and soil moisture. Thermal imagery is particularly valuable for the following:

1. the detection of plant water stress before visible symptoms appear;
2. mapping evapotranspiration;
3. optimizing irrigation schedules;
4. monitoring canopy temperature;
5. evaluating drought impact through temperature anomalies over time; and
6. high-throughput phenotyping, especially for traits related to water-use efficiency.

In addition to the imaging sensors discussed earlier, 3D sensors, including LiDAR, multicamera arrays, and millimeter-wave radar, provide valuable spatial depth and elevation data, enabling precise modeling of terrain and structures. LiDAR uses laser pulses to generate high-resolution 3D maps, which are particularly useful for terrain mapping and vegetation analysis. Multicamera arrays allow for robust 3D reconstruction through photogrammetry, while millimeter-wave radar delivers accurate distance and velocity measurements, even under challenging weather conditions.

6.2.3 Data preprocessing

UAV data preprocessing is essential for enhancing drone-captured images, ensuring they are accurate and ready for detailed analysis. The process involves several key techniques:

1. Noise reduction, which removes unwanted elements caused by atmospheric conditions, sensor limitations, or motion blur, thereby improving image clarity.
2. Geometric correction (or orthorectification), which eliminates distortions resulting from camera angles and the Earth's curvature, ensuring spatial accuracy.
3. Mosaic generation, which stitches overlapping images together to create a seamless and comprehensive view of the area.
4. Enhanced feature extraction, which makes faint or low-contrast details more visible, aids in accurate object detection.

Various software tools are available for preprocessing, and the integration of ML and DL algorithms has significantly improved the speed and efficiency of these processes.

6.3 DL—Concept and Architecture

DL, introduced by Hinton *et al.* (2006), is a subset of ML, which itself comes under the broader domain of AI. Within this hierarchy, DL represents an advanced AI function that simulates human cognitive processes to analyze data, recognize patterns, and make informed decisions. DL algorithms are inspired by the human brain's neural architecture and are implemented through artificial neural networks (ANNs). While both ML and DL aim to enable machines to learn from data, they differ significantly in their approach. Traditional ML typically involves manual feature engineering and performs best with structured datasets. In contrast, DL automatically learns hierarchical features directly from raw data—such as images, audio, or text—removing the need for manual extraction. Although DL models often require longer training times due to their complexity and dense parameter configuration, they typically perform faster and more efficiently during the testing or inference phase (Xin *et al.*, 2018). This performance advantage has driven rapid advancements in fields such as computer vision, object detection, image classification, and semantic segmentation. DL models are especially effective at identifying intricate patterns—such as edges, color gradients, textures, and shapes—that may be missed by traditional algorithms or even human experts. In agriculture, the ability to process aerial imagery from drones, satellites, or field cameras has made detection and classification tasks far more efficient. As a result, DL assumes a critical role in modern agriculture.

6.3.1 Artificial neural networks (ANNs)

DL relies on ANNs, which are modeled after the behavior of neurons in the human brain. A standard neural network is made up of three primary layers:

1. *Input layer.* Receives raw data inputs (e.g. pixel values from an image).

2. *Hidden layer.* Performs computations for feature extraction.
3. *Output layer.* Final prediction or classification.

Each connection between neurons is given a weight, which is modified during training to minimize the discrepancy between predicted and actual outputs. Non-linear activation functions, such as ReLU, Sigmoid, or SoftMax, are employed to capture complex, non-linear patterns in the data.

The architecture of a neural network includes several key components, each playing a vital role in its learning and classification processes.

6.3.2 Learning process

The training process involves two key steps:

1. *Forward propagation.* The observed data is passed through the network to generate the output.
2. *Back propagation.* The error is determined, and optimization algorithms such as stochastic gradient descent (SGD) or Adam are used to adjust the weights.

Through repeated iterations, the model progressively refines its ability to make predictions. After training, it can generate quick and accurate predictions on new, unseen data.

6.4 DL—Detection and Classification in Crop Production

Detection and classification tasks using spatial data play a vital role in agriculture. Traditional methods involving GIS tools and ML algorithms, such as RF, SVMs, and k-nearest neighbors (k-NNs), often require significant human intervention and are time-intensive, particularly when processing extensive datasets.

In contrast, DL algorithms offer automated feature recognition with minimal or no human input. They are capable of processing vast amounts of data efficiently while achieving high levels of accuracy. This makes DL techniques particularly valuable for a range of agricultural applications, including the following:

1. crop detection;
2. weed detection;
3. pest and disease identification;
4. fruit detection in orchards; and
5. crop classification.

Among the various DL architectures, CNNs are especially designed for computer vision and image processing tasks. Their capability to automatically capture spatial hierarchies of features makes them ideal for the detection and classification of objects in UAV-based imagery.

This section provides a detailed discussion on CNNs along with other key DL techniques used for detection and classification in crop production.

6.4.1 Convolutional neural network (CNN)

CNNs or ConvNets (LeCun *et al.*, 1998) are a distinct category of DL models specifically developed for handling and interpreting visual data (O'Shea and Nash, 2015; Gu *et al.*, 2017; Dai, 2021). They are particularly well suited for detection and classification tasks involving large and complex image datasets, such as crop monitoring, weed identification, pest detection, and disease classification in agricultural applications.

These are feedforward neural networks that utilize convolutional operations to extract hierarchical features from input data. CNNs are discriminative DL architectures, which learn directly from raw inputs, bypassing the need for manual feature extraction (Sarker, 2021). CNNs are specifically designed to handle various 2D shapes and structures in image data. Each layer within a CNN is responsible for learning optimal parameters that contribute to meaningful feature extraction while simultaneously reducing model complexity (Fig. 6.1). The basic components of CNNs include convolution layers, pooling layers, activation layers, fully connected layers, and dropout layers.

Convolutional layer

The convolutional layer is the foundational block of CNNs. It applies a set of learnable filters (or kernels) that move across the input image to produce feature maps. These feature maps highlight essential local features, such as shape, edges, texture, and so on, which are crucial for distinguishing different objects in the image.

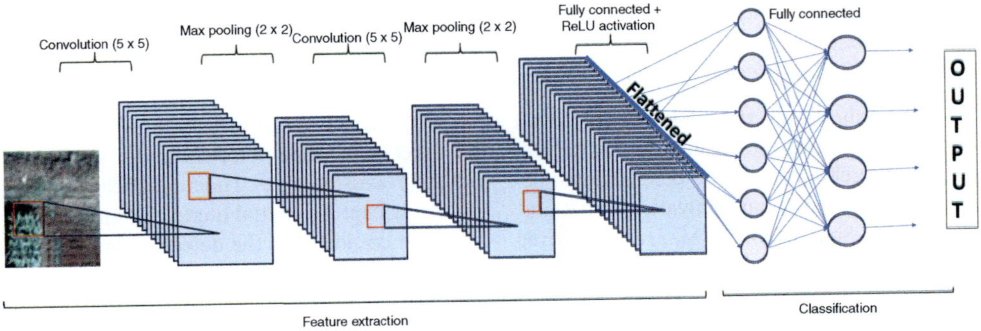

Fig. 6.1. Structure of a convolution neural network. (Author's own image.)

In precision agriculture, the convolutional layer enables models to detect subtle visual cues, such as early signs of disease, pest infestation, or differences between crop species.

Key features:

1. Filters use the same weights across the input, reducing the number of trainable parameters and increasing model efficiency.
2. Each neuron connects only to a localized portion of the input, allowing the model to capture the spatial relationships between adjacent pixels.
3. The network can detect objects irrespective of their position in the image, which is important for detection tasks in agriculture.

Pooling layer

The pooling layer follows the convolutional layers and is used to downscale the feature maps, which often contain large amounts of data. This process reduces the spatial scale of the data while preserving essential information, making the model computationally efficient and robust to small variations in the input, all while decreasing complexity. Pooling helps generalize the features extracted during convolution and reduces the risk of overfitting.

In agricultural applications, pooling enhances the model's proficiency in recognizing patterns consistently under varying lighting conditions, scales, and perspectives such as those found in aerial or ground-level imagery.

The pooling techniques commonly utilized are as follows:

1. *Max pooling.* Extracts the maximum value from each section of the feature map, retaining the most prominent features.
2. *Average pooling.* Averages the values within each section, often used to reduce noise and smooth the data.

Activation layer

The activation layer introduces non-linearity into the CNN, enabling it to learn complex patterns. The most commonly used activation function in CNNs is the rectified linear unit (ReLU), given as the following:

$$f(x) = \max\{0, x\}.$$

This function hastens the training by alleviating the vanishing gradient problem and helps the network converge faster.

In classification tasks, especially at the output layer, a SoftMax activation function is often used to convert the final scores into class probabilities.

Fully connected layer

Following convolutional and pooling layers, the CNN transitions into one or more fully connected layers, also called as dense layers. These layers flatten the extracted feature maps into a single vector, serving as input for classification or detection tasks.

Key functions:

1. Combines features learned in previous layers to make predictions.
2. Assigns the final class label based on the extracted features (e.g. diseased vs healthy leaf and weed vs crop).

Although fully connected layers contain many parameters, they enable the network to make informed decisions by considering all the extracted features.

Dropout layer

To mitigate the problem of overfitting, CNNs often incorporate dropout layers. By randomly deactivating a portion of neurons during training, this technique encourages the network to learn generalized patterns instead of memorizing the training data.

In scenarios with limited labeled datasets, which are most common in agriculture, dropout provides a natural regularization mechanism and improves the model's performance on unseen data.

Object detection

Object detection plays a pivotal role in agricultural applications, supporting critical tasks such as weed identification, disease recognition, stress detection, and pest damage assessment. As a foundational aspect of computer vision, object detection is integral to tackling more complex vision problems like image segmentation (Minaee *et al.*, 2021), object tracking (Luo *et al.*, 2014), and behavior recognition (Hu *et al.*, 2022).

At its core, the object detection process typically involves two main stages:

1. *Localization:* pinpointing the potential locations of target objects within an image.
2. *Classification:* categorizing the detected objects into predefined classes.

Unlike general image classification, which processes the entire image holistically, object detection narrows its focus to specific regions containing relevant features. This localized attention allows for more accurate recognition of multiple objects within a single frame, which is especially beneficial in agricultural environments where scenes may contain complex, cluttered backgrounds and diverse objects.

Significant advancements in this field began with the resurgence of DL, particularly following the success of CNNs in the 2012 ImageNet Challenge. This breakthrough sparked a wave of research and development in DL-based object detection, greatly enhancing detection accuracy and computational efficiency.

With continued in-depth exploration by researchers, a wide array of CNN-based object detection algorithms has emerged. These algorithms generally follow two main technical paths:

1. *Two-stage detectors (region-based).* This approach was developed earlier, and it involves a two-step process. First, a set of candidate regions is generated, often through methods such as sliding windows or region proposal networks. These regions are then further refined by classifying objects and adjusting their boundary boxes. Although this approach tends to be computationally intensive and slower in training, it offers high accuracy. Some of the known models in this category include R-CNN, fast R-CNN, and faster R-CNN.
2. *One-stage detectors (regression-based).* Developed later, this method bypasses the region proposal stage and performs detection in a single step. By directly generating object boundary boxes and class labels from the input image, they achieve rapid detection without major compromises in accuracy. Representative algorithms include the You Only Look Once (YOLO) series and Single Shot MultiBox Detector (SSD).

The choice between one-stage and two-stage detectors depends on the specific demands of the task, whether it is the need for real-time detection in drone-based field monitoring or high-accuracy analysis for disease diagnosis and so on.

WEED DETECTION. Weed detection stands out as a prominent and impactful application of DL in agriculture. Weeds are invasive and undesirable plant species that compete with crops for critical resources such as sunlight, moisture, and nutrients, and certain types may even be toxic. Therefore, the ability to accurately distinguish weeds from crops in agricultural settings is crucial for optimizing crop health and yield. Currently, DL models—particularly CNNs—have proven highly effective in automating weed identification across various stages of crop growth. Among the advanced techniques, object detection algorithms have been widely adopted for weed classification and localization. These models enable precise herbicide application, reducing the overuse of chemicals and promoting environmentally sustainable smart farming practices. The leading DL frameworks utilized for this task include YOLO series, Vision Transformers, U-Net,

VGG16, and ResNet101, among others (Zhang *et al.*, 2024).

A study by Ong *et al.* (2023) deployed CNN models on UAV-based aerial imagery captured using 20-MP cameras. They achieved a commendable accuracy of 92%, outperforming the RF classifier, and showed greater resilience to class imbalance in the dataset. Similarly, Haq (2022) conducted a comparative analysis using RGB images taken from UAVs flying at a height of 4 m above ground level (AGL). The results indicated that CNNs consistently outperformed traditional ML algorithms such as SVM, decision trees (DT), RF, and AdaBoost, achieving a peak accuracy of 99%.

Gallo *et al.* (2023) introduced a novel dataset known as the Chicory Plant (CP) dataset, which consists of over 3000 RGB images captured using UAVs during various growth stages of weeds and crops. This dataset features 12,113 manually annotated bounding boxes, specifically labeled to identify *Mercurialis annua*, a common weed species found in chicory plantations. To examine the performance of contemporary object detection algorithms, several state-of-the-art DL models were tested, among which YOLOv7 demonstrated the highest effectiveness. On the CP dataset, YOLOv7 achieved notable results, recording a mean average precision at 0.5 (mAP@0.5) intersection over union (IoU) of 56.6%, alongside a recall of 62.1% and a precision of 61.3%. To further validate the robustness and generalizability of the model, YOLOv7 was also applied to the publicly available Lincoln Beet (LB) dataset, which had previously been analyzed using earlier versions of the YOLO algorithm. Remarkably, YOLOv7 demonstrated substantial improvements across all evaluation metrics—raising the total mAP from 51% to 61%, the weed-specific mAP from 67.5% to 74.1%, and the sugar beet-specific mAP from 34.6% to 48%.

In addition to RGB and multispectral imagery, vegetation indices have proven to be valuable tools for improving weed detection accuracy. An example of this is the study by Osorio *et al.* (2020), who employed multispectral UAV imagery that included green, red, RE, and NIR bands. To further enhance discrimination between vegetation and soil background, the normalized difference vegetation index (NDVI) was integrated. In a comparative performance analysis, YOLOv3 and Mask R-CNN significantly outperformed SVM, achieving F1-scores as high as 94%.

DISEASE DETECTION. Pests and diseases pose serious threats to agricultural productivity, making timely and accurate detection essential for effective management. Thanks to advancements in DL, especially CNNs, disease detection has become both more accurate and efficient.

Wu *et al.* (2024) utilized drone-based remote sensing and an improved YOLOv3 model optimized with the CIoU loss function for detecting tree diseases and pests in pine forests. The CIoU loss, which accounts for overlapping area, center distance, and aspect ratio consistency, enhanced both accuracy and speed. The model achieved over 95% accuracy with an average processing time of less than 0.5 s per image, demonstrating its potential for forest inspection and protection. Similarly, the YOLOv4 algorithm showed impressive results in plant disease detection, recording an accuracy of 99.99% on the Plant Village dataset (Aldakheel *et al.*, 2024). Liang *et al.* (2022) proposed an Enhanced CenterNet (ECenterNet), an anchor-free detector that improved detection accuracy without increasing computational load. Trained on 5281 tiles and tested on 3842 images, ECenterNet achieved a mAP of 54.7% under COCO Challenge metrics, compared to 49.8% from the original CenterNet model.

Sangaiah *et al.* (2024) have developed the UAV T-YOLO-Rice network using Tiny YOLOv4 as the base detector, enhanced with modules such as spatial pyramid pooling (SPP), convolutional block attention module (CBAM), sand clock feature extraction module (SCFEM), ghost modules, and additional convolutional layers. Using an augmented dataset of rice leaf diseases (bacterial leaf blight, rice blast, and brown spot), the model achieved a mAP of 86%, outperforming previous models, including YOLOv7.

Image classification

Image classification involves the process of distinguishing and categorizing various object classes within visual data (Chandra and Bedi, 2018). While traditional classification algorithms can perform adequately on simple datasets, their effectiveness diminishes in more complex scenarios that require nuanced feature discrimination. On the other hand, CNNs excel in image classification tasks, outperforming other models. CNNs utilize convolutional kernels to extract spatial and contextual features, significantly improving model generalization across large and complex datasets.

As DL technologies advance, CNN-based image classification architectures have seen continuous development and refinement (Panda *et al.*, 2021). Notable models, such as ConNeXt, GoogLeNet, AlexNet, and ResNet, have shown strong performance across a wide range of image classification benchmarks. These models differ in-depth, architecture design, and parameter optimization strategies but consistently outperform traditional methods, especially in large-scale and high-dimensional image analysis tasks.

CROP CLASSIFICATION. CNNs have emerged as a potential tool for crop classification, playing a vital role in monitoring land use, assessing crop rotations, and supporting crop planning. Crop classification encompasses the identification and categorization of various crop types, typically based on remote sensing imagery.

For instance, Galodha *et al.* (2022) utilized a UAV equipped with a terrestrial hyperspectral spectroradiometer to acquire aerial imagery of different crop types. Their study evaluated the performance of CNN architectures with varying depths and kernel sizes. They reported that CNNs comprising three or five layers, when applied with 7 × 7 convolutional kernels, achieved a comparable classification accuracy of approximately 87%. In another study, Kwak *et al.* (2021) introduced a hybrid approach that combines CNNs with RF classifiers for early-stage crop mapping using limited training data. The hybrid CNN–RF model outperformed standalone CNNs and RFs, demonstrating improved feature extraction and classification capabilities by leveraging the complementary strengths of both algorithms.

Similarly, Pandey and Jain (2021) proposed an intelligent crop classification system based on CNNs using UAV-acquired RGB imagery, and several other studies have successfully demonstrated the application of CNNs for crop classification using UAV-based imagery (Agilandeeswari *et al.*, 2022; Xi-hong *et al.*, 2025).

In essence, the integration of CNNs with UAV-based remote sensing data, particularly multispectral and hyperspectral imagery, has significantly advanced the precision and reliability of crop classification.

General workflow of CNN-based models using UAV imagery

The application of CNNs for detection and classification in crop production using UAV imagery follows a structured pipeline, as shown in Fig. 6.2, comprising the following key stages.

DATA ACQUISITION. UAVs equipped with high-resolution sensors (RGB, multispectral, hyperspectral, or thermal cameras) capture aerial imagery with high resolution. Flight planning is critical to data quality and involves considerations such as altitude, speed, image overlap (typically 70–90%), ground sampling distance (GSD), and lighting conditions. Optimal flights are usually conducted mid-morning to early afternoon to minimize shadows and lighting variations.

DATASET PREPARATION AND AUGMENTATION. Raw UAV data undergoes preprocessing steps, including orthomosaic generation, georeferencing, noise reduction, radiometric calibration, and normalization. The dataset is then prepared as follows:

1. *Data splitting*. Typically, 70% training, 15% validation, and 15% testing.
2. *Augmentation*: Flipping, rotation, and scaling techniques are used to augment the data diversity and enhance model robustness.
3. *Image tiling*. Large images are broken down into smaller patches (e.g. 256 × 256 or 512 × 512 pixels), as per the required input size of the CNN models.

CNN MODEL TRAINING AND FEATURE EXTRACTION. Once prepared, the dataset is fed into CNN models for feature learning and classification:

1. *Feature extraction*: CNNs automatically extract features through convolutional, activation (e.g. ReLU), and pooling layers, capturing both low-level and abstract image features.
2. *Model selection*:
   - Classification. VGGNet, ResNet, DenseNet, and EfficientNet.
   - Object Detection. YOLO, SSD, and Faster R-CNN.

3. *Training Process*: The model applies iterations to minimize prediction error using loss functions (e.g. cross-entropy and IoU loss) and optimizers (e.g. Adam and SGD).

Since the collection and labeling of large datasets are a complex task, many studies utilize publicly available benchmark datasets for training purposes (Table 6.2).

MODEL EVALUATION. Various metrics are used to evaluate the model's performance:

1. *Classification*. Recall, accuracy, F1-score, and precision.
2. *Object Detection*. mAP and IoU.

6.5 Challenges and Limitations

While the integration of DL models for UAV imagery analysis in agriculture offers several advantages, it is also essential to be mindful of the associated challenges and limitations for responsible application and progressive research.

1. Despite the increased use of UAVs as remote sensing platforms, most of them have limitations in their payload capacity and endurance.
2. Images captured by UAVs are often affected by various weather factors, illumination, shadow, occlusion, and so on, especially in diversified agricultural settings.

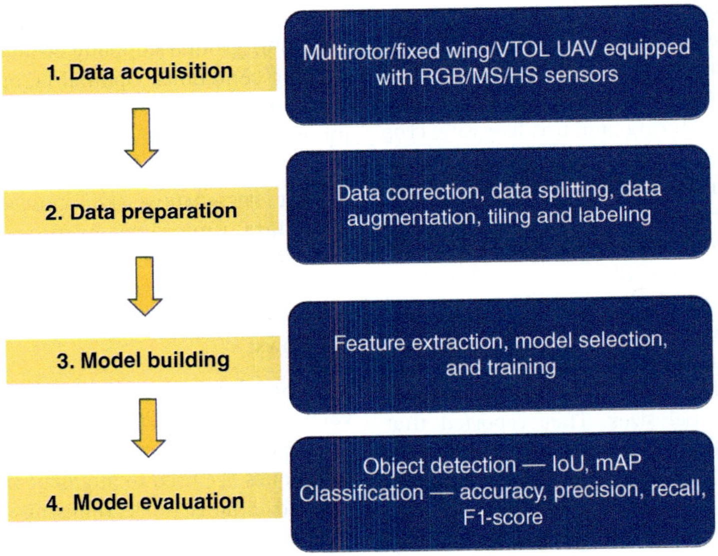

Fig. 6.2. General workflow of CNN-based models using UAV imagery for object detection and classification in agriculture. CNNs, convolutional neural networks; UAVs, unmanned aerial vehicles (Author's own image.)

Table 6.2. Some examples of publicly available benchmark datasets. Author's own table.

Dataset	Purpose	Reference	Available at
WHU-Hi dataset	Crop classification	Zhong *et al.* (2020)	http://rsidea.whu.edu.cn/e-resource_WHUHi_sharing.htm
Weed25	Weed detection	Wang *et al.* (2022)	https://pan.baidu.com/s/1rnUoDm7lxxmX1n1LmtXNXw (Password: rn5h)
CoFly-WeedDB	Weed detection and species identification	Krestenitis *et al.* (2022)	https://zenodo.org/record/6697343#.YrQpwHhByV4
MH-SoyaHealthVision	Disease detection and classification	Shinde and Attar (2025)	https://data.mendeley.com/datasets/hkbgh5s3b7/1
WeedsGalore	Weed detection and classification	Celikkon *et al.* (2025)	https://github.com/GFZ/weedsgalore

3. It is crucial to capture imagery during the appropriate stage to ensure that it is representative and suitable for analysis.
4. DL models require a large volume of data for training. However, in agricultural settings, data collection and data labeling are highly challenging, given the dynamic nature of the diversified and complex environments.
5. Training DL models is a time-consuming process, owing to the higher number of parameters in the model compared to traditional ML models.
6. Powerful processing hardware, preferably graphics processing units (GPUs) over central processing units (CPUs), is often required for efficient handling of large and complex datasets.
7. Small changes in the hyperparameters (such as stride and kernel) can significantly influence models' performance. Thus, the selection of appropriate hyperparameters is crucial, and it requires considerable expertise and domain knowledge.
8. Overfitting is the most common issue in DL models, where the model becomes excessively specialized to the training data and fails to generalize to new data.
9. DL models cannot register qualitative relationships among the variables; they work on numbers and their statistical regularities (Hall *et al.*, 2019). Thus, they are often referred to as "black boxes," as it is difficult to interpret and verify how they arrive at specific outputs. In contrast, ML models are typically based on rule-based logic.
10. DL models still lack strong theoretical and mathematical foundations (Zhao *et al.*, 2024).

6.6 Conclusion

UAVs are emerging as powerful tools with a wide range of applications in agriculture. As remote sensing platforms, UAVs have substantially simplified the task of data collection and enhanced the availability of high-quality spatial data at relatively low costs. DL, an advanced form of AI, has made significant strides in the processing and analysis of UAV-based aerial imagery. In the context of digital agriculture, the integration of UAVs with DL algorithms has proved to be a timely breakthrough, enabling efficient crop monitoring, detection, and classification of pests, diseases, weeds, crop types, tree species, and so on. Despite these advancements, research to date has been relatively limited, especially with regard to food grain crops. Moving forward, the seamless integration of technological innovations in UAVs and DL methods holds tremendous potential for detection and classification tasks in agriculture. By strengthening both the theoretical foundations and practical applications of DL methods in these tasks, we can unlock new avenues for intelligent, digital, smart, and sustainable farming practices.

References

Adithya, K., Ramesh, T., Rathika, S., Vanniarajan, C. and Kalimuthu, R. (2024) Exploring the feasibility of using drones for seeding of rice (*Oryza sativa*. L.). *Plant Science Today* 11 (sp4). https://doi.org/10.14719/pst.4826

Agilandeeswari, L., Manoharan, P., Radhesyam, V., Phaneendra, K. L. N. B. and Farhan, A. (2022) Crop classification for agricultural applications in hyperspectral remote sensing images. *Applied Sciences* 12(3), 1670. https://doi.org/10.3390/app12031670

Aldakheel, E. A., Zakariah, M. and Alabdalall, A. H. (2024) Detection and identification of plant leaf diseases using YOLOv4. *Frontiers in Plant Science* 15. https://doi.org/10.3389/fpls.2024.1355941

Celikkan, E., Kunzmann, T., Yeskaliyev, Y., Itzerott, S., Klein, N. and Herold, M. (2025) WeedsGalore: A multispectral and multitemporal UAV-based dataset for crop and weed segmentation in agricultural maize fields. *arXiv*. https://doi.org/10.48550/arxiv.2502.13103

Chandra, M. A. and Bedi, S. S. (2018) Survey on SVM and their application in image classification. *International Journal of Information Technology* 13(5), 1. https://doi.org/10.1007/s41870-017-0080-1

Dai, D. (2021) An introduction of CNN: models and training on neural network models. In: *International Conference on Big Data, Artificial Intelligence and Risk Management (ICBAR)* Shanghai, China, 2021, pp. 135-138. https://doi.org/10.1109/icbar55169.2021.00037

Feng, F., Gao, M., Liu, R., Yao, S. and Yang, G. (2023) A deep learning framework for crop mapping with reconstructed Sentinel-2 time series images. *Computers and Electronics in Agriculture* 213, 108227. https://doi.org/10.1016/j.compag.2023.108227

Gallo, I., Rehman, A. U., Dehkordi, R. H., Landro, N., Grassa, R. L. and Boschetti, M. (2023) Deep object detection of crop weeds: performance of YOLOv7 on a real case dataset from UAV Images. *Remote Sensing* 15(2), 539. https://doi.org/10.3390/rs15020539

Galodha, A., Vashisht, R., Nidamanuri, Rama R. and Ramiya, A. M. (2022) Convolutional Neural Network (CNN) for crop - classification of drone acquired hyperspectral imagery. In: *IGARSS 2022 - 2022 IEEE International Geoscience and Remote Sensing Symposium* 16, 7741. https://doi.org/10.1109/igarss46834.2022.9883141

Gu, J., Wang, Z., Kuen, J., Ma, L., Shahroudy, A., Shuai, B., Liu, T., Wang, X., Wang, G., Cai, J. and Chen, T. (2017) Recent advances in convolutional neural networks. *Pattern Recognition* 77, 354. https://doi.org/10.1016/j.patcog.2017.10.013

Hafeez, A., Husain, M. A., Singh, S., Chauhan, A., Khan, M. T., Kumar, N., Chauhan, A. and Soni, S. K. (2022) Implementation of drone technology for farm monitoring and pesticide spraying: A review. *Information Processing in Agriculture* 10(2), 192. https://doi.org/10.1016/j.inpa.2022.02.002

Hall, M., Harborne, D., Tomsett, R., Galetic, V., Quintana-Amate, S., Nottle, A. and Preece, A. (2019) A systematic method to understand requirements for explainable AI (XAI) Systems.

Haq, M. A. (2022) CNN based automated weed detection system using UAV imagery. *Computer Systems Science and Engineering* 42(2), 837. https://doi.org/10.32604/csse.2022.023016

Hinton, G. E., Osindero, S. and Teh, Y. (2006) A fast-learning algorithm for deep belief nets. *Neural Computation* 18(7), 1527. https://doi.org/10.1162/neco.2006.18.7.1527

Hu, K., Jin, J., Zheng, F., Weng, L. and Ding, Y. (2022) Overview of behaviour recognition based on deep learning. *Artificial Intelligence Review* 56(3), 1833. https://doi.org/10.1007/s10462-022-10210-8

Karmakar, P., Teng, S. W., Murshed, M., Pang, S., Li, Y. and Lin, H. (2023) Crop monitoring by multimodal remote sensing: A review. *Remote Sensing Applications Society and Environment* 33, 101093. https://doi.org/10.1016/j.rsase.2023.101093

Katsigiannis, P., Misopolinos, L., Liakopoulos, V., Alexandridis, T. and Zalidis, G. (2016) An autonomous multi-sensor UAV system for reduced-input precision agriculture applications. In: *24th Mediterranean Conference on Control and Automation (MED),* Athens, Greece, pp. 60-64. https://doi.org/10.1109/med.2016.7535938

Kim, J., Kim, S. C., Ju, C. and Son, H. I. (2019) Unmanned aerial vehicles in agriculture: a review of perspective of platform, control, and applications. *IEEE Access* 7, 105100. https://doi.org/10.1109/access.2019.2932119

Krestenitis, M., Raptis, E. K., Kapoutsis, A. Ch., Ioannidis, K., Kosmatopoulos, E. B., Vrochidis, S. and Kompatsiaris, I. (2022) CoFly-WeedDB: A UAV image dataset for weed detection and species identification. *Data in Brief* 45, 108575. https://doi.org/10.1016/j.dib.2022.108575

Wang, P., Tang, Y., Luo, F., Wang, L., Li, C., Niu, Q. and Li, H. (2022) Weed25: A deep learning dataset for weed identification. *Frontiers in Plant Science* 13. https://doi.org/10.3389/fpls.2022.1053329

Kwak, G.-H., Park, C., Lee, K., Na, S.-I., Ahn, H. and Park, N.-W. (2021) Potential of hybrid CNN-RF model for early crop mapping with limited input data. *Remote Sensing* 13(9), 1629. https://doi.org/10.3390/rs13091629

Laghari, A. A., Jumani, A. K., Laghari, R. A. and Nawaz, H. (2022) Unmanned aerial vehicles: A review. *Cognitive Robotics* 3, 8. https://doi.org/10.1016/j.cogr.2022.12.004

Lambertini, A., Mandanici, E., Tini, M. A. and Vittuari, L. (2022) Technical challenges for multi-temporal and multi-sensor image processing surveyed by UAV for mapping and monitoring in precision agriculture. *Remote Sensing* 14*(19)*, 4954. https://doi.org/10.3390/rs14194954

LeCun, Y., Bottou, L., Bengio, Y. and Haffner, P. (1998) Gradient-based learning applied to document recognition. *Proceedings of the IEEE* 86(11), 2278. https://doi.org/10.1109/5.726791

Liang, D., Liu, W., Zhao, L., Zong, S. and Luo, Y. (2022) An improved convolutional neural network for plant disease detection using unmanned aerial vehicle images. *Nature Environment and Pollution Technology* 21(2), 899. https://doi.org/10.46488/nept.2022.v21i02.053

Liu, X., Zhai, H., Shen, Y., Lou, B., Jiang, C., Li, T., Hussain, S. B. and Shen, G. (2020) Large-Scale Crop Mapping from Multisource Remote Sensing Images in Google Earth Engine. *IEEE Journal of Selected Topics in Applied Earth Observations and Remote Sensing* 13, 414. https://doi.org/10.1109/jstars.2019.2963539

Luo, W., Xing, J., Milan, A., Zhang, X., Liu, W. and Kim, T.-K. (2014) Multiple object tracking: a literature review. *arXiv*. https://doi.org/10.48550/ARXIV.1409.7618

Machichi, M. A., Mansouri, L. E., Imani, Y., Bourja, O., Lahlou, O., Zennayi, Y., Bourzeix, F., Houmma, I.H. and Hadria, R. (2023) Crop mapping using supervised machine learning and deep learning: a systematic literature review. *International Journal of Remote Sensing* 44(8), 2717. https://doi.org/10.1080/01431161.2023.2205984

Makam, S., Komatineni, B. K., Meena, S. S. and Meena, U. J. (2024) Unmanned aerial vehicles (UAVs): an adoptable technology for precise and smart farming. *Discover Internet of Things* 4(1). https://doi.org/10.1007/s43926-024-00066-5

Minaee, S., Boykov, Y., Porikli, F., Plaza, A., Kehtarnavaz, N. and Terzopoulos, D. (2021) Image segmentation using deep learning: a survey. *IEEE Transactions on Pattern Analysis and Machine Intelligence* 1. https://doi.org/10.1109/tpami.2021.3059968

Nex, F., Armenakis, C., Cramer, M., Cucci, D. A., Gerke, M., Honkavaara, E., Kukko, A., Persello, C. and Škaloud, J. (2022) UAV in the advent of the twenties: Where we stand and what is next. *ISPRS Journal of Photogrammetry and Remote Sensing* 184, 215. https://doi.org/10.1016/j.isprsjprs.2021.12.006

O'Shea, K., and Nash, R. (2015) An introduction to convolutional neural networks. *arXiv*. https://doi.org/10.48550/arxiv.1511.08458

Ong, P., Teo, K. S., and Kiong, S. C. (2023) UAV-based weed detection in Chinese cabbage using deep learning. *Smart Agricultural Technology* 4, 100181. https://doi.org/10.1016/j.atech.2023.100181

Osorio, K., Lara, A., Pedraza, C., Jamaica, D. and Rodríguez, L. (2020) A deep learning approach for weed detection in lettuce crops using multispectral images. *AgriEngineering*, 2(3) 471. https://doi.org/10.3390/agriengineering2030032

Panda, B., Mishra, M. K., Mishra, B. S. P. and Tiwari, A. K. (2021) Parametric analysis on crop/weed classification via optimized convolutional neural network. In: *Third International Conference on Intelligent Communication Technologies and Virtual Mobile Networks (ICICV)*, 2021, Tirunelveli, India, pp. 1477-1484. https://doi.org/10.1109/icicv50876.2021.9388462

Pandey, A. and Jain, K. (2021) An intelligent system for crop identification and classification from UAV images using conjugated dense convolutional neural network. *Computers and Electronics in Agriculture* 192, 106543. https://doi.org/10.1016/j.compag.2021.106543

Sangaiah, A. K., Yu, F.-N., Lin, Y., Shen, W.-C. and Sharma, A. (2024) UAV T-YOLO-Rice: an enhanced tiny Yolo networks for rice leaves diseases detection in paddy agronomy. *IEEE Transactions on Network Science and Engineering* 11(6), 5201. https://doi.org/10.1109/tnse.2024.3350640

Sarker, I. H. (2021) Deep learning: a comprehensive overview on techniques, taxonomy, applications and research directions. *SN Computer Science* 2(6). https://doi.org/10.1007/s42979-021-00815-1

Sharma, V., Tripathi, A. K. and Mittal, H. (2022) Technological revolutions in smart farming: current trends, challenges and future directions. *Computers and Electronics in Agriculture*, 201, 107217. https://doi.org/10.1016/j.compag.2022.107217

Shen, X., Jin, Y., Zhou, K. and Zhang, Y. (2020) Application analysis and prospect of deep learning in remote sensing image classification. In: *Eleventh International Symposium on Multispectral Image Processing and Pattern Recognition (MIPPR2019), 2019*, Wuhan, China, 114320A. https://doi.org/10.1117/12.2538161

Shinde, S. K. and Attar, V. (2025) An Indian UAV and leaf image dataset for integrated crop health assessment of soyabean crop. *Data in Brief* 111517. https://doi.org/10.1016/j.dib.2025.111517

Toscano, F., Fiorentino, C., Capece, N., Erra, U., Travascia, D., Scopa, A., Drosos, M. and D'Antonio, P. (2024) Unmanned aerial vehicle for precision agriculture: a review. *IEEE Access* 12, 69188. https://doi.org/10.1109/access.2024.3401018

Wu, Y., Yang, H. and Mao, Y. (2024). Detection of the pine wilt disease using a joint deep object detection model based on drone remote sensing data. *Forests* 15(5), 869. https://doi.org/10.3390/f15050869

Xi-hong, G., Feng, Q. and Guo, F. (2025) CMTNet: a hybrid CNN-transformer network for UAV-based hyperspectral crop classification in precision agriculture. *Scientific Reports* 15(1). https://doi.org/10.1038/s41598-025-97052-w

Xin, Y., Kong, L., Liu, Z., Chen, Y., Li, Y., Zhu, H., Gao, M., Hou, H. and Wang, C. (2018) Machine learning and deep learning methods for cybersecurity. *IEEE Access* 6, 35365. https://doi.org/10.1109/access.2018.2836950

Zhang, J., Feng, Y., Zhang, Q., Wang, M., Yu, J. and Tan, Y. (2024) Advancements of UAV and deep learning technologies for weed management in farmland. *Agronomy* 14(3), 494. https://doi.org/10.3390/agronomy14030494

Zhang, Z. and Zhu, L. (2023) A Review on unmanned aerial vehicle remote sensing: platforms, sensors, data processing methods, and applications. *Drones* 7(6), 398. https://doi.org/10.3390/drones7060398

Zhao, X., Wang, L., Zhang, Y., Han, X., Deveci, M. and Parmar, M. (2024). A review of convolutional neural networks in computer vision. *Artificial Intelligence Review* 57(4). https://doi.org/10.1007/s10462-024-10721-6

Zhong, Y., Hu, X., Luo, C., Wang, X., Zhao, J. and Zhang, L. (2020) WHU-Hi: UAV-borne hyperspectral with high spatial resolution (H2) benchmark datasets and classifier for precise crop identification based on deep convolutional neural network with CRF. *Remote Sensing of Environment* 250, 112012. https://doi.org/10.1016/j.rse.2020.112012

7 Deep Learning for Plant High-Throughput Phenotyping and Phenomics

Amarjeet Singh Bhogal, Ellapu Govardhana Rao, Dikshita Hazarika, Dikshita Gogoi, Gayatree Hazarika, Niharika Saharia, Dibosh Bordoloi, and Debojit Sarma*

Department of Plant Breeding and Genetics, Assam Agricultural University, Jorhat, Assam, India

Abstract

Deep learning (DL), a specialized subset of artificial intelligence (AI) and machine learning (ML), holds immense potential in the field of plant high-throughput phenotyping (HTP) and phenomics. It enables computational frameworks utilizing multiple processing layers to gain insight into data representations at numerous degrees of abstraction, recognizing complexities in big datasets by the backpropagation strategy, and instructing a system where to modify its internal features that calculate subsequent layer's depiction of the preceding layer. In today's era of climate change, DL is a beacon of hope, enabling the automatic assessment of intricate plant traits such as developmental cycles, disease resistance, and stress responses. This technology improves crop enhancement and oversight efficiency by precisely evaluating massive amounts of image and sensor data. DL simplifies complex phenotyping and significantly minimizes human error. This chapter details various deep learning strategies for HTP and plant phenomics, offering a glimpse into the exciting future of plant science.

Keywords: climate change, crop improvement, deep learning, food security, phenomics

7.1 Introduction

Phenotype is said to be the culmination of all the observable characteristics, which is often influenced by both the genetic makeup (genotype) and its respective environment, which includes both microenvironment and macro-environment. In the context of plants, these attributes often include a wide range of physiological parameters such as transpiration rate, stomatal conductance, water-use efficiency, and photosynthetic performance, as well as morphological traits such as plant height, leaf characteristics, biomass, and flower color (Akhtar *et al.*, 2024, Li *et al.*, 2014). According to Fiorani and Schurr (2013), plant phenotyping encompasses a broad range of techniques and procedures designed to measure plant composition, architecture, and biomass at different scales, from the cellular to the whole-plant level, enabling the identification of features for breeding high-yielding, stress-resilient cultivars, essential for crop development. Given the population expansion, climate change, and the loss of arable land, accurate phenotyping has

*E-mail: debojit.sarma@aau.ac.in

DOI: 10.1079/9781836991120.0007

become a crucial aspect in realizing a sustainable and secure community based on agricultural produce (Tardieu *et al.*, 2017). Integrating phenotypic and genotypic data speeds the production of plant varieties that are resistant to both abiotic and biotic stress factors (Yang *et al.*, 2020).

Traditional plant phenotyping relies on manual equipment and visual scoring to assess traits such as plant height, leaf area index (LAI), biomass, disease severity, area of plant structure under diseased conditions, and a measure of reproductive structures. These methods frequently involve the use of rulers, manual counts, and subjective rating scales (Catchpole and Wheeler, 1992; Liu *et al.*, 2018). Traditional phenotyping exhibits some clear advantages, such as a simple and easy assessment of the phenotype, but these techniques are time-consuming, labor-intensive, and inappropriate for large-scale or multi-seasonal studies (Xie and Yang, 2020). Destructive sampling, such as harvesting for biomass, prevents repeated observations and limits our understanding of dynamic plant responses, while observer bias and variability compromise accuracy and repeatability. Modern breeding efforts require quick, large-scale phenotyping to take advantage of genetic discoveries for producing high-yielding, stress-resilient cultivars that are severely hampered by low-throughput methods (Yang *et al.*, 2020).

In the field of research pertaining to plant sciences itself, high-throughput phenotyping (HTP) has emerged as a transformative approach that has proven to be successful over multiple instances and overcomes the limitations of subjective and time-consuming conventional phenotyping techniques. HTP has paved the way for advanced imaging technologies such as RGB (red–green–blue), hyperspectral, multispectral, panchromatic, thermal, and fluorescence-based along with more efficient drones, satellites, and automated equipment capable of cross-platform rapid, non-destructive, and high-resolution data generation, collection over large plant populations, and interpretation in a more swift and precise manner (Yang *et al.*, 2020). To reduce human bias and increase scalability, remote sensing and image-based HTP provide real-time monitoring of characteristics such as growth modulation, stress responses, and yield potential (Furbank and Tester, 2011). By capturing intricate morphological data such as canopy volume and branch angles, recent advances in three-dimensional (3D) imaging, including light detection and ranging (LiDAR) and structure-from-motion, significantly enhance trait quantification and get over the occlusion issues faced during 2D approaches (Harandi *et al.*, 2023). These innovations have considerably sped up breeding programs and helped in a more desired goal of sustainable agriculture amid climate change and population expansion (Araus and Cairns, 2014).

HTP generates an astounding amount of high-quality data, which may pose some significant problems despite its transformational promise. Terabytes of data, comprising millions of photos, sensor readings, and environmental observations, may be produced by a single HTP experiment (Singh *et al.*, 2016). Traditional data processing methods, such as manual picture grading and simple statistical models, are ineffective at handling the complexity, and diversity, provided by the HTP data. This barrier emphasizes how complex computing techniques are required to transform unprocessed data into useful insights. With the ability to fully utilize HTP data through automated feature extraction, pattern identification, and predictive modeling, machine learning (ML)—particularly deep learning (DL)—has emerged as a powerful alternative (Singh *et al.*, 2016; Gill *et al.*, 2022). As environmental and demographic concerns increase, HTP can help progress plant science and agriculture by integrating many computational methods, which will help to create sustainable agricultural solutions (Araus *et al.*, 2018).

In artificial intelligence (AI), DL is a subset of ML that transforms plant phenotyping using multilayered, complex neural networks with a significant credit assignment path (CAP) depth to achieve high-level abstraction from unstructured HTP data, including raw RGB and multispectral images (Lecun *et al.*, 2015; Arya *et al.*, 2022). An important aspect of DL is its ability to learn, thereby improving performance in both classification and regression tasks. It enables models to automatically extract hierarchical features from raw data (LeCun *et al.*, 2015), and if provided with sufficient labeled data, DL models deliver high parallelization that allows for quick and precise analysis of complex features. Convolutions, pooling, memory cells, and activation functions are some of the layers that DL integrates to handle different phenotyping tasks, depending on the architecture—whether convolutional neural networks (CNNs), recurrent neural networks (RNNs), or recursive neural networks (RvNNs) (Kamilaris and Prenafeta-Boldú, 2018).

For tasks such as plant species classification, disease detection, weed identification, and quantification of traits such as leaf count and canopy structure, DL models, in particular CNNs, use convolution operations with flexible kernels (e.g. 3 × 3) and non-linear activation functions to automatically extract features, in contrast to traditional ML, which necessitates manual feature selection (Mohanty *et al.*, 2016; Sandhu *et al.*, 2020; Arya *et al.*, 2022). Convolutional, pooling, and fully connected CNNs have outperformed traditional approaches in the segmentation of damaged cucumber leaves, achieving up to 96% accuracy (Lin *et al.*, 2019). As per Tsaftaris *et al.* (2016) and Yang *et al.* (2020), DL speeds up breeding programs that are essential for sustainable agriculture by facilitating multitask phenotyping, for instance, detecting *Arabidopsis* genotypes and forecasting their growth stages. Data annotation bottlenecks and overfitting risks, which require strategies such as dropout and backpropagation to mitigate, make DL dependent on sizable, labeled datasets, high-performance computational resources (e.g. GPUs), and optimal hyperparameter tuning (e.g. learning rate and layer count) hazardous. Notwithstanding these obstacles, DL's ability to handle large HTP datasets makes it a fundamental component of plant phenomics, propelling improvements in food security and agricultural development. This chapter provides an overview of several DL approaches for plant phenomics and HTP, providing insight into the fascinating future of plant science.

7.2 Fundamentals of DL

An artificial neural network (ANN) is a flexible ML technique that is modeled after biological neural networks. It is made up of layers of "neurons" that receive inputs and send outputs to the next layer. These layers are connected by weighted connections (Zou *et al.*, 2008). ANNs are devices that mimic the functioning of the human brain to carry out particular tasks. They are composed of a multitude of artificial neurons or processors. The ANN is achieved by building a cognitive learning algorithm that does not require programming all of the rules because it is capable of developing independent norms of behavior through what is commonly referred to as "experience." Since neural networks are massively parallel computing systems composed of countless interconnected basic processing units (neurons) that learn from their surroundings and synaptic weights that store and retrieve the abilities of the interconnected neurons, they can be used in real-world scenarios (Haykin, 2009). Deep neural networks (DNNs) are an advanced version of common ANNs. DNNs are different from traditional neural networks in two key ways. CNNs include only one or two hidden layers, which are shallow. DNNs, however, have a large number of hidden layers. DNNs, CNNs, RNNs, and long short-term memory (LSTM) are just a few of the many models for DNNs. Even attention-based networks that concentrate on particular regions of a DNN have been developed recently. The complexity, resources, and training time of a network increase with its size and number of layers. It takes less time to train DNNs on GPU-based architectures than on traditional CPUs, and new advancements have significantly reduced training times (Sarkar *et al.*, 2018).

A hierarchy is formed by the many neurons that make up each layer of a DNN. The subsequent layer uses the preceding layer's output as its input, and the process continues. The input data's patterns become more intricate with each succeeding layer. Abstractions at the highest level in the data are usually learned by deeper layers, while lower layers usually learn low-level features. Because of DNN's multilayered feed-forward neural network strategy, they have the simplest structure (Ahmad *et al.*, 2018). CNNs are composed of three distinct types of layers: fully connected, pooling, and convolutional. With the use of filters and kernels, whose coefficients are modified during the training phase, the convolution layers try to identify important features that might be present in the data. To create a feature map, each filter is independently convolved over the input; greater activation values indicate the presence of features (Ahmad *et al.*, 2017). Despite being extremely effective frameworks for analyzing non-sequential data, CNN and DNN are not appropriate for identifying patterns in time-series data. RNNs are a new family of architectures that were developed for this purpose. The recurrent connections in each

RNN unit allow the network to retain information for a longer amount of time. Thus, RNNs can discern patterns in sequential voice, movies, and text (Zaremba *et al.*, 2014). A RNN type called LSTM has a high capacity for learning and making predictions about sequential data. According to research, RNN's ability to preserve long-term memory is restricted. Thus, the LSTM was developed to overcome this limitation by including a memory structure that can retain its configuration over time, along with gates to determine what should be the outcome, what should be discarded, and what should be preserved. Several applications that are intrinsically sequential, such as handwriting recognition, language modeling and translation, speech recognition, and speech synthesis, demonstrate the effectiveness of the LSTM (Al-Jabery *et al.*, 2019).

7.3 HTP Platforms

Traditional methods of plant phenotyping are less efficient, costly, destructive, and time-consuming. Handheld and ground vehicle-based mounted sensors are crop-limited, whereas satellite-based phenotyping is not reliable due to its poor resolution and dependency on weather parameters. Thus, efficient phenotyping platforms ease the process of phenotypic data collection, processing, and interpretation. A few revolutionary platforms have been discussed in the following section.

7.3.1 Imaging systems

Micro-phenotypes of crop organs or tissues are gaining importance in the phenomics study of crops. Various scientists and researchers have proposed robust methods for accurate measurement of crop dynamics in different crops such as wheat and corn. Non-destructive methods of HTP are of great importance in elucidating the relationships between organ or tissue traits and their functions (Fahlgren *et al.*, 2015).

RGB, multispectral, and hyperspectral imaging (HSI)

RGB imaging is a low-cost, sensor-based spectral visualization of an original crop image that is capable of creating a 3D crop model and detecting plant information at three wavelengths. HTP using RGB imaging is an additive color-based model that involves three primary colors, namely red, green, and blue in the visible region of the spectrum (640, 550, and 460 nm, respectively). These primary colors are combined in various proportions to reproduce a wide range of colors, thus converting real-world colors into digital signals through 8- or 24-bit color encoding systems. The limitation of this system is that it does not provide a continuous array of spectrum information as it relies on three primary colors. Using RGB imaging, phenotypic data, such as plant height, LAI, biomass, and yield, can be calculated (Song *et al.*, 2021). Multispectral imaging (MSI) is based on several wavelengths in the visible and near-infrared (NIR) regions. It offers seamless resolution refinement, achieving precision as fine as 0.25 m. On the other hand, HSI is carried out using hundreds or more than hundreds, even thousands, of continuous wavelengths. The spectral resolution is highest among all these three sensors as it involves numerous wavebands.

Thermal imaging and 3D imaging

HTP using thermal printing and 3D imaging such as LiDAR is categorized as an unmanned aerial vehicle (UAV) that can be used for taking data in a non-destructive method for traits such as yield, biomass, height, and LAI (Xei and Yang, 2020). 3D imaging tools enable accurate measurement of plant features. Cameras based on these technologies can detect long-wave infrared (LWIR) light emitted by plant leaves (Fahlgren *et al.*, 2015). These imaging techniques are cost-effective, providing high-resolution images.

X-ray computed tomography for root system analysis

Plant root system architecture (RSA) is crucial for water and nutrient uptake, stability of plant, and interaction with soil microorganisms. Traditional methods of studying root system involve destruction of root biota, thus limiting the complete analysis of root system. HTP using X-ray computed tomography (CT) is a non-destructive imaging method that gives cross-sectional images from different angles. Root traits, such as root length, volume, diameter, branching patterns, and root angles, can be measured accurately using X-ray CT (Song *et al.*, 2021).

7.3.2 Non-imaging systems

Sensor-based HTP tools can be used to detect plant physiological parameters such as chlorophyll content and CO_2 assimilation rate. Controlled environments can be created through automation, and data can be collected through sensor-based approaches such as ground robots or rovers and UAVs. Robotic tools may be of fixed-structure model, movable between two points located in a short distance, and may be employed with a conveyor belt-based phenotyping system. These tools help in precision farming, for trait and stress monitoring.

7.3.3 Integration with IoT and big data

Acquisition of high-resolution data is very important for selecting the best performing genome line. HTP platforms generate massive, heterogeneous datasets, often terabytes per day, whose manual interpretation is infeasible. An interdisciplinary technique using statistics, decision theory, data visualization, optimization, and neural networks is needed to analyze these data (Gill *et al.*, 2022). The assistance of ML and DL programs is important for managing such large datasets.

7.4 DL in Plant Image Analysis

7.4.1 Growth monitoring

Monitoring plant growth stands is an essential aspect when implementing precision agriculture along with plant breeding programs. Traditional plant size and height measurement methods require expensive resources and considerable human involvement, thus creating potential for human error in the results. HTP became possible due to DL technology advancements, enabling automated plant image analysis. Event detection algorithms, when used in conjunction with bounding box regression, serve as essential tools in plant image analysis systems, enabling accurate plant detection and precise measurement of their dimensions (Tong *et al.*, 2022). The object detection process in DL involves the automated recognition of objects within images, along with their spatial location information. Several classification categories exist for different DL object detection approaches. Two-stage detector techniques work through two sequential detection steps, where the initial phase generates area proposals, while the second phase classifies these proposals. The region-based CNN (R-CNN) system functions by obtaining features from the proposed regions before classifying the results (Sun *et al.*, 2024). Fast R-CNN boosts the R-CNN framework through a single procedure that examines images completely. Faster R-CNN adds a region proposal network (RPN) to make the proposal generation process more efficient. One-step detectors execute their detection process at once, which leads to quicker processing but yields lower accuracy than two-stage detection approaches. You Only Look Once (YOLO) serves as an illustration of a single forward network pass detection system that provides opening boxes alongside class probability predictions (Diwan *et al.*, 2023). The Single Shot MultiBox Detector (SSD) system operates in a manner similar to YOLO by deploying numerous aspect ratios and scales across multiple feature maps for object detection functions. The anchor-based detector system makes use of predefined anchor boxes that serve as predictors for objects appearing at various scales together with aspect ratios. The architectures of Faster R-CNN and SSD are anchor-based, as they rely on predefined anchor boxes, whereas anchor-free detectors instead predict object positions directly using key points or center points. Such detectors function without anchors because they directly establish object positions based on key points or centers. A method, which uses a heatmap without predefined anchors to predict object centers serves as an example. Instance segmentation deals with both recognizing objects and separating each class instance while generating pixel-level segmentation results. Mask R-CNN represents an enhanced version of Faster R-CNN, which adds a new branch to generate mask predictions from each Region of Interest (RoI). Key point detection operates in a manner different from regular object detection because it focuses on finding particular points inside objects, which includes human body joint detection. The real-time detection system Open Pose is used for multi-person 2D pose recognition. Through DL object detection methods, researchers can achieve multiple applications, including plant growth monitoring by identifying fruits, leaves, and different plant components to measure exact growth metrics.

The U-Net and DeepLabV3+ systems serve as semantic segmentation methods that detect leaves effectively among complex visual scenes to enable accurate measurements of leaf size (Surehli *et al.*, 2025). Such advanced models go beyond standard image processing because they acquire hierarchical characteristics from extensive datasets, thus achieving enhanced accuracy throughout different plant species and environmental situations. Leaf area can be accurately measured through semantic segmentation of images that separates leaves from the surrounding objects. Scientists can use this method to measure leaf area because it serves as a key indicator for evaluating plant photosynthetic functions and wellness status. Persistent image segmentation throughout plant development enables scientists to make better decisions about plant fertilization by monitoring leaf morphological changes at various growth stages. The evaluation of leaf morphology depends critically on the combination of shape descriptors, including aspect ratio convexity along with circularity because they provide valuable information about growth patterns according to Polly and Devi (2024). The incorporation of DL-based segmentation enables these descriptors to execute automated HTP processes that eliminate manual workload in agricultural research. Researchers use morphological analysis to evaluate leaf shapes through shape descriptors, which help identify stress or disease symptoms. The shape of a leaf indicates how well the plant adapts to its natural environment as well as the genetic differences between plant species. Shape descriptors operate as an automated tool for effective large-scale analysis of plants. Such type of analysis serves breeding programs and agricultural research most efficiently. Progress in both CNNs and transformer-based models has increased the accuracy and efficiency of plant picture analysis, establishing it as a crucial management tool in precision agriculture. Research in the subject should attempt to improve model applicability across different plant kinds while also using numerous data sources for better growth parameter evaluation. Analysis of canopy structure and architecture occurs through studies of 3D point cloud data and voxel-based representations. The collection of data points, called a 3D point cloud, exists in 3D areas that LiDAR and photogrammetry techniques typically measure. Every spatial point in the dataset shows the position through its x, y, and z coordinates; thus, it retains the detailed forms of plant structures. Agricultural crop canopy structure assessment is improved through the integration of DL methods in plant image analysis. Plant morphological characteristics can be accurately measured through 3D point cloud data and voxel-based models because these models are essential for plant health evaluation, stress response assessment, and growth dynamic tracking. Research has proven that DL networks achieve top results when processing 3D point cloud information for plant applications. Researchers applied LSTM and gated recurrent unit (GRU) network combinations to LiDAR-derived point clouds for vegetable crop height and crown dimension estimation, which produced results with an accuracy of about 80% (Nidamanuri, 2024). Point voxel CNNs serve to segment cotton plant components in three dimensions and produce structural traits (Saeed *et al.*, 2023). The analysis of 3D point clouds enables experts to measure canopy volume together with surface area and leaf density, which provides detailed information about plant structural characteristics (Tong *et al.*, 2022). Research on altering canopy structures through time provides knowledge about environmental impacts that affect plant growth patterns. Through voxel-based methodology, scientists acquire better insights into plant organizational patterns. Researchers use voxel grids to build plant structures, which enable them to obtain detailed 3D characteristics. Choudhury *et al.* (2020) created a method to grid-mesh corn plants into voxel shapes for extracting complex morphological data that analyzed plant progression. The voxel-based representation technique breaks 3D space into small grid blocks called voxels, which serves as an efficient method to analyze plant structures through computational analysis. The reconstruction process of detailed plant canopy structures occurs through voxel models, while these models enable researchers to study leaf and branch spatial positions. Optimal space utilization, in combination with understanding plant architecture, forms the foundation for comprehensively analyzing plant structures and their functions within a canopy. DL models accept voxel grid data for direct classification of all plant canopy elements, including leaf surfaces and leaf interfacial angles. The DL methods optimize processing capabilities for 3D data structures to deliver substantial information from voxel-based complex images (Tong *et al.*, 2022). These advances highlight the promising role of DL along with 3D data in transforming plant phenotyping, allowing for efficient, accurate, and non-invasive ways of in-depth growth monitoring in agricultural research.

7.4.2 Plant stress detection

Training a computer algorithm involves ML methods, called image classification, which can sort different images into defined categories. Initially, a large assortment of plant images, including disease stages together with clean plant pictures, needs to be collected. A suitable labeling system must accompany the image collection, with each photograph identifying the diseases or conditions it presents. The training process of CNNs makes use of this dataset. Through weight adjustment during training, the network develops proficiency at recognizing patterns, which correspond to its predefined classes. The CNN utilizes multiple processing layers to evaluate input pictures by separating features that start from basic visual elements all the way to sophisticated structural shapes consistent with particular diseases (Tong *et al.*, 2022). Images of leaves processed with CNNs achieve successful identification of plant diseases. A deep CNN processed 54,306 images, which contained 14 crop species and 26 diseases, with 99.35% accuracy (Mohanty *et al.*, 2016). The CNN identified crop diseases with 99.35% accuracy (Mohanty *et al.*, 2016). Researchers conducted an examination using modern DL models to detect plant diseases while achieving detection accuracy between 99% and 99.2% (Shoaib *et al.*, 2023). The YOLO family of object detection models has been deployed for identifying and localizing pests present in images. YOLOv3 and YOLOv4 models succeeded in detecting fruit leaf diseases in a research study that demonstrated their speed and success, as stated in the study of Alhwaiti *et al.* (2025).

One of the most common methods for detecting plant diseases is image classification, which employs DL-based supervised learning on labeled datasets. According to research, CNNs are extremely effective at precisely identifying plant illnesses. The research of Soumia *et al.* presented a straightforward CNN, which obtained 97.04% accuracy for tomato disease detection. The CNN-based model developed by Gnanave *et al.* achieved a training efficiency of 98% and a testing accuracy of 88.17% using 3,000 PlantVillage images, outperforming the pretrained ResNet152 and VGG19 models in this task. Abbas *et al.* applied conditional generative adversarial networks (c-GANs) for synthetic tomato leaf creation to train a DenseNet121 model, which performed exceptionally well with 99.51% accuracy in identifying five disease types (Odounfa *et al.*, 2024).

The ability of image classification models to distinguish healthy from diseased plants remains successful, but they face limitations when identifying the exact position of pathological lesions in images. Research models, including Faster R-CNN, YOLO, and SSD, address the need for accurate disease lesion localization by performing simultaneous disease detection in single images. Research demonstrates that real-time plant disease and pest identification becomes practical through YOLO-based frameworks, which have proven to be highly effective. Rishabh *et al.* developed a multi-classification CNN model, which classified mustard crop images into healthy and mildew-infected categories with a 96.66% success rate (Odounfa *et al.*, 2024).

The identification of abiotic stress conditions (drought and salinity, along with nutrient deficits) depends on spectral as well as multispectral image data analysis. The stability of crop health and agricultural sustainability suffers from abiotic stressors, which include drought stress, salt damage, harsh temperatures, and essential nutrient shortages. Detecting stresses in the early stages is crucial so that proper actions can be taken to reduce the negative effects on crop production. Visual and chemical assessment of stress often demands excessive manual work and takes extended periods of time while frequently causing human errors to occur. New advances in spectral and MSI have revolutionized plant stress detection through the development of non-invasive automatic monitoring systems. Plant stress-related changes can be identified through these imaging methods, which detect different wavelengths of plant reflectance before stress symptoms appear to the human eye (Almouctar *et al.*, 2024). Advanced ML techniques and spectral and MSI create significant improvements in detection accuracy as well as classification and quantification of crop stress.

The fundamental aspects of spectral along with MSI can be understood. Spectral imaging captures extensive unseen spectral information by photographing objects at many wavelengths without destroying them. It is broadly classified into MSI, which captures data in a small number of discrete spectral bands (often 3–20), and HSI, which gathers data by scanning through hundreds (sometimes up to thousands) of adjacent spectral bands without interruption. Both methods target plant physiological functions, thus enabling researchers to uncover early signs of abiotic stress through wavelength-specific light measurement of reflectance and absorption. Spectral as well as MSI techniques perform essential functions in detecting crop abiotic stressors by tracking the changes in plant reflectance across different wavelength values. The assessment of particular stress situations utilizes spectral indices, which include normalized difference vegetation index (NDVI), photochemical reflectance index (PRI), normalized difference water index (NDWI), and chlorophyll Index (CI_red-edge).

The research develops new vegetation indices that operate with ML methods to analyze wheat drought stress under variable nitrogen nutrient levels. Hyperspectral data proved successful in agricultural crop drought stress monitoring and prediction according to the new proposed indices, which achieved accuracy levels above 94% across all model instances (Okyere *et al.*, 2024). A research team used HSI and photosystem evaluation to detect early indicators of drought stress in durum wheat plants. Researchers attempted to use this novel technique to evaluate and select drought-tolerant wheat genotypes for breeding purposes (Roy *et al.*, 2023). The research team is investigating a unique remote sensing approach to evaluate combined drought and salinity effects on vegetative growth by adopting image processing approaches that estimate drought levels and salinity exposure while also evaluating crop development (Wen *et al.*, 2022).

7.4.3 Phenotypic trait extraction

The exact assessment of leaf colors and chlorophyll composition is critical in agricultural applications for assessing plant health and detecting nutritional deficiencies, which aids in the development of better farming practices. The current monitoring methods, based on spectrophotometry and soil plant analysis development (SPAD) meters, maintain high accuracy while being difficult to scale up for extensive plant assessments at reasonable costs. Progress in digital imaging and DL technology now enables effective non-invasive extraction of phenotypic traits according to Barman (2021).

To quantify leaf colors and evaluate chlorophyll concentrations, digital image processing makes use of three basic color spaces: RGB, HSV (hue, saturation, and value), and LAB. Leaf photos obtained under controlled illumination settings exhibit distinct color characteristics that scientists can associate with chlorophyll measurement levels. According to Barman (2021), the assessment of chlorophyll concentration in micropropagated potato plants is based on RGB picture component analysis for research objectives. The methodology uses leaf color intensity to quantify chlorophyll content, giving a non-damaging approach to assessing plant health. The prediction of chlorophyll quantities through leaf images gets boosted by CNNs in DL, which show exceptional performance for chlorophyll analysis. Barman (2021) carried out a study where digital image color features were used to estimate tea leaf chlorophyll concentrations through a 1D CNN model. CSA along with its DL regression models surpassed multiple linear regression (MLR) as well as k-nearest neighbor (KNN) to achieve 81% prediction accuracy. Through DL technology, models have shown the ability to boost agricultural chlorophyll prediction accuracy and effectiveness. The combination of color space analysis and ML techniques creates an effective plant health assessment method, according to the research by Askey *et al.* (2019). The technology allows researchers to assess plant health indicators using anthocyanin monitoring, which is directly related to stress situations. Predictions about *Arabidopsis thaliana* leaf anthocyanin levels may now be made non-invasively using regression models that process digital picture color data. Early plant stress detection through these models allows scientists to take immediate action for reducing possible crop yield reduction.

RSA analysis remains essential in agriculture because it helps scientists understand plant nutrient and water intake, which directly affects crop yield and stress tolerance (Weihs *et al.* (2024). Current RSA processing methods take too much time to analyze roots and often fail to show root structures accurately enough. Thanks to modern DL techniques with enhanced image processing capabilities, researchers can now execute automated, precise HTP of RSA systems. The process of acquiring images represents the initial step needed to conduct accurate RSA analysis. X-ray CT, together with minirhizotrons and rhizotrons, constitutes established methods for obtaining data. The observation systems of minirhizotrons and rhizotrons allow *in situ* monitoring of root development, through their transparent viewing components, without harming the roots. The study of RSA becomes complete through X-ray CT imaging since it provides 3D views of root structures in soil (Shen *et al.*, 2020). The obtained images undergo processing techniques that separate roots from their surrounding environmental materials. The edge detection and thresholding methods show limitations in application because of variations between different soil types and changing light environments. DL models, especially CNNs, serve as modern methods for more dependable root segmentation. U-Net stands as a CNN model that successfully executes root segmentation tasks, particularly in biomedical image segmentation. The symmetrical design of this method allows precise location of complex root patterns, making it suitable for their definition. RootPainter operates according to the description provided by Smith *et al.* (2022). An open-source software introduced by Smith *et al.* (2022) employs deep learning techniques to segment biological imagery, with a particular emphasis on plant root structures. The system enables fast neural network learning, thus making root length study and analysis more efficient. Quantification of RSA traits through DL models allows scientists to measure factors such as root length and diameter together with root branching patterns as well as overall architecture. The research conducted by Shen *et al.* (2020) achieved a perfect F1-score of 0.9773 using DeepLabv3+ architecture to segment cotton roots in 2020, enabling precise measurements of root diameters and lengths according to Gaggion *et al.* (2021). DL models operating through automated imaging systems have enabled the development of high-throughput RSA phenotyping. Gaggion *et al.* (2021) developed ChronoRoot as a study, which uses deep segmentation networks and 3D-printed open hardware to track root growth dynamics at high time intervals, thereby providing essential knowledge about root developmental changes.

7.5 DL for Plant Phenomics

7.5.1 Time-series analysis

Growth curve prediction using RNNs and LSTMs to model plant growth dynamics

DL plant growth curve prediction is a novel concept utilizing advanced ML techniques that estimate crop yields, biomass accumulation, flowering dates, maturity dates, and all types of plant growth parameters. This field has gained significance in advancing agricultural production systems and resource management strategies recently, particularly with the increased demand for food on a global scale, and the deterioration of our natural resources. DL algorithms perform much better than traditional methods for predicting outcomes because they can incorporate large datasets with variables that account for environmental conditions, management practices, and plant genomics. DL architectures such as ANNs, CNNs, and RNNs, which process complex datasets with spatial and temporal dimensions, have gained popularity in the recent years (Khaki *et al*, 2020, Yasrab, 2021).

ANN is a potent tool for evaluating non-linear systems and is largely based on the idea of the biological neuron system. A class of ANN is used for sequential data analysis in RNN (Lin *et al.*, 2024), which is of immense importance in the agricultural sector and has been used in areas such as crop growth prediction, water-use efficiency, and energy efficiency (Küçükönder *et al.*, 2016). RNNs and LSTM networks are advanced ML frameworks used in modeling plant growth dynamics. These models are good at processing sequential data (Jung *et al.*, 2020) and can represent complex temporal relationships between different environments and plant physiological responses to those environments over time. Agricultural applications of these models are particularly important due to their ability to analyze complex time-series data, and they are particularly relevant for improving predictions regarding biomass yield and growth efficiency in different conditions. Standard RNNs struggle with the identification of long-term dependencies (Khaki *et al.*, 2020). Standard long-term dependencies with RNNs are difficult due to the vanishing gradient problem. To address this, LSTMs were built to incorporate memory cells, and blocking mechanisms improve the learning on large data sequences as well as being able to predict plant responses better, irrespective of environmental conditions (Sherstinsky, 2020). Lack of comprehensive standardized datasets, particularly those that integrate multimodal data, generally poses a major barrier to developing robust models, thereby limiting their generalizability across different plant species and growth scenarios (Malashin *et al.*, 2024). Further, complex biological processes may lead to predictions that do not fully capture the dynamics of plant growth, and therefore RNNs and LSTMs represent a promising frontier in the modeling of plant growth dynamics, offering nuanced insights into the interactions between environmental variables and plant development.

RNNs are effective at processing sequential data and are hence used in modeling plant growth dynamics. This category of neural networks can be classified into different types, such as fully RNNs (FRNNs), RvNN, recurrent multilayer perceptron (RMLP), bidirectional RNN, and GRUs (Katte, 2018). FRNNs were created in the 1980s and are capable of learning temporal sequences in both online and batch environments. They are made up of two layers, one for input and one for output, each of which has linear and non-linear units. RvNN is a type of RNN in which the same set of weights is applied to the network recursively in topological order to construct a network with a structure resembling a differentiable graph. RMLPs are defined as a series of cascaded, feedforward, fully connected subnetworks. Each subnetwork consists of node layers, and feedback connections are allowed only in the last layer. The RMLP neural network structure is valuable for incorporating temporal behavior during the learning process and, importantly, for the detection and management of non-linear dynamic processes. Bidirectional RNNs can process a sequence both forward and backward. The bidirectional structure has the advantage of integrating information from both past and future contexts and consequently increases our understanding of the dependencies associated with the data. Tasks that are complex and where the future context can influence the current predictions benefit from this structure. This is especially useful in dynamic modeling contexts, for example, plant growth. GRUs are designed to capture dependencies in sequential data but with a simpler architecture. They are similar to LSTMs but consist of fewer items. RNNs, particularly those with LSTM architectures, have become effective methods for the prediction of plant growth under greenhouse conditions. Specifically, RNN models predict various growth parameters of plants from time-series data and allow growers to modify conditions for optimal yield and resource usage. RNNs have been utilized to predict the production of crops such as *Ficus benjamina* and tomato based on microclimate conditions and previous growth data (Alhnaity *et al.*, 2019; Naidu, 2023). A DNN based on autoencoder wavelets and featuring an attention mechanism was created to predict changes in plant stem diameter. It significantly outperformed the existing models in terms of error criteria such as root mean squared error (RMSE), mean absolute error (MAE), and mean absolute percentage error (MAPE) by employing an LSTM-based encoder–decoder framework for feature extraction and wavelet decomposition to reduce noise (Alhnaity *et al.*, 2021). RNNs and ANNs more broadly have been used to predict plant growth using *in vitro* conditions of a plant, estimating biomass, classifying somatic embryos, and predicting optimal *in vitro* culture conditions. This suggests the broad applicability of neural networks and their usefulness for providing predictive modeling across a variety of biological processes.

Stress response modeling: identifying temporal patterns and predicting the onset and severity of stress events

Plant stress response is an important field of study that focuses on how plants withstand different environmental stresses. Stress can be categorized into two primary types: biotic stress, which arises from living organisms such as pathogens, and abiotic stress, which is induced by non-living environmental factors. Abiotic stress includes a wide range of environmental conditions, including drought, flooding, extreme temperatures, salinity, light variations, air pollution, and nutrient imbalances. DL has emerged as a powerful tool for predicting and identifying the different types of stress and how plants respond to them. It also offers solutions that are both accurate and efficient (Ali *et al*, 2024). DL and ML are mainly utilized in plants for predicting stress results based on different environmental parameters. Researchers can use data collected from different environmental conditions to create models that predict how plants will respond to abiotic stresses such as drought, temperature variations, and salinity (Gou *et al.*, 2024). Generation of stress-resistant crop varieties and the implementation of effective stress management strategies rely significantly on this predictive capacity.

Drought stress is a major obstacle in agriculture, impacting growth and yield within crops. DL models have been utilized to successfully identify drought stress in crops such as potato and corn. One study using UAV-based imaging and a pretrained DenseNet121 model achieved a 97% precision in identifying drought stress in potatoes (Patra and Sahoo, 2024). Another study used a deep CNN (DCNN) for drought stress classification in corn plants and found that it achieved an accuracy of 98.14% in distinguishing stress conditions (An *et al.*, 2019). These models utilize features such as texture differences and color changes in capturing plant images to identify plant stress early and accurately. Biotic stress, such as insect infestation and diseases, also poses a major threat to crop health and its yield. To identify biotic stresses in plants, various DL models have been developed, and one such is the EfficientNetB3 model, which achieved 98.2% accuracy in identifying various biotic stresses, including insect infestations and diseases, by employing a diverse collection of plant images using transfer learning (Soni and Taneja, 2024). The accuracy of the pretrained ResNet-50 model in identifying biotic stressors in rice crops was 92.61% (Malvade *et al.*, 2022). High accuracy was attained by the Fast Fourier convolutional neural network (FFDNN) in differentiating between wheat yellow rust and nitrogen deficiency (Shi *et al.*, 2023). In a study, water stress in corn was classified with 91.30% accuracy using a deformable attention-based network (StressNet) (Nampally *et al.*, 2023). Using 30,000 RGB images, VGG16 architectures were found to be an effective system for identifying and classifying distinct abiotic stresses in diverse rice crop kinds with an accuracy of 92.13% and 95.08%, respectively (Anami *et al.*, 2020). Pretrained CNN models, such as AlexNet, VGG16, VGG19, GoogLeNet, ResNet50, ResNet101, and DenseNet201, are mostly used for the identification and classification of different diseases (Chakraborty *et al.*, 2022).

7.5.2 Genotype–phenotype mapping

DL for quantitative trait locus (QTL) mapping: integrating DL with genetic data for high-resolution mapping

QTL mapping is a crucial statistical technique that integrates phenotypic data, which encompasses observable traits, with genotypic data derived from molecular markers (Rani *et al.*, 2023). This approach is employed to identify specific genetic regions, known as QTLs, that are associated with complex traits in particular populations. A variety of QTL mapping methods have been used over the years (e.g. standard interval mapping (SIM), multiple imputation (IMP) unlinked QTL, composite interval mapping (CIM) linked and unlinked QTL/gene across chromosomes). The application of DL in QTL mapping has received a great deal of attention, particularly in terms of working with large, high-dimensional datasets, or datasets often associated with plant genomics. For example, CNNs have differentiated plant phenotypes and identified diseases with more than 99% accuracy. Unlike traditional methods that required expert visual inspection and human knowledge, ML- and DL-based technologies classified plants and diseases with greater speed, adaptability, and reduced labor costs (Miao *et al.*, 2021).

Various tools and methodologies have been developed to improve the precision and efficiency of QTL mapping, marking a shift from traditional methods using DL technologies. DeepBSA is a novel software developed for bulked segregant analysis (BSA), which is essential for mapping mutations and QTLs in both plants and animals. This tool incorporates two innovative algorithms that have shown superior performance compared to existing methods, particularly in terms of absolute bias and signal-to-noise ratio (Li *et al.*, 2022). DeepBSA's design is particularly focused on high-resolution mapping and functional gene cloning, thereby accelerating the identification of novel quantitative trait genes and reducing the workload associated with candidate gene testing. QTG-Finder2 utilizes ML to prioritize candidate causal genes in QTLs, significantly reducing the time required for fine mapping. The algorithm's ability to prioritize genes based on known causal genes was tested and demonstrated a recall rate of approximately 64% for *Arabidopsis* and 83% for rice when orthologs of known causal genes were utilized as the training set. The versatility of this algorithm has increased its usability by taking advantage of orthologs of known causative genes to adapt to specific plant species (Lin *et al.*, 2020). DL models such as SegFormer are being utilized to extract multiple features associated with yield in phenotypic data from high-throughput sources. For example, an illustration of DL's potential with respect to trait quantification is the identification of 3586 time-specific QTLs related to rice panicle traits (Geng *et al.*, 2024).

Predicting phenotypes from genotypes using DL models

Predicting phenotypic effects from genotypes is a key challenge in genetics, and one that seeks to understand the way in which genotypes, and single-nucleotide polymorphisms (SNPs) in particular, impact phenotypes, or observable traits (Li *et al.*, 2019). Several conventional statistical techniques have been used in genetics, such as best linear unbiased prediction (BLUP) and Bayesian models, to model the effects of genotypes and predict phenotypes. These methods often incorporate unknown genotype effects (random effects) assumed to follow prior distributions, typically Gaussian, and treat contributions from each genotype as independent features. Each treatment of genotype data as independent features would likely require large numbers of samples to effectively capture population structure as well as the effect of unknown individual genotypes and epistasis, which refers to the interaction of SNPs, which might make predicting complex traits more difficult (Danilevicz *et al.*, 2022). There has been greater acceptance of non-linear methods, including ML or DL methods, to predict phenotypes of plants to build on traditional assumptions of linear models in genomic selection. DL models can better distinguish and leverage complex patterns in the data, and there are now ample data collected on plants. For example, deep learning models can adaptively learn complex representations that connect genotype information to phenotypic responses, effectively handling the high dimensionality of genomic data to improve trait quantification.

Different types of architectures have been developed to improve prediction accuracy and performance for phenotype prediction tasks. One notable architecture is DeepAT, which exploits several important features for phenotype prediction. First, DeepAT captures representative feature vectors from high-dimensional SNP data well, and second, it utilizes the rectified linear unit (ReLU) activation function to optimally represent non-linear information in the architecture, which accelerates convergence during training (Li *et al.*, 2024). An interpretable neural network model, DeepGAMI, enhances genotype and phenotype prediction within multimodal data. DeepGAMI creates directed neural network connections based on functional genomic data, such as gene regulation, and eQTLs (Chandrashekhar *et al.*, 2023). DeepGS is an emerging model that employs DCNNs to predict phenotypes from genotype data. It includes hidden variables assumed to jointly represent the features of genotypic markers and to improve prediction accuracy (Wenlon *et al.*, 2017).

7.5.3 Precision agriculture

Amid resource constraints and ever-increasing global food demands, there has been an urgent need to advance crop management practices; usage of AI, especially DL, has proved to have significantly transformed agricultural practices by optimizing fertilizer application, irrigation schedules, and enhancing detection of diseases and pests in crops. DL uses complex algorithms and datasets from various sources, including weather and soil nutrient datasets and imaging technologies, and allows farmers and researchers to make decisions based on combinations of data that can improve crop yield, reduce cost, and promote sustainable farming practices. Sensors are used to track a range of parameters, including light intensity, temperature, humidity, nutrient, and moisture levels. DL algorithms enable actionable insights regarding the most effective use of resources such as water, fertilizer, and pesticides by analyzing high-resolution real-time data and improving efficiency and production of agricultural operations.

Different approaches have emerged that improve efficiency, forecast water needs, and incorporate real-time data. These approaches both conserve water and enable/advocate sustainable agricultural practices. The deep reinforcement learning irrigation controller (DRLIC) uses deep reinforcement learning (DRL) to optimize an irrigation decision by predicting future soil moisture loss. This approach can achieve up to 9.52% savings in water when compared to traditional methods (Du and Ding, 2024). The application of DL in fertilizer application and management is revolutionizing agricultural practices by enhancing soil fertility and optimizing fertilizer recommendations, thereby reducing labor cost and improving crop growth as well as yield. Various studies have demonstrated the effectiveness of DL models in accurately assessing soil conditions and predicting the necessary fertilizer types and amounts, ultimately leading to improved crop yields. The 1D-CNN model (Sujatha *et al.*, 2023) developed to investigate soil fertility classification exhibited impressive accuracy at 97.9% and surpassed conventional DL-based methods such as extreme learning machines (ELMs) (Sirat *et al.*, 2018) and multilayer perceptrons (MLPs). A hybrid DCNN–LSTM model showed high performance in predicting soil nitrogen levels in muskmelon (Chang *et al.*, 2021).

A system using DL was developed to model granular fertilizer distribution methods, which achieved a high precision and recall rate in making the fertilizer application process more precise (Liu *et al.*, 2023). Another model, UNet-ConvLSTM, combined remote sensing and DL architecture to improve prediction accuracy of agricultural water demand forecasting. The results were also impressive with the model achieving R^2 values of 0.927 and 0.935 based on the MODIS and GLDAS datasets, respectively (Ye *et al.*, 2024). This model is able to effectively represent the spatial and temporal complexity in water management. A hybrid model particle swarm optimization-long short-term memory (PSO-LSTM) was created to predict evapotranspiration (Jia *et al.*, 2023) and optimize hyperparameters, thereby improving the accuracy of predictions and facilitating the generation of appropriate irrigation schedules. Multimodal DL approach was utilized to detect water status in tomato ((Zuo *et al.*, 2023), and green beans (Mokhtar *et al.*, 2023).

DL algorithms (such as CNNs) detect early signs of pest infestation through image analysis by applying filters to an image and identifying variation in textures, shapes, and so on (Kowalska and Asraf, 2023). The ultra-lightweight efficient network (ULEN), which contains a deep feature extraction module and a multiscale feature module, has been developed for an efficient deployment on devices with limited computational power, achieving high accuracy with lower parameters aiding in detecting pests and diseases (Zhang *et al.*, 2023). Integrated frameworks, for example, DIANA (Ilyas *et al.*, 2022), detect and assess the severity of plant diseases in paprika. PlantInfoCMS enables collecting and managing high-quality datasets for training DL models and saves time and money for researchers, addressing the issue of data scarcity in agricultural applications (Jin *et al.*, 2023).

7.6 Challenges

7.6.1 Data availability and preparation

DL might seem advantageous in many circumstances; yet, it proves challenging when the volume of training data is low. When the data is insufficient or inconsistent, DL models may struggle to generalize and provide reliable results. Occlusion is a necessary challenge when phenotyping canopy structures, especially as the foliage of the crop becomes denser in later growth stages, making it challenging for sensors and cameras to obtain precise images of leaves and other plant elements (Ninomiya, 2022). The generation of the training dataset, specifically for feature localization, counting, and picture segmentation, is a labor- and time-intensive operation and requires specialized tools to arrange the dataset in the right format (Arya *et al.*, 2022).

7.6.2 Big data management

After collecting massive volumes of data, managing and handling the large datasets in terms of processing, accessibility, and storage can be challenging. Unlike genomic and other omics data, phenotypic data is rarely reused or made openly accessible (Yang *et al.*, 2020). The findable, accessible, interoperable, and reusable (FAIR) principles help standardize data management across diverse persons or groups (Wilkinson *et al.* 2016). Although the FAIR principles offer strategies for handling big datasets, their application differs from one another (Yang *et al.*, 2020). Several initiatives for data management and analysis have also been built in accordance with the FAIR principle, such as PHENOPSIS DB (Fabre *et al.*, 2011) and CropSight (Reynolds *et al.*, 2019), but data sharing and uniformity within communities remain to be a huge challenge. Another difficulty is the limited funding and data infrastructure for effective data management (Bolger *et al.*, 2019).

7.6.3 Computational costs

Computability expenditure is another challenging problem in DL. It requires a certain level of computability for both online and offline computation, which elevates the cost. It is difficult to implement the expensive DL method in practical agricultural applications (Jin *et al.*, 2025).

7.6.4 Algorithmic limitations

For DL models to produce reliable results, clean, high-quality data is necessary. Preprocessing activities such as picture segmentation, contrast improvement, and noise removal are crucial but can be difficult, particularly when the original data quality is low. The method is made more difficult by object detection, feature extraction, and trait selection (Yang *et al.*, 2020). CNNs demonstrate sensitivity to geometric transformations such as rotation and scaling while being slightly robust to translation. This means that some degree of preprocessing might be necessary to ensure model stability. On the contrary, noise and interference from other sequences greatly reduce RNN prediction accuracy, especially when irrelevant data is present. Moreover, the strong performance of RNNs on short sequences is countered by their inability to deal with training instabilities, leading to information loss. In addition, due to the large amount of data needed within the extensive training phase, transformers are also less suited for time-relevant tasks (Jin *et al.*, 2025).

7.6.5 Technical and computing skill requirements

One of the biggest hurdles for new researchers in the field of plant breeding and agriculture is the high degree of technical and computer skills required for the application of DL in plant phenotyping. Some studies aimed to ease the adoption of technology with the development of graphical user interface (GUI) tools, for example, RootNav2 (Yasrab *et al.*, 2019). However, there is a shortage of such easy-to-use GUI tools for speedy construction and deployment of DL models. Hence, the transfer of research to mass application in phenomics facilities around the world for real-time inference is still lagging behind. It is further complicated by the fact that DL model deployment for inference has multiple dependencies, which are related to both hardware and software. However, application of such ready-to-use devices and other solutions must be further researched for potential real-time application to obtain higher throughput in plant phenotyping (Arya *et al.*, 2022).

7.6.6 Diverse sensor data handling

DL research mainly addresses RGB images, with many algorithms modified for their efficiency and accuracy. Plant phenomics, however, is based on multispectral, hyperspectral, and fluorescence imaging, each producing distinct data compared to conventional RGB images (Signoroni *et al.*, 2019). Hyperspectral data offer both spectral and spatial information, while fluorescence imaging traces physiological processes over time. Harnessing such valuable datasets requires special CNN architectures for both better accuracy and efficiency. Phenomics data is collected across the entire development of the crop, which implies that temporal datasets play an essential role during training. Implementing a uniform image analysis procedure is challenging because different imaging sensors produce different data formats. Every kind of sensor offers different information; therefore, it might be difficult to combine data from several sources (Yang *et al.*, 2020).

7.6.7 Challenges in field conditions

The primary challenge is that most studies take place in a controlled environment, avoiding the issues such as occlusion and light variability (Fahlgren *et al.*, 2015). Parameters estimated using data from specific locations or crop varieties may not generalize well to other environments or species. This limits the broader applicability of DL models for phenotyping. Acquiring manual measurements of data is labor-intensive and time-consuming, making large-scale phenotyping projects challenging (Ninomiya, 2022). The lack of appropriate technology has made non-destructive root phenotyping in the field challenging. Underground root imaging is still difficult since there are not enough appropriate technologies, in contrast to canopy phenotyping, which has greatly profited from developments in drone (UAV) and sensor technology (Yang *et al.*, 2020).

7.7 Future Research Trends

7.7.1 Development of universal datasets

For the plant community to train reliable ML models for practical applications in decision-making, uniform datasets for a range of stresses are needed. Realistic and perhaps deteriorated sensing conditions (such as foggy, cloudy, gloomy, and saturated sunlight) should be included in the datasets to guarantee strong in-field performance of ML (Gill *et al.*, 2022).

7.7.2 Enhancement of decision support

DL has the potential to help with decision-making in plant research in the future. This includes predicting which genomic region should be altered to produce a desired phenotype and ensuring ideal local growth conditions by evaluating crop performance *in vivo* in the field or greenhouse (Gill *et al.*, 2022).

7.7.3 Advancements in transfer learning

The adaptation of pretrained models to various biotic and abiotic stresses, such as diseases, pests, floods, droughts, and salinity, will be enhanced through advances in AI and transfer learning. This will reduce the requirement of large datasets so that researchers who have limited data can utilize DL. Also, the development of community-labeled datasets, such as MNIST and ImageNet benchmarks, will increase model deployment, decrease duplication of data generation, and enhance collaboration among ML experts and plant scientists. DL will soon revolutionize plant phenotyping through the convergence of sophisticated imaging systems, increased processing power, and effective storage solutions (Gill *et al.*, 2022).

7.7.4 Easy-to-use phenotyping tools

In order to ease the complexity in acquiring and processing field images developed using unmanned aerial systems (UAS), enhancement of user-friendly phenotyping tools is crucial for managing these processes. Few examples of such tools are Easy Intermediate Data Processor (EasyIDP) (Wang *et al.*, 2021), Easy Microplot Extraction (EasyMPE) (Tresch *et al.*, 2019), Easy Pixel Color Classification (EasyPCC) (Guo *et al.*, 2017), and Easy Drone-based Canopy Profiling (EasyDCP) (Feldman *et al.*, 2021). The integration of these tools into a single pipeline on a shared data exchange platform would then be crucial to provide easy access to genotypic data (Ninomiya, 2022).

7.7.5 Training the agricultural research workforce

Training in both computer science and biology is necessary for the future research workforce. To properly utilize DL techniques and computer science advancements to address new questions in agriculture, biological science trainees need to be trained in additional computer science concepts, similar to bioinformatics training programs that were established after high-throughput sequencing technologies were developed (Murphy *et al.*, 2024).

7.8 Conclusion

Food security and sustainable agriculture are being changed by the inclusion of DL into plant phenotyping and phenomics. DL becomes more vital as climate change presents more and more issues for resilience and crop production. This technology is very accurate and quick at analyzing high-dimensional, sophisticated data, notably from sensors and photos. The potential of DL to automate the examination of complex plant features, including growth patterns, disease resistance, and stress tolerance, which are currently analyzed using arduous and error-prone manual approaches, is one of its key advantages. DL models can improve the selection of high-performing cultivars and speed up breeding by finding subtle patterns that are typically overlooked by human observation when using massive datasets. Furthermore, DL promotes HTP, which allows thousands of plants to be studied at once. Developing climate-adaptive crops demands improved knowledge of genotype-by-environment interactions, a capability strengthened by this approach. Predictive modeling also aids in the early diagnosis of crop disease and stress, enabling rapid treatments that protect yields and minimize the need for chemical inputs, and supporting more ecologically friendly farming methods. DL increases the effectiveness, precision, and scalability of crop development operations by bridging the gap between enormous amounts of phenotypic data and usable insights. DL becomes a formidable instrument to guarantee resilient and successful agricultural systems in the future as climatic circumstances become more variable and global food demands increase.

References

Ahmad, J., Farman, H., & Jan, Z. (2018). Deep learning methods and applications. In *SpringerBriefs in computer science* (pp. 31–42). https://doi.org/10.1007/978-981-13-3459-7_3

Ahmad, J., Muhammad, K., & Baik, S. W. (2017). Medical Image Retrieval with Compact Binary Codes Generated in Frequency Domain Using Highly Reactive Convolutional Features. *Journal of Medical Systems*, 42(2). https://doi.org/10.1007/s10916-017-0875-4

Akhtar, M. S., Zafar, Z., Nawaz, R., & Fraz, M. M. (2024). Unlocking plant secrets: A systematic review of 3D imaging in plant phenotyping techniques. *Computers and Electronics in Agriculture*, 222, 109033.

Alhnaity, B., Kollias, S., Leontidis, G., Jiang, S., Schamp, B., & Pearson, S. (2021). An autoencoder wavelet based deep neural network with attention mechanism for multi-step prediction of plant growth. *Information Sciences*, 560, 35-50.

Alhnaity, B., Pearson, S., Leontidis, G., & Kollias, S. (2019, June). Using deep learning to predict plant growth and yield in greenhouse environments. In *International Symposium on Advanced Technologies and Management for Innovative Greenhouses: GreenSys2019 1296* (pp. 425-432).

Alhwaiti, Y., Khan, M., Asim, M., Siddiqi, M. H., Ishaq, M., & Alruwaili, M. (2025). Leveraging YOLO deep learning models to enhance plant disease identification. *Scientific Reports*, 15(1), 7969.

Ali, T., Rehman, S. U., Ali, S., Mahmood, K., Obregon, S. A., Iglesias, R. C., Khursaid, T & Ashraf, I. (2024). Smart agriculture: utilizing machine learning and deep learning for drought stress identification in crops. *Scientific Reports*, 14(1), 1-16. https://doi.org/10.1038/s41598-024-74127-8

Al-Jabery, K. K., Obafemi-Ajayi, T., Olbricht, G. R., & Wunsch, D. C., II. (2019). Selected approaches to supervised learning. In *Elsevier eBooks* (pp. 101–123). https://doi.org/10.1016/b978-0-12-814482-4.00004-8

Almouctar, M. A. S., Wu, Y., Zhao, F., & Qin, C. (2024). Drought analysis using normalized difference vegetation index and land surface temperature over Niamey region, the southwestern of the Niger between 2013 and 2019. *Journal of Hydrology: Regional Studies*, 52, 101689.

An, J., Li, W., Li, M., Cui, S., & Yue, H. (2019). Identification and classification of maize drought stress using deep convolutional neural network. *Symmetry*, 11(2), 256.

Anami BS, Malvade NN, Palaiah S (2020) Deep learning approach for recognition and classification of yield affecting paddy crop stresses using field images. *AI Agri* 4:12–20. https://doi.org/10.1016/j.aiia.2020.03.001

Araus, J. L., & Cairns, J. E. (2014). Field high-throughput phenotyping: the new crop breeding frontier. *Trends in plant science*, 19(1), 52-61.

Araus, J. L., Kefauver, S. C., Zaman-Allah, M., Olsen, M. S., & Cairns, J. E. (2018). Translating high-throughput phenotyping into genetic gain. *Trends in plant science*, 23(5), 451-466.

Arya, S., Sandhu, K. S., Singh, J., & Kumar, S. (2022). Deep learning: as the new frontier in high-throughput plant phenotyping. *Euphytica*, 218(4). https://doi.org/10.1007/s10681-022-02992-3

Askey, B. C., Dai, R., Lee, W. S., & Kim, J. (2019). A noninvasive, machine learning–based method for monitoring anthocyanin accumulation in plants using digital color imaging. *Applications in Plant Sciences*, 7(11), e11301.

Barman, U. (2021). Deep Convolutional neural network (CNN) in tea leaf chlorophyll estimation: A new direction of modern tea farming in Assam, India. *Journal of Applied & Natural Science*, 13(3).

Bolger, A. M., Poorter, H., Dumschott, K., Bolger, M. E., Arend, D., Osorio, S., Gundlach, H., Mayer, K. F. X., Lange, M., Scholz, U., & Usadel, B. (2018). Computational aspects underlying genome to phenome analysis in plants. *The Plant Journal*, 97(1), 182–198. https://doi.org/10.1111/tpj.14179

Caiming Gou, Sara Zafar, Fatima , Zuhair Hasnain, Nazia Aslam, Naeem Iqbal, Sammar Abbas, Hui Li, Jia Li, Bo Chen, Arthur J. Ragauskas, Manzar Abbas. Machine and Deep Learning: Artificial Intelligence Application in Biotic and Abiotic Stress Management in Plants. *Front. Biosci. (Landmark Ed)* **2024**, 29(1), 20. https://doi.org/10.31083/j.fbl2901020

Catchpole, W. R., & Wheeler, C. J. (1992). Estimating plant biomass: a review of techniques. *Australian Journal of Ecology*, 17(2), 121-131.

Chakraborty, S. K., Chandel, N. S., Jat, D., Tiwari, M. K., Rajwade, Y. A., & Subeesh, A. (2022). Deep learning approaches and interventions for futuristic engineering in agriculture. *Neural Computing and Applications*, 34(23), 20539-20573.

Chandrashekar, P. B., Alatkar, S., Wang, J., Hoffman, G. E., He, C., Jin, T., Khullar, S., Bendl, J., Fullard, F. J., Roussos, P. and Wang, D. (2023). DeepGAMI: deep biologically guided auxiliary learning for multimodal integration and imputation to improve genotype–phenotype prediction. *Genome Medicine*, 15(1), 88.

Chang, L., Li, D., Hameed, M. K., Yin, Y., Huang, D., & Niu, Q. (2021). Using a hybrid neural network model DCNN–LSTM for image-based nitrogen nutrition diagnosis in muskmelon. *Horticulturae*, 7(11), 489.

Danilevicz, M. F., Gill, M., Anderson, R., Batley, J., Bennamoun, M., Bayer, P. E., & Edwards, D. (2022). Plant genotype to phenotype prediction using machine learning. *Frontiers in Genetics*, 13, 822173.

Das Choudhury, S., Maturu, S., Samal, A., Stoerger, V., & Awada, T. (2020). Leveraging image analysis to compute 3D plant phenotypes based on voxel-grid plant reconstruction. *Frontiers in plant science*, 11, 521431.

Diwan, T., Anirudh, G., & Tembhurne, J. V. (2023). Object detection using YOLO: Challenges, architectural successors, datasets and applications. *multimedia Tools and Applications*, 82(6), 9243-9275.

Du, W., & Ding, X. (2024). Optimizing Irrigation Efficiency using Deep Reinforcement Learning in the Field. *ACM Transactions on Sensor Networks.* https://doi.org/10.1145/3662182

Fabre, J., Dauzat, M., Nègre, V., Wuyts, N., Tireau, A., Gennari, E., Neveu, P., Tisné, S., Massonnet, C., Hummel, I., & Granier, C. (2011). PHENOPSIS DB: an Information System for Arabidopsis thaliana phenotypic data in an environmental context. *BMC Plant Biology*, 11(1). https://doi.org/10.1186/1471-2229-11-77

Fahlgren, N., Feldman, M., Gehan, M. A., Wilson, M. S., Shyu, C., Bryant, D. W., Hill, S. T., McEntee, C. J., Warnasooriya, S. N., Kumar, I., Ficor, T., Turnipseed, S., Gilbert, K. B., Brutnell, T. P., Carrington, J. C., Mockler, T. C., & Baxter, I. (2015). A versatile phenotyping system and analytics platform reveals diverse temporal responses to water availability in Setaria. *Molecular Plant*, 8(10), 1520–1535. https://doi.org/10.1016/j.molp.2015.06.005

Fahlgren, N., Gehan, M. A., & Baxter, I. (2015). Lights, camera, action: high-throughput plant phenotyping is ready for a close-up. *Current Opinion in Plant Biology*, 24, 93–99. https://doi.org/10.1016/j.pbi.2015.02.006

Feldman, A., Wang, H., Fukano, Y., Kato, Y., Ninomiya, S., & Guo, W. (2021). EasyDCP: An affordable, high-throughput tool to measure plant phenotypic traits in 3D. *Methods in Ecology and Evolution*, 12(9), 1679–1686. https://doi.org/10.1111/2041-210x.13645

Fiorani, F., & Schurr, U. (2013). Future scenarios for plant phenotyping. *Annual review of plant biology*, 64(1), 267-291.

Franceschi, L., Donini, M., Perrone, V., Klein, A., Archambeau, C., Seeger, M., ... & Frasconi, P. (2024). Hyperparameter Optimization in Machine Learning. *arXiv preprint arXiv:2410.22854*.

Furbank, R. T., & Tester, M. (2011). Phenomics–technologies to relieve the phenotyping bottleneck. *Trends in plant science*, 16(12), 635-644.

Gaggion, N., Ariel, F., Daric, V., Lambert, E., Legendre, S., Roulé, T., ... & Ferrante, E. (2021). ChronoRoot: High-throughput phenotyping by deep segmentation networks reveals novel temporal parameters of plant root system architecture. *GigaScience*, 10(7), giab052.

Gill, T., Gill, S. K., Saini, D. K., Chopra, Y., De Koff, J. P., & Sandhu, K. S. (2022). A comprehensive review of high throughput phenotyping and machine learning for plant stress phenotyping. *Phenomics*, 2(3), 156–183. https://doi.org/10.1007/s43657-022-00048-z

Harandi, N., Vandenberghe, B., Vankerschaver, J., Depuydt, S., & Van Messem, A. (2023). How to make sense of 3D representations for plant phenotyping: a compendium of processing and analysis techniques. *Plant Methods*, 19(1), 60.

Haykin S (2009) Neural networks and learning machines, 3rd edn. Pearson Prentice Hall, New York.

Ilyas, T., Jin, H. J., Siddique, M., Lee, S.-J., Kim, H., & Chua, L. O. (2022). DIANA: A deep learning-based paprika plant disease and pest phenotyping system with disease severity analysis. *Frontiers in Plant Science*, 13. https://doi.org/10.3389/fpls.2022.983625

Jia, W., Zhang, Y., Wei, Z., Zheng, Z., & Xie, P. (2023). Daily reference evapotranspiration prediction for irrigation scheduling decisions based on the hybrid PSO-LSTM model. *PLOS ONE*, 18(4), e0281478. https://doi.org/10.1371/journal.pone.0281478

Jin, C., Zhou, L., Pu, Y., Zhang, C., Qi, H., & Zhao, Y. (2025). Application of deep learning for high-throughput phenotyping of seed: a review. *Artificial Intelligence Review*, 58(3). https://doi.org/10.1007/s10462-024-11079-5

Jin, D., Yin, H., Zheng, R., Yoo, S. J., & Gu, Y. H. (2023). PlantInfoCMS: Scalable Plant Disease Information Collection and Management System for Training AI Models. *Sensors*, 23(11), 5032. https://doi.org/10.3390/s23115032

Jung, D., Kim, H. S., Jhin, C., Kim, H., & Park, S. H. (2020). Time-serial analysis of deep neural network models for prediction of climatic conditions inside a greenhouse. *Computers and Electronics in Agriculture*, 173, 105402. https://doi.org/10.1016/j.compag.2020.105402

Kamilaris, A., & Prenafeta-Boldú, F. X. (2018). Deep learning in agriculture: A survey. *Computers and electronics in agriculture*, 147, 70-90.

Katte, T. (2018). Recurrent neural network and its various architecture types. *International Journal of Research and Scientific Innovation*, 5(3), 124-129.

Khaki, S., Wang, L., & Archontoulis, S. V. (2020). A CNN-RNN framework for crop yield prediction. *Frontiers in Plant Science*, 10, 1750.

Kowalska, A., & Ashraf, H. (2023). Advances in deep learning algorithms for agricultural monitoring and management. *Applied Research in Artificial Intelligence and Cloud Computing*, 6(1), 68-88.

Kranz, J. (1988). Measuring plant disease. In *Experimental techniques in plant disease epidemiology* (pp. 35-50). Berlin, Heidelberg: Springer Berlin Heidelberg.

Küçükönder, H., Boyaci, S., & Akyüz, A. (2016). A modeling study with an artificial neural network: developing estimationmodels for the tomato plant leaf area. TURKISH JOURNAL OF AGRICULTURE AND FORESTRY, 40, 203–212. https://doi.org/10.3906/tar-1408-28

LeCun, Y., Bengio, Y., & Hinton, G. (2015). *Deep learning. nature*, 521(7553), 436-444.

Li, J., He, Z., Zhou, G., Yan, S., & Zhang, J. (2024). DeepAT: A Deep Learning Wheat Phenotype Prediction Model Based on Genotype Data. *Agronomy*, 14(12), 2756.

Li, L., Zhang, Q., & Huang, D. (2014). A review of imaging techniques for plant phenotyping. *Sensors*, 14(11), 20078-20111.

Li, Z., Chen, X., Shi, S., Zhang, H., Wang, X., Chen, H., ... & Li, L. (2022). DeepBSA: A deep-learning algorithm improves bulked segregant analysis for dissecting complex traits. *Molecular Plant*, 15(9), 1418-1427.

Lin, F., Lazarus, E. Z., & Rhee, S. Y. (2020). QTG-Finder2: a generalized machine-learning algorithm for prioritizing QTL causal genes in plants. *G3: Genes, Genomes, Genetics*, 10(7), 2411-2421.

Lin, K., Gong, L., Huang, Y., Liu, C., & Pan, J. (2019). Deep learning-based segmentation and quantification of cucumber powdery mildew using convolutional neural network. *Frontiers in plant science*, 10, 155.

Lin, Y., Fang, S., Kang, L., Chen, C., Yao, M., & Kuo, B. (2024). Combining recurrent neural network and sigmoid growth models for Short-Term temperature forecasting and tomato growth prediction in a plastic greenhouse. *Horticulturae*, 10(3), 230. https://doi.org/10.3390/horticulturae10030230

Liu, H. M., Wang, L., Shi, Y., Wang, X., Chang, F., & Wu, Y. (2023). A deep learning-based method for detecting granular fertilizer deposition distribution patterns in centrifugal variable-rate spreader fertilization. *Computers and Electronics in Agriculture*, 212, 108107. https://doi.org/10.1016/j.compag.2023.108107

Liu, J., Liu, Y., & Doonan, J. (2018, August). Point cloud based iterative segmentation technique for 3d plant phenotyping. In *2018 IEEE International Conference on Information and Automation (ICIA)* (pp. 1072-1077). IEEE.

Liu, Y., Wang, D., He, F., Wang, J., Joshi, T., & Xu, D. (2019). Phenotype prediction and genome-wide association study using deep convolutional neural network of soybean. *Frontiers in genetics*, 10, 1091.

Ma, W., Qiu, Z., Song, J., Cheng, Q., & Ma, C. (2017). DeepGS: Predicting phenotypes from genotypes using Deep Learning. *BioRxiv*, 241414.

Malashin, I., Tynchenko, V., Gantimurov, A., Nelyub, V., & Borodulin, A. (2024). Applications of Long Short-Term Memory (LSTM) Networks in Polymeric Sciences: A review. *Polymers*, 16(18), 2607. https://doi.org/10.3390/polym16182607

Malvade, N. N., Yakkundimath, R., Saunshi, G., Elemmi, M. C., & Baraki, P. (2022). A comparative analysis of paddy crop biotic stress classification using pre-trained deep neural networks. *Artificial Intelligence in Agriculture*, 6, 167–175.

Miao, C., Guo, A., Thompson, A. M., Yang, J., Ge, Y., & Schnable, J. C. (2021). Automation of leaf counting in maize and sorghum using deep learning. *The Plant Phenome Journal*, 4(1), e20022.

Mohanty, S. P., Hughes, D. P., & Salathé, M. (2016). Using deep learning for image-based plant disease detection. *Frontiers in plant science*, 7, 215232.

Mokhtar, A., Al-Ansari, N., El-Ssawy, W., Graf, R., Aghelpour, P., He, H., Hafez, S. M., & Abuarab, M. E. (2023). Prediction of Irrigation Water Requirements for Green Beans-Based Machine Learning Algorithm Models in Arid Region. *Water Resources Management*, 37(4), 1557–1580. https://doi.org/10.1007/s11269-023-03443-x

Murphy, K. M., Ludwig, E., Gutierrez, J., Gehan, M. A., & Donald Danforth Plant Science Center. (2024). Deep Learning in Image-Based Plant Phenotyping. In *Annual Review of Plant Biology* (pp. 771–795). https://doi.org/10.1146/annurev-arplant-070523-042828

Naidu, D. S. (2023). Prediction of tomato plant growth and yield in green house environment using ann and deep learning. *International Research Journal Of Modernization In Engineering Technology And Science. https://doi.org/10.56726/irjmets43938*

Nampally, T., Kumar, K., Chatterjee, S., Pachamuthu, R., Naik, B., & Desai, U. B. (2023). StressNet: a spatial-spectral-temporal deformable attention-based framework for water stress classification in maize. *Frontiers in Plant Science*, 14, 1241921.

Nidamanuri, R. R. (2024). Deep learning-based prediction of plant height and crown area of vegetable crops using LiDAR point cloud. *Scientific Reports*, 14(1), 14903.

Ninomiya, S. (2022). High-throughput field crop phenotyping: current status and challenges. *Breeding Science*, 72(1), 3–18. https://doi.org/10.1270/jsbbs.21069

Odounfa, M. G. F., Gbemavo, C. D., Tahi, S. P. G., & Kakaï, R. L. G. (2024). Deep learning methods for enhanced stress and pest management in market garden crops: A comprehensive analysis. *Smart Agricultural Technology*, 100521.

Okyere, F. G., Cudjoe, D. K., Virlet, N., Castle, M., Riche, A. B., Greche, L., ... & Hawkesford, M. J. (2024). Hyperspectral imaging for phenotyping plant drought stress and nitrogen interactions using multivariate modeling and machine learning techniques in wheat. *Remote Sensing*, 16(18), 3446.

Patra, A. K., & Sahoo, L. (2024). Explainable light-weight deep learning pipeline for improved drought stress identification. *Frontiers in Plant Science*, 15, 1476130.

Polly, R., & Devi, E. A. (2024). Semantic segmentation for plant leaf disease classification and damage detection: A deep learning approach. *Smart Agricultural Technology*, 9, 100526.

Rani, K., Kumar, M., Razzaq, A., Ajay, B. C., Kona, P., Bera, S. K., & Wani, S. H. (2023). Recent advances in molecular marker technology for QTL mapping in plants. *QTL Mapping in Crop Improvement*, 1-15.

Reynolds, D., Ball, J., Bauer, A., Davey, R., Griffiths, S., & Zhou, J. (2019). CropSight: a scalable and open-source information management system for distributed plant phenotyping and IoT-based crop management. *GigaScience*, 8(3). https://doi.org/10.1093/gigascience/giz009

Roy, B., Sagan, V., Haireti, A., Newcomb, M., Tuberosa, R., LeBauer, D., & Shakoor, N. (2023). Early detection of drought stress in durum wheat using hyperspectral imaging and photosystem sensing. *Remote Sensing*, 16(1), 155.

Saeed, F., Sun, S., Rodriguez-Sanchez, J., Snider, J., Liu, T., & Li, C. (2023). Cotton plant part 3D segmentation and architectural trait extraction using point voxel convolutional neural networks. *Plant methods*, 19(1), 33.

Sandhu, K. S., Lozada, D. N., Zhang, Z., Pumphrey, M. O., & Carter, A. H. (2021). Deep learning for predicting complex traits in spring wheat breeding program. *Frontiers in plant science*, 11, 613325.

Sarkar, D., Bali, R., & Sharma, T. (2018). Practical machine learning with Python. *Book" Practical Machine Learning with Python*, 25-30.

Shen, C., Liu, L., Zhu, L., Kang, J., Wang, N., & Shao, L. (2020). High-throughput in situ root image segmentation based on the improved DeepLabv3+ method. *Frontiers in Plant Science*, 11, 576791.

Sherstinsky, A. (2020). Fundamentals of Recurrent Neural Network (RNN) and Long Short-Term Memory (LSTM) network. *Physica D Nonlinear Phenomena*, 404, 132306. https://doi.org/10.1016/j.physd.2019.132306

Shi, Y., Han, L., González-Moreno, P., Dancey, D., Huang, W., Zhang, Z., ... & Dai, M. (2023). A fast Fourier convolutional deep neural network for accurate and explainable discrimination of wheat yellow rust and nitrogen deficiency from Sentinel-2 time series data. *Frontiers in Plant Science*, 14, 1250844.

Shoaib, M., Shah, B., Ei-Sappagh, S., Ali, A., Ullah, A., Alenezi, F., ... & Ali, F. (2023). An advanced deep learning models-based plant disease detection: A review of recent research. *Frontiers in Plant Science*, 14, 1158933.

Signoroni, A., Savardi, M., Baronio, A., & Benini, S. (2019). Deep Learning Meets Hyperspectral Image Analysis: A Multidisciplinary review. *Journal of Imaging*, 5(5), 52. https://doi.org/10.3390/jimaging5050052

Singh, A., Ganapathysubramanian, B., Singh, A. K., & Sarkar, S. (2016). Machine learning for high-throughput stress phenotyping in plants. *Trends in plant science*, 21(2), 110-124.

Sirsat, M. S., Cernadas, E., Fernández-Delgado, M., & Barro, S. (2018). Automatic prediction of village-wise soil fertility for several nutrients in India using a wide range of regression methods. *Computers and electronics in agriculture*, 154, 120-133.

Smith, A. G., Han, E., Petersen, J., Olsen, N. A. F., Giese, C., Athmann, M., ... & Thorup-Kristensen, K. (2022). RootPainter: deep learning segmentation of biological images with corrective annotation. *New Phytologist*, 236(2), 774-791.

Song, P., Wang, J., Guo, X., Yang, W., & Zhao, C. (2021). High-throughput phenotyping: Breaking through the bottleneck in future crop breeding. *The Crop Journal*, 9(3), 633–645. https://doi.org/10.1016/j.cj.2021.03.015

Soni, D. K., & Taneja, A. (2024, November). Leveraging EfficientNetB3 for Automated Plant Stress Identification. In *2024 8th International Conference on Electronics, Communication and Aerospace Technology (ICECA)* (pp. 1580-1584). IEEE.

Sujatha, M., Jaidhar, C. D., & Lingappa, M. (2023). 1D convolutional neural networks-based soil fertility classification and fertilizer prescription. *Ecological Informatics.* https://doi.org/10.1016/j.ecoinf.2023.102295

Sun, Y., Sun, Z., & Chen, W. (2024). The evolution of object detection methods. *Engineering Applications of Artificial Intelligence*, 133, 108458.

Surehli, M. K., Aggarwal, N., Joshi, G., & Nayyar, H. (2025). Semantic Segmentation of Plant Structures with Deep Learning and Channel-wise Attention Mechanism. *Journal of Telecommunications and Information Technology*, 56-66.

Tardieu, F., Cabrera-Bosquet, L., Pridmore, T., & Bennett, M. (2017). Plant phenomics, from sensors to knowledge. *Current Biology*, 27(15), R770-R783.

Tong, Y. S., Lee, T. H., & Yen, K. S. (2022). Deep Learning for Image-Based Plant Growth Monitoring: A Review. *International Journal of Engineering & Technology Innovation*, 12(3).

Tresch, L., Mu, Y., Itoh, A., Kaga, A., Taguchi, K., Hirafuji, M., Ninomiya, S., & Guo, W. (2019). Easy MPE: Extraction of quality microplot images for UAV-Based High-Throughput field phenotyping. *Plant Phenomics, 2019*. https://doi.org/10.34133/2019/2591849

Tsaftaris, S. A., Minervini, M., & Scharr, H. (2016). Machine learning for plant phenotyping needs image processing. *Trends in plant science*, 21(12), 989-991.

Wang, H., Duan, Y., Shi, Y., Kato, Y., Ninomiya, S., & Guo, W. (2021). EasyIDP: a Python package for intermediate data processing in UAV-Based plant phenotyping. *Remote Sensing*, 13(13), 2622. https://doi.org/10.3390/rs13132622

Weihs, B. J., Heuschele, D. J., Tang, Z., York, L. M., Zhang, Z., & Xu, Z. (2024). The state of the art in root system architecture image analysis using artificial intelligence: a review. *Plant Phenomics*, 6, 0178.

Wen, W., Timmermans, J., Chen, Q., & van Bodegom, P. M. (2022). Monitoring the combined effects of drought and salinity stress on crops using remote sensing. *Hydrology and Earth System Sciences Discussions*, 2022, 1-15.

Wilkinson, M. D., Dumontier, M., Aalbersberg, I. J., Appleton, G., Axton, M., Baak, A., Blomberg, N., Boiten, J., Da Silva Santos, L. B., Bourne, P. E., Bouwman, J., Brookes, A. J., Clark, T., Crosas, M., Dillo, I., Dumon,

O., Edmunds, S., Evelo, C. T., Finkers, R., . . . Mons, B. (2016). The FAIR Guiding Principles for scientific data management and stewardship. *Scientific Data*, 3(1). https://doi.org/10.1038/sdata.2016.18

Xie, C., & Yang, C. (2020). A review on plant high-throughput phenotyping traits using UAV-based sensors. *Computers and Electronics in Agriculture*, 178, 105731. https://doi.org/10.1016/j.compag.2020.105731

Yang, W., Feng, H., Zhang, X., Zhang, J., Doonan, J. H., Batchelor, W. D., Xiong, L., & Yan, J. (2020). Crop Phenomics and High-Throughput Phenotyping: past decades, current challenges, and future Perspectives. *Molecular Plant*, 13(2), 187–214. https://doi.org/10.1016/j.molp.2020.01.008

Yasrab, R., Atkinson, J. A., Wells, D. M., French, A. P., Pridmore, T. P., & Pound, M. P. (2019). RootNav 2.0: Deep learning for automatic navigation of complex plant root architectures. *GigaScience*, 8(11). https://doi.org/10.1093/gigascience/giz123

Yasrab, R., Zhang, J., Smyth, P., & Pound, M. P. (2021). Predicting plant growth from time-series data using deep learning. *Remote Sensing*, 13(3), 331.

Ye, Z., Yin, S., Cao, Y., & Wang, Y. (2024). AI-driven optimization of agricultural water management for enhanced sustainability. *Dental Science Reports*, 14(1). https://doi.org/10.1038/s41598-024-76915-8

Zaremba, W., Sutskever, I., & Vinyals, O. (2014). Recurrent neural network regularization. *arXiv (Cornell University)*. https://doi.org/10.48550/arxiv.1409.2329

Zhang, C., Cao, C., Huang, D., & Gong, Y. (2023). An ultra-lightweight efficient network for image-based plant disease and pest infection detection. *Precision Agriculture.* https://doi.org/10.1007/s11119-023-10020-0

Zou, J., Han, Y., & So, S. (2008). Overview of artificial neural networks. *Methods in Molecular Biology*, 14–22. https://doi.org/10.1007/978-1-60327-101-1_2

Zuo, Z.-T., Bu, Q., Mao, H. P., Zhang, X., Han, L., & Ni, J. (2023). Study on the detection of water status of tomato (Solanum lycopersicum L.) by multimodal deep learning. *Frontiers in Plant Science*, 14. https://doi.org/10.3389/fpls.2023.1094142

8 Deep Learning-Enhanced Plant Genomic Prediction

Sarfraz Ahmad[1]*, Saniya[2], Mohammad Ghani Raghib[3], Vikas Belwal[4], Shaukat Ali[5], Rubina Khan[6], Sunil Gocher[7], and Sumer Singh Punia[8]

[1]Department of Genetics and Plant Breeding, College of Agriculture Mandawa, Swami Keshwanand Rajasthan Agriculture University, Bikaner, Rajasthan, India; [2]Department of Horticulture, ICAR-National Research Centre for Grapes, Pune, Maharashtra, India; [3]Department of Horticulture, Faculty of Agriculture, Nangarhar University, Jalalabad, Nangarhar, Afghanistan; [4]Department of Tree Improvement, College of Forestry Ranichauri, VCSG Uttarakhand University of Horticulture and Forestry, Bharsar, Uttarakhand, India; [5]Department of Agronomy, Krishi Vigyan Kendra Chandgothi, Swami Keshwanand Rajasthan Agriculture University, Bikaner, Rajasthan, India; [6]Department of Genetics and Plant Breeding, College of Agriculture Jobner, Sri Karan Narendra Agriculture University, Jobner, Rajasthan, India; [7]Department of Entomology, College of Agriculture Jobner, Sri Karan Narendra Agriculture University, Jobner, Rajasthan, India; [8]Department of Genetics and Plant Breeding, College of Agriculture Jobner, Sri Karan Narendra Agriculture University, Jobner, Rajasthan, India

Abstract

Genomic prediction (GP) has become a crucial component of modern plant breeding, enabling early selection of high-performance genotypes based on molecular markers. While traditional statistical models, such as genomic best linear unbiased prediction (GBLUP) and Bayesian approaches, have facilitated genetic gain, they often fall short in capturing the non-linear and high-dimensional interactions inherent in plant genomes. Deep learning (DL), with its hierarchical architecture and representation learning capabilities, has emerged as a powerful alternative, offering improved accuracy, scalability, and adaptability in GP. This chapter provides a comprehensive overview of DL applications in plant GP, discussing model architectures (convolutional neural networks (CNNs), recurrent neural networks (RNNs), transformers), data integration strategies, and performance comparisons with conventional methods. It further explores case studies across key crops (e.g. corn, rice, wheat), examines challenges such as overfitting and interpretability, and discusses future prospects, including multi-omics integration and explainable artificial intelligence (XAI). The integration of DL and genomics holds the potential to revolutionize data-driven breeding for climate-resilient and high-yielding crops.

Keywords: convolutional neural networks, deep learning, genotype–phenotype mapping, genomic selection, machine learning, multi-omics integration, non-linear prediction models, plant breeding

*E-mail: sarfraz_firoz@rediffmail.com

DOI: 10.1079/9781836991120.0008

8.1 Introduction

Plant breeding has undergone a transformative evolution with the advent of molecular genetics and high-throughput technologies. Traditional phenotypic selection, although foundational, is time-consuming, labor-intensive, and heavily influenced by environmental variability. The integration of genotyping technologies, especially high-density single-nucleotide polymorphism (SNP) arrays and genotyping-by-sequencing, has enabled breeders to incorporate molecular data into selection decisions. Genomic prediction (GP), particularly, has emerged as a cornerstone in modern plant breeding programs. It allows breeders to predict the genetic merit of individuals based on genome-wide marker data, expediting the selection cycle and improving genetic gain per unit time (Sinha *et al.*, 2023).

Traditional GP models, such as genomic best linear unbiased prediction (GBLUP), ridge regression BLUP (RR-BLUP), and Bayesian approaches (e.g. BayesA, BayesB), are primarily linear models. These methods assume additive genetic effects and typically use marker-derived genomic relationship matrices. While effective in many scenarios, these models have limitations in capturing complex genetic architectures, including non-additive effects such as dominance, epistasis, and genotype × environment (G × E) interactions (Jadhav and Jumbo, 2024). Moreover, as omics data scales in volume and complexity, these linear models struggle to exploit the full potential of available datasets.

Recent advances in artificial intelligence (AI), and more specifically deep learning (DL), present a paradigm shift in the modeling of complex biological systems. DL models, as a subset of machine learning, are characterized by their ability to learn hierarchical representations from raw data through multiple layers of non-linear transformation. These models are especially suited for high-dimensional, unstructured, and multimodal data features increasingly common in plant genomic datasets. The potential of DL to model non-linear and interactive genetic effects without prior assumptions makes it a powerful tool for GP (Montesinos-López *et al.*, 2021).

One of the key advantages of DL lies in its flexibility and scalability. Architectures such as convolutional neural networks (CNNs), recurrent neural networks (RNNs), autoencoders (AEs), and transformer models have been successfully adapted to GP tasks. CNNs are particularly effective at detecting spatial and local patterns in data, such as those found in ordered SNP arrays, by convolving filters over genetic marker matrices. RNNs, and more advanced versions such as long short-term memory (LSTM) networks, are adept at modeling sequential dependencies and have been explored for temporal phenotyping and longitudinal GP (Mienye *et al.*, 2024). AEs have demonstrated utility in feature extraction and dimensionality reduction, offering a path to mitigate overfitting in genomic data. More recently, transformer-based models, originally developed for natural language processing, have been employed for modeling long-range interactions in genomic sequences and integrating heterogeneous data types (Consens *et al.*, 2025).

As the size and complexity of breeding datasets continue to grow, driven by advancements in phenotyping platforms (e.g. UAVs, hyperspectral imaging), multi-omics profiling (e.g. transcriptomics, metabolomics), and environmental data acquisition, DL's capacity to integrate and model heterogeneous inputs becomes even more critical. Integrative frameworks that combine genomic data with phenotypic, environmental, and omics data using DL architectures are showing promise in predicting complex traits more accurately than conventional models (Wu and Xie, 2021).

Despite its strengths, DL in plant GP is still an emerging field with unique challenges. DL models are often data-hungry and require large, high-quality labeled datasets for effective training. In many crop species, particularly minor or orphan crops, such data is scarce. Moreover, the "black-box" nature of deep neural networks presents challenges in biological interpretation and model transparency. To address these concerns, explainable AI (XAI) techniques are being developed to identify influential markers, genes, or genomic regions contributing to trait predictions (Novielli *et al.*, 2024).

Importantly, the application of DL in plant breeding is not intended to replace traditional methods but rather to augment and enhance them. By enabling breeders to exploit non-linear patterns, latent feature representations, and multidimensional data integration, DL models offer a complementary toolset to traditional quantitative genetics. The integration of DL with GP not only holds promise for improving prediction accuracy but also facilitates data-driven decision-making in breeding pipelines.

In conclusion, DL offers unprecedented opportunities to improve the accuracy and efficiency of GP in plant breeding. Its ability to handle complex, non-linear, and high-dimensional data makes it uniquely suited for the genomic era of plant science. However, realizing its full potential will require interdisciplinary collaboration among geneticists, breeders, data scientists, and computer scientists. As computational resources become more accessible and data generation continues to expand, DL-enhanced GP will likely become an indispensable component of next-generation breeding programs focused on sustainability, productivity, and resilience.

8.2 Fundamentals of GP in Plants

GP refers to the use of genome-wide molecular marker data to predict the genetic value of individuals, facilitating the selection of superior genotypes without exhaustive phenotyping. Unlike traditional marker-assisted selection (MAS), which relies on a few statistically significant markers linked to major quantitative trait loci (QTL), GP simultaneously utilizes all available markers across the genome, regardless of their statistical significance (Wang *et al.*, 2018). This holistic approach enhances the prediction of complex quantitative traits that are influenced by many loci of small effect.

GP has revolutionized modern plant breeding by enabling early selection of elite individuals, reducing reliance on large-scale field trials, and shortening the breeding cycle (Sinha *et al.*, 2023). These improvements are particularly valuable in long-duration crops such as wheat, corn, and forest trees, where traditional breeding timelines span multiple years.

8.2.1 Principles of genomic selection

The underlying rationale of GP is based on the premise that dense molecular markers (typically SNPs) can capture most of the additive genetic variance through linkage disequilibrium (LD) with causal variants. When individuals are genotyped with high-density markers and their phenotypes are available for one generation (training population), this information is used to build a predictive model. The trained model is then applied to predict the genomic estimated breeding values (GEBVs) of new, unphenotyped individuals (test population), thereby guiding selection decisions.

The success of GP depends on several factors:

- *Trait heritability.* Higher heritability generally leads to better prediction accuracy.
- *Marker density.* Adequate marker coverage is essential to capture the genome-wide variance.
- *Training population size and composition.* Larger and genetically diverse training sets yield more robust models.
- *Genetic architecture.* Traits controlled by additive effects are better predicted than those with strong epistatic interactions using linear models.

8.2.2 Conventional models in GP

Several statistical models form the foundation of GP, each tailored to different trait architectures and assumptions about marker effects.

Genomic best linear unbiased prediction (GBLUP)

GBLUP is one of the most widely used methods in GP and is a natural extension of traditional best linear unbiased prediction (BLUP) . It models the genetic relationship matrix using SNP data instead of pedigree information. Assuming equal contribution of all markers, GBLUP estimates GEBVs through mixed linear models:

$$\mathbf{y} = \mathbf{X}\boldsymbol{\beta} + \mathbf{Z}\mathbf{g} + \boldsymbol{\varepsilon}$$

where

- $\mathbf{y}$ is the vector of observed phenotypes.
- $\mathbf{X}$ is the design matrix for fixed effects.
- $\boldsymbol{\beta}$ is the vector of fixed effects (e.g. mean, environmental effects).
- $\mathbf{Z}$ is the design matrix for random genetic effects.
- $\mathbf{g}$ is the vector of random additive genetic effects (breeding values).
- $\boldsymbol{\varepsilon}$ is the vector of residuals.

GBLUP performs well for traits primarily governed by additive genetic variance and has been successfully applied in crops such as wheat, rice, and barley (Budhlakoti *et al.*, 2022).

Ridge regression BLUP (RR-BLUP)

RR-BLUP is a frequentist approach that minimizes the residual sum of squares with a penalty on the size of the regression coefficients, effectively shrinking all marker effects toward zero. It is computationally efficient and particularly suitable for traits with many small-effect QTLs. However, its uniform shrinkage assumption may limit its utility for traits with large-effect loci or non-additive effects.

Bayesian models

Bayesian approaches provide a flexible framework by incorporating prior distributions on marker effects. Common models include the following:

- *BayesA* has its own variance, allowing for heteroscedasticity.
- *BayesB* assumes that many markers have zero effect, better for sparse traits.
- *BayesC* estimates the proportion of markers with non-zero effects from the data.

Bayesian models are useful when prior biological knowledge is available or when trait architecture is complex, involving a mix of large and small effect loci (Howard *et al.*, 2022).

Machine learning methods

Recent years have seen the introduction of classical machine learning algorithms, such as support vector machines, random forests, and gradient boosting machines into GP. These models can capture some non-linear relationships and interactions among markers but often require hyperparameter tuning and may not scale well with very high-dimensional data (Zhou *et al.*, 2024).

8.2.3 Factors influencing GP accuracy

Several factors influence the performance of GP models:

- Genetic relatedness between training and testing populations affects how well marker effects transfer.
- LD patterns affect how comprehensively markers can tag causal variants.
- Trait architecture, including the number of QTLs, their effect sizes, and interactions, dictates the suitability of a modeling approach.
- Phenotyping precision is critical since noisy trait measurements can reduce model accuracy.
- Population size and structure influence the power to detect marker–trait associations and generalize predictions.

A well-designed training population with high diversity and accurate phenotyping is critical for building robust GP models.

8.2.4 Evaluation of model performance

GP models are typically evaluated using cross-validation, where the dataset is partitioned into training and validation sets. The following metrics are commonly used:

- *Predictive correlation (r)* measures the Pearson correlation between observed and predicted values.
- *Root mean squared error (RMSE)* reflects the average deviation of predictions from true values.
- *Mean absolute error (MAE)* offers an interpretable measure of average error.

Multi-environment prediction strategies, including G × E interaction modeling, are increasingly used in crops exposed to diverse growing conditions (Malosetti *et al.*, 2013).

8.2.5 Transitioning to DL

While linear and Bayesian methods have served as the foundation for GP, they often fall short in accounting for the complexity of genetic architectures, especially for traits with epistatic interactions, non-additive effects, and G × E interactions. DL offers a powerful alternative by learning complex, non-linear patterns from high-dimensional data without predefined assumptions.

Moreover, DL models are more adept at integrating multimodal data sources, including

- genotypic data (SNPs, haplotypes);
- phenotypic records;

- environmental variables; and
- multi-omics data (transcriptomics, metabolomics).

This transition from traditional statistical learning to DL in GP is a pivotal step toward the implementation of predictive breeding systems that are adaptive, scalable, and data-driven.

8.3 DL Architectures for GP

DL, a subset of machine learning, has gained substantial traction in the genomics domain due to its capacity to automatically extract intricate features from large-scale, high-dimensional datasets. In plant breeding, DL provides an alternative to conventional GP models by modeling complex non-linear relationships and capturing higher-order interactions among genetic markers that traditional linear models struggle to detect (Abdollahi-Arpanahi *et al.*, 2020). High-level workflows showing how DL is used in plant GP are presented in Fig. 8.1.

Unlike traditional methods that require prespecifying interaction terms or kernel functions, DL models are data-driven and hierarchical in nature. This means they can adapt to discover latent genomic structures without manual feature engineering, making them particularly suitable for polygenic traits governed by intricate genetic architectures, including epistasis, G × E interactions, and non-additive effects.

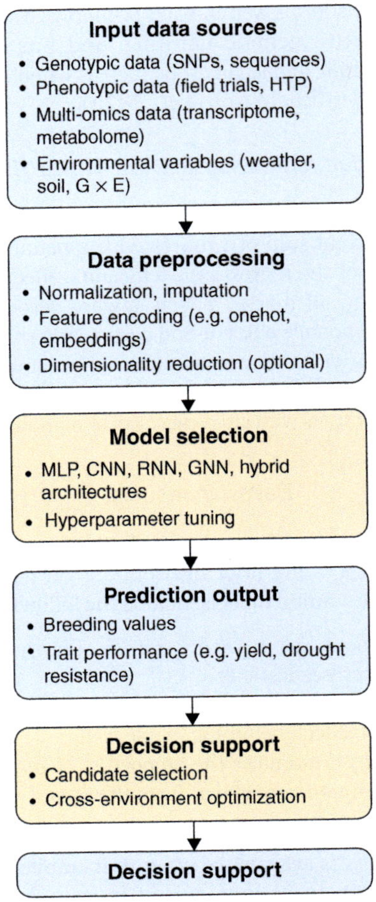

Fig. 8.1. General workflow of deep learning in plant genomic prediction. CNN, convolutional neural network; GNN, graph neural network; MLP, multilayer perceptron; RNN, recurrent neural network (Author's own image.)

8.3.1 Common DL architectures in GP

Several DL architectures have been explored for plant GP, each offering unique advantages depending on data structure and target traits. A brief summary of DL architectures used in plant GP is presented in Table 8.1.

Fully connected neural networks (FCNN or multilayer perceptron (MLP))

The most straightforward DL model applied in GP is the FCNN, also known as an MLP. These networks consist of an input layer (e.g. SNPs), one or more hidden layers with non-linear activation functions (e.g. rectified linear unit), and an output layer predicting phenotypic traits or breeding values.

APPLICATIONS.

- Jubair *et al.* (2023) demonstrated the use of FCNNs in wheat for yield prediction.
- Zhou *et al.* (2024) reported that MLPs outperformed GBLUP for certain traits in soybean, especially when complex marker interactions were involved.

ADVANTAGES.

- Captures non-linear relationships and interactions
- Easy to implement with moderate computing resources.

LIMITATIONS.

- Struggles with overfitting when training data are limited
- Less effective at modeling sequential or spatial dependencies in marker data.

Convolutional neural networks (CNNs)

CNNs are designed to capture local dependencies and hierarchical patterns, originally popularized in image processing. In GP, the SNP sequence can be modeled as a one-dimensional array, where convolutional filters detect short-range marker interactions (e.g. haplotypes or LD blocks).

APPLICATIONS.

- Washburn *et al.* (2020) applied CNNs to corn genotypic data and observed improved predictive accuracy compared to traditional linear models.
- Waldmann *et al.* (2020) evaluated CNNs for multiple traits across species, finding that CNNs captured local SNP interactions more effectively than RR-BLUP.

ADVANTAGES.

- Effective at handling high-dimensional input data
- Capable of detecting local interaction patterns between markers.

LIMITATIONS.

- May overlook long-range interactions unless combined with deeper layers or dilated convolutions
- Requires substantial training data to generalize well.

Recurrent neural networks (RNNs) and long short-term memory networks (LSTMs)

RNNs are designed for sequence data and are especially useful when the order of inputs carries meaning. In GP, the sequence of SNPs can be treated in a manner similar to time-series data. LSTM networks, a variant of RNNs, are capable of learning long-range dependencies and avoiding the vanishing gradient problem.

APPLICATIONS.

- Wang *et al.* (2025) demonstrated that LSTM networks could capture long-distance epistatic interactions in wheat.

Table 8.1. DL architectures used in plant GP. Author's own table.

DL architecture	Description	Use case	References
MLP (fully connected NN)	Learns global patterns from genotypic data	Trait prediction (e.g. grain yield)	Sandhu *et al.* (2021b)
CNN	Extracts local feature patterns from genomic sequences or images	Marker pattern learning, image-based traits	Zhang and Imoto (2024)
RNN/LSTM	Models temporal data and sequential dependencies	Time-series traits, G × E dynamics	Waqas and Humphries (2024)
GNNs	Operates on structured genomic graphs/ networks	Multi-omics and gene networks	Li *et al.* (2024b)
Autoencoder	Learns compressed representation for feature reduction	Dimensionality reduction, data integration	Petscharnig *et al.* (2017)
Hybrid (CNN + LSTM)	Combines spatial and temporal pattern recognition	Climate-resilient trait prediction	Khalilzadeh *et al.* (2024)

CNN, convolutional neural network; GNN, graph neural network; LSTM, long short-term memory; MLP, multilayer perceptron; NN, neuron network; RNN, recurrent neural network.

- Shook *et al.* (2021) integrated LSTMs with image phenotyping and genotyping data for soybean yield prediction.

ADVANTAGES.

- Models sequential dependencies and long-range marker interactions
- Particularly suitable for genome sequences with conserved patterns.

LIMITATIONS.

- More computationally demanding
- Requires careful tuning of hyperparameters to prevent overfitting.

Autoencoders (AEs)

AEs are unsupervised DL architectures used to reduce the dimensionality of genomic data by learning compressed latent representations. The encoder transforms the input SNP data into a lower-dimensional feature space, while the decoder reconstructs the original input. The latent features can then be fed into a downstream prediction model.

APPLICATIONS.

- Li *et al.* (2024a) used stacked AEs for wheat GP and found that latent representations improved trait prediction accuracy.
- Tsimenidis *et al.* (2022) highlighted the usefulness of AEs in extracting biologically meaningful features from large-scale omics data.

ADVANTAGES.

- Effective at denoising and compressing high-dimensional SNP data
- Reduces overfitting risk by filtering redundant features.

LIMITATIONS.

- Reconstruction loss may not always align with predictive objectives
- Interpretability of latent features is limited.

Graph neural networks (GNNs)

GNNs extend DL to graph-structured data, where SNPs or genes can be treated as nodes and their biological or statistical relationships as edges. GNNs offer an advanced way to model structured data and leverage prior biological knowledge (e.g. gene regulatory networks or metabolic pathways).

APPLICATIONS.

- Liu (2024) applied GNNs to integrate genotype and expression data for gene–trait prediction.
- Some studies propose combining GNNs with GBLUP to form hybrid prediction models that account for both statistical and biological interactions.

ADVANTAGES.

- Capture both local and global interactions among genomic features
- Integrate multi-omics and network information.

LIMITATIONS.

- Still emerging in plant GP applications
- Complex to implement and require biological network data.

8.3.2 Comparison with traditional models

DL methods generally outperform linear models (e.g. GBLUP, RR-BLUP) when:

- The trait has a complex architecture with non-additive effects.
- Large, diverse training populations are available.
- Multiple data modalities (e.g. genomics, phenomics, environment) are integrated.

However, traditional models are still competitive for traits with high heritability and additive variance, especially in small datasets, where DL models risk overfitting. Moreover, DL models often lack interpretability—a key drawback in breeding applications, where biological understanding is critical.

8.3.3 Transfer learning and pretraining

Transfer learning, using a model trained on one task or crop and fine-tuning it for another, has emerged as a promising approach to overcome data scarcity in minor crops. Pre-trained models on major crops can be reused, reducing the need for large-scale training datasets.

Example:

- A CNN model trained on corn genotypes may be fine-tuned for predicting drought tolerance in sorghum if the marker architecture is sufficiently conserved.

8.3.4 Integration with multi-omics and environmental data

Advanced DL architectures can fuse SNP data with environmental, transcriptomic, and phenotypic data for G × E interaction modeling and trait dissection. Multimodal DL models are being developed to process heterogeneous inputs simultaneously, leveraging attention mechanisms and encoder–decoder structures.

Example:

- Togninalli *et al.* (2023) proposed a multimodal DL framework that integrates genomic and environmental covariates for predicting grain yield across environments.

8.3.5 Future architectures and innovations

Emerging architectures such as:

- *Transformer models*, adapted from natural language processing, are showing promise for modeling long-range genomic dependencies.
- *Attention mechanisms* offer better interpretability by highlighting influential SNPs or genomic regions.
- *Hybrid models* combining statistical and DL components may balance accuracy and interpretability.

8.4 Data Integration in Deep GP

In plant genomics, accurate trait prediction often requires the integration of multiple types of data beyond genomic markers. While single-omic approaches based solely on SNPs or genotypic data have shown promise, they may not fully capture the complexity of plant phenotypes, which are the cumulative result of interactions among genes, environmental factors, and various molecular processes. Integrating multi-omics data (e.g. transcriptomics, metabolomics, proteomics) along with environmental and high-throughput phenotyping (HTP) information has become increasingly important to develop robust, accurate, and biologically meaningful GP models (Misra *et al.*, 2019).

DL frameworks are particularly well suited to handle such integration tasks due to their ability to model non-linear relationships and accommodate diverse, high-dimensional data modalities. This section outlines the strategies, architectures, and challenges associated with integrating heterogeneous datasets in DL-based plant GP.

8.4.1 The rationale for multimodal integration

Phenotypic variation in plants arises from a cascade of molecular processes that begin at the genome level and are influenced by environmental cues. Single-level data often lack sufficient resolution to capture this complexity, especially for traits influenced by epistasis, gene expression regulation, and G × E interactions. Integrating multimodal data helps:

- *Capture biological context*. For example, integrating SNPs with transcriptomic data can help identify gene expression signatures underlying trait variation (Zhang *et al.*, 2019).
- *Improve predictive power*. Combining omics data can reduce noise and redundancy, allowing models to focus on biologically relevant features.
- *Model complex interactions*. DL architectures can learn intricate hierarchical relationships among omics layers and external covariates.

8.4.2 Types of data integrated in deep GP

Genomic data (SNPs and haplotypes)

SNPs form the backbone of most GP models. In DL approaches, SNP arrays are typically used as input vectors or matrices, allowing the model to learn non-linear relationships with phenotypes. Haplotypes and LD blocks can also be included to capture more stable genomic segments.

Transcriptomics

Gene expression profiles reflect the intermediate functional state of the genome. Transcriptomic data can be particularly informative in modeling traits such as drought tolerance, flowering time, and disease resistance.

Example:

- Yuan *et al.* (2019) used DL to integrate RNA-Seq data with SNPs in corn, improving prediction accuracy for flowering time.

Epigenomics and proteomics

Although less commonly used, DNA methylation, histone modifications, and protein abundance profiles provide additional regulatory and post-translational information.

- These layers are vital for modeling developmental and stress-responsive traits.

Metabolomics

Metabolite profiles are often closely linked to phenotypes, especially yield and quality traits. Integrating metabolomics can help identify biochemical pathways influencing target traits.

Environmental and management data

Environmental covariates, such as temperature, precipitation, and soil properties, are crucial in modeling G × E interactions. DL models can dynamically integrate such data using spatio-temporal or recurrent architectures.

- Jackson *et al.* (2023) developed a DL model incorporating weather data with SNPs, improving yield prediction across environments in wheat.

HTP data

HTP technologies, such as hyperspectral imaging, UAVs, and thermal sensors, generate large-scale phenotypic data. These can be used as features or even directly modeled using CNNs.

- Larue *et al.* (2024) demonstrated a CNN model predicting crop biomass using image and genomic data.

8.4.3 DL strategies for data integration

Early fusion

In early fusion, multiple data types are concatenated at the input layer. This method assumes that all data are of similar scale and importance.

- Example: Concatenating SNP and transcriptome vectors into a single input for an FCNN.

ADVANTAGES.

- Simple to implement
- Allows direct modeling of cross-modal interactions.

LIMITATIONS.

- May lead to dominance of high-dimensional data types (e.g. SNPs over metabolites)
- Requires careful preprocessing and normalization.

Late fusion

In late fusion, each data type is processed through a separate subnetwork, and the outputs are merged at a later stage.

- For instance, a CNN may process SNPs, while an RNN handles environmental time-series data. Their outputs are then concatenated and passed to a shared dense layer.

ADVANTAGES.

- Preserves modality-specific feature learning
- More robust to noise in individual modalities.

LIMITATIONS.

- More complex architecture requiring more parameters and training time.

Hybrid and attention-based models

Hybrid fusion combines early and late strategies, using attention mechanisms to weight each modality based on its contribution to prediction accuracy.

- Transformers and multihead attention networks are being explored to dynamically select relevant features from each omics layer.

Example:

- Huang *et al.* (2023) implemented a transformer-based genomic model that integrated SNPs and transcriptomic data, achieving state-of-the-art results in *Arabidopsis* seed germination.

8.4.4 Challenges in data integration

Data heterogeneity

Different omics platforms generate data with varying dimensions, scales, and distributions. Normalization and feature transformation are necessary before integration.

Missing data

Missing values are common in large-scale multi-omics studies due to cost, technical limitations, or sample dropout. DL models must be robust to partial input, and imputation strategies may be required.

Overfitting and model complexity

As more data types are integrated, the risk of overfitting increases. Techniques such as dropout, batch normalization, and transfer learning are essential to regularize complex models.

Interpretability

Integrated DL models often become "black boxes," making it difficult to determine which data types or features are most influential. This poses challenges in breeding applications where explainability is essential.

8.4.5 Future directions in multi-omics integration

- *Graph-based integration.* Using GNNs to model biological networks (e.g. gene regulatory or metabolic networks) as part of the learning framework.
- *Federated learning.* Distributed DL frameworks that allow data integration across institutions without sharing sensitive data.
- *Multitask learning.* Simultaneously predicting multiple traits using shared representations learned from integrated data.
- *Causal modeling.* Combining DL with causal inference techniques to identify drivers of phenotypic variation.

8.5 Application of DL in Enhanced Plant GP

The application of DL in plant GP represents a transformative shift in modern crop improvement strategies. As plant breeders face challenges such as climate variability, resource limitations, and rising global food demands, the need for predictive models that can capture complex genetic and environmental interactions has never been more urgent. DL models provide an advanced analytical framework capable of leveraging vast, multidimensional datasets, including genomics, phenomics, transcriptomics, and environmental data, to predict complex traits with improved accuracy and robustness.

This section outlines key applications of DL in GP, spanning across crops, traits, and breeding objectives, and demonstrates how DL has been used to enhance prediction accuracy, accelerate selection, and enable more informed decision-making in plant breeding pipelines. Table 8.2 presents a summary comparing key DL applications across different crops and traits for plant GP.

Table 8.2. Comparing key DL applications across different crops and traits for plant genomic prediction. Author's own table.

Crop	Trait(s)	DL architecture	Data integrated	Key findings	References
Corn	Plant height, grain yield	MLP, CNN	Genomic image-based prediction in corn hybrids	MLP and CNN outperformed GBLUP.	Galli *et al.* (2022)
Soybean	Yield prediction	RNN, LSTM	Image phenotyping and genotyping data	RNN, LSTM integrated with image phenotyping and genotyping data for yield prediction.	Shook *et al.* (2021)
Wheat	Yield prediction	CNN, LSTM	Weather data combined with Sentinel-2 data to establish the deep learning network	Weather data can improve the accuracy of crop yield predictions.	Peng *et al.* (2024)
Sorghum	Biomass	LSTM, RNN	Genomic, time-series remote sensing and weather data	LSTM-based RNN model showed high accuracy for biomass prediction.	Wang *et al.* (2023)
Rice	Rice blast disease assessment	CNN, LSTM	Robust dataset of over 15,000 distinctly labeled images	DL models achieved a high accuracy rate of 97.97%.	Kukreja *et al.* (2024)
Cassava	Plant disease	Transfer learning, CNN	Dataset of 5656 images of cassava	Deep transfer learning model attains a promising accuracy of 88%.	Akinpelu *et al.* (2025)
Arabidopsis	Seed germination	Transformer-based genomic model	SNPs and transcriptomic data	Achieving state-of-the-art results in *Arabidopsis* seed germination.	Huang *et al.* (2023)
Tomato	Fruit size, disease resistance, yield	CNN, MLP, LSTM, ANN	Genomic data, phenotypic data	DL models were used to predict fruit size and disease resistance more effectively than conventional models.	Gerakari *et al.* (2025)
Barley	Flowering time and grain yield estimation	CNN, LSTM	Genomic and phenomic data	LSTM outperformed the other baseline methods in handling complex, high-dimensional agricultural datasets and enhancing the accuracy of crop phenotype prediction.	Wang *et al.* (2024)

ANN, artificial neural network; CNN, convolutional neural network; DL, deep learning; LSTM, long short-term memory; MLP, multilayer perceptron; NN, neuron network; RNN = recurrent neural network.

8.5.1 GEBV predictions

A central application of DL in GP is the prediction of GEBVs, which quantify the genetic potential of individuals for a given trait. Traditional methods, such as GBLUP and Bayesian approaches, perform well for additive genetic architectures but fall short when modeling non-linear or epistatic interactions.

DL models, such as MLPs, CNNs, and RNNs, have shown improved performance by capturing high-order marker interactions.

- Rice and Lipka (2021) applied DL models to predict GEBVs in corn for grain yield, showing higher accuracy than GBLUP in scenarios involving non-additive effects.
- Singh *et al.* (2021) found that DL models, particularly CNNs, yielded better GEBV predictions for soybean seed composition traits.

These studies demonstrate the suitability of DL for traits governed by complex genetic architectures and for leveraging high-dimensional marker data.

8.5.2 Multi-trait GP

Multi-trait GP is essential in breeding programs where selection indices are based on multiple correlated traits (e.g. yield, drought tolerance, and disease resistance). DL models can simultaneously learn patterns across traits, exploiting genetic correlations to enhance predictive accuracy.

- Montesinos-López *et al.* (2019) used a multi-output DL model to predict five agronomic traits in corn simultaneously. Their model outperformed univariate models, especially when traits exhibited moderate-to-high genetic correlations.

By incorporating multitask learning strategies, DL models support integrated selection strategies, helping breeders make more holistic decisions.

8.5.3 Modeling G × E interactions

Environmental heterogeneity poses a major challenge in GP. Accurately predicting trait performance across diverse environments requires models that can dynamically incorporate environmental variables and their interactions with genotype.

DL offers several advantages in modeling G × E interactions:

- Peng *et al.* (2024) used LSTM networks to model temporal environmental data in wheat breeding trials, integrating genotype and weather sequences to predict yield stability.
- Srivastava *et al.* (2022) developed a CNN-based G × E model that included weather covariates (e.g. rainfall, temperature) as parallel inputs alongside genomic data. The model showed improved accuracy for grain yield prediction across environments in wheat.

These applications underscore the capability of DL to model non-linear G × E effects, supporting genotype selection for specific environments or across multiple locations.

8.5.4 Incorporating HTP

HTP technologies such as hyperspectral imaging, light detection and ranging, and UAV-based platforms—generate massive phenotypic datasets that are often underutilized in conventional GP frameworks. DL models, particularly CNNs and AEs, are adept at extracting informative features from HTP data.

- Ranđelović *et al.* (2023) developed a DL framework combining hyperspectral images and genotypic data to predict biomass in soybean. Their CNN architecture improved prediction by learning spatial and spectral features relevant to plant vigor.
- Jurado-Ruiz *et al.* (2023) used deep AEs to compress image-based phenotyping data and integrate it with SNP data.

By integrating phenomic and genomic data streams, DL enhances the resolution and predictive power of genomic selection models.

8.5.5 Enhancing predictions in low-heritability traits

Traits with low heritability, such as complex stress responses or quality traits, pose a challenge for GP due to their large environmental influence and polygenic nature. DL models are better equipped to capture the subtle, non-linear patterns that drive these traits.

- Sandhu *et al.* (2021a) compared multiple DL architectures with linear models for several complex traits across corn and wheat and found that DL consistently improved accuracy for traits with low-to-moderate heritability.
- Bhat *et al.* (2023) integrated transcriptomic data using DL to improve predictions for traits with complex regulatory architectures, including flowering time and disease susceptibility.

These findings suggest that DL is particularly valuable in tackling prediction challenges associated with low signal-to-noise ratios.

8.5.6 Prediction in understudied and orphan crops

One of the more recent applications of DL in GP is in orphan or underrepresented crops, which often suffer from data scarcity and less investment in breeding infrastructure. DL models combined with transfer learning or data augmentation techniques can enable meaningful prediction in such crops.

- Gogoi *et al.* (2023) implemented transfer learning with CNN-based architectures for the classification of rice diseases.
- Selvaraj *et al.* (2020) applied DL to cassava, a major tropical root crop, demonstrating improved root yield prediction despite sparse genomic resources.

This highlights DL's potential in democratizing breeding technologies for neglected crops, contributing to global food security and biodiversity.

8.5.7 GP with multi-omics data

The integration of multi-omics data (e.g. SNPs, transcriptomics, metabolomics) into DL models allows for a more holistic representation of the genotype-to-phenotype continuum.

- Huang *et al.* (2023) employed a transformer-based genomic model combining SNP and RNA-Seq data to predict seed germination in *Arabidopsis*, achieving higher accuracy.
- Talukder *et al.* (2021) emphasized the utility of DL in integrating epigenomic and gene expression data for more biologically grounded predictions.

As more multi-omics datasets become available, DL will continue to be the primary framework for their integration in predictive breeding.

8.5.8 Decision support and breeding strategy optimization

Beyond raw prediction, DL models can support decision-making in breeding pipelines, including:

- parent selection through GEBVs and trait dissection;
- trait prioritization using attention mechanisms to identify influential markers; and
- risk assessment by quantifying prediction uncertainty.

Some DL approaches incorporate explainability tools such as SHAP (SHapley Additive exPlanations) values or saliency maps to provide insights into model behavior and guide breeding decisions (Hosain *et al.*, 2024).

8.6 Challenges and Limitations

Despite the promising applications of DL in plant GP, several challenges and limitations must be addressed before widespread adoption in practical breeding programs. These limitations are primarily related to data constraints, model interpretability, computational costs, and integration into traditional plant breeding workflows.

8.6.1 Data availability and quality

A fundamental limitation of DL models in GP is the requirement for large, high-quality datasets. Deep neural networks often require thousands of training examples to generalize effectively. However, in plant breeding, especially for minor or orphan crops, genomic and phenotypic datasets are often small and fragmented.

- For example, Dong *et al.* (2022) highlighted that in small datasets (fewer than 1000 samples), DL models frequently underperform compared to traditional linear models due to overfitting.
- HTP data, while voluminous, may suffer from noise and inconsistencies due to varying environmental and imaging conditions, further complicating DL training (Galli *et al.*, 2022).

Moreover, missing data common in field trials due to logistical constraints may introduce bias or require imputation, which can degrade the model's predictive power if not handled properly.

8.6.2 Overfitting and model generalization

DL models possess high flexibility, allowing them to capture complex, non-linear relationships. However, this same flexibility can lead to overfitting, especially when the model complexity is high relative to dataset size.

- Overfitting occurs when a DL model learns noise or irrelevant patterns in the training data, resulting in poor generalization to new, unseen data (Dong *et al.*, 2022).
- In GP, this can lead to overestimated predictive ability in internal validation but degraded performance in independent environments or years, which limits utility in real breeding scenarios.

Regularization techniques—such as dropout, L1/L2 penalties, or early stopping—are often used, but the risk of overfitting remains a concern, particularly when cross-environment robustness is required.

8.6.3 Interpretability and biological insight

Another significant limitation of DL models is their black-box nature, which hinders biological interpretability.

- While DL may improve predictive accuracy, breeders often need to understand the biological underpinnings of predictions, such as identifying key genetic markers or pathways associated with trait variation.
- Traditional GP methods, such as GBLUP or Bayesian models, provide interpretable marker effects or posterior distributions, whereas DL models typically lack such transparency.

Recent advances in model interpretability tools—such as SHAP, Local Interpretable Model-Agnostic Explanations, and saliency maps—have been introduced to partially overcome this limitation (Hosain *et al.*, 2024), but these tools are still underutilized in the plant science domain.

8.6.4 Computational demands and resource requirements

DL models, particularly those with many layers or using large multi-omics datasets, are computationally intensive.

- Training such models often requires high-performance computing resources, including graphics processing unit or tensor processing unit, and extensive memory.
- Small breeding programs, especially in developing countries, may lack access to such infrastructure, creating an accessibility barrier.

Additionally, tuning hyperparameters (e.g. learning rate, number of layers, activation functions) is non-trivial and often requires iterative experimentation, further increasing the time and computational cost.

8.6.5 Model selection and hyperparameter optimization

Selecting the most appropriate DL architecture for a specific GP problem is still more of an art than a science. Unlike classical methods that rely on established statistical principles, DL models have numerous hyperparameters that can drastically affect performance.

- The optimal architecture varies depending on the trait architecture, data modality, sample size, and crop species.

- For example, CNNs are well suited for spatially structured inputs such as images, whereas MLPs may suffice for vectorized genomic data. Hybrid architectures combining CNNs, RNNs, and attention mechanisms are emerging but require expert design.

Automated machine learning (AutoML) and neural architecture search offer potential solutions (Salehin *et al.*, 2024), but these methods are resource-intensive and not widely adopted in plant science.

8.6.6 Limited transferability across environments and populations

GP models, including those based on DL, may have limited transferability across populations or environmental conditions.

- A model trained on data from one breeding population or location may not perform well in others due to differences in genetic architecture, LD patterns, or G × E interactions.
- This restricts the scalability of DL models in multilocation or global breeding programs, where generalizability is crucial.

While domain adaptation and transfer learning techniques show promise (Kamath *et al.*, 2019), their application in plant genomics is still in the early stages and requires further validation.

8.6.7 Integration into breeding pipelines

Finally, the integration of DL tools into practical breeding pipelines remains a challenge.

- Breeders need user-friendly tools and decision support systems, not just accurate predictions. Many DL models remain in academic prototypes and lack robust interfaces for end users.
- Moreover, integrating DL predictions with existing tools (e.g. pedigree-based selection, MAS) and decision-making frameworks requires careful adaptation.

Efforts to develop XAI platforms and DL-assisted decision support tools are ongoing, but widespread adoption will require capacity building, cross-disciplinary collaboration, and investment in digital breeding infrastructure.

8.6.8 Ethical and reproducibility concerns

As with other AI applications, issues of reproducibility, data sharing, and ethical considerations are gaining attention.

- Lack of standardized datasets and benchmarking protocols makes it difficult to compare model performance across studies.

Proprietary data, particularly from private breeding companies, limits open science and reproducibility.

- Furthermore, the ethical use of AI in agriculture must consider equity and access, especially in regions where technological resources are scarce.

Establishing community benchmarks, open-access data repositories, and fair-use policies is essential for the responsible development of DL-enhanced GP.

8.7 Future Perspectives

The integration of DL in plant GP is still in its developmental stages, yet its transformative potential is increasingly evident. As both methodological innovations and agricultural challenges converge, DL is poised to play a central role in accelerating genetic gains, improving crop resilience, and optimizing resource use in plant breeding programs. This section outlines key directions and emerging opportunities that will shape the future of DL-enhanced GP.

8.7.1 Multi-omics integration and systems biology

One of the most promising frontiers is the integration of multi-omics datasets, including genomics, transcriptomics, proteomics, metabolomics, and epigenomics, into DL models. Such integration allows for a systems-level understanding of trait architecture and biological regulation.

- By combining genomic markers with transcriptomic data, for example, DL models can predict not only phenotypes but also intermediate molecular states that mediate trait expression (Washburn *et al.*, 2020).
- GNNs and attention mechanisms are being explored to handle complex multilayered omics data structures (Lee *et al.*, 2020), offering enhanced capacity to uncover gene–trait relationships.

8.7.2 Climate-smart and environmentally adaptive breeding

As climate change intensifies abiotic and biotic stresses, GP models must evolve to include environmental variables and climate projections. DL architectures such as RNNs and temporal convolutional networks are particularly suitable for modeling G × E interactions over time.

- Coupling DL models with spatiotemporal weather data, soil conditions, and management practices can yield climate-resilient phenotypic predictions (Khan *et al.*, 2022).
- Developing climate-smart DL prediction tools will be vital for sustainable agriculture, especially in regions prone to drought, flooding, or temperature extremes.

8.7.3 Transfer learning and domain adaptation

DL models trained on large, well-characterized crops (e.g. corn, wheat) can be transferred to understudied crops (e.g. sorghum, millet) using transfer learning techniques.

- Transfer learning enables model reuse, reducing the need for large training datasets and computational costs (Iman *et al.*, 2023).
- Domain adaptation approaches are being explored to adjust models trained in one environment or population to perform effectively in new, diverse conditions.

These advances can democratize the benefits of DL across global agriculture, particularly in resource-limited settings.

8.7.4 XAI and model interpretability

As breeders demand more transparent decision support tools, XAI methods will become essential.

- Tools such as SHAP, attention mechanisms, and saliency maps are being tailored for genomic data to identify key loci, genes, or interactions driving trait variation (Hosain *et al.*, 2024).
- Increasing model interpretability will also improve biological insight and foster trust and adoption among domain experts.

Future DL pipelines may combine high predictive accuracy with interactive visual explanations and gene network maps.

8.7.5 Federated and collaborative learning

The future of GP may also be shaped by federated learning frameworks, allowing breeding institutions to collaboratively train DL models without sharing sensitive data.

- This is particularly useful for private–public partnerships, where data privacy and intellectual property are major concerns.
- Early frameworks, such as federated averaging and split neural network, offer avenues for secure, distributed DL training in agriculture (Hafi *et al.*, 2024).

Federated DL could unlock the potential of large-scale, cross-population GP across continents while preserving data sovereignty.

8.7.6 Automation and integration in digital breeding pipelines

To realize operational impact, DL models must be embedded into automated breeding workflows.

- Integration with decision support systems, robotic phenotyping, and genomic selection software will streamline candidate evaluation and accelerate breeding cycles.

- Advances in AutoML will also allow breeders with limited computational expertise to deploy effective models with minimal manual tuning (Salehin *et al.*, 2024).

These developments align with the broader goals of digital agriculture and precision breeding, where real-time data and predictive modeling drive faster, informed decisions.

8.7.7 Education, capacity building, and open resources

A major challenge moving forward is the capacity gap in DL expertise among plant scientists and breeders. Addressing this requires:

- developing open-source DL libraries tailored to plant breeding (e.g. DeepGP, PlantDeep);
- launching interdisciplinary training programs, workshops, and curriculum that bridge genomics, statistics, and AI; and
- encouraging FAIR (findable, accessible, interoperable, reusable) data sharing and open benchmarks to foster reproducibility and innovation.

Collaborative platforms such as PlantGenIE, GenomicSelection.org, and AI4Ag could serve as hubs for community-driven tool development and knowledge exchange.

8.8 Conclusion

The integration of DL into plant GP marks a transformative step in modern computational breeding. With increasing demands for climate-resilient and high-yield crops, DL offers the capacity to model complex, non-linear genotype–phenotype relationships and effectively utilize large, heterogeneous biological datasets. This chapter discussed foundational DL architectures, including MLPs, convolutional, and RNNs, and highlighted their improved prediction accuracy compared to traditional statistical models. While DL brings significant advantages, challenges such as limited data availability, interpretability, computational demands, and generalizability across environments remain. Bridging the gap between algorithmic development and field application requires user-friendly, explainable tools, and strong collaboration among plant scientists, breeders, and data experts. Emerging approaches such as multi-omics integration, federated learning, and climate-smart prediction are poised to enhance DL's impact on breeding. Ultimately, DL has the potential to enable faster, more accurate, and sustainable genomic selection, driving the future of precision agriculture.

References

Abdollahi-Arpanahi, R., Gianola, D., & Peñagaricano, F. (2020). Deep learning versus parametric and ensemble methods for genomic prediction of complex phenotypes. *Genetics Selection Evolution*, 52, 1-15.

Akinpelu, S. A., Olasoji, O. E., Akindolani, A., Adeyanju, K. I., Ajagbe, S. A., & Taiwo, G. A. (2025). Detection of Cassava Plant Disease using Deep Transfer Learning Approach. *ParadigmPlus*, 6(1), 1-12.

Bhat, J. A., Feng, X., Mir, Z. A., Raina, A., & Siddique, K. H. (2023). Recent advances in artificial intelligence, mechanistic models, and speed breeding offer exciting opportunities for precise and accelerated genomics-assisted breeding. *Physiologia Plantarum*, 175(4), e13969.

Budhlakoti, N., Kushwaha, A. K., Rai, A., Chaturvedi, K. K., Kumar, A., Pradhan, A. K., & Kumar, S. (2022). Genomic selection: A tool for accelerating the efficiency of molecular breeding for development of climate-resilient crops. *Frontiers in Genetics*, 13, 832153.

Consens, M. E., Dufault, C., Wainberg, M., Forster, D., Karimzadeh, M., Goodarzi, H., & Wang, B. (2025). Transformers and genome language models. *Nature Machine Intelligence*, 1-17.

Dong, Y., Zhou, S., Xing, L., Chen, Y., Ren, Z., Dong, Y., & Zhang, X. (2022). Deep learning methods may not outperform other machine learning methods on analyzing genomic studies. *Frontiers in Genetics*, 13, 992070.

Galli, G., Sabadin, F., Yassue, R. M., Galves, C., Carvalho, H. F., Crossa, J., & Fritsche-Neto, R. (2022). Automated machine learning: a case study of genomic "image-based" prediction in maize hybrids. *Frontiers in Plant Science*, 13, 845524.

Gerakari, M., Katsileros, A., Kleftogianni, K., Tani, E., Bebeli, P. J., & Papasotiropoulos, V. (2025). Breeding of Solanaceous Crops Using AI: Machine Learning and Deep Learning Approaches—A Critical Review. *Agronomy*, 15(3), 757.

Gogoi, M., Kumar, V., Begum, S. A., Sharma, N., & Kant, S. (2023). Classification and detection of rice diseases using a 3-stage CNN architecture with transfer learning approach. *Agriculture*, 13(8), 1505.

Hafi, H., Brik, B., Frangoudis, P. A., Ksentini, A., & Bagaa, M. (2024). Split federated learning for 6G enabled-networks: Requirements, challenges, and future directions. *IEEe Access*, 12, 9890-9930.

Hosain, M. T., Jim, J. R., Mridha, M. F., & Kabir, M. M. (2024). Explainable AI approaches in deep learning: Advancements, applications and challenges. *Computers and electrical engineering*, 117, 109246.

Howard, R., Jarquin, D., & Crossa, J. (2022). Overview of Genomic Prediction Methods and the Associated Assumptions on the Variance of Marker Effect, and on the Architecture of the Target Trait. In *Genomic prediction of complex traits: methods and protocols* (pp. 139-156). New York, NY: Springer US.

Huang, K., Tian, J., Sun, L., Xie, P., Zhou, S., Deng, A., & Jiang, X. (2023). TransGeneSelector: A Transformer-based Approach Tailored for Key Gene Mining with Small Plant Transcriptomic Datasets. *bioRxiv*, 2023-09.

Iman, M., Arabnia, H. R., & Rasheed, K. (2023). A review of deep transfer learning and recent advancements. *Technologies*, 11(2), 40.

Jackson, R., Buntjer, J. B., Bentley, A. R., Lage, J., Byrne, E., Burt, C., & Ober, E. S. (2023). Phenomic and genomic prediction of yield on multiple locations in winter wheat. *Frontiers in genetics*, 14, 1164935.

Jadhav, Y., & Jumbo, M. B. (2024). Genomic Selection in Plant Breeding. In *Plant Molecular Breeding in Genomics Era: Applications* (pp. 145-181). Cham: Springer Nature Switzerland.

Jubair, S., Tremblay-Savard, O., & Domaratzki, M. (2023). Gxenet: Novel fully connected neural network based approaches to incorporate gxe for predicting wheat yield. *Artificial Intelligence in Agriculture*, 8, 60-76.

Jurado-Ruiz, F., Rousseau, D., Botía, J. A., & Aranzana, M. J. (2023). GenoDrawing: an autoencoder framework for image prediction from SNP markers. *Plant Phenomics*, 5, 0113.

Kamath, U., Liu, J. & Whitaker, J. (2019) Introduction. In: *Deep learning for NLP and speech recognition,* 3-38, Springer, Cham, Switzerland.

Khalilzadeh, Z., Kashanian, M., Khaki, S., & Wang, L. (2024). A hybrid deep learning-based approach for optimal genotype by environment selection. *Frontiers in Artificial Intelligence*, 7, 1312115.

Khan, M. H. U., Wang, S., Wang, J., Ahmar, S., Saeed, S., Khan, S. U., & Feng, X. (2022). Applications of artificial intelligence in climate-resilient smart-crop breeding. *International Journal of Molecular Sciences*, 23(19), 11156.

Kukreja, V., Sharma, V., & Sharma, R. (2024). Utilizing a Hybrid CNN-LSTM Model for Automated Severity Classification of Rice Panicle Blast Disease. In *2024 International Conference on Intelligent and Innovative Technologies in Computing, Electrical and Electronics (IITCEE)* (pp. 1-5). IEEE.

Larue, F., Rouan, L., Pot, D., Rami, J. F., Luquet, D., & Beurier, G. (2024). Linking genetic markers and crop model parameters using neural networks to enhance genomic prediction of integrative traits. *Frontiers in Plant Science*, 15, 1393965.

Lee, B., Zhang, S., Poleksic, A., & Xie, L. (2020). Heterogeneous multi-layered network model for omics data integration and analysis. *Frontiers in genetics*, 10, 1381.

Li, J., He, Z., Zhou, G., Yan, S., & Zhang, J. (2024a). DeepAT: A Deep Learning Wheat Phenotype Prediction Model Based on Genotype Data. *Agronomy*, 14(12), 2756.

Li, Q., Hu, Z., Wang, Y., Li, L., Fan, Y., King, I. & Li, Y. (2024b). Progress and opportunities of foundation models in bioinformatics. *Briefings in Bioinformatics*, 25(6), bbae548.

Liu, R. (2024). *Network Biology Powered by Graph Deep Learning: Understanding the Molecular Basis of Human Genetics*. Michigan State University.

Malosetti, M., Ribaut, J. M., & van Eeuwijk, F. A. (2013). The statistical analysis of multi-environment data: modeling genotype-by-environment interaction and its genetic basis. *Frontiers in physiology*, 4, 44.

Mienye, I. D., Swart, T. G., & Obaido, G. (2024). Recurrent neural networks: A comprehensive review of architectures, variants, and applications. *Information*, 15(9), 517.

Misra, B. B., Langefeld, C., Olivier, M., & Cox, L. A. (2019). Integrated omics: tools, advances and future approaches. *Journal of molecular endocrinology*, 62(1), R21-R45.

Montesinos-López, O. A., Montesinos-López, A., Crossa, J., Cuevas, J., Montesinos-López, J. C., Gutiérrez, Z. S., & Singh, R. (2019). A Bayesian genomic multi-output regressor stacking model for predicting multi-trait multi-environment plant breeding data. *G3: Genes, Genomes, Genetics*, 9(10), 3381-3393.

Montesinos-López, O. A., Montesinos-López, A., Pérez-Rodríguez, P., Barrón-López, J. A., Martini, J. W., Fajardo-Flores, S. B., & Crossa, J. (2021). A review of deep learning applications for genomic selection. *BMC genomics*, 22, 1-23.

Novielli, P., Romano, D., Pavan, S., Losciale, P., Stellacci, A. M., Diacono, D., & Tangaro, S. (2024). Explainable artificial intelligence for genotype-to-phenotype prediction in plant breeding: a case study with a dataset from an almond germplasm collection. *Frontiers in Plant Science*, 15, 1434229.

Peng, D., Cheng, E., Feng, X., Hu, J., Lou, Z., Zhang, H., & Zhang, B. (2024). A Deep–Learning Network for Wheat Yield Prediction Combining Weather Forecasts and Remote Sensing Data. *Remote Sensing*, 16(19), 3613.

Petscharnig, S., Lux, M., & Chatzichristofis, S. (2017). Dimensionality reduction for image features using deep learning and autoencoders. In *Proceedings of the 15th international workshop on content-based multimedia indexing* (pp. 1-6).

Ranđelović, P., Đorđević, V., Miladinović, J., Prodanović, S., Ćeran, M., & Vollmann, J. (2023). High-throughput phenotyping for non-destructive estimation of soybean fresh biomass using a machine learning model and temporal UAV data. *Plant Methods*, 19(1), 89.

Rice, B.R. & Lipka, A.E. (2021) Diversifying maize genomic selection models. *Molecular Breeding*, 41(5), 33.

Salehin, I., Islam, M. S., Saha, P., Noman, S. M., Tuni, A., Hasan, M. M., & Baten, M. A. (2024). AutoML: A systematic review on automated machine learning with neural architecture search. *Journal of Information and Intelligence*, 2(1), 52-81.

Sandhu, K. S., Lozada, D. N., Zhang, Z., Pumphrey, M. O., & Carter, A. H. (2021a). Deep learning for predicting complex traits in spring wheat breeding program. *Frontiers in plant science*, 11, 613325.

Sandhu, K., Patil, S. S., Pumphrey, M., & Carter, A. (2021b). Multitrait machine-and deep-learning models for genomic selection using spectral information in a wheat breeding program. *The Plant Genome*, 14(3), e20119.

Selvaraj, M. G., Valderrama, M., Guzman, D., Valencia, M., Ruiz, H., & Acharjee, A. (2020). Machine learning for high-throughput field phenotyping and image processing provides insight into the association of above and below-ground traits in cassava (Manihot esculenta Crantz). *Plant methods*, 16, 1-19.

Shook, J., Gangopadhyay, T., Wu, L., Ganapathysubramanian, B., Sarkar, S., & Singh, A. K. (2021). Crop yield prediction integrating genotype and weather variables using deep learning. *Plos one*, 16(6), e0252402.

Singh, A. K., Singh, A., Sarkar, S., Ganapathysubramanian, B., Schapaugh, W., Miguez, F. E., & Zhang, J. (2021). High-throughput phenotyping in soybean. *High-throughput crop phenotyping*, 129-163.

Sinha, D., Maurya, A. K., Abdi, G., Majeed, M., Agarwal, R., Mukherjee, R., & Chen, J. T. (2023). Integrated genomic selection for accelerating breeding programs of climate-smart cereals. *Genes*, 14(7), 1484.

Srivastava, A. K., Safaei, N., Khaki, S., Lopez, G., Zeng, W., Ewert, F., & Rahimi, J. (2022). Winter wheat yield prediction using convolutional neural networks from environmental and phenological data. *Scientific reports*, 12(1), 3215.

Talukder, A., Barham, C., Li, X., & Hu, H. (2021). Interpretation of deep learning in genomics and epigenomics. *Briefings in Bioinformatics*, 22(3), bbaa177.

Togninalli, M., Wang, X., Kucera, T., Shrestha, S., Juliana, P., Mondal, S., & Poland, J. (2023). Multi-modal deep learning improves grain yield prediction in wheat breeding by fusing genomics and phenomics. *Bioinformatics*, 39(6), btad336.

Tsimenidis, S., Vrochidou, E., & Papakostas, G. A. (2022). Omics data and data representations for deep learning-based predictive modeling. *International Journal of Molecular Sciences*, 23(20), 12272.

Waldmann, P., Pfeiffer, C., & Mészáros, G. (2020). Sparse convolutional neural networks for genome-wide prediction. *Frontiers in Genetics*, 11, 25.

Wang, C., Zhang, D., Ma, Y., Zhao, Y., Liu, P., & Li, X. (2025). WheatGP, a genomic prediction method based on CNN and LSTM. *Briefings in Bioinformatics*, 26(2), bbaf191.

Wang, G., Xuan, J., Wang, P., Li, C., & Lu, J. (2024). Lstm autoencoder-based deep neural networks for barley genotype-to-phenotype prediction. In *Australasian Joint Conference on Artificial Intelligence* (pp. 342-353). Singapore: Springer Nature Singapore.

Wang, T., Crawford, M. M., & Tuinstra, M. R. (2023). A novel transfer learning framework for sorghum biomass prediction using UAV-based remote sensing data and genetic markers. *Frontiers in Plant Science*, 14, 1138479.

Wang, X., Xu, Y., Hu, Z., & Xu, C. (2018). Genomic selection methods for crop improvement: Current status and prospects. *The Crop Journal*, 6(4), 330-340.

Waqas, M., & Humphries, U. W. (2024). A critical review of RNN and LSTM variants in hydrological time series predictions. *MethodsX*, 102946.

Washburn, J. D., Burch, M. B., & Franco, J. A. V. (2020). Predictive breeding for maize: Making use of molecular phenotypes, machine learning, and physiological crop models. *Crop Science*, 60(2), 622-638.

Wu, Y., & Xie, L. (2025). AI-driven multi-omics integration for multi-scale predictive modeling of genotype-environment-phenotype relationships. *Computational and Structural Biotechnology Journal.*

Yuan, Y., Cairns, J. E., Babu, R., Gowda, M., Makumbi, D., Magorokosho, C., & Zhang, X. (2019). Genome-wide association mapping and genomic prediction analyses reveal the genetic architecture of grain yield and flowering time under drought and heat stress conditions in maize. *Frontiers in plant science*, 9, 1919.

Zhang, W., Voloudakis, G., Rajagopal, V. M., Readhead, B., Dudley, J. T., Schadt, E. E., & Roussos, P. (2019). Integrative transcriptome imputation reveals tissue-specific and shared biological mechanisms mediating susceptibility to complex traits. *Nature communications*, 10(1), 3834.

Zhang, Y. Z., & Imoto, S. (2024). Genome analysis through image processing with deep learning models. *Journal of Human Genetics*, 69(10), 519-525.

Zhou, W., Yan, Z., & Zhang, L. (2024). A comparative study of 11 non-linear regression models highlighting autoencoder, DBN, and SVR, enhanced by SHAP importance analysis in soybean branching prediction. *Scientific Reports*, 14(1), 5905.

9 Deep Learning for Advancing Synthetic Biology in Plants

Mokhtar Guerfel[1]* and J. Essemine[2]

[1]Applied College, Shaqra University, Afif, Saudi Arabia; [2]Bioinformatics Section, Department of Informatics, University of Quebec at Montreal, Pavillon President-Kennedy, Montreal, Canada

Abstract

Plant synthetic biology (SynBio) represents a rapidly evolving field focused on optimizing crop traits and bioproduction by applying engineering methodologies, particularly through the design and implementation of genetic circuits for precise modifications. This emerging platform applies cyclic design principles to engineer novel biological components, devices, and systems, aiming to enhance yield, nutritional content, nutrient uptake, and stress resilience. Nowadays, deep learning has demonstrated significant strengths in learning and design, offering new approaches to study the mechanisms underlying complex biological phenomena in plants. This chapter investigates the integration of deep learning methodologies with emerging advancements in SynBio in plants, focusing on approaches such as genome editing and design, gene regulatory networks, phenotype prediction, protein structure prediction, and metabolic pathway engineering. This chapter also delineates key challenges that pose significant barriers to the continued advancement of deep learning and synthetic biology.

Keywords: convolutional neural network, crop improvement, deep learning, engineering strategies, plant synthetic biology

9.1 Introduction

Given the challenges posed by global warming and the increasing frequency of extreme weather events, the urgent need to develop new technologies to ensure global food security has become more pressing than ever. Researchers have been persistently advancing new methodologies to ensure food security, with widespread consensus that the incorporation of technology into agricultural practices is essential to achieving this objective. Plant synthetic biology (SynBio) has rapidly developed as a promising yet intrinsically uncertain platform for the generation of novel agricultural innovations. Future research trajectories are expected to critically advance efforts to optimize crop yields and fortify the nutritional composition of plant-derived food sources. According to Tiwari *et al.* (2021), plant SynBio is an emerging field that applies engineering principles to biological systems. It focuses on designing artificial genetic components and reconstructing synthetic biological systems in plants, supported by mathematical modeling and genome engineering technologies.

SynBio has been widely applied across industrial and academic domains, fostering innovation in areas such as food production, biofuels,

*E-mail: mokhtar_guerfel@su.edu.sa

DOI: 10.1079/9781836991120.0009

bioremediation, and pharmaceuticals. Among these advances, plant SynBio has emerged as a particularly effective approach, addressing food security challenges while contributing to environmentally sustainable food systems (Shi *et al.*, 2022). Additionally, plant SynBio has gained significant attention for its broad range of applications, including the production of specialized metabolites, crop breeding, biosensor development, and the enhancement of stress tolerance, often through constitutive design strategies (Gupta *et al.*, 2021).

Deep learning holds significant potential for the detection and analysis of physiological traits in plants (Sahraeian *et al.*, 2019; Yang *et al.*, 2020). Furthermore, to make accurate predictions, deep learning is highly effective at uncovering patterns within large datasets, offering fresh insights into complex biological processes. For example, these models can identify key genomic regions involved in regulating gene expression (Washburn *et al.*, 2019) and can even design ligands that selectively bind to specific protein sites without prior knowledge of the ligand's structure (Lin *et al.*, 2022).

The design process in SynBio encompasses a systematic sequence of stages—conceptualization, design, modeling, construction, and testing (Liu and Stewart, 2015)—that collectively facilitates the development of genetic systems characterized by reliability, reproducibility, and predictability. This approach has also markedly advanced directed protein engineering, facilitating the rational incorporation of novel phenotypic traits to enhance crop performance.

Crop SynBio is emerging as a comprehensive and integrative strategy for crop design and breeding. Unlike conventional transgenic methods that primarily involve introducing foreign DNA (originating from other plant species or synthetically constructed sequences) into the host genome, crop SynBio also enables precise manipulation of the crop's own genome through genome editing. Furthermore, it facilitates the redesign of metabolic pathways to optimize the synthesis of desirable traits in crop improvement (Kaul *et al.*, 2021).

Ongoing advancements in DNA synthesis, high-throughput sequencing technologies, and laboratory automation are increasingly empowering synthetic biologists to integrate deep learning methodologies into their workflows. SynBio, grounded in engineering principles, focuses on the rational design, assembly, and optimization of biological components and systems, and is extensively applied across industrial, agricultural, pharmaceutical, and environmental domains (Meng and Ellis, 2020). Deep learning, a subfield of machine learning characterized by multilayered artificial neural networks (ANNs) such as convolutional and recurrent architectures, offers powerful tools for modeling complex, nonlinear relationships within biological datasets. In SynBio, deep learning applications have included the prediction of genetic element activity—such as promoter strength and specificity—from primary sequence data, the design of novel regulatory elements, and the generation of highly accurate protein structure predictions through end-to-end learning frameworks.

Deep learning offers powerful capabilities for analyzing genetic data, identifying candidate genes associated with specific traits, and accurately predicting phenotypic outcomes based on genotype information (Wang *et al.*, 2023). Moreover, deep learning plays a critical role in directed protein evolution by facilitating the prediction and prioritization of protein variants that meet predefined engineering objectives, leveraging advanced machine learning algorithms (Cadet *et al.*, 2022).

By leveraging deep neural networks composed of multiple interconnected layers, deep learning excels at automatically extracting and learning complex, high-dimensional features from large-scale datasets. This capability allows it to uncover complex patterns and relationships that often elude traditional machine learning methods, particularly in contexts where high predictive accuracy is critical. The emergence of deep learning represents a significant milestone in the evolution of machine learning, bringing computational models closer to aspects of human-like intelligence. Compared to conventional analytical approaches, deep learning has demonstrated superior performance, especially in its ability to generalize from and predict outcomes using large and complex datasets. Its effectiveness has been widely validated across diverse scientific domains, including molecular biology, proteomics, genetics, and genomics (Zhou and Troyanskaya, 2015; Chen *et al.*, 2019; Luo *et al.*, 2019; Zou *et al.*, 2019; Liu *et al.*, 2020; Jumper *et al.*, 2021; de Almeida *et al.*, 2022).

SynBio aims to develop models capable of generalizing from training data to accurately predict outputs for previously unseen inputs.

A defining characteristic of deep learning is its ability to progressively extract complex features through layered processing within ANNs (Goodfellow, 2016). Deep learning has demonstrated high accuracy in predicting epigenetic modification sites, offering significant potential for synthetic epigenetics when integrated with targeted editing technologies (Yang *et al.*, 2022). For instance, deep learning models have achieved precise predictions of dynamic modifications such as 5mC, 6mA, m6A, and histone marks in plants (Wang *et al.*, 2021). The neural network model Deep Learning Explore Nanopore (DENA) predicts ribonucleic acid (RNA) m6A modifications in *Arabidopsis* with an accuracy of 90% (Qin *et al.*, 2022), while Smart Model for Open Chromatin (SMOC) region prediction identifies open chromatin regions within rice genomes with over 80% accuracy (Guo *et al.*, 2022). These deep learning tools are advancing the field of synthetic epigenetics and providing valuable insights to guide crop breeding strategies.

This chapter reviews recent developments in the application of deep learning to plant SynBio, emphasizing case studies wherein synthetic regulatory circuits have been engineered to enhance crop yield, metabolic efficiency, stress resistance, product quality, and nutritional value.

9.2 Applications of DL in Crop SynBio

Crop SynBio is rapidly emerging as a comprehensive and integrative framework for the rational design and improvement of crop species through advanced breeding strategies. It goes beyond traditional transgenic approaches that introduce foreign DNA, sourced from other species or synthetically constructed, into the host genome. It also enables precise genome manipulation through genome editing and the redesign of metabolic pathways to enhance desirable traits (Rao and Wan, 2021). The overarching goal is to deepen our understanding of plant genomics, predict gene–gene interactions, elucidate genotype–phenotype relationships, and provide informed recommendations for breeding. With this knowledge, breeders can implement molecular-designed breeding strategies more efficiently and accurately using genome-editing tools.

As deep learning and genome-editing technologies continue to advance, our capacity to interpret and engineer plant genomes will expand significantly. These advancements will enable breeders to design and develop optimized crop varieties with an unprecedented degree of precision, efficiency, and adaptability.

Crop SynBio, as a cutting-edge interdisciplinary science, applies engineering principles to the design and manipulation of biological systems. It focuses on constructing artificial genetic components and reconstructing synthetic biological circuits in plants, supported by genome engineering technologies and mathematical modeling frameworks (Braguy and Zurbriggen, 2016; Tiwari *et al.*, 2021). When implemented in crop species, this approach enables the development of varieties with enhanced yield potential, increased resilience to biotic and abiotic stresses, and improved nutritional profiles.

Figure 9.1 shows some of the ways in which deep learning is being applied in SynBio of plants.

9.2.1 Genome editing and design

Advances in crop genome editing offer considerable opportunities to improve agronomic traits, including yield enhancement, nutritional fortification, and increased resilience to biotic and abiotic challenges. Deep learning is emerging as a disruptive tool for automating and refining genome-editing workflows in plants. By enhancing the precision and efficiency of editing systems, deep learning facilitates the identification and optimization of functional genomic regions, target sequences, and off-target minimization, thereby accelerating the implementation of precision breeding strategies. Through advanced predictive models, deep learning also aids in the design of high-fidelity genetic modifications and the strategic selection of molecular targets for enhanced trait expression.

SynBio involves the design, engineering, and *de novo* synthesis of biological systems with defined functions, leveraging fundamental principles of life sciences in conjunction with advanced engineering methodologies. Its applications span the creation of novel genes and, in some cases, entirely new organisms. Genome editing serves as a pivotal technique within SynBio, enabling

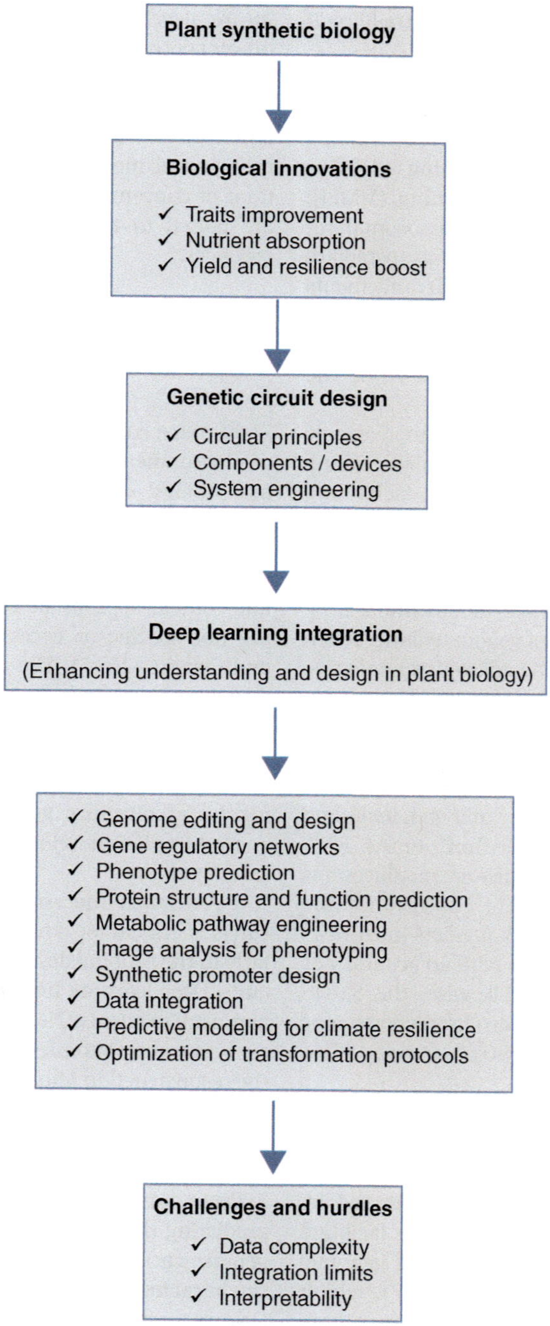

Fig. 9.1. Integration of deep learning and synthetic biology in plant improvement. (Author's own image.)

precise and rapid genetic modifications for the controlled manipulation of living systems. For example, prime editing has been utilized to generate the synthetic allele Xa23^SW14, which confers resistance to bacterial blight in rice (Gupta *et al.*, 2023). Moving forward, genome-editing technologies are anticipated to play a critical role in the engineering of novel genes with tailored functionalities, specifically designed to meet targeted breeding objectives, thereby addressing the limitations inherent in traditional breeding methodologies.

In certain cases, genome editing induces multiple DNA double-strand breaks simultaneously. These DNA breaks may be repaired through non-homologous end joining (NHEJ) or microhomology-mediated end joining (MMEJ), processes that can lead to chromosomal rearrangements and potential genomic instability (Schmidt *et al.*, 2020). Such rearrangements can substantially alter chromosome structure—possibly even reducing chromosome number—leading to reproductive isolation from the wild-type progenitor.

Deep learning algorithms are increasingly applied to predict the outcomes of genome-editing techniques, such as CRISPR-Cas9, facilitating the design of guide RNAs with enhanced specificity and efficiency to minimize off-target effects. Furthermore, deep learning models enable precise prediction of genomic modification sites, enhancing the accuracy of genome editing. When combined with epigenetic editing technologies, these models hold significant promise for advancing the nascent field of synthetic epigenetics (Yang *et al.*, 2022). Various deep learning frameworks have demonstrated high predictive accuracy for a range of epigenetic markers, including dynamic DNA methylation (5mC, 6mA), RNA modifications (m6A), and histone modifications in plants (Wang *et al.*, 2021). For instance, the DENA m6A neural network predicts m6A RNA modifications in *Arabidopsis* with an accuracy of 90% (Qin *et al.*, 2022). Likewise, the SMOC model accurately predicts open chromatin regions in *Oryza sativa* with over 80% accuracy (Guo *et al.*, 2022).

Gene editing technologies have already made substantial contributions to the progression of plant breeding (Li *et al.*, 2018). Modern intelligent breeding strategies, empowered by big data and artificial intelligence (AI), facilitate targeted breeding through integrated genomic and environmental prediction (iGEP), which incorporates both genomic data and population–environment interactions to optimize breeding outcomes (Yin *et al.*, 2008). These tools are also crucial in designing effective synthetic epigenetic strategies for plant improvement.

The integration of genome editing with deep learning, SynBio, and advanced biotechnological tools presents immense potential for the future of crop improvement. This interdisciplinary approach is set to broaden the scope and capabilities of genome editing, providing breeders with a sophisticated toolkit for precise, efficient, and targeted crop enhancement. As these novel methodologies exceed the limitations of conventional breeding techniques, they are poised to transform modern agricultural practices.

9.2.2 Gene regulatory networks

Plants must continuously integrate a wide array of environmental signals, including both biotic and abiotic stress cues, alongside intrinsic signals such as those derived from developmental processes and hormonal pathways. These responses are precisely regulated to enable rapid and context-appropriate reactions while avoiding non-specific or excessive activation (Swift and Coruzzi, 2017). The coordination of such responses is mediated by a complex network of transcription factors (TFs), which orchestrate gene expression through gene regulatory networks (GRNs). These GRNs represent the fundamental architecture governing transcriptional control in plant development and environmental adaptation.

A detailed understanding and modeling of GRNs is critical for advancing SynBio, particularly in the rational design of synthetic gene circuits. Deep learning has emerged as a powerful approach to infer GRNs from large-scale datasets such as transcriptomics. These models facilitate the reconstruction of transcriptional regulatory landscapes and the prediction of interactions between genes and regulatory elements.

For example, the deep learning framework ExPecto has demonstrated high accuracy in predicting the transcriptional consequences of genetic variants, including rare or previously uncharacterized mutations, based solely on DNA sequence context (Zhou *et al.*, 2018). This capability allows for preliminary evaluations of evolutionary constraints on gene regulation, as well as the potential impact of deleterious mutations associated with disease. Notably, deep learning models trained on one species have exhibited transferable predictive power in closely related species, owing to the conservation of molecular regulatory mechanisms across species (Kelley, 2020).

Recent advancements underscore the versatility of deep learning in genomics. Chen *et al.* (2023) introduced an unsupervised clustering approach integrated with a deep learning model to predict gene mutations. In a similar vein, Ghahramani *et al.* (2018) utilized generative adversarial networks (GANs) to simulate gene expression and predict cellular perturbations at the single-cell level. This approach facilitates the identification of genes essential for determining biological states and enables the inference of gene regulatory relationships.

9.2.3 Phenotype prediction

Deep learning has emerged as a powerful approach for solving complex problems by leveraging large-scale data. In the field of crop breeding, images continue to serve as the main source of data for selecting desirable traits (Araus and Cairns, 2014). Recent advances in deep learning, especially the use of convolutional neural networks (CNNs), have significantly accelerated progress in analyzing plant phenotyping images. These techniques offer major benefits in processing and interpreting complex, high-dimensional phenotypic information (Chang *et al.*, 2016). With their powerful abilities in feature extraction and predictive analysis, CNNs are creating new possibilities for overcoming the challenges associated with large-scale phenotyping data.

Plant phenotyping entails a detailed evaluation of complex traits, including growth patterns, developmental stages, stress resilience (both abiotic and biotic), structural characteristics, physiological processes, environmental adaptability, yield performance, and other essential quantitative traits necessary for thorough trait analysis (Li *et al.*, 2014). Progress in this area has been significantly enhanced by the adoption of smart farming innovations (Wolfert *et al.*, 2017) and precision agriculture techniques (Gebbers and Adamchuk, 2010), which have enhanced traditional methodologies to incorporate autonomous, intelligent, and remotely operated systems. Central to this transformation are big data technologies, with deep learning models playing a pivotal role in managing and interpreting the vast amounts of phenotypic information.

There has been a notable rise in the use of deep learning techniques for plant phenotyping, driven by their superiority over traditional analytical methods (Chandra *et al.*, 2020; Jiang and Li, 2020; Ren *et al.*, 2020; Kolhar and Jagtap, 2021; Arya *et al.*, 2022). These approaches allow for the direct prediction of plant phenotypes based on genotypic information, reducing the reliance on time-consuming and resource-intensive field experiments. Deep learning techniques have been utilized across various plant phenomics applications, such as image classification (Ramcharan *et al.*, 2017), object detection (Ghosal *et al.*, 2019), and semantic segmentation (Aich and Stavness, 2017). Their use has shown promise in forecasting growth patterns, estimating crop yields, recognizing developmental phases, and detecting plant responses to stress factors.

Enhancing crop varieties to withstand abiotic stresses is crucial for boosting agricultural productivity while reducing environmental harm. SynBio offers powerful tools for engineering stress-resilient plants, contributing to higher yields and advancing more sustainable farming methods.

Over the last 20 years, machine learning has had a profound impact on agriculture, with deep learning—an advanced branch of machine learning—emerging as a fundamental element of contemporary smart farming practices (Kamilaris and Prenafeta-Boldú, 2018). Although research in this field is expanding, there is still a shortage of comprehensive reviews specifically addressing the role of deep learning in plant phenotyping. Notable contributions by Jiang and Li (2020), Arya *et al.* (2022), Chandra *et al.* (2020), Ren *et al.* (2020), and Kolhar and Jagtap (2021) have offered valuable analyses of the latest developments in this domain.

Although deep learning offers high accuracy and the ability to uncover complex patterns in plant phenotypic data, its successful deployment often requires specialized knowledge of machine learning methodologies. The wide range of available models and their diverse applications present challenges for plant scientists, who may struggle with selecting appropriate architectures, implementing algorithms, and integrating results into biological studies.

9.2.4 Protein structure and function prediction

The engineering of plant proteins and the practice of directed evolution often rely on the heterologous production of modified proteins originating from both plant and non-plant organisms (Engqvist and Rabe, 2019). A critical consideration in these processes is the context-dependent nature of protein stability, particularly in heterologous plant expression systems. Protein stability is influenced by three-dimensional (3D) structure, and recent advancements in deep learning have revolutionized structural prediction. AlphaFold, a deep learning-based framework, offers highly accurate protein structure predictions, surpassing previous computational methods and presenting a viable alternative to time- and resource-intensive techniques such as X-ray crystallography (Jumper *et al.*, 2021).

Prior to the advent of deep learning, predicting the functional impact of individual amino acid substitutions required empirical measurements, which were laborious and time-consuming. Deep learning has since enabled the prediction of both protein structure and function from sequence data, facilitating the rational design of synthetic enzymes and novel proteins with tailored characteristics. These predictive capabilities are essential for advancing SynBio and protein engineering.

One prominent example is the development of a homology model for the multifunctional cytochrome P450 enzyme, CYP8D2, which was based on the crystal structure of *Saccharomyces cerevisiae* CYP51 (Li *et al.*, 2019). Nevertheless, the natural diversity of available protein structures is still relatively narrow (Opuu and Simonson, 2023), highlighting the critical need to broaden the catalog of known protein architectures to better facilitate directed evolution efforts.

AlphaFold and its successor, AlphaFold2, represent significant breakthroughs in protein science, employing geometric deep learning to predict protein conformations with unprecedented accuracy, even in the absence of homologous templates (Jumper *et al.*, 2021). These models have inspired the development of additional deep learning frameworks for protein structure prediction (Mirdita *et al.*, 2022; Baek *et al.*, 2023). Furthermore, deep learning has also been applied to epigenetic studies, such as predicting methylation states at cysteine (C) loci, thereby extending its utility in functional proteomics and regulatory biology.

9.2.5 Metabolic pathway engineering

Altering plant-specific synthetic metabolic pathways offers considerable promise for improving both crop yields and nutritional content. Synthetic metabolic engineering in plants focuses on the precise modification of key genes, particularly those that encode rate-limiting enzymes within plant metabolic networks. Protein engineering and directed evolution serve as potent techniques for optimizing these pathways. These approaches have been successfully used to boost the catalytic activity of native plant enzymes and to enable the expression of foreign enzymes within plant systems.

Recent advances in deep learning have further enabled the rational design of metabolic pathways by leveraging embedded representations of enzymatic reactions. These models utilize substrate–product molecular structures in conjunction with amino acid sequence data of the corresponding enzymes. By employing integrative methods, plant synthetic metabolic engineering can be utilized to develop new metabolic pathways in higher plants, customized to meet particular human demands. Specifically, the combination of directed evolution and protein engineering enables the creation of enzymes that are not naturally found in plant systems, thus broadening the range of the synthetic pathways available.

For example, the long-chain polyunsaturated fatty acids eicosapentaenoic acid (EPA) and docosahexaenoic acid (DHA), which are typically not produced in higher plants, have been successfully synthesized in species such as *Arabidopsis thaliana*, *Brassica juncea*, *Camelina sativa*, and canola. This was achieved by introducing genes that encode desaturase, elongase, or subunits of unsaturated fatty acid synthase (Zhu *et al.*, 2020). Additionally, the incorporation of a phosphopantetheinyl transferase gene into cotton, via SynBio techniques, has enabled the reprogramming of genetic circuits to produce multicolored fibers (Zhu *et al.*, 2020; Ge *et al.*, 2023).

Deep learning approaches further contribute to the design and optimization of plant metabolic pathways by predicting the impact of genetic modifications on metabolic flux. These models are increasingly being employed to identify and enhance biosynthetic routes for the production of high-value compounds, including biofuels, pharmaceuticals, and specialty chemicals. By integrating large-scale datasets pertaining to plant metabolism, deep learning algorithms offer a powerful tool for pathway optimization, facilitating improved efficiency and output of target metabolites.

9.2.6 Image analysis for phenotyping

The application of deep learning to extract meaningful information from images and videos has become an essential technique for identifying phenotypic variations in plants. High-throughput phenotyping, enhanced by machine learning-driven image analysis, is expected to play a crucial role in enhancing crop yields. Specifically, deep learning methodologies—particularly those within the field of computer vision, such as image processing and pattern recognition—allow for the precise extraction of phenotypic traits, including color, shape, and texture, from plant imagery. These traits enable the effective detection and classification of plant diseases, supporting early and accurate diagnosis and facilitating timely, targeted management interventions (Shrivastava and Pradhan, 2021; Khan *et al.*, 2022).

In addition to disease diagnostics, deep learning algorithms have proven to be effective in predicting crop yields with enhanced accuracy by integrating and analyzing large-scale agricultural datasets, including meteorological data, soil properties, and field management practices (van Klompenburg *et al.*, 2020). Moreover, deep learning has been successfully applied to various plant phenotyping tasks in conjunction with a range of sensor platforms. These applications include crop counting for harvest optimization, precision weed management, and the detection of crop stress (Schumann *et al.*, 2018; Bauer *et al.*, 2019).

In the context of crop stress detection, deep learning-based image analysis methods vary depending on the specific task—ranging from image classification and segmentation to object detection—and on the stress type and sensor modality employed. The unprecedented advancement of computer vision techniques, powered by deep learning, has significantly enhanced the automation of imaging applications in SynBio as well. These include tasks such as colony detection on growth plates and microscopy image analysis. Two common types of image analysis in this context are classification (e.g. identifying the presence of microbial colonies) and segmentation (e.g. distinguishing individual cells within an image). Early CNN architectures, such as AlexNet (Krizhevsky *et al.*, 2017), LeNet-5 (LeCun *et al.*, 1998), and ResNets (He *et al.*, 2016), were foundational in advancing these capabilities.

CNNs, in particular, have revolutionized image-based plant phenotyping, enabling the automated assessment of plant growth, health status, and other phenotypic traits. In the context of global climate change, which poses growing risks to agricultural productivity and crop quality (Praveen and Sharma, 2019), the application of deep learning models for plant image analysis offers a powerful approach for identifying superior genotypes with desirable traits. This includes the selection of crop varieties that are high-yielding, disease-resistant, and resilient to climate-related stresses. Furthermore, these models can evaluate plant responses to environmental stressors such as drought and salinity, thereby aiding in the development of more robust cultivars (Sun *et al.*, 2019).

9.2.7 Synthetic promoter design

The synergistic integration of SynBio and deep learning is transforming our capacity to decode the regulatory grammar of DNA and to predict gene expression with high accuracy across multiple species. SynBio, employing a bottom-up design strategy, enables the construction of synthetic regulatory elements with expanded sequence diversity and a broader range of TF binding combinations compared to those found in natural genomic sequences (Davis *et al.*, 2020). Deep learning techniques, especially CNNs, have proven to be pivotal in the rational design of synthetic promoters, facilitating the precise and predictable modulation of gene expression.

The composition of DNA sequences and their methylation status are closely associated with gene expression activity, including the differential expression levels of orthologous gene pairs. Washburn *et al.* (2019) demonstrated, using a CNN-based model, that promoter and terminator sequences can be used to accurately predict gene expression status. Moreover, their study revealed that these sequence features could also be used to determine which gene in a pair of orthologs—such as between corn and sorghum—would exhibit higher expression levels.

In addition to sequence-based features, epigenetic modifications, such as DNA methylation, play a critical role in regulating gene expression. Hollister and Gaut (2009) reported a strong association between the methylation status of flanking regions and gene expression activity, with increased methylation generally correlating with reduced expression levels. This inverse relationship is attributed to the decreased binding affinity of TFs to methylated DNA regions (Busslinger *et al.*, 1983).

9.2.8 Data integration

Deep learning possesses substantial potential for the interpretation and analysis of genetic data, enabling the identification of candidate genes associated with specific traits and providing accurate trait predictions based on genotype information (Wang *et al.*, 2023). By integrating diverse data types—including genomic, transcriptomic, proteomic, and metabolomic datasets—deep learning can offer a more holistic understanding of plant biology and support advancements in SynBio. While the integration of multi-omics data presents significant computational and methodological challenges, it is essential for elucidating the complexity of biological systems. Such integrative approaches can uncover novel biological insights, identify previously unrecognized relationships among molecular components, and facilitate the discovery of biomarkers and therapeutic targets.

A variety of methodologies have been developed for the integration of omics data, including correlation-based approaches, machine learning models, and network-based frameworks. The holistic integration of genomics, metabolomics, phenomics, proteomics, and transcriptomics enables the analysis of biological changes and regulatory mechanisms, supporting the identification of key genes and regulatory elements crucial for crop improvement and breeding (Yang *et al.*, 2021).

Moreover, the integration of deep learning, epigenetics, and SynBio offers transformative opportunities for advancing agricultural practices.

Deep learning enables the analysis of vast and complex datasets to elucidate complex interactions among genetic elements, epigenetic modifications, and phenotypic traits in crops. This in-depth understanding paves the way for precision engineering through SynBio approaches to enhance crop performance. In particular, deep learning enhances the screening and optimization of biological chassis and components, thereby streamlining breeding processes. By leveraging automation and high-throughput computation, it enables the efficient identification of optimal genomic combinations for traits such as yield, disease resistance, and environmental adaptability.

9.2.9 Predictive modeling for climate resilience

The impact of abiotic environmental stresses on agricultural yields is both significant and well-documented. Rising awareness of the harmful effects of these stresses, which are exacerbated by climate change, has driven global initiatives aimed at adapting agricultural practices to increasingly hostile environmental conditions (Lobell *et al.*, 2008; Ortiz *et al.*, 2008; Wassmann *et al.*, 2009). Strategies proposed to address these challenges include the implementation of advanced agronomic management practices and the creation of novel crop genotypes with improved tolerance to environmental stressors.

Deep learning models provide powerful tools for predicting plant responses to various environmental stresses, thus aiding in the design and selection of crop varieties that are more resilient to climate variability. Notably, the integration of deep learning with epigenetic data and SynBio offers a promising path for advancing agricultural innovation. These models enable the analysis of large, complex datasets to reveal intricate relationships among genomic sequences, epigenetic changes, and phenotypic traits. Such a comprehensive understanding facilitates the targeted engineering of genetic and regulatory components through SynBio, with the goal of optimizing plant performance under stress conditions. Additionally, deep learning can significantly expedite the screening and evaluation of synthetic biological chassis and components, thereby improving the efficiency of breeding programs.

By leveraging automation and high-throughput computational capabilities, deep learning supports the rapid identification of optimal genome combinations associated with key agronomic traits such as yield potential, disease resistance, and environmental adaptability.

9.2.10 Optimization of transformation protocols

Plant transformation represents a powerful tool for advancing our understanding of plant responses to environmental changes, enabling the design of novel traits and the development of plants as sustainable bio-factories for high-quality agricultural products. While considerable challenges persist, recent technological advancements have significantly improved the feasibility and scope of plant transformation. A key priority for the future is expanding access to transformation technologies for a broader research community and investing in the education and training of the next generation of plant scientists. Ongoing innovation in this field holds substantial promise for both fundamental plant biology and agricultural improvement.

Since its inception, plant transformation has played a pivotal role in elucidating plant gene function and enhancing agronomic traits. However, even after more than four decades of development, major bottlenecks remain, particularly in achieving efficient transformation across a wide range of plant species at medium- to high-throughput levels. One of the primary limitations lies in the regeneration of transformed plants from explants, a process that is often hindered by genotype dependency and low regeneration capacity. To address these issues, researchers have implemented various strategies, including the optimization of hormone treatments in culture media, the selection of explants with reduced genotype sensitivity, and the ectopic expression of morphogenetic TFs (MTFs) to enhance regeneration potential (Lee and Wang, 2023).

New approaches, including the use of deep learning, open up fresh possibilities for advancing transformation protocols. Deep learning models can be applied to optimize the insertion of synthetic constructs into plant cells and improve regeneration efficiency by forecasting the most effective conditions and parameter combinations, thus streamlining and expediting the transformation process.

9.3 Challenges

Although the integration of deep learning with SynBio offers significant potential for enhancing crop improvement, several challenges persist (Fig. 9.1). SynBio is a rapidly evolving discipline focused on engineering biological systems for diverse applications, including biofuel production, pharmaceutical development, and crop trait enhancement. However, the inherent complexity and incomplete understanding of biological systems present significant data-related limitations for the field. Despite recent progress in improving the predictability of synthetic systems, key gaps in biological knowledge persist, which can hinder the effective application of AI techniques.

One critical challenge in applying AI to SynBio is model interpretability. Many deep learning models function as "black boxes," offering limited insight into the underlying rationale behind their predictions. This lack of transparency can pose obstacles for hypothesis-driven research and regulatory acceptance. Moreover, substantial sociological and technological barriers continue to exist between the domains of SynBio and deep learning, further complicating interdisciplinary collaboration.

The development and deployment of deep learning models are also constrained by their high demand for large, high-quality datasets. Data acquisition remains a major bottleneck in SynBio, particularly in areas where deep learning models are known to be data-intensive. Generating large-scale biological datasets is often expensive, time-consuming, and technically demanding. Additional challenges, including imbalanced training data, uncertainties in data labeling, catastrophic forgetting, overfitting, and vanishing gradients, further complicate the development and performance of models. In agriculture, deep learning has been successfully applied to various plant phenomics tasks, such as image classification (Ramcharan *et al.*, 2017), object detection (Ghosal *et al.*, 2019), and semantic segmentation (Aich and Stavness, 2017).

These applications have demonstrated the potential of deep learning to predict plant growth, estimate yield, determine maturity, and detect both biotic and abiotic stresses. However, the effectiveness of these models is highly dependent on the availability of extensive and well-labeled datasets (Cordts *et al.*, 2016). Data collection can be particularly costly when models must distinguish between numerous or subtly different categories (Tong *et al.*, 2022; Taghavi Namin *et al.*, 2018). Phenotypic data acquisition is further hampered by challenges such as severe occlusion in plant canopies and varying lighting conditions, both of which increase the difficulty and time required for accurate annotation (Scharr *et al.*, 2016).

Genotypic, phenotypic, and environmental datasets are essential resources for the advancement of intelligent breeding strategies (Talbot *et al.*, 2017). However, a notable gap persists in the availability and accessibility of high-quality phenotypic data. Traditional manual phenotyping methods are increasingly seen as inadequate, underscoring the need for the creation of high-throughput, cost-effective, and precise alternatives (Liu and Wang, 2021). Overcoming these challenges will be crucial to fully leverage the combined potential of deep learning and SynBio in advancing agricultural innovation.

9.4 Conclusion

The integration of deep learning with SynBio represents a transformative shift in the way biological systems are designed, understood, and optimized. Deep learning models, with their ability to uncover complex, non-linear relationships within vast biological datasets, are increasingly being applied across diverse areas of SynBio—from metabolic engineering and protein design to plant phenotyping and crop improvement. These models have shown great potential in accelerating the design–build–test cycle, enhancing predictive capabilities, and supporting high-throughput and precise engineering of biological systems.

However, despite significant progress, several challenges must be addressed to fully harness the power of deep learning in SynBio. Data scarcity, especially in high-quality, labeled datasets, remains a major bottleneck. Additionally, the interpretability of deep learning models continues to be a concern, limiting their utility in hypothesis-driven research. Issues such as model overfitting, vanishing gradients, and data imbalance further complicate model development and application.

Looking ahead, a key priority will be the development of more interpretable, data-efficient, and robust models tailored to the specific complexities of biological systems. Equally important is the need to foster interdisciplinary collaboration, improve data sharing infrastructure, and train researchers with combined expertise in biology and computational methods. By addressing these challenges, the integration of deep learning with SynBio is poised to drive significant innovation across biotechnology, agriculture, and environmental sustainability.

Overall, deep learning has already begun to reshape the landscape of SynBio. As the field continues to mature, we anticipate a new era of intelligent biological design—where machine learning not only assists in discovery but also becomes an integral component of the engineering process.

References

Aich, S. and Stavness, I. (2017) Leaf counting with deep convolutional and deconvolutional networks. In: *Proceedings of the IEEE International Conference on Computer Vision Workshops*. IEEE, Piscataway, NJ, USA, 2080–2089.

Almahairi, A., Rajeshwar, S., Sordoni, A., Bachman, P. and Courville, A. (2018) Augmented cyclegan: learning many-to-many mappings from unpaired data. In: *International Conference on Machine Learning*. PMLR, Sweden, 195–204.

Araus, J.L. and Cairns, J.E. (2014) Field high-throughput phenotyping: the new crop breeding frontier. *Trends in Plant Science* 19, 52–61. DOI: 10.1016/j.tplants.2013.09.008
Arya, S., Sandhu, K.S., Singh, J. and Kumar, S. (2022) Deep learning: as the new frontier in high-throughput plant phenotyping. *Euphytica* 218, 1–22. DOI: 10.1007/s10681-022-02992-3
Baek, M., Dimaio, F., Anishchenko, I., Dauparas, J., Ovchinnikov, S., Lee, G.R. *et al.* (2023) Accurate prediction of protein structures and interactions using a three-track neural network. *Science* 373, 871–876.
Bauer, A., Bostrom, A.G., Ball, J., Applegate, C., Cheng, T., Laycock, S. and Zhou, J. (2019) Combining computer vision and deep learning to enable ultra-scale aerial phenotyping and precision agriculture: A case study of lettuce production. *Horticulture Research* 6, 1–12.
Braguy, J. and Zurbriggen, M.D. (2016) Synthetic strategies for plant signalling studies: molecular toolbox and orthogonal platforms. *Plant Journal* 87, 118–138. DOI: 10.1111/tpj.13218
Busslinger, M., Hurst, J. and Flavell, R.A. (1983) DNA methylation and the regulation of globin gene expression. *Cell* 34, 197–206.
Cadet, X.F., Gelly, J.C., van Noord, A., Cadet, F. and Acevedo-Rocha, C.G. (2022) Learning strategies in protein directed evolution. In: Currin, A. and Swainston, N. (eds) *Directed Evolution: Methods and Protocols*. Springer US, New York, NY, pp. 225–275. DOI: 10.1007/978-1-0716-2152-3_15
Chandra, A.L., Desai, S.V., Guo, W. and Balasubramanian, V.N. (2020) Computer vision with deep learning for plant phenotyping in agriculture: a survey. *arXiv*. Available online at: https://arxiv.org/abs/2006.11391v1
Chang, L., Deng, X.-M., Zhou, M.-Q., Wu, Z.-K., Yuan, Y., Yang, S. *et al.* (2016) Convolutional neural networks in image understanding. *Acta Automatica Sinica* 42, 1300–1312. DOI: 10.16383/j.aas.2016.c150800
Chen, W.-Y., Liu, Y.-C., Kira, Z., Wang, Y.-C.F. and Huang, J.-B. (2019) A closer look at few-shot classification. *arXiv preprint arXiv*:1904.04232, 1–17. DOI: 10.48550/arXiv.1904.04232
Chen, Z., Li, X., Yang, M., Zhang, H. and Xu, X.S. (2023) Optimization of deep learning models for the prediction of gene mutations using unsupervised clustering. *Journal of Pathology: Clinical Research* 9, 3–17. DOI: 10.1002/cjp2.302
Cordts, M., Omran, M., Ramos, S., Rehfeld, T., Enzweiler, M., Benenson, R. *et al.* (2016) The cityscapes dataset for semantic urban scene understanding. In: *Proceedings of the IEEE Conference on Computer Vision and Pattern Recognition*. IEEE Computer Society, Los Alamitos, CA, USA, 3213–3223.
Davis, J.E., Insigne, K.D., Jones, E.M., Hastings, Q.A., Boldridge, W.C. and Kosuri, S. (2020) Dissection of c-AMP response element architecture by using genomic and episomal massively parallel reporter assays. *Cell Systems* 11, 75–85.
de Almeida, B.P., Reiter, F., Pagani, M. and Stark, A. (2022) DeepSTARR predicts enhancer activity from DNA sequence and enables the de novo design of synthetic enhancers. *Nature Genetics* 54, 613–624.
Engqvist, M.K.M. and Rabe, K.S. (2019) Applications of protein engineering and directed evolution in plant research. *Plant Physiology* 179, 907–917.
Ge, X., Wang, P., Wang, Y., Wei, X., Chen, Y. and Li, F. (2023) Development of an eco-friendly pink cotton germplasm by engineering betalain biosynthesis pathway. *Plant Biotechnology Journal* 21, 674–676.
Gebbers, R. and Adamchuk, V.I. (2010) Precision agriculture and food security. *Science* 327, 828–831. DOI: 10.1126/science.1183899
Ghahramani, A., Watt, F.M. and Luscombe, N.M. (2018) Generative adversarial networks simulate gene expression and predict perturbations in single cells. *bioRxiv*, 262501. DOI: 10.1101/262501
Ghosal, S., Zheng, B., Chapman, S.C., Potgieter, A.B., Jordan, D.R., Wang, X. *et al.* (2019) A weakly supervised deep learning framework for sorghum head detection and counting. *Plant Phenomics*. DOI: 10.34133/2019/1525874
Goodfellow, I., Bengio, Y. and Courville, A. (2016) Deep convolutional feature fusion model for multispectral maritime imagery ship recognition. In: *Deep Learning*. MIT Press, Cambridge, MA, USA.
Guo, T., Liu, C., Meng, F., Hu, L., Fu, X., Yang, Z., Wang, N., Jiang, Q., Zhang, X. and Ma, F. (2022) The m6A reader MhYTP2 regulates MdMLO19 mRNA stability and antioxidant genes translation efficiency conferring powdery mildew resistance in apple. *Plant Biotechnology Journal* 20, 511–525.
Gupta, A., Liu, B., Chen, Q.J. and Yang, B. (2023) High-efficiency prime editing enables new strategies for broad-spectrum resistance to bacterial blight of rice. *Plant Biotechnology Journal* 21(7), 1454–1464. DOI: 10.1111/pbi.14049
Gupta, D., Sharma, G., Saraswat, P. and Ranjan, R. (2021) Synthetic biology in plants, a boon for coming decades. *Molecular Biotechnology* 63, 1138–1154.
He, K., Zhang, X., Ren, S. *et al.* (2016) Deep residual learning for image recognition. In: *Proceedings of the IEEE Conference on Computer Vision and Pattern Recognition*, pp. 770–778.

Hollister, J.D. and Gaut, B.S. (2009) Epigenetic silencing of transposable elements: a trade-off between reduced transposition and deleterious effects on neighboring gene expression. *Genome Research* 19, 1419–1428.

Jiang, Y. and Li, C. (2020) Convolutional neural networks for image-based high-throughput plant phenotyping: a review. *Plant Phenomics* 2020. DOI: 10.34133/2020/4152816

Jumper, J., Evans, R., Pritzel, A., Green, T., Figurnov, M., Ronneberger, O. *et al.* (2021) Highly accurate protein structure prediction with AlphaFold. *Nature* 596, 583–589. DOI: 10.1038/s41586-021-03819-2

Kamilaris, A. and Prenafeta-Boldú, F.X. (2018) Deep learning in agriculture: a survey. *Computers and Electronics in Agriculture* 147, 70–90. DOI: 10.1016/j.compag.2018.02.016

Kaul, S., Choudhary, M., Gupta, S. and Dhar, M.K. (2021) Engineering host microbiome for crop improvement and sustainable agriculture. *Frontiers in Microbiology* 12, 635917. DOI: 10.3389/fmicb.2021.635917

Kelley, D.R. (2020) Cross-species regulatory sequence activity prediction. *PLoS Computational Biology* 16, e1008050. DOI: 10.1371/journal.pcbi.1008050

Khan, H., Ul Haq, I., Munsif, M., Mustaqeem, K.S.U. and Lee, M.Y. (2022) Automated wheat diseases classification framework using advanced machine learning technique. *Agriculture* 12(8), 1226. DOI: 10.3390/agriculture12081226

Kolhar, S. and Jagtap, J. (2021) Plant trait estimation and classification studies in plant phenotyping using machine vision – a review. *Information Processing in Agriculture* 10, 114–135. DOI: 10.1016/j.inpa.2021.02.006

Krizhevsky, A., Sutskever, I. and Hinton, G.E. (2017) ImageNet classification with deep convolutional neural networks. *Communications of the ACM* 60, 84–90. DOI: 10.1145/3065386

LeCun, Y., Bottou, L., Bengio, Y. *et al.* (1998) Gradient-based learning applied to document recognition. *Proceedings of the IEEE* 86(11), 2278–2324. DOI: 10.1109/5.726791

Lee, K. and Wang, K. (2023) Strategies for genotype-flexible plant transformation. *Current Opinion in Biotechnology* 79, 102848. DOI: 10.1016/j.copbio.2022.102848

Li, D.W., Ma, Y.S., Zhou, Y., Gou, J.B., Zhong, Y., Zhao, L.L., Han, L.D., Ovchinnikov, S., Ma, L., Huang, S.W. *et al.* (2019) A structural and data-driven approach to engineering a plant cytochrome P450 enzyme. *Science China Life Sciences* 62, 873–882.

Li, L., Zhang, Q. and Huang, D. (2014) A review of imaging techniques for plant phenotyping. *Sensors* 14, 20078–20111. DOI: 10.3390/s141120078

Li, T., Yang, X., Yu, Y., Si, X., Zhai, X., Zhang, H. *et al.* (2018) Domestication of wild tomato is accelerated by genome editing. *Nature Biotechnology* 36, 1160–1163. DOI: 10.1038/nbt.4273

Lin, H., Huang, Y., Liu, M., Li, X., Ji, S. and Li, S.Z. (2022) Diffbp: Generative diffusion of 3D molecules for target protein binding. *arXiv preprint* arXiv:2211.11214.

Liu, J. and Wang, X. (2021) Plant diseases and pests detection based on deep learning: a review. *Plant Methods* 17, 1–18. DOI: 10.1186/s13007-021-00722-9

Liu, J., Li, J., Wang, H. and Yan, J. (2020) Application of deep learning in genomics. *Science China Life Sciences* 63, 1860–1878.

Liu, W. and Stewart, C.N. Jr. (2015) Plant synthetic biology. *Trends in Plant Science* 20, 309–317.

Lobell, D.B., Burke, M.B., Tebaldi, C., Mastrandrea, M.D., Falcon, W.P. and Naylor, R.L. (2008) Prioritizing climate change adaptation needs for food security in 2030. *Science* 319, 607–610.

Luo, R., Sedlazeck, F.J., Lam, T.W. and Schatz, M.C. (2019) A multi-task convolutional deep neural network for variant calling in single molecule sequencing. *Nature Communications* 10, 998.

Meng, F. and Ellis, T. (2020) The second decade of synthetic biology: 2010–2020. *Nature Communications* 11, 5174.

Mirdita, M., Schütze, K., Moriwaki, Y., Heo, L., Ovchinnikov, S. and Steinegger, M. (2022) ColabFold: making protein folding accessible to all. *Nature Methods* 19, 679–682.

Mortensen, A.K., Skovsen, S., Karstoft, H. and Gislum, R. (2019) The oil radish growth dataset for semantic segmentation and yield estimation. In: *2019 IEEE/CVF Conference on Computer Vision and Pattern Recognition Workshops (CVPRW)*. IEEE, Long Beach, CA, USA, 2703–2710.

Opuu, V. and Simonson, T. (2023) Enzyme redesign and genetic code expansion. *Protein Engineering, Design and Selection* 36, gzad017.

Ortiz, R., Sayre, K.D., Govaerts, B., Gupta, R., Subbarao, G.V., Ban, T., Hodson, D., Dixon, J.A., Ortiz-Monasterio, J.I. and Reynolds, M. (2008) Climate change: can wheat beat the heat? *Agriculture, Ecosystems & Environment* 126, 46–58.

Praveen, B. and Sharma, P. (2019) A review of literature on climate change and its impacts on agriculture productivity. *Journal of Public Affairs* 19, e1960. DOI: 10.1002/pa.1960

Qin, H., Ou, L., Gao, J., Chen, L., Wang, J.W., Hao, P. and Li, X. (2022) DENA: training an authentic neural network model using Nanopore sequencing data of Arabidopsis transcripts for detection and quantification of N(6)-methyladenosine on RNA. *Genome Biology*, 23, p.25.

Ramcharan, A., Baranowski, K., McCloskey, P., Ahmed, B., Legg, J. and Hughes, D.P. (2017) Deep learning for image-based cassava disease detection. *Frontiers in Plant Science* 8, 1852. DOI: 10.3389/fpls.2017.01852

Ren, C., Kim, D.-K. and Jeong, D. (2020) A survey of deep learning in agriculture: techniques and their applications. *Journal of Information Processing Systems* 16, 1015–1033. DOI: 10.3745/JIPS.04.0187

Sahraeian, S.M.E., Liu, R., Lau, B., Podesta, K., Mohiyuddin, M. and Lam, H.Y.K. (2019) Deep convolutional neural networks for accurate somatic mutation detection. *Nature Communications* 10, 1041.

Scharr, H., Minervini, M., French, A.P., Klukas, C., Kramer, D.M., Liu, X. *et al.* (2016) Leaf segmentation in plant phenotyping: a collation study. *Machine Vision and Applications* 27, 585–606. DOI: 10.1007/s00138-015-0737-3

Schmidt, C., Schindele, P. and Puchta, H. (2020) From gene editing to genome engineering: restructuring plant chromosomes via CRISPR/Cas. *aBIOTECH* 1, 21–31. DOI: 10.1007/s42994-019-00002-0

Schumann, A., Waldo, L., Holmes, W., Test, G. and Ebert, T. (2018) Artificial intelligence for detecting citrus pests, diseases and disorders. In: *Citrus Industry News, Technology*. AgNet Media, Inc., Gainesville, FL, USA, 2 July.

Shi, S., Wang, Z., Shen, L. and Xiao, H. (2022) Synthetic biology: a new frontier in food production. *Trends in Biotechnology* 40(7), 781–803. DOI: 10.1016/j.tibtech.2022.01.002

Shrivastava, V.K. and Pradhan, M.K. (2021) Rice plant disease classification using color features: a machine learning paradigm. *Journal of Plant Pathology* 103, 17–26. DOI: 10.1007/s42161-020-00683-3

Sun, D., Zhu, Y., Xu, H., He, Y. and Cen, H. (2019) Time-series chlorophyll fluorescence imaging reveals dynamic photosynthetic fingerprints of sos mutants to drought stress. *Sensors* 19, 2649. DOI: 10.3390/s19122649

Swift, J. and Coruzzi, G.M. (2017) A matter of time – how transient transcription factor interactions create dynamic gene regulatory networks. *Biochimica et Biophysica Acta (BBA) - Gene Regulatory Mechanisms* 1860, 75–83. DOI: 10.1016/j.bbagrm.2016.08.007

Taghavi Namin, S., Esmaeilzadeh, M., Najafi, M., Brown, T.B. and Borevitz, J.O. (2018) Deep phenotyping: deep learning for temporal phenotype/genotype classification. *Plant Methods*, 14, pp.1–14. doi: 10.1186/s13007-018-0333-4.

Talbot, B., Chen, T.-W., Zimmerman, S., Joost, S., Eckert, A.J., Crow, T.M., *et al.* (2017). Combining genotype, phenotype, and environment to infer potential candidate genes. *Journal of Heredity*, 108, pp.207–216. doi: 10.1093/jhered/esw077.

Tiwari, V., Joshi, R.C. and Dutta, M.K. (2021) Dense convolutional neural networks based multiclass plant disease detection and classification using leaf images. *Ecological Informatics*, 63, p.101289.

Tong, Y.-S., Tou-Hong, L. and Yen, K.-S. (2022) Deep learning for image-based plant growth monitoring: A review. *International Journal of Engineering and Technology Innovation*, 12, p.225. doi: 10.46604/ijeti.2022.8865.

van Klompenburg, T., Kassahun, A. and Catal, C. (2020) Crop yield prediction using machine learning: a systematic literature review. *Computers and Electronics in Agriculture*, 177, p.105709. doi: 10.1016/j.compag.2020.105709.

Wang, K., Abid, M.A., Rasheed, A., Crossa, J., Hearne, S. and Li, H. (2023) Dnngp, a deep neural network-based method for genomic prediction using multi-omics data in plants. *Molecular Plant*, 16, pp.279–293. doi: 10.1016/j.molp.2022.11.004.

Wang, Y., Zhang, P., Guo, W., Liu, H., Li, X., Zhang, Q., Du, Z., Hu, G., Han, X., Pu, L., *et al.* (2021) A deep learning approach to automate whole-genome prediction of diverse epigenomic modifications in plants. *New Phytologist*, 232, pp.880–897.

Washburn, J.D., Mejia-Guerra, M.K., Ramstein, G., Kremling, K.A., Valluru, R., Buckler, E.S., *et al.* (2019) Evolutionarily informed deep learning methods for predicting relative transcript abundance from DNA sequence. *Proceedings of the National Academy of Sciences of the United States of America*, 116, pp.5542–5549.

Wassmann, R., Jagadish, S.V.K., Heuer, S., Ismail, A., Redona, E., Serraj, R., Singh, R.K., Howell, G., Pathak, H. and Sumfleth, K. (2009) Climate change affecting rice production: the physiological basis for possible adaptation strategies. *Advances in Agronomy*, 101, pp.59–122.

Wolfert, S., Ge, L., Verdouw, C. and Bogaardt, M.-J. (2017) Big data in smart farming - a review. *Agricultural Systems*, 153, pp.69–80. doi: 10.1016/j.agsy.2017.01.023.

Yang, L., Zhang, P., Wang, Y., Hu, G., Guo, W., Gu, X. and Pu, L. (2022) Plant synthetic epigenomic engineering for crop improvement. *Science China Life Sciences*, 65, pp.2191–2204.

Yang, Y., Saand, M.A., Huang, L., Abdelaal, W.B., Zhang, J., Wu, Y., *et al.* (2021) Applications of multi-omics technologies for crop improvement. *Frontiers in Plant Science*, 12, p.563953. doi: 10.3389/fpls.2021.563953.

Yang, Z., Gao, S., Xiao, F., Li, G., Ding, Y., Guo, Q., *et al.* (2020) Leaf to panicle ratio (LPR): a new physiological trait indicative of source and sink relation in japonica rice based on deep learning. *Plant Methods*, 16, pp.1–15. doi: 10.1186/s13007-020-00660-y.

Yin, J., Huo, L., Guo, L. and Hu, J. (2008) Short-term load forecasting based on improved gene expression programming. In: *7th World Congress on Intelligent Control and Automation*. Piscataway: IEEE, pp.5647–5650.

Zhou, J. and Troyanskaya, O.G (2015) Predicting effects of noncoding variants with deep learning-based sequence model. *Nature Methods*, 12, pp.931–934.

Zhou, J., Theesfeld, C.L., Yao, K., Chen, K.M., Wong, A.K. and Troyanskaya, O.G., (2018) Deep learning sequence-based ab initio prediction of variant effects on expression and disease risk. *Nature Genetics*, 50, pp.1171–1179. doi: 10.1038/s41588-018-0160-6.

Zhu, Q., Wang, B., Tan, J., Liu, T., Li, L. and Liu, Y.G. (2020) Plant synthetic metabolic engineering for enhancing crop nutritional quality. *Plant Communications*, 1, p.100017.

Zou, J., Huss, M., Abid, A., Mohammadi, P., Torkamani, A. and Telenti, A. (2019) A primer on deep learning in genomics. *Nature Genetics*, 51, pp.12–18.

10 Deep Learning in Managing Crop Production Under a Changing Climate

Ankan Das[1], Rajashree Mitra Thakur[1], Sucharita Mondal[1], Shruti Mukhopadhyay[1], and Sandip Debnath[2*]

[1]Palli Siksha Bhavana (Institute of Agriculture), Visva-Bharati, Sriniketan, West Bengal, India; [2]Department of Genetics and Plant Breeding, Palli Siksha Bhavana (Institute of Agriculture), Visva-Bharati, Sriniketan, West Bengal, India

Abstract

Climate change presents escalating challenges to global agriculture, necessitating advanced strategies to enhance resilience in crop production. This chapter explores how deep learning (DL) technologies are revolutionizing climate-adaptive farming by integrating diverse data streams—such as satellite imagery, Internet of Things (IoT) sensors, and climate models—to predict and optimize crop yields amid environmental volatility. Autonomous systems, equipped with computer vision and robotics, further enable real-time monitoring and precision tasks, such as irrigation and pest management, adapting dynamically to shifting weather patterns. Complementing these innovations, edge computing brings lightweight artificial intelligence (AI) models directly to farms, allowing rapid, localized decision-making without dependence on centralized infrastructure, while generative AI addresses data gaps by simulating realistic climate scenarios to improve predictive accuracy. A critical focus is placed on explainable AI (XAI), which demystifies complex algorithms to foster trust among farmers and policymakers, ensuring transparent and actionable insights. Emerging quantum-inspired DL methods are also examined for their potential to tackle large-scale challenges, such as optimizing water distribution and creating digital twins (DTs) of agroecosystems. However, these advancements are not without hurdles: high costs, interoperability issues, and ethical concerns about equitable access pose significant barriers. Looking ahead, interdisciplinary collaboration is essential to refine AI models, democratize their adoption, and align technological progress with sustainable, agroecological practices. By synthesizing cutting-edge innovations and persistent challenges, this chapter outlines a roadmap for leveraging DL to build adaptive, equitable, and climate-resilient agricultural systems in an uncertain future.

Keywords: climate-resilient crop management, deep learning, explainable AI, predictive agricultural analytics, quantum-inspired optimization

10.1 Introduction to Deep Learning in Climate-Adaptive Agriculture

The Indian agricultural industry is suffering due to the effects of climate change. The negative consequences of climate change are having an increasing impact on crop productivity. Farmers are finding it more difficult to guarantee steady production and food security as a result of climate change, especially in the form of storms, droughts, floods, and heat waves. These extreme weather events affect water availability, soil health, and

*E-mail: sandip.debnath@visva-bharati.ac.in

 Deep Learning in Plant Science and Agricultural Biotechnology (Jen-Tsung Chen ed.)
DOI: 10.1079/9781836991120.0010

plant physiology, which in turn leads to lower yields and the possibility of crop failure. Reduced agricultural output is a result of a number of factors, including higher global temperatures, unpredictable rainfall patterns, dwindling water supplies, and extreme weather events, including heat waves, droughts, and floods.

Artificial intelligence (AI) is now being used in the agricultural sector to help confront the growing issues of climate change, food, and shelter. AI can transform precision farming and make data-driven, climate-resilient crop management possible. With its advanced algorithmic structures, real-time data analytics, and machine learning (ML) models, AI has the potential to revolutionize agriculture by reducing risks, maximizing resource use, and improving agricultural operations. Advancements such as big data analytics, the Internet of Things (IoT), and precision farming enable farmers to track and supervise their crops with greater accuracy and efficacy. Farmers are making data-driven decisions to optimize planting schedules, water use, and pest control with the use of remote sensing, drones, and geographic information systems (GIS) (Patil *et al.*, 2024).

These technologies help mitigate key threats to agriculture by enabling farmers to shift from vulnerable practices to more adaptive, resilient farming systems. Predictive data analysis aids farmers in anticipating weather patterns, disease and insect attacks, and soil moisture deficits. Additionally, it aids in scheduling irrigation and water conservation by forecasting droughts and heavy precipitation. Planting schedules, climate-resilient crop selection, and precautions against severe weather events are all made possible by ML algorithms that forecast future weather conditions based on historical weather data and current conditions. To predict crop outcomes and propose strategies, it also makes use of real-time, data-driven systems and satellite imagery. Technologies such as satellite-based weather forecasting systems and soil moisture sensors can assist farmers in maximizing water use in drought-prone areas. In many areas, soil degradation is being driven by climate change, which lowers productivity. In order to maximize yields and improve endurance to changing climate conditions, such as extreme weather events and variations in temperature, farmers can implement more effective mitigation techniques to minimize the effects of climate change on farming practices (Shafi *et al.*, 2019). Deep learning (DL) plays a critical role in climate-adaptive agriculture by using sophisticated algorithms to analyze massive databases from sensors, drones, and satellites (Logeshwaran *et al.*, 2024).

By using AI algorithms to evaluate and predict agricultural outcomes and offer recommendations to enhance crop output, DL can be used to boost crop yields. The purpose of this chapter is to provide an overview of the ways in which DL can be used to combat climate change. We will discuss how to use a variety of AI tools and ML techniques to adapt to climatic unpredictability and prevent the negative effects of the climate.

10.2 Multimodal Predictive Analytics for Dynamic Yield Optimization

Accurate prediction of crop yields has become increasingly important for supporting farmers and policymakers in making better decisions, especially under the growing challenges of climate change. As crops react differently to changing weather patterns in different regions, it becomes essential to create adaptive strategies that are specific to local conditions. Climate change has placed agriculture under serious pressure worldwide, threatening food security and making the need for reliable yield forecasting even more urgent (Kumar *et al.*, 2022). Understanding the complex relationship between climatic factors, such as temperature and rainfall, and their effects on crop productivity is necessary to build strong adaptation plans. In this context, DL models offer a powerful solution by capturing complex patterns hidden in large agricultural datasets (Wang *et al.*, 2022). Unlike traditional methods, DL models can process large amounts of raw data, learn intricate relationships, and make accurate predictions without the need for manual feature selection (Ahmed *et al.*, 2023). In this study, the influence of important climatic elements, such as rising temperatures and reduced rainfall, on crop production is analyzed. Before building the models, the dataset is carefully prepared by filling missing values, removing duplicates, converting categorical information into numerical formats using one-hot encoding, and identifying and removing outliers through clustering techniques. Key variables that influence yield are selected, and then DL models are applied to predict future yields with greater precision. A hybrid modeling approach, combining long short-term memory (LSTM) networks and transformer structures, is used to enhance the model's ability to capture both temporal trends and hidden feature relationships, though the main focus remains on DL's overall predictive strength (Garg *et al.*, 2024). DL, a specialized area of ML, uses multiple hidden layers to uncover deep and complex patterns from raw data, leading to highly accurate predictions (Sarker, 2021). Models such as deep neural networks (DNNs), convolutional neural networks (CNNs), recurrent neural networks (RNNs), and LSTM networks have shown strong abilities in different areas of crop-yield forecasting (Oikonomidis *et al.*, 2022). DNNs handle complex, non-linear data patterns; CNNs automatically extract meaningful features; RNNs manage time-based data efficiently; and LSTM models specialize in learning long-term dependencies that are crucial for seasonal crop predictions. Although DL models have brought remarkable improvements, challenges still exist, mainly due to the diversity of crop types, data sources, and regional conditions. A systematic review was performed to gather insights into the different ML and DL approaches applied to crop-yield prediction (Table 10.1). Overall, integrating DL models into agriculture provides a promising path toward dynamic yield optimization, helping farming systems stay productive and resilient under the pressures of a changing climate. A detailed study about DL models in dynamic crop-yield prediction is discussed below.

The illustration Fig. 10.1 shows a smooth flow of how DL helps in optimizing crop yield, even as the climate keeps changing. First, different types of data are gathered from many sources. Satellite-based remote sensing tools collect broad meteorological information, such as temperature, rainfall patterns, and cloud movements. On the ground, special instruments such as rainfall loggers and compact weather stations like the Vaisala WXT520 capture local or "microclimate" conditions, including real-time temperature, humidity, wind speed, and rainfall at a very detailed level. After the collection, all the data—both from space and from field sensors—is sent to cloud computing platforms. Here, strong computer systems store, manage, and organize this huge amount of information securely and efficiently. In the next step, DL models come into action. These smart computer programs are trained to find hidden patterns and relationships inside the collected data. They estimate important agricultural factors, such as soil moisture, plant health, crop growth stages, and potential yield. Finally, all these estimations are combined. The DL models such as DNN, CNN, RNN, and LSTM dynamically calculate the best strategies for maximizing crop yield. They take into account not just the current weather and soil conditions but also future climate predictions. In this way, the entire cycle—from gathering raw data to fine-tuning farming decisions—is completed in a seamless, intelligent system powered by DL, offering farmers a smart way to adapt and thrive even under shifting climate conditions.

10.3 Autonomous Climate-Smart Farming Systems

The integration of DL into modern agriculture has revolutionized the way crop production is managed under the ever-changing climate. DL, a branch of ML inspired by the structure of the human brain, excels at processing vast volumes of complex data and identifying patterns that are often too subtle for traditional algorithms or human analysis (Vaissnave *et al.*, 2024). These capabilities have enabled the development of autonomous, climate-resilient farming systems that adapt in real time to environmental fluctuations, optimize resource use, and significantly enhance crop productivity.

One of the most notable applications of DL in autonomous farming is in precision weed management (Upadhyay *et al.*, 2024). Conventional manual methods, such as hand weeding, hoeing, or chemical spraying, are labor-intensive, time-consuming, and can inadvertently harm desirable crops. To address these inefficiencies, DL-driven robotic systems have been developed. A notable example is AGRIBOT, an autonomous four-wheeled robot equipped with navigational sensors such as GPS, compass, inertial measurement units (IMU), and vision-based cameras. AGRIBOT leverages DL algorithms, particularly CNNs, to perform real-time weed and crop classification in complex field environments. The

Table 10.1. Multimodal deep learning algorithms for crop-yield prediction and optimization. Author's own table.

Algorithm/model	Type of input data handled	Core functional role	Multimodal strategy	Key use cases in agriculture	References
Deep neural network (DNN)	Structured tabular data (area, year, rainfall, pesticide use)	Learns complex, non-linear relationships between inputs and outputs	Acts as a fusion model to combine soil, climate, and crop metadata	General yield estimation from categorical and numerical input data	Bisht (2023)
Convolutional neural network (CNN)	Spatial data (e.g. soil depth layers, satellite images)	Extracts local patterns and spatial dependencies	Handles soil maps and rainfall patterns as image-like inputs	Soil texture recognition, rainfall region mapping, crop-zone classification	Hosseini *et al.* (2023)
Recurrent neural network (RNN)	Time series (e.g. rainfall over months, seasonal trends)	Maintains sequence memory for time-dependent data	Fuses temporal features, such as crop-cycle stages, with yield output	Modeling seasonal yield variation and climate-shift impact	Wang *et al.* (2022)
Long short-term memory (LSTM)	Long-term sequences (multiyear rainfall, temperature trends)	Remembers long dependencies, solves RNN limitations	Integrates long-range climatic sequences, crop history, and inputs	Highly accurate yield prediction across fluctuating seasons	Wang and Karimi (2024)
CLSTMT (CNN + LSTM + transformer)	Mixed-modal inputs: images, time series, tabular data	Combines feature detection, temporal memory, and attention mechanisms	Synchronizes multiple data types: sensor input, climate, and crop features	Advanced climate-responsive yield modeling under dynamic scenarios	Viel *et al.* (2023)
Hybrid (LSTM + SVM/RF)	Tabular and time series	Uses DL for feature extraction and ML for classification	Extracted deep features from LSTM passed to SVM or RF for prediction	Real-time crop advisories and disease-risk alerts	Gao *et al.* (2022)
IoT-integrated LSTM model	Sensor data (real-time temperature, humidity, soil pH)	Continuously updates predictions based on live data streams	Multimodal flow from hardware (sensors) to software (DL model)	Live yield forecasting, irrigation control, pesticide timing	Saha *et al.* (2025)
Transformer-based models	Sequential and contextual features	Captures global dependencies using self-attention	Blends contextual crop cycles, rainfall, and fertilizer scheduling	Forecasting under erratic climate shifts, adaptive planning	Liu *et al.* (2025)

RF, random forest; SVM, support vector machine.

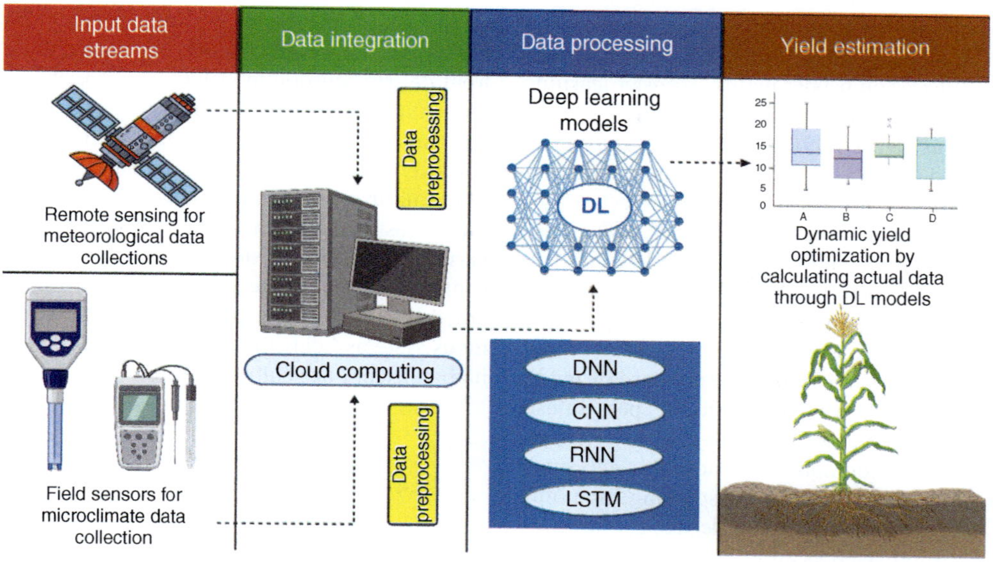

Fig. 10.1. Flow diagram of dynamic crop-yield prediction by deep learning algorithms. (Author's own image.)

CNN is trained using thousands of labeled images, enabling it to accurately distinguish between different plant species based on color, shape, texture, and edge features at the pixel level (Patel *et al.*, 2022).

This pixel-wise classification capability, which CNN architectures such as VGG16 and U-Net excel at, allows for precise targeting of weeds without disturbing the crops. Models trained on multispectral data—beyond just RGB channels—can capture additional wavelength information that enhances object separation, especially under variable natural lighting. In comparative evaluations, DL models using combined RGB and NIR (near-infrared) input achieved higher accuracy than those relying solely on RGB inputs (Sun *et al.*, 2020). Some advanced weed recognition systems even integrate encoder–decoder architectures that refine spatial information during up-sampling, which improves boundary detection between crops and weeds.

Beyond open-field applications, DL has also been instrumental in the creation of intelligent greenhouse systems, which simulate and regulate microclimatic conditions to ensure year-round, climate-smart cultivation. These smart greenhouses rely on a network of IoT-enabled sensors—including soil moisture detectors, ambient temperature probes, humidity sensors, and CO_2 monitors—that collect real-time environmental data. This multidimensional data is processed and stored in cloud platforms, where DL models are trained to predict and control variables critical to plant health (Wang *et al.*, 2022). For instance, RNNs or LSTM models are particularly suited for time-series forecasting, enabling the prediction of temperature fluctuations and moisture loss patterns (Waqas and Humphries, 2024). Based on these forecasts, automated systems can adjust irrigation, shading, and ventilation dynamically to maintain optimal conditions for plant growth.

DL also aids in the automation of irrigation systems, where models process live soil moisture data and correlate it with plant water requirements, historical climate data, and weather forecasts (Sami *et al.*, 2022). Decision-making within these models is powered by activation functions such as ReLU, sigmoid, or softmax, which help the system decide when and how much water to apply. These functions allow the models to interpret environmental signals probabilistically, facilitating precise and efficient water usage even in the presence of incomplete or noisy data.

To overcome the challenge of limited training datasets in some agricultural regions, techniques such as transfer learning are widely employed. In this approach, pretrained DL models—initially developed on large, generic image datasets—are fine-tuned with smaller, domain-specific datasets, such as images of pests on tomato leaves or diseases on pepper plants. Transfer learning has proven remarkably effective, with models such as DenseNet201, MobileNet, and InceptionV3 achieving classification accuracies exceeding 94% in real-time pest detection systems (Islam and Rahman, 2024). These systems are often integrated with low-cost embedded devices, such as Raspberry Pi units equipped with high-resolution cameras. These devices scan plants in the field, and the images are instantly analyzed by the onboard DL model to detect disease symptoms or pest infestations. If abnormalities are detected, the system can autonomously activate mitigation actions, such as localized spraying, or notify farm operators via mobile alerts.

Another application is the use of autoencoders in greenhouse management (Adkisson *et al.*, 2021). Unlike supervised CNNs that require labeled data, autoencoders can learn patterns from unlabeled data by compressing and reconstructing input features. This makes them ideal for anomaly detection, where they can identify deviations from normal temperature, humidity, or light levels. By learning the "normal" pattern of greenhouse operation, autoencoders can detect subtle signs of malfunction or external stress—such as failing cooling systems or sudden humidity drops—allowing early intervention before crop damage occurs.

Field experiments and case studies further validate the effectiveness of these autonomous, DL-based systems. In smart greenhouse trials involving crops such as tomatoes, cucumbers, beans, and bell peppers, systems integrated with CNN-based controllers demonstrated significant improvements in yield, water savings, and energy efficiency compared to manually controlled systems (Akbar *et al.*, 2024). Moreover, the synergy between edge computing, cloud platforms, and real-time sensor networks allows these DL models to be deployed in scalable architectures. For instance, edge devices handle real-time processing tasks, such as object recognition and immediate control actions, while cloud systems manage longer-term data analytics, model training, and performance evaluation. This distributed framework ensures low-latency responses for critical operations while enabling robust learning and decision-making over time.

DL is at the heart of autonomous, climate-smart farming systems. By fusing data from diverse sensors, modeling complex environmental relationships, and automating control mechanisms, these systems offer a sustainable, adaptive, and intelligent approach to agriculture. They minimize human intervention, reduce input waste, and optimize crop growth across variable climatic conditions. As climate change continues to challenge conventional farming practices, the convergence of DL and autonomous systems holds the promise of a resilient, high-efficiency agricultural future.

10.4 Edge AI and IoT–enabled Resource Management

Continuous spatiotemporal fluctuations in weather, conditions of the soil, and plant development status define the changing agricultural environment. As a result of these differences, different input demands, such as those for fertilizer and water, are shifted. Applying these inputs precisely and on time has increased agricultural output alongside efficient usage of resources. In a similar vein, uncontrolled pests and weeds can seriously impair crop development and even cause total crop failure.

Thus, employing IoT and edge AI has greatly increased operational efficiency through continuous and autonomous monitoring and evaluation of field conditions (Abioye *et al.*, 2022). IoT's primary goals are to streamline operations, decrease human intervention, and bring everything online. The IoT was first introduced by Kevin Ashton in 1999, and since then, it has grown in prevalence and improved all facets of people's lives (Ashton, 2009). The IoT is growing in importance in the current digital era, which is why the array of connected IoT devices is rising quickly. These extensively used IoT devices at the edge of the network will generate enormous volumes of data.

It is anticipated that processing, examining, and storing this enormous volume of data in a centralized cloud will result in higher network traffic, delays, and bandwidth usage. Furthermore, cloud computing cannot satisfy the demands of IoT applications that need real-time responses, since cloud servers are situated far from IoT end-user devices. To collect extensive data from the field, *in situ* sensors, unmanned aerial vehicles (UAVs), and remote sensing devices are used (Nawaz *et al.*, 2024).

With the help of IoT and AI, real-time analysis of fertilization and irrigation needs, as well as early weed and pest detection capacities, is made possible by the incorporation of multisensor data, which involves both micro- and macronutrients, soil pH, moisture level, and optical sensors such as near-infrared (NIR) reflectance spectroscopy. Moreover, variable rate technologies (VRTs) for irrigation and fertilization control offer real-time modifications, improving the distribution of water and nutrients according to the circumstances. Additional uses of IoT and AI in agriculture include weather prediction, yield estimation, crop selection, and determining the best time to plant (Sharma *et al.*, 2020).

The remote sensing technologies in the IoT control the entire operation. In addition to measuring plant growth and height by taking pictures with a high-resolution camera utilizing an IoT-based drone, field monitoring periodically senses the environmental soil, water level, and wetness for watering. This approach is used for lowering network load and enabling real-time IoT applications.

In addition, implementing algorithms based on AI on edge or local devices (Fig. 10.2) rather than depending on centralized cloud servers is known as edge AI. By lowering the response time, this procedure enables data processing within the local edge device. This method is frequently employed in ML model applications. By bringing computing power and AI closer to the agricultural field, edge computing enables real-time data processing and decision-making. The method starts with gathering data from sensors and IoT devices positioned throughout the fields of agriculture (Alonso *et al.*, 2020). These tools gather information on a variety of environmental factors, including soil moisture. Data processing then occurs immediately on edge devices. Mini-servers, edge gateways, or embedded systems with computing power can all be considered edge devices (Tuli *et al.*, 2014). In order to derive significant facts pertaining to plant health and environmental circumstances, these devices preprocess and analyze data in real time. In the last few decades, edge computing—which offers services near IoT end users without requiring strong internet connectivity—has become a prevalent application. Climate-adaptive agriculture relies heavily on a wireless sensing network. After detecting the surroundings, the

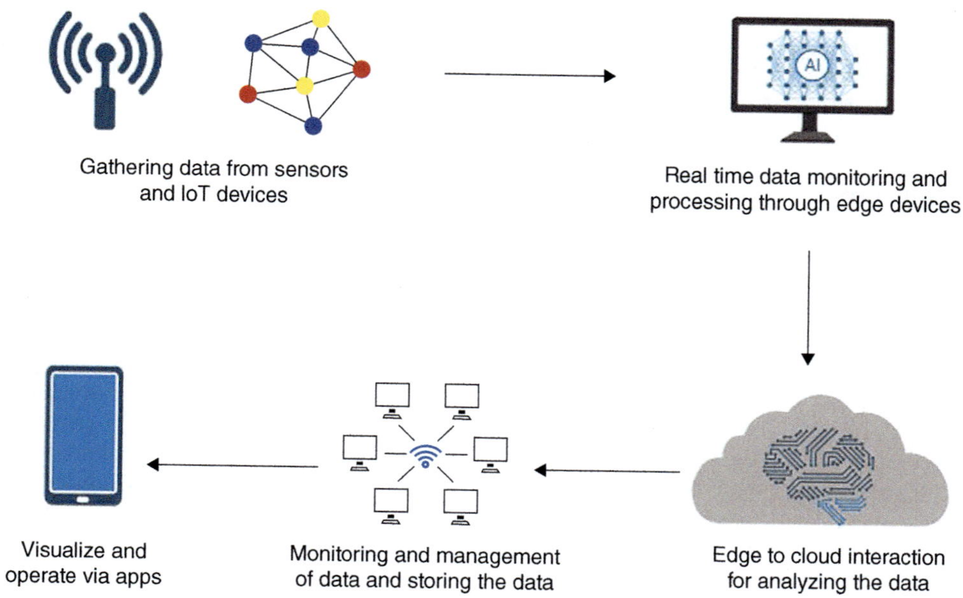

Fig. 10.2. Edge computing design for climate-adaptive farming. (Author's own image.)

sensor transmits the data to the activator for additional analysis, enabling efficient irrigation scheduling. The sensor in a smart irrigation system senses the moisture threshold level to determine when watering is necessary. IoT can monitor the field conditions in addition to irrigation (Avanija *et al.*, 2024).

Through IoT, tiny, low-cost sensors are made possible to continuously evaluate the environment. Developments in sensor, microcontroller, and wireless communication technologies have enabled the mass manufacturing of inexpensive IoT platforms with a wide range of connectivity possibilities. This trend of TinyML makes it possible to implement ML models in small, low-power, and low-latency IoT devices. The goal of this TinyML framework entails offering IoT devices greater intelligence so that new apps may be developed that incorporate ML tasks into them and permit certain interpretation and evaluation of data to be done in real time at the place of collection, which involves major savings in costs and improved safeguarding of privacy. TinyML models, such as decision trees, k-nearest neighbors, or simple neural networks (CNNs for image-based pest detection that are integrated into edge nodes, like mobile devices, and some deep CNNs—AlexNet, MobileNet, and VGG16—to detect diseases automatically) are frequently used for soil moisture and pest detection on devices with limited resources because they require little memory and computational capacity yet still produce precise categorization according to data collected by sensors from soil moisture probes or camera traps. TinyML holds several benefits that are real-time observation, minimal power intake, and inexpensiveness.

10.5 Generative AI for Synthetic Agricultural Data

Generative AI is used to counteract climate change by creating synthetic data. Generative AI is a field of AI that uses data from input in order to generate new material, including text, graphics, molecular structures, and other data (Sinha, 2024). Instead of merely examining or categorizing preexisting data, these trained algorithms synthesize enormous datasets and produce extremely detailed meteorological models by creating distinct patterns and structures. In order to provide optimal climate-resilient farming, it forecasts unpredictable weather, variations in the climate, and how various crops will respond to various climatic scenarios. By improving yield estimating systems' proficiency and dependability, generative AI contributes significantly to precision agricultural improvements. Insufficient or bare datasets are a common problem for traditional crop cultivation predicting models, and a number of sources, including environmental change, seasonal changes, and unreliable verification processes, cause data gaps. By mimicking various agricultural scenarios and producing complementary artificial data, generative AI overcomes this constraint. A potent method for enhancing agricultural output predictions is the combination of generative AI with remote sensing. Data on crop performance, soil moisture, and growth patterns may be obtained in real time using remote sensing through satellites or drones. However, there may be discrepancies in such information, particularly at certain periods like harvest (Li *et al.*, 2020). By creating synthetic data that closely resembles actual distant images, generative AI fills in those blanks and develops more comprehensive data. Data augmentation is the primary function of generative AI. In order to supplement actual datasets and facilitate real-time farm condition observation, synthetic samples are developed. In addition to identifying risk zones prone to unfavorable weather alterations, it may offer ways to reduce production losses and give data concerning soil water content, nutritional levels of difficulty, plant diseases, and insect populations. Without causing agricultural harm, it offers a basis for future weather forecasting for crop design.

Generative AI models, such as generative adversarial networks (GANs), are employed to predict future climatic occurrences and their consequences on agriculture. A hybrid structure of GANs is essential for generating crop production and growth prospects in a variety of scenarios while gradually improving the model's accuracy. When real data is inadequate or limited, GANs' capacity to deliver superior-quality, varied data improves the modeling process and results in more accurate and thorough yield projections. Two neural networks make up GANs: one is a generator that creates artificial data, and the other is a discriminator that assesses the veracity of the created data. The generator creates synthetic datasets from random noise in order to provide data that is almost equivalent to true samples. In order to provide accurate results, it uses many DNN layers to modify noise, accepting both fluctuating noise and previous meteorological information. In contrast, the discriminator checks the datasets to ascertain whether they are real or fraudulent (from the generator) (Leonila *et al.*, 2024). Employing activation functions such as ReLU or sigmoid, it takes in real and created meteorological data series and generates a probability value indicating whether the incoming data is true or artificial. The capacity of deep convolutional generative adversarial networks (DCGANs) to produce incredibly accurate information makes them popular in image-analysis applications such as flexible context production, image synthesis, sharpness augmentation, and image recovery from fragmented images.

Variational autoencoders (VAEs) are another generative model that combines concepts from statistical graphical models and autoencoders. They are especially helpful for image processing and data decoding services because they are programmed to acquire the variance of data in the space known as latent, which allows them to sample from it and produce a new one (Asperti *et al.*, 2021). Artificial neural networks (ANNs) are crucial in generative AI in increasing reliability in predictions or closing data collection gaps. They do this by creating new synthetic data and assessing historical information such as weather and soil characteristics (Ray, 2024). The input, hidden, and output layers are its three main components. The input layer gets the initial data, the hidden layer carries out the real programming, and the ultimate prediction is produced by the output layer. More data typically improves ANN performance, enabling them to discover undetected trends and correlations (Sapkota *et al.*, 2024).

In particular, in science and technology, physics-informed neural networks (PINNs) are a specific kind of neural network that incorporates concepts from physics in their learning procedures. PINNs may be used to construct physically robust and scientifically viable predictive frameworks, and along with this, they can produce precise projections by boosting the applicability of models for hidden circumstances, involving relevance to fluctuating platforms like climate shifts (Mowlavi *et al.*, 2023). They play a major role in the field of soil physics to represent intricate processes such as soil–pile relationships, soil microbial growth, and soil heat and moisture transfer. They have benefits when data are scarce and can deal with data with noise. Furthermore, if there are disturbances in measurements, PINNs can accurately model changes in field soil temperature (Kashinath *et al.*, 2021). By taking into account climatic factors including temperature, solar radiation, air humidity, soil hydration state, and outside weather, they may simulate the progress of bacterial and fungal communities.

10.6 XAI for Transparent Decision-Making

Agriculture is undergoing a digital transformation driven by AI and ML. These technologies are being implemented for tasks such as predicting crop yields, analyzing soil conditions, managing irrigation, and detecting pests. A majority of AI models are employed as "black boxes," which have limited transparency and give little insight into how decisions are made (Martin *et al.*, 2024; Shams *et al.*, 2024). Explainable AI (XAI) addresses this issue by making AI models more interpretable, enhancing trust, usability, and decision-making in agricultural applications, as shown in Fig. 10.3 (Mohan *et al.*, 2025). The algorithms, which include SHapley Additive exPlanations (SHAP) and local interpretable model-agnostic explanations (LIME), aim at making the use of AI-based agro solutions intelligible. For instance, SHAP has been used in assessing feature importance in crop-yield prediction models to predict how factors such as temperature, rainfall, and soil nutrients influence expected yields (Martin *et al.*, 2024). Similarly, in smart farming systems, LIME has been applied to understand local AI-generated recommendations so that the decisions are compatible with agronomic best practices (Singh *et al.*, 2022).

The importance of XAI in agriculture goes far beyond transparency to sustainability and climate change resilience that will truly matter. AI models can quantify climate variability impacts on crop yields; such models' recommendations

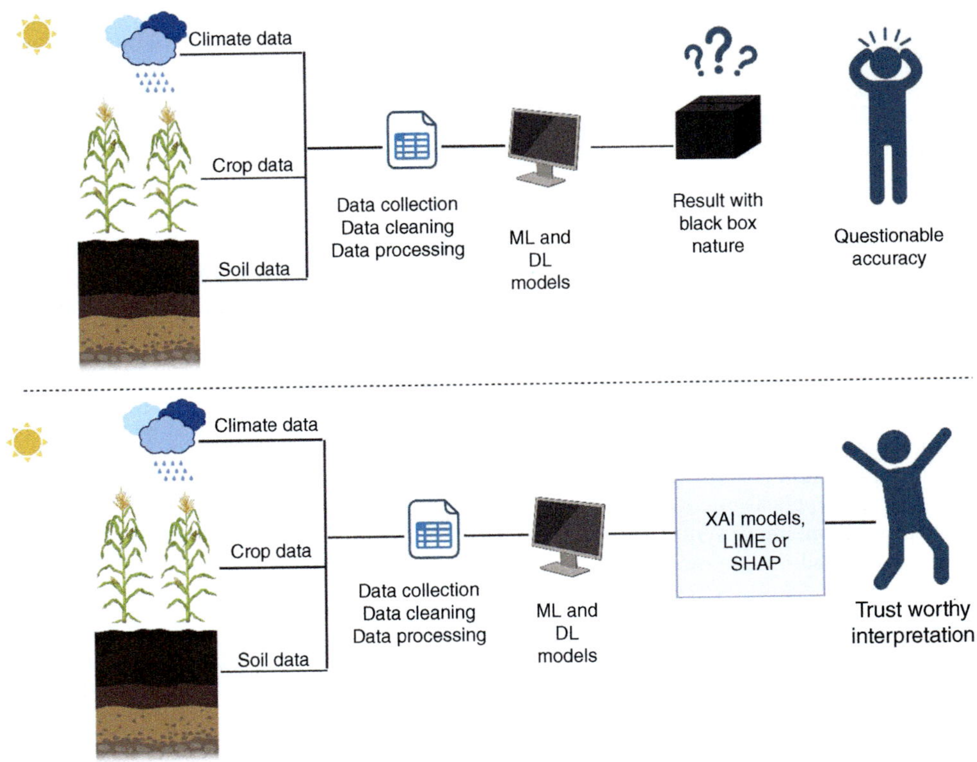

Fig. 10.3. Comparison between normal machine learning, deep learning model, and explainable AI-assisted models. (Author's own image.)

would be difficult to trust if interpreted as failures, since impacts affect any action (Malashin *et al.*, 2024). It is thus through integration into XAI that predictions become actionable expressions, guiding farmers in addressing emerging environmental conditions and utilizing resources optimally (Maraveas *et al.*, 2024).

Recent studies have shown that XAI in digital agriculture is needed, wherein AI-led decision support systems dictate irrigation, fertilization, and pest-control strategies. XAI-based smart agriculture frameworks have demonstrated increased productivity in food and sustainability through the combination of AI with interpretable decision models (Cartolano *et al.*, 2024). XAI techniques have also been applied to satellite and hyperspectral imaging data, enhancing precision agriculture by identifying key growth patterns and indicators of crop health (Mohan *et al.*, 2025).

XAI renders AI-based recommendations meaningful by bridging the gap between highly scientific, human-directed agriculture and the needs of beneficiaries, thus guaranteeing a transparent, sustainable, and efficient agricultural ecosystem.

As it pertains to making smart decisions about agricultural practices, neural networks are fundamental in understanding the climate determinants that drive crop production, such as temperature, precipitation, and soil moisture (Martin *et al.*, 2024). Explainable AI techniques such as SHAP and LIME indicate which of the different climate factors are weighted by these models (Singh *et al.*, 2022). By assigning a value of importance to the features, SHAP enables researchers and farmers to see whether a rise in temperature or a variation in rainfall contributes more to predicting crop yield (Cartolano *et al.*, 2024).

Studies indicate that temperature changes are frequently the strongest influence, followed by alterations in precipitation and soil quality (Maraveas *et al.*, 2024). Those visualizations assist farmers in adjusting their irrigation and fertilization strategies through AI-based modeling toward better climate change resilience (Mohan *et al.*, 2025).

10.6.1 User-friendly XAI interfaces for non-technical stakeholders

Despite the development of AI in agriculture, one of the most persistent challenges remains the interpretability of AI-driven findings by farmers and policymakers (Martin *et al.*, 2024). The majority of AI models have strong predictive ability but lack clear explanations, which makes it difficult for non-technical stakeholders to accept or act upon AI insights (Singh *et al.*, 2022).

To address this problem, researchers have designed user-friendly XAI interfaces for AI interpretations (Cartolano *et al.*, 2024). These include interactive dashboards, color-coded risk indicators, and voice-based AI assistants that enable farmers to make data-driven decisions even without technical expertise (Malashin *et al.*, 2024). In this context, a case study of farming frameworks that employ XAI demonstrated how visual supports such as real-time climate trend graphs and alerts generated by AI can increase accessibility for farmers relying on experiential knowledge instead of data analytics (Maraveas *et al.*, 2024).

Moreover, researchers further stress the need for tailored XAI interfaces based on specific agricultural regions to ensure that AI-driven recommendations are localized and suited for a particular crop or climate (Mohan *et al.*, 2025). Such adaptation will ensure that AI-oriented solutions gain the much-needed traction within agricultural communities (Table 10.2) that were traditionally averse to technology (Martin *et al.*, 2024).

SHapley Additive exPlanations

SHAP is one of the most common XAI techniques, offering quantification of the contribution of each feature in an ML model toward its predictions, assuring transparency in AI-driven decision-making (Cartolano *et al.*, 2024). An offspring of cooperative game theory, SHAP gives theoretically sustainable directions on feature attribution by fairly distributing prediction output among input features according to their marginal contributions (Singh *et al.*, 2022).

Making SHAP a top-tier tool in agriculture involves analysis of climate-related variables such as temperature, precipitation, and soil moisture, delivering clear explanations of how such factors influence crop-yield predictions (Martin *et al.*, 2024). SHAP thus builds interpretable insights and, in this regard, enhances trust and usability of AI-driven agricultural systems, making such interventions accessible for farmers, agronomists, and policymakers (Malashin *et al.*, 2024). SHAP calculates the marginal contribution of a feature by collating all possible subsets of features, thus ensuring the allocation of importance is fair and consistent (Singh *et al.*, 2022).

The average contribution of each feature is calculated by KernelExplainer of SHAP (Cartolano *et al.*, 2024); it is used in PLSR and SVR models, whereas TreeExplainer has been used in models such as RF and GBR (Singh *et al.*, 2022).

Local Interpretable Model-agnostic Explanations

LIME ensures the transparency of ML models to users, thus making those models more meaningful to stakeholders such as farmers and agricultural experts (Martin *et al.*, 2024). LIME aims to explain the prediction function of the original complex model through a simpler linear model. Two very common problems of ML are: a) trust in a single prediction and b) trust in the model as a whole. The authors of LIME tried to find a solution to these two problems (Cartolano *et al.*, 2024). LIME can provide visualization of the contribution of different features to a prediction; for example, in a recommendation for irrigation in a field, LIME shows the relationship and effect of soil moisture and historical yield data on this particular decision (Martin *et al.*, 2024).

Comparative experiments between LIME and SHAP increase the strength of explanations and improve the model's credibility, which is essential for building trust in critical domains of ML where understanding the model's behavior is vital (Malashin *et al.*, 2024).

10.7 Quantum-Inspired DL for Scalable Solutions

10.7.1 Quantum annealing

Quantum annealing (QA) is a computational paradigm that leverages quantum-mechanical principles to solve combinatorial optimization problems prevalent in agriculture, such as resource

Table 10.2. XAI applications in crop production. Author's own table.

Sl. no.	XAI applications	Techniques used	Key outcomes	References
1	Connective AI and XAI techniques for precision crop yield predictions	CNN, TEXTLIP, SHAP, LIME	Utilized satellite and drone imagery to analyze crop health and predict yield. XAI methods improved interpretability, helping stakeholders understand how environmental factors (e.g. soil moisture, temperature) affect crop yields.	Mohan *et al.* (2025)
2	Enrichment of crop recommendation systems with XAI	XAI-CROP framework	Enhanced transparency in crop recommendations based on environmental conditions and historical yield data. Improved decision-making for farm owners by explaining why specific crops are recommended.	Shams *et al.* (2024)
3	Wheat and grape yield prediction case studies	Machine learning with XAI methods	Provided insights into key factors influencing yield gains in wheat and grape production. Ensured machine learning models were interpretable and actionable for agricultural stakeholders.	Porfirio *et al.* (2024)

CNN, convolutional neural network; LIME, local interpretable model-agnostic explanation; SHAP, SHapley Additive exPlanations; XAI, explainable AI.

allocation, crop-yield prediction, and disease detection. Unlike classical optimization methods constrained by local minima, QA exploits quantum tunneling and superposition to navigate complex energy landscapes, enabling efficient exploration of near-optimal solutions (Setiadi *et al.*, 2024). Rooted in the adiabatic theorem, QA operates by gradually evolving a quantum system from a simple initial Hamiltonian—with an easily preparable ground state—to a final Hamiltonian encoding the optimization problem (Singh and Khan, 2023). This adiabatic evolution minimizes transitions away from the ground state, theoretically ensuring convergence to the optimal solution.

The process begins with a "driver" Hamiltonian that introduces quantum fluctuations, allowing the system to tunnel through energy barriers. As the system transitions to the final "problem" Hamiltonian—which mathematically represents the agricultural challenge (e.g. minimizing water usage while maximizing crop yield)—quantum effects are harnessed to identify global minima (Attri *et al.*, 2024). For instance, optimizing fertilizer distribution across a farm could be modeled using the Ising model, where qubits represent discrete decisions (e.g. fertilizing vs not fertilizing a specific plot), and their interactions encode cost–benefit trade-offs (Singh *et al.*, 2023).

Performance enhancement of quantum annealing

The efficacy of QA in agricultural applications depends on suppressing non-adiabatic transitions and refining parameter tuning. Recent advances focus on optimizing annealing schedules—the rate at which the Hamiltonian evolves—and improving qubit-embedding strategies to reduce computational complexity (Setiadi *et al.*, 2024). For example, mapping a crop-rotation optimization problem onto a quantum annealer requires embedding spatial and temporal constraints into qubit-connectivity patterns, a process that can be streamlined using hybrid quantum-classical algorithms (Singh and Khan, 2023).

Griffiths singularities—energy landscape irregularities that trap classical optimizers—pose challenges but can be mitigated through adaptive control of quantum fluctuations. In one hypothetical scenario, dynamically adjusting annealing parameters enabled QA to outperform classical algorithms in optimizing irrigation schedules for drought-prone regions, reducing water waste by 30% (Attri *et al.*, 2024). Such enhancements highlight QA's potential to transform resource-intensive agricultural workflows into polynomial-time-solvable problems.

Applications of quantum annealing

- Machine Learning and Artificial Intelligence. QA augments classical ML by accelerating feature selection and hyperparameter tuning in agricultural datasets. Hybrid models, such as quantum-enhanced XGBoost, have demonstrated superior accuracy in predicting rice yields by identifying non-linear relationships between soil pH, rainfall, and crop health (Setiadi *et al.*, 2024). In another case, QA-driven clustering algorithms categorized pest-infestation patterns in wheat fields, enabling targeted pesticide application and reducing ecological harm.
- Image Processing and Agriculture. QA underpins advanced image-processing techniques for crop monitoring. The entangled quantum image descriptor (EQID) model, which integrates quantum entanglement and superposition, achieved 99% accuracy in detecting early-stage blight in tomato plants—surpassing classical models such as VGGNet-16 (Attri *et al.*, 2024). By encoding images into quantum states, EQID captures subtle visual features (e.g. leaf discoloration) while exploiting entanglement for parallel processing. For example, entangled qubits could simultaneously analyze multiple spectral bands in satellite imagery, identifying water-stressed regions faster than classical convolutional networks (Singh *et al.*, 2023).
- Quantum Image Descriptors and Quantum Entanglement. Quantum image descriptors (QIDs) leverage superposition to represent high-resolution agricultural imagery compactly. A QID-based system could map a 10-terabyte drone-captured field image into a 100-qubit state, enabling real-time analysis of crop health. Entanglement further enhances processing: if two qubits encoding soil moisture and chlorophyll data are entangled, measuring one instantly reveals the state of the other, accelerating multispectral analysis (Klusch *et al.*, 2024).

Challenges and future directions

Despite its promise, QA faces obstacles in agricultural deployment. Quantum decoherence—the loss of quantum states due to environmental interference—limits computation time, particularly in field-deployable sensors (Singh and Khan, 2023). Qubit-connectivity constraints and hardware noise further hinder scalability; current annealers struggle with problems exceeding 5,000 variables, such as continent-scale supply-chain optimization (Maraveas *et al.*, 2024).

Future research aims to hybridize QA with classical DL, creating robust frameworks for real-world agricultural challenges. For instance, a quantum-classical neural network could predict locust swarm migration by combining quantum-optimized weather data with classical satellite imagery. Innovations in error-corrected qubits and modular quantum architectures may soon enable QA to tackle grand challenges such as global food security and climate-resilient farming (Setiadi *et al.*, 2024).

10.7.2 Digital twin

Digital twin (DT) technology is a transformative framework that creates dynamic, quantum-enhanced virtual replicas of physical agricultural systems, enabling real-time synchronization between crop fields and computational models. Originally developed for industrial applications, DT now integrates quantum computing principles to address complex challenges in crop production, such as yield optimization, disease prediction, and resource management (Purcell *et al.*, 2023). By merging real-time data streams with quantum simulations, DT systems offer predictive insights and adaptive control for sustainable agriculture (Table 10.3). The core architecture of a quantum-augmented DT includes six interconnected components, each enhanced by quantum computational advantages (Qi *et al.*, 2022):

- Physical Entity. The agricultural system being mirrored—such as a wheat field, irrigation network, or greenhouse—equipped with quantum-enabled sensors. For example, spin-based quantum sensors could measure soil moisture at subatomic resolution, capturing micronutrient gradients undetectable by classical devices (Pylianidis *et al.*, 2021).
- Virtual Model. A quantum computational representation of the physical entity, built using hybrid quantum-classical neural networks. This model simulates crop growth under variable conditions (e.g. drought, pest outbreaks) by leveraging quantum parallelism. A hypothetical DT for corn production might encode plant genetics and environmental variables into entangled qubits, enabling simultaneous evaluation of 10,000 growth scenarios in a single simulation (Qi *et al.*, 2022).
- Data Integration and Connectivity. Real-time data fusion via quantum communication protocols, such as quantum key distribution (QKD), ensures secure, low-latency transmission between IoT sensors and cloud platforms. In a rice-paddy DT, quantum-entangled photon pairs could synchronize soil pH and water-salinity data across distributed sensors, reducing data loss by 40% compared to classical networks (Huang *et al.*, 2021; Cesco *et al.*, 2023).
- Analytics and AI Processing. Quantum machine learning (QML) algorithms analyze hyperspectral crop imagery and sensor data to detect anomalies. For instance, a quantum support vector machine (QSVM) could identify early-stage fungal infections in soybean leaves with 98% accuracy—20% higher than classical CNNs—by exploiting quantum-kernel methods (Pylianidis *et al.*, 2021).
- Feedback Mechanism and Actuation. Quantum optimization algorithms, like Grover-adaptive search, generate actionable recommendations. A DT for citrus orchards might use QA to allocate pesticides optimally, reducing chemical usage by 35% while maintaining yield (Qi *et al.*, 2022).
- User Interface and Visualization. Quantum-inspired holographic dashboards provide 3D visualizations of crop health, overlaying quantum simulation results with real-time drone imagery. Farmers could interactively test irrigation strategies via quantum circuit-based risk assessment tools (Huang *et al.*, 2021).

Table 10.3. Case studies in digital twin. Author's own table.

Case study	Location	Key features	Economic impact	References
Zero-energy farm	Thailand	Remote soil and crop-health monitoring; automated irrigation scheduling.	Water savings; cost-effective resource use.	Li Wang (2024)
Mandarins twin	Jeju Island, South Korea	Data aggregation; multiscale analysis.	15% higher fruit-market value.	Kim and Seong Heo (2024)
Smart agriculture system	Not specified	Collects crop and environmental data in real time using sensors; generates digital twins modeled with machine learning algorithms; facilitates easy data collection and analysis; enables quick responses to problems and monitoring of pesticide use, crop growth, and disease outbreaks.	Improved crop productivity by approximately 20% compared with physical greenhouses; aims to reduce costs.	Kim *et al.* (2023)

By combining real-time data, AI, and the IoT for predictive modeling and resource optimization, DT technology is propelling notable breakthroughs in sustainability and Agriculture 5.0, as shown in Table 10.3 (Purcell *et al.*, 2023). In terms of sustainability, DTs play a crucial role in precision agriculture, urban planning, environmental monitoring, and climate modeling. They also help with energy efficiency, waste reduction, and resource conservation (Purcell *et al.*, 2023). DTs improve biodiversity conservation and sustainable farming methods while reducing the effects of climate change by offering data-driven insights (Purcell *et al.*, 2023).

Operational workflow of quantum-enhanced Digital Twin technology

The workflow integrates quantum computing at each stage to resolve agricultural bottlenecks, as shown in Fig. 10.4.

- Data Acquisition and Processing. Quantum sensors embedded in soil and drones collect tera-scale datasets on microclimate conditions. For example, nitrogen-vacancy (NV) center sensors in a potato farm could map root-zone temperature fluctuations with nanokelvin precision. Quantum error correction ensures data fidelity even in noisy field environments (Qi *et al.*, 2022; Purcell *et al.*, 2023).
- Model Synchronization and Simulation. Hybrid quantum-classical models update the virtual twin using variational quantum algorithms (VQAs). A coffee plantation DT might simulate shade management strategies by solving Schrödinger-like equations for canopy light distribution, achieving 100× faster convergence than finite-element methods (Huang *et al.*, 2021).
- Decision-Making and Optimization. Quantum Bayesian networks predict crop outcomes under uncertainty. In a drought-prone sorghum farm, the DT could evaluate 50,000 irrigation schedules in milliseconds using quantum amplitude amplification, identifying the optimal balance between water conservation and yield (Pylianidis *et al.*, 2021; Qi *et al.*, 2022).
- Implementation of Corrective Actions. Quantum reinforcement learning agents autonomously adjust farm equipment. For instance, a DT-controlled robotic harvester might use quantum Thompson sampling to prioritize ripe tomatoes while avoiding damage, improving harvest efficiency by 25% (Cesco *et al.*, 2023).

By embedding quantum principles into DT architecture, agriculture gains a scalable tool to tackle climate volatility and food security. Future advancements in photonic quantum sensors and fault-tolerant qubits will further bridge the gap between theoretical models and real-world farming systems (Purcell *et al.*, 2023).

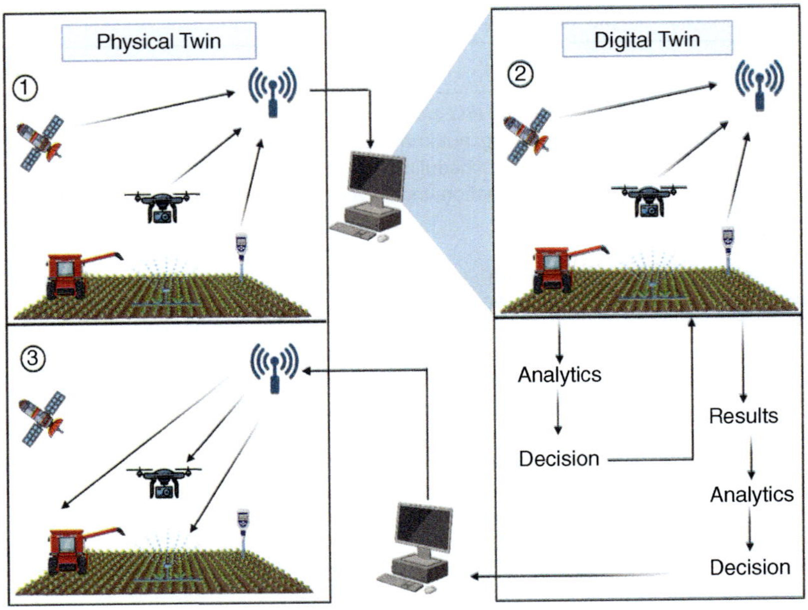

Fig. 10.4. Working principle of digital twin. (Author's own image.)

10.8 Future Directions and Challenges

10.8.1 Challenges

DL has emerged as a transformative tool for addressing climate-driven agricultural challenges, offering capabilities in predicting yield variability, detecting climate-resilient crop traits, and optimizing resource use. However, the dynamic and unpredictable nature of climate change introduces unique obstacles to deploying DL systems effectively in real-world farming scenarios.

Data scarcity and climate variability

DL models require large, annotated datasets to capture the complex interactions between crops and shifting climatic conditions. However, agricultural data under climate change scenarios are often fragmented, as extreme weather events (e.g. droughts, floods) disrupt consistent data collection (Albahar, 2023; Araújo *et al.*, 2023). For instance, a DL model trained to predict wheat yields in temperate zones may fail in regions experiencing sudden aridification due to insufficient training data on soil moisture dynamics under prolonged droughts (Mukherjee *et al.*, 2025). Additionally, labeling datasets for novel climate-induced diseases—such as heat-stress blight in corn—is costly and time-consuming, limiting model readiness for emerging threats (Ojo and Zahid, 2022).

Computational demands for real-time climate adaptation

The urgency of climate adaptation necessitates real-time DL processing, particularly in controlled environment agriculture (CEA) systems like vertical farms. However, simulating microclimate adjustments (e.g. optimizing LED light spectra during heat waves) demands intensive GPU resources, which are often inaccessible to smallholder farmers in climate-vulnerable regions (Albahar, 2023; Mukherjee *et al.*, 2025). A hypothetical DL system designed to regulate irrigation in response to erratic monsoon patterns might require cloud-based quantum simulators to achieve scalability—a barrier for communities lacking high-speed internet (Ojo and Zahid, 2022).

Generalization across climate-region shifts

DL models trained on historical climate data struggle to generalize under rapidly changing conditions. For example, a CNN trained to detect nutrient deficiencies in rice paddies using decade-old satellite imagery may misclassify drought-induced leaf discoloration as fungal infection, leading to incorrect fertilizer recommendations (Ojo and Zahid, 2022; Mukherjee *et al.*, 2025). Such discrepancies highlight the need for adaptive architectures that integrate real-time climate projections, such as hybrid models combining DL with regional climate model (RCM) outputs.

Interpretability for climate-resilient decision-making

The "black-box" nature of DL models undermines trust in critical climate adaptation strategies. A convolutional neural network might recommend reducing irrigation in a drought-stricken soybean field based on latent features unrelated to soil hydrology, leaving farmers skeptical of its logic (Araújo *et al.*, 2023; Shoaib *et al.*, 2023). Explainable AI techniques, such as attention maps highlighting heat-stressed root zones, could bridge this gap by aligning model decisions with agronomic principles.

Interdisciplinary knowledge gaps

Developing DL solutions for climate-resilient agriculture requires expertise in climatology, agronomy, and data science—a rare combination. A DL model predicting pest migrations under warmer winters, for instance, might overlook critical bioclimatic variables like diurnal temperature ranges if trained solely by computer scientists (Albahar, 2023; Mukherjee *et al.*, 2025).

Ethical concerns in climate data governance

Climate data ownership disputes complicate DL deployments. For example, sensor data from smallholder farms in flood-prone Bangladesh could be exploited by agribusinesses to develop proprietary flood-resistant crop varieties without compensating local communities (Araújo *et al.*, 2023). Equitable data-sharing frameworks are essential to ensure farmers benefit from climate adaptation insights.

Sensor reliability under environmental stress

Climate extremes degrade sensor performance, corrupting DL inputs. A soil moisture network in a heat wave-affected vineyard might report erroneous data due to sensor drift, leading a DL system to over-irrigate and deplete aquifers (Ojo and Zahid, 2022). Robust sensor fusion techniques, such as integrating quantum-resistant encryption for data integrity, could mitigate these risks.

10.8.2 Future directions

Future directions of DL for crop production under a changing climate emphasize a scientific and data-driven approach to improving yield prediction and farm management. As climate variability intensifies, enhancing model robustness becomes essential, which involves refining DL architectures through deeper networks, advanced activation functions, optimized hyperparameters, and better loss functions to capture complex, non-linear relationships among environmental, soil, and crop parameters (Jabed and Murad, 2022). Precision in crop-yield forecasting depends heavily on the integration of heterogeneous data—such as multispectral satellite imagery, UAV-based vegetation indices, ground-based IoT sensor data (e.g. soil moisture, temperature), and meteorological records—through multimodal data fusion techniques (Wang *et al.*, 2024). These approaches allow models to generate spatiotemporally rich representations of agricultural systems, enhancing both granularity and accuracy in predictions. Given the challenge of limited labeled agricultural data, especially for certain crops or regions, the application of transfer learning enables models to adapt pre-trained feature extractors from domains with abundant data (such as wheat or corn) to target crops with sparse datasets (e.g. millets or pulses). Domain adaptation further fine-tunes these models to the specific distribution of local data, increasing prediction reliability across diverse agroecological zones. Reinforcement learning introduces dynamic feedback loops between models and real-time environmental data, enabling autonomous decision-making systems that adjust planting schedules, irrigation patterns, or nutrient delivery in response to changing field conditions. Explainability remains a scientific priority, with ongoing efforts to incorporate methods like SHAP, saliency maps, and attention mechanisms that reveal which features—such as chlorophyll content, canopy structure, or rainfall variability—most influence model outputs. This is crucial for gaining the trust of end users and ensuring practical utility. Data preprocessing pipelines involving normalization, outlier handling, and dimensionality reduction (e.g. PCA or t-SNE) are critical for improving model training efficiency and reducing noise. Robustness is further improved through ensemble learning, where multiple models vote or average predictions, and through data augmentation strategies that synthetically expand datasets via transformations or generative approaches. Looking ahead, AI-powered decision support systems will integrate DL outputs with economic and climatic models, providing actionable insights not only for yield prediction but also for optimizing input use, market planning, and risk mitigation. These systems will also play a pivotal role in climate-smart agriculture by enabling early detection of abiotic stress (e.g. drought, heat) and biotic threats (e.g. disease or pest outbreaks), thereby reducing losses and ensuring resilient crop production. Overall, the scientific evolution of DL in agriculture is positioned to drive transformative improvements in sustainability, productivity, and climate adaptation across farming systems globally (Raza *et al.*, 2023).

10.9 Conclusion

DL has emerged as a pivotal tool in addressing the complexities of crop production under climate change, leveraging architectures such as CNNs, LSTMs, and hybrid models to decode intricate agro-climatic interactions through multimodal data fusion—spanning satellite imagery, IoT sensors, and climate simulations. These models enable precise yield forecasting, autonomous climate-resilient farming systems, and dynamic resource optimization, while innovations like quantum-inspired algorithms and DTs enhance scalability and scenario modeling. However, challenges such as data scarcity under extreme weather, computational demands, model generalizability across shifting climates, and ethical concerns in data governance necessitate adaptive frameworks, transfer learning, and interdisciplinary collaboration. XAI bridges transparency gaps, fostering trust in algorithmic decisions, while generative AI mitigates data limitations by synthesizing realistic agricultural scenarios. Future advancements must prioritize robust, real-time reinforcement learning systems, equitable data-sharing protocols, and integration with regional climate models to align predictive accuracy with agronomic realities. Ultimately, the convergence of DL with IoT, edge computing, and quantum technologies offers a transformative pathway toward sustainable agriculture, balancing productivity with ecological resilience, contingent on collaborative efforts to democratize access and translate computational insights into actionable, context-specific strategies for global food security.

References

Adkisson, M., Kimmell, J. C., Gupta, M., & Abdelsalam, M. (2021, December). Autoencoder-based anomaly detection in smart farming ecosystem. In *2021 IEEE International Conference on Big Data (Big Data)* (pp. 3390-3399). IEEE.

Ahmed, S. F., Alam, M. S. B., Hassan, M., Rozbu, M. R., Ishtiak, T., Rafa, N., ... & Gandomi, A. H. (2023). Deep learning modelling techniques: current progress, applications, advantages, and challenges. *Artificial Intelligence Review*, 56(11), 13521-13617.

Akbar, J. U. M., Kamarulzaman, S. F., Muzahid, A. J. M., Rahman, M. A., & Uddin, M. (2024). A comprehensive review on deep learning assisted computer vision techniques for smart greenhouse agriculture. *IEEE Access*, 12, 4485-4522.

Albahar, M. (2023). A survey on deep learning and its impact on agriculture: challenges and opportunities. *Agriculture*, 13(3), 540.

Alibabaei, K., Gaspar, P. D., Lima, T. M., Campos, R. M., Girão, I., Monteiro, J., & Lopes, C. M. (2022). A review of the challenges of using deep learning algorithms to support decision-making in agricultural activities. *Remote Sensing*, 14(3), 638.

Araújo, S. O., Peres, R. S., Ramalho, J. C., Lidon, F., & Barata, J. (2023). Machine learning applications in agriculture: current trends, challenges, and future perspectives. *Agronomy*, 13(12), 2976.

Attri, I., Awasthi, L. K., & Sharma, T. P. (2025). EQID: Entangled quantum image descriptor an approach for early plant disease detection. *Crop Protection*, 188, 107005.

Avanija, J., Rajyalakshmi, C., Madhavi, K. R., & Rao, B. N. K. (2024). Enabling Smart Farming Through Edge Artificial Intelligence (AI). In *Agriculture and Aquaculture Applications of Biosensors and Bioelectronics* (pp. 69-82). IGI Global Scientific Publishing.

Bisht, B. (2023). *Yield prediction using spatial and temporal deep learning algorithms and data fusion* (Doctoral dissertation, Université d'Ottawa/University of Ottawa).

Cao, X., Yao, Y., Li, L., Zhang, W., An, Z., Zhang, Z., ... & Luo, D. (2022, June). igrow: A smart agriculture solution to autonomous greenhouse control. In *Proceedings of the AAAI Conference on Artificial Intelligence* (Vol. 36, No. 11, pp. 11837-11845).

Cartolano, A., Cuzzocrea, A., & Pilato, G. (2024). Analyzing and assessing explainable AI models for smart agriculture environments. *Multimedia Tools and Applications*, 83(12), 37225-37246.

Cesco, S., Sambo, P., Borin, M., Basso, B., Orzes, G., & Mazzetto, F. (2023). Smart agriculture and digital twins: Applications and challenges in a vision of sustainability. *European Journal of Agronomy*, 146, 126809.

Deep Learning in Agriculture: Challenges and Opportunities – A Comprehensive Review. (2025). *African Journal of Biomedical Research*, 28(1S), 2397-2415. https://doi.org/10.53555/AJBR.v28i1S.6701

Farea, A., Yli-Harja, O., & Emmert-Streib, F. (2024). Understanding physics-informed neural networks: techniques, applications, trends, and challenges. *AI*, 5(3), 1534-1557.

Gao, W., Li, Z., Chen, Q., Jiang, W., & Feng, Y. (2022). Modelling and prediction of GNSS time series using GBDT, LSTM and SVM machine learning approaches. *Journal of Geodesy*, 96(10), 71.

Geli, H., Prihodko, L., Randall, J., Tran, S. C., Cao, H., Misra, S., ... & Hoffman, D. J. (2019). Climate adaptive smart systems for future agricultural and rangeland production. A White Paper on Artificial Intelligence Applications in Agriculture. Web access: https://aces. nmsu. edu/cesfas/documents/artificial-intelligence-in-agriculture---white-paper-with-summary---v6. pdf. Accessed Jan, 10, 2020.

Hosseini, F. S., Razavi-Termeh, S. V., Sadeghi-Niaraki, A., Choi, S. M., & Jamshidi, M. (2023). Spatial prediction of physical and chemical properties of soil using optical satellite imagery: a state-of-the-art hybridization of deep learning algorithm. *Frontiers in Environmental Science*, 11, 1279712.

Huang, Z., Shen, Y., Li, J., Fey, M., & Brecher, C. (2021). A survey on AI-driven digital twins in industry 4.0: Smart manufacturing and advanced robotics. *Sensors*, 21(19), 6340.

Islam, M. T., & Rahman, M. S. (2024, May). An efficient deep learning approach for jute pest classification using transfer learning. In *2024 6th International Conference on Electrical Engineering and Information & Communication Technology (ICEEICT)* (pp. 1473-1478). IEEE.

Jabed, M. A., & Murad, M. A. A. (2024). Crop yield prediction in agriculture: A comprehensive review of machine learning and deep learning approaches, with insights for future research and sustainability. *Heliyon*.

Kashinath, K., Mustafa, M., Albert, A., Wu, J. L., Jiang, C., Esmaeilzadeh, S., ... & Prabhat, N. (2021). Physics-informed machine learning: case studies for weather and climate modelling. *Philosophical Transactions of the Royal Society A*, 379(2194), 20200093.

Kim, H. J., Lee, M. H., & Yoe, H. (2023, July). Research on the Design and Application of Digital Twin-Based Smart Agricultural Systems. In 2023 Congress in Computer Science, Computer Engineering, & Applied Computing (CSCE) (pp. 1770-1773). IEEE.

Kim, S., & Heo, S. (2024). An agricultural digital twin for mandarins demonstrates the potential for individualized agriculture. *Nature Communications*, 15(1), 1561.

Klusch, M., Lässig, J., Müssig, D., Macaluso, A., & Wilhelm, F. K. (2024). Quantum Artificial Intelligence: A Brief Survey. *KI-Künstliche Intelligenz*, 1-20.

Kumar, L., Chhogyel, N., Gopalakrishnan, T., Hasan, M. K., Jayasinghe, S. L., Kariyawasam, C. S., ... & Ratnayake, S. (2022). Climate change and future of agri-food production. In *Future foods* (pp. 49-79). Academic Press.

Liu, S., Liu, K., Wang, Z., Liu, Y., Bai, B., & Zhao, R. (2025). Investigation of a transformer-based hybrid artificial neural networks for climate data prediction and analysis. *Frontiers in Environmental Science*, 12, 1464241.

Logeshwaran, J., Srivastava, D., Kumar, K. S., Rex, M. J., Al-Rasheed, A., Getahun, M., & Soufiene, B. O. (2024). Improving crop production using an agro-deep learning framework in precision agriculture. *BMC bioinformatics*, 25(1), 341.

Malashin, I., Tynchenko, V., Gantimurov, A., Nelyub, V., Borodulin, A., & Tynchenko, Y. (2024). Predicting sustainable crop yields: Deep learning and explainable AI tools. *Sustainability*, 16(21), 9437.

Maraveas, C., Konar, D., Michopoulos, D. K., Arvanitis, K. G., & Peppas, K. P. (2024). Harnessing quantum computing for smart agriculture: Empowering sustainable crop management and yield optimization. *Computers and Electronics in Agriculture*, 218, 108680.

Martin, R. J., Mittal, R., Malik, V., Jeribi, F., Siddiqui, S. T., Hossain, M. A., & Swapna, S. L. (2024). *XAI-Powered Smart Agriculture Framework for Enhancing Food Productivity and Sustainability*. IEEE Access.

Mohan, R. J., Rayanoothala, P. S., & Sree, R. P. (2025). Next-gen agriculture: integrating AI and XAI for precision crop yield predictions. *Frontiers in Plant Science*, 15, 1451607.

Nawaz, M., & Babar, M. I. K. (2024). IoT and AI: a panacea for climate change-resilient smart agriculture. *Discover Applied Sciences*, 6(10), 517.

Oikonomidis, A., Catal, C., & Kassahun, A. (2022). Hybrid deep learning-based models for crop yield prediction. *Applied artificial intelligence*, 36(1), 2031822.

Ojo, M. O., & Zahid, A. (2022). Deep learning in controlled environment agriculture: A review of recent advancements, challenges and prospects. *Sensors*, 22(20), 7965.

Paparella, A. Diffusion generative models for weather forecasting.

Patel, D., Gandhi, M., Shankaranarayanan, H., & Darji, A. D. (2022). Design of an autonomous agriculture robot for real-time weed detection using CNN. In *Advances in VLSI and Embedded Systems: Select Proceedings of AVES 2021* (pp. 141-161). Singapore: Springer Nature Singapore.

Patil, D. (2024). Artificial Intelligence Innovations In Precision Farming: Enhancing Climate-Resilient Crop Management. Available at SSRN 5057424.

Porfírio, R. P., Santos, P. A., & Madeira, R. N. (2024, November). Enhancing digital agriculture with XAI: Case studies on tabular data and future directions. In Companion Proceedings of the 26th International Conference on Multimodal Interaction (pp. 211-217).

Purcell, W., & Neubauer, T. (2023). Digital Twins in Agriculture: A State-of-the-art review. *Smart Agricultural Technology*, 3, 100094.

Pylianidis, C., Osinga, S., & Athanasiadis, I. N. (2021). Introducing digital twins to agriculture. *Computers and Electronics in Agriculture*, 184, 105942.

Qi, Y., Li, H., Pang, Z., Gao, W., & Liu, C. (2022). A case study of the relationship between vegetation coverage and urban heat island in a coastal city by applying digital twins. *Frontiers in Plant Science*, 13, 861768.

Raza, A., Shahid, M. A., Safdar, M., Zaman, M., & Sabir, R. M. (2023, November). The Role of Artificial Intelligence in Climate-Smart Agricul-ture: A Review of Recent Advances and Future Directions. In *Presented at the 2nd International Electronic Conference on Agriculture* (Vol. 1, p. 15).

Rolnick, D., Donti, P. L., Kaack, L. H., Kochanski, K., Lacoste, A., Sankaran, K., ... & Bengio, Y. (2022). Tackling climate change with machine learning. *ACM Computing Surveys (CSUR)*, 55(2), 1-96.

S, Y. P., Garg, N., Arora, R., Singh, S., & S, S. S. (2024). Predictive Modeling of Crop Yield Using Deep Learning Based Transformer with Climate Change Effects. International Research Journal of Multidisciplinary *Technovation*, 223–240. https://doi.org/10.54392/irjmt24616

Saha, G., Shahrin, F., Khan, F. H., Meshkat, M. M., & Azad, A. A. M. (2025). Smart IoT-driven precision agriculture: Land mapping, crop prediction, and irrigation system. *PloS one*, 20(3), e0319268.

Sami, M., Khan, S. Q., Khurram, M., Farooq, M. U., Anjum, R., Aziz, S., ... & Sadak, F. (2022). A deep learning-based sensor modeling for smart irrigation system. *Agronomy*, 12(1), 212.

Sarker, I. H. (2021). Deep learning: a comprehensive overview on techniques, taxonomy, applications and research directions. *SN computer science*, 2(6), 1-20.

Setiadi, D. R. I. M., Susanto, A., Nugroho, K., Muslikh, A. R., Ojugo, A. A., & Gan, H. S. (2024). Rice yield forecasting using hybrid quantum deep learning model. *Computers*, 13(8), 191.

Shams, M. Y., Gamel, S. A., & Talaat, F. M. (2024). Enhancing crop recommendation systems with explainable artificial intelligence: a study on agricultural decision-making. *Neural Computing and Applications*, 36(11), 5695-5714.

Shoaib, M., Shah, B., Ei-Sappagh, S., Ali, A., Ullah, A., Alenezi, F., ... & Ali, F. (2023). An advanced deep learning models-based plant disease detection: A review of recent research. *Frontiers in Plant Science*, 14, 1158933.

Singh, H., Roy, A., Setia, R. K., & Pateriya, B. (2022). Estimation of nitrogen content in wheat from proximal hyperspectral data using machine learning and explainable artificial intelligence (XAI) approach. *Modeling Earth Systems and Environment*, 8(2), 2505-2511.

Singh, R. K., & Khan, A. (2023). A comparative study of quantum and classical deep learning for intelligent agriculture. *Journal of Information and Computational Science*, 13.

Sinha, M., Maurya, H., & Soni, G. (2024). Leveraging Generative AI for Enhancing Climate Resilience and Plant Breeding Techniques. Available at SSRN 5153865.

Suganthi, S. U., Prinslin, L., Selvi, R., & Prabha, R. (2025). Generative AI in Agri: Sustainability in Smart Precision Farming Yield Prediction Mapping System Based on GIS Using Deep Learning and GPS. *Procedia Computer Science*, 252, 365-380.

Sun, C., Zhang, C., & Xiong, N. (2020). Infrared and visible image fusion techniques based on deep learning: A review. *Electronics*, 9(12), 2162.

Upadhyay, A., Zhang, Y., Koparan, C., Rai, N., Howatt, K., Bajwa, S., & Sun, X. (2024). Advances in ground robotic technologies for site-specific weed management in precision agriculture: A review. *Computers and Electronics in Agriculture*, 225, 109363.

Vaissnave, V., Nandhini, S., Davamani, K. A., Malathi, P., & Pothumani, S. (2024). Advancements in Deep Learning Algorithms.

Viel, F., Maciel, R. C., Seman, L. O., Zeferino, C. A., Bezerra, E. A., & Leithardt, V. R. Q. (2023). Hyperspectral image classification: An analysis employing CNN, LSTM, transformer, and attention mechanism. *IEEE access*, 11, 24835-24850.

Wang, D., Cao, W., Zhang, F., Li, Z., Xu, S., & Wu, X. (2022). A review of deep learning in multiscale agricultural sensing. *Remote Sensing*, 14(3), 559.

Wang, J., Li, X., Li, J., Sun, Q., & Wang, H. (2022). NGCU: A new RNN model for time-series data prediction. *Big Data Research*, 27, 100296.

Wang, L. (2024). Digital Twins in Agriculture: A Review of Recent Progress and Open Issues. *Electronics*, 13(11), 2209.

Wang, Y., & Karimi, H. A. (2024). Advanced deep learning models and algorithms for spatial-temporal data. In *Big Data* (pp. 227-253). CRC Press.

Wang, Y., Zhang, Q., Yu, F., Zhang, N., Zhang, X., Li, Y., ... & Zhang, J. (2024). Progress in Research on Deep Learning-Based Crop Yield Prediction. *Agronomy*, 14(10), 2264.

Waqas, M., & Humphries, U. W. (2024). A critical review of RNN and LSTM variants in hydrological time series predictions. *MethodsX*, 102946.

11 Deep Learning for Accelerating Crop Breeding

Siddhartha Das[1]* and Sweety Maity[2]

[1]Department of Plant Pathology, M.S. Swaminathan School of Agriculture, Centurion University of Technology and Management, Paralakhemundi, Odisha, India; [2]Department of Agricultural Biotechnology, College of Agriculture, Odisha University of Agriculture and Technology, Bhubaneswar, Odisha, India

Abstract

One of the most effective strategies for increasing crop yields and improving quality is crop breeding. The breeding procedure encounters various obstacles, such as intricate data complexities, challenges in data collection, and low accuracy in predictions, which collectively hinder breeding efficiency and extend the duration of the process. Deep learning has been a revolutionary tool in crop breeding in recent years, offering the unmatched potential to accelerate the development of robust, sustainable, and high-yielding crop types. Recurrent neural networks (RNNs) and convolutional neural networks (CNNs) can quickly and accurately assess large datasets, such as hyperspectral images, genetic sequences, and environmental data. This approach facilitates the development of more productive, adaptable, and disease-resistant cultivars for agricultural use, while also accelerating crop improvement and increasing breeding efficiency. It also makes genome-wide association studies (GWAS) more successful. Global studies using deep learning approaches illustrate the contrast between traditional breeding and transformative breeding, as well as their future agricultural challenges.

Keywords: crop breeding, deep learning, genotyping, phenotyping

11.1 Introduction

Modern technology in food and agriculture sciences is needed due to resource depletion, climate change, and rising food demands (Ali *et al.*, 2018). A vibrant area of agricultural research, plant breeding has developed from the straightforward selection of striking plants to contemporary techniques based on genomics and biotechnology. Hybrid breeding programs, yield-aspect analysis, yield stability evaluation, and genetic diversity assessment are examples of traditional techniques. Molecular breeding methods based on *in vitro* include *Agrobacterium*-mediated gene transformation, artificial polyploidy induction, twofold haploid production, and *in vitro* micropropagation. Conventional statistical techniques, which mostly rely on variable and regression-type models, are used to evaluate and analyze the results for both conventional and *in vitro* enhancement studies. Various models, however, have trouble deciphering intricate correlations between independent and dependent variables. Plant breeding machine learning (ML) has been widely used in research, the field that teaches programs to learn from

*E-mail: siddhartha.das10@gmail.com

DOI: 10.1079/9781836991120.0011

data, and to comprehend how information flows from the DNA sequence to observed phenotypes. Regression, classification, and the interpretation of formless datasets, such as sequencing or image date data, can all be accomplished with ML methods. Probabilistic models with hidden layers, known as artificial neural networks (ANNs), are more predictive than support vector machines (SVMs). The principles of natural visual perception serve as the basis for the most advanced deep learning architecture known as convolutional neural networks (CNNs). CNNs are employed for a number of purposes, such as motion identification, image classification, object tracking, object recognition, position calculation, visual saliency recognition, text detection, action identification, science labeling, voice, and natural language processing. Their automatic feature extraction makes them appropriate for classification studies. SVMs are more efficient for regression analysis, classification, and clustering because they employ a method of supervised learning to determine linear and non-linear correlations in data (Zheng *et al.*, 2018). Random forest (RF) regression, which combines multiple independent predictors, can control overfitting and is appropriate for spectral data processing. Abiotic as well as biotic stress can be diagnosed and categorized using deep CNNs. CNNs are additionally useful for recognizing objects and finding patterns in remote sensing. A nine-layer deep CNN algorithm was created to detect plant leaf diseases, and it had a 96.46% classification accuracy (Geetharamani and Pandian, 2019). For diagnosing plant diseases, ANN-based techniques are superior to SVM-based techniques because the latter employ fewer samples and characteristics. To determine which algorithm is optimal for a given dataset, it is appropriate to compare the performance of various algorithms.

11.2 Conventional Plant Breeding

11.2.1 Evaluation and categorization of genetic variation

A key prerequisite for plant breeding operations is genetic diversity, which makes it possible to identify important accessions and employ them in breeding initiatives (Srivastava *et al.*, 2020). Physiological, biochemical, and morphological indicators have been examined to look into the genetic diversity of plants. Measurement of morphological traits is the most straightforward and does not call for specialized equipment or methods. A statistical evaluation of these indicators can show that the genotypes under evaluation exhibit genetic diversity. Eight different ecotypes of ajwain (*Carum copticum* L.) with therapeutic value were examined using analysis of variance (ANOVA) to determine the coefficient of variation (CV) of several agro-morphological characteristics, including number of branches, plant height, the number of umbellets within an inflorescence, biological yield, number of umbels, and yield of a single plant (Niazian *et al.*, 2017). Genetic and molecular markers are another class of markers that allow for genotype discrimination and evaluation of genetic variation. The most popular molecular markers for examining species designation and genetic diversity in various host plants are amplified fragment length polymorphism (AFLP), restriction-fragment length polymorphism (RFLP), single-nucleotide polymorphism (SNP), randomly amplified polymorphic DNA (RAPD), simple sequence repeats (SSR), and inter-simple sequence repeats (ISSR). These genetic markers are more accurate and efficient than physical markers at estimating evolutionary relationships and identifying varieties (Boonsrangsom, 2020). Molecular markers are superior to morphological markers in identifying different plant genotypes, discriminating among them, and assessing genetic diversity; nonetheless, there are certain technical and financial constraints. Conventional multivariate methods such as cluster analysis, principal component analysis (PCA), and analysis of discriminant functions have been utilized to classify and group different genotypes in a variety of plant species using molecular, physiological, morphological, and biochemical indicators (Poletto *et al.*, 2020). It is possible to apply deep learning algorithms for detecting objects, which may result in plant genotype classification and efficient genetic diversity assessment. CNN is an effective alternative for classifying morphological characteristics since it is less time-consuming and feature extraction-dependent than traditional classification methods, such as probabilistic k-nearest neighbor, neural networks, SVMs, PCA, and evolutionary algorithms. The genetic composition of 90 accessions of soybeans (*Glycine max* (L.) Merr.) was determined to be diversified through a thorough assessment of stomatal density. In *Cinnamomum osmophilum* (Lauraceae), CNN classifiers in terms of performance (96.7% vs 74.6%) when a deep CNN was employed to differentiate between morphologically identical species. In order to evaluate genetic diversity and categorize 10 plant populations, Sant'Anna *et al.* compared the effectiveness of ANN with Anderson's discriminant functions and Fisher's conventional multivariate statistical technique. Because ANN-classified populations with equal levels of divergence had fewer incorrectly classified individuals, they were found to be superior to traditional approaches. Ten potato types were identified and distinguished using morphological data gathered through image processing, non-linear ANN techniques, and linear discriminant evaluation. The classification accuracy of the ANN approach was 100% (Azizi *et al.*, 2016). In many plant species, ANN algorithms have also been used for genomic selection (GS) and genomic prediction. Based on genomic information gathered from 33,709 DArT markers, a deep CNN was utilized to predict the characteristics of 2000 Iranian bread wheat landrace collections. When comparing observed and estimated parameters (grain width, grain length, test weight, grain hardness, sodium dodecyl sulfate precipitation, plant height, and grain protein), the authors discovered that deep CNN had a higher Pearson's correlation coefficient than other GS methods (Ma *et al.*, 2018).

11.2.2 Analysis of yield components and indirect selection (projection)

Most breeding programs always aim to increase the economic yield, which includes the production of seeds, oil, sugar, essential oils, biomass, straw, lint percentage, and other products. However, yield is a quantitative feature that is extremely complex, typically controlled by multiple genes, and heavily influenced by environmental factors. Thus, direct selection does not enhance such traits, and yield traits have poor heritability. To have more control over production, plant breeders typically choose to use simpler, highly linked features. Future research will employ the chosen yield component or components as "selection criteria," or indirect selection (Ghffari *et al.*, 2020). Multiple regression evaluations (forward, reverse, and stepwise), path analysis, PCA, correlation coefficient evaluation, and multivariate linear techniques to enhance critical quantitative features such as yield and resilience to biotic and abiotic stressors have been employed in traditional plant breeding. Correlation coefficient analysis and path analysis have been utilized to establish causal linkages between correlated variables and to assess a straightforward relationship between two traits, respectively. Regression-based multivariate statistical techniques are the most suited for indirect selection. Their basis is the linear correlation between multiple independent factors and a single dependent variable (Y) (Zou *et al.*, 2020). Regression analysis, including stepwise, forward, and reverse, has been used to ascertain how different economic quantitative traits of different crops are affected by yield components. Utilizing backward stepwise regression, the link between variations in rice (*Oryza sativa* L.) yield components and grain yield was determined in terms of the absolute sensitivity ratio to increased CO_2 (Lv *et al.*, 2020). Stepwise regression was used to determine the variables affecting sugarbeet (*Beta vulgaris* L.) productivity under foliar jasmonic acid application and limited water management. Stepwise regression analysis was employed by Zou *et al.* to identify the yield parameters that influence cotton (*Gossypium hirsutum* L.) drought resistance (Niazian *et al.*, 2018). Non-linear algorithms for ML can be used in components of yield assessment and studies that use indirect selection to predict economic quantitative characteristics, interpret non-linear interactions between independent and dependent variables, and assess the importance of yield components to yield. ANNs fared better than multiple linear regression (MLR) in forecasting the oil content and seed yield of ajwain (*Trachyspermum ammi* L.). In their study, Emamgholizadeh and associates found that ANN performed better than MLR in forecasting sesame (*Sesamum indicum* L.) seed output (Emamgholizadeh *et al.*, 2015). The ANN model had a lower root mean square error (RMSE) and a higher determination coefficient (R^2). The flowering period and the number of capsular structures per plant were the most and least significant factors, respectively, according to the ANN model's sensitivity analysis. The production of pears, apples, chives, and onions has been accurately predicted by an ANN that accounts for crop diseases, current temperature, harvest date, humidity, precipitation (including snowfall), atmospheric pressure, ground temperature, sunshine, and moisture evaporation in the ground (Lee *et al.*, 2019). Additionally, ANNs have been employed to forecast yields of rapeseed and winter wheat using meteorological data (such as precipitation and air temperature) as well as details on the application of mineral fertilizers. Wheat production prediction through deep neural networks (DNNs; long short-term memory) has been shown to be more accurate than the autoregressive integrated moving average (ARIMA) model (Haider *et al.*, 2019). The qualitative traits of several plants have also been estimated and predicted using neural networks. Estimation of oil content in sesame seeds, tested through an ANN model, was performed with greater precision and effectiveness compared to the MLR model (Abdipour *et al.*, 2018). Using a multilayer perceptron (MLP)-ANN, Parsaeian *et al.* (2020) estimated the oil and protein content in sesame based on 138 morphological parameters from 125 genotypes of sesame seeds. Image processing was employed to gauge the morphological traits of the seeds. MLP outperformed the extended M5-Rule, extended radial basis function RBF (ERBF), radial basis function (RBF), SVM regression, M5-Tree, GRNN, and linear regression models, based on qualitative standards of protein and oil content in sesame seeds calculated using R^2 and RMSE statistics. In order to assess how weather and cultivar affect ferulic acid concentration, researchers used a multilayer perceptron ANN to connect the ferulic acid ratio with nivalenol and deoxynivalenol content in 23 different winter wheat strains with varying levels of *Fusarium* resistance. A total of 14 features were independent variables, consisting of two qualitative and 12 quantitative components. Neural network sensitivity analysis revealed that the winter wheat cultivar and experiment variant were the primary factors influencing the levels of nivalenol, deoxynivalenol, and ferulic acid in winter wheat seeds (Niedbała *et al.*, 2020). To evaluate the impact of topography, soil, and outside influences (18 input parameters, including soil nutrients and climate conditions) on the concentration of the active ingredient coronarin D in white ginger lilies (*Hedychium coronarium*), Ray *et al.* (2020) applied ANN modeling. According to the ANN's sensitivity analysis, altitude, manganese, and zinc had the greatest effects on coronarin D content.

11.2.3 Yield consistency and genotype–environment interaction

The elements that affect crop production and productivity are plant genotype, agricultural practices (sowing, cultivation, and harvesting), and the environment (soil and climate). Environmental factors and the genotype–environment interaction (GEI) give rise to variations in a given genotype's productivity and phenotypic traits from year to year. Due to the GEI, selecting a particular genotype for a desired attribute is a difficult and complex process because several genotypes react differently to different ecosystem factors. The ideal remedy for these annual fluctuations is to estimate genotypes' relative performance across environments using stability analysis (Ajay *et al.*, 2020). Finlay and Wilkinson's regression analysis and CV, Eberhart and Russell's coefficient of regression (S^2di), Wricke's ecovalence (Wi), Shukla's procedure of stability variance, and the CV and Lin and Binns cultivar performance measure are examples of traditional univariate methods for assessing the GEI. The primary components of these techniques are variance components and linear regression analysis. The sustainability yield index (SYI), in addition to the statistical techniques indicated above, is used to assess how agricultural practices affect the sustainability of crop output. Since all of these techniques are parametric, it is important to take into account the data distribution and variance homogeneity assumptions before using them. The stability parameters Si_1, Si_2, Si_3, and Si_6, as well as the Kang parameter, Ketrank and Ketyield charts, Fox-rank, and Star, are non-parametric univariate techniques for assessing the GEI. These analytical clustering methods, which are free from modeling assumptions and rely on ranks rather than data, are known as non-parametric stability statistics. If a genotype remains largely constant in ranking across environments, it is said to be stable.

11.2.4 Evaluation of biotic and abiotic stress

A variety of biotic and abiotic stressors are experienced by plants. Various techniques have been used to assess plant genotypes' resilience and tolerance to stressors to identify superior genotypes. Breeding efforts to withstand drought stress have been many. Tolerance (TOL), mean productivity (MP), stress susceptibility index (SSI), geometric mean productivity (GMP), harmonic mean (HARM), relative drought index (RDI), stress tolerance index (STI), yield stability index (YSI), yield index (YI), and modified stress tolerance index (K1STI and K2STI) are some statistical measures that have been used to assess a plant's ability to withstand drought (Han *et al.*, 2020). The yield obtained under non-stress (Yp) and stress (Ys) conditions is the main focus of these conventional techniques, which are based on morphological information. However, in addition to physical characteristics, plants respond to environmental stressors through a variety of physiological and biochemical mechanisms. Cellular antioxidants, secondary metabolites, plant growth regulators, compatible solutes, and polyamines all play a role in plant response to biotic and abiotic stresses. Plant reactions to abiotic and biotic stressors can be properly quantified using a combination of phenomic, metabolomic, and genomic data. This enormous volume of data (one or more unique variables against many dependent variables) cannot be efficiently managed by conventional multivariate statistical techniques. Linear regression is the most applied technique to identify nutritional deficiencies through RGB images (Barbedo, 2019).

Still, a linear model of regression cannot cover up variables that were obtained through digital images, having non-linear relationships with the nutritional composition. Through the processing of all the omics and phenomic information, ML models combined with digital images are applied to model and predict responses of genotypes under stressful conditions, as well as discover the most stress-resistant and non-stress environments. High-throughput stress phenotyping can be done by applying ML to assess large data, including imaging and remote sensing data. Ravari *et al.* reported that 41 Iranian wheat cultivars (*Triticum aestivum* L.) were assessed for salinity tolerance employing an MLP-ANN and the MP, TOL, GMP, SSI, HM, STI, YI, and YSI indices. To predict cultivars that would be tolerant to salinity, YSI, MP, GMP, and STI were the best-performing indices. Arabidopsis (*Arabidopsis thaliana*) decision tree (DT), SVM, least-square support vector machine (LSSVM), and Naïve Bayes (NB) ML models took miRNA expressions as input features to predict plant responses to abiotic stressors of salinity, drought, heat, and cold.

A variety of biotic stress-caused plant diseases have been identified and categorized using deep CNNs. Image identification and classification of corn leaf diseases, including gray leaf spot (*Cercospora*), common rust (*Puccinia sorghi*), and northern corn leaf blight (*Exserohilum*), were performed with a 93.35% accuracy rate using a deep CNN. To segment powdery mildew disease on leaf images at the pixel level in cucumbers (*Cucumis sativus*), a CNN-based semantic segmentation model was established. The CNN model's pixel accuracy (96.08%) was higher than that of k-means, RF, and GBDT segmentation methods (Lin *et al.*, 2019). DNNs have also been utilized to identify mildew disease in pearl millet (*Pennisetum glaucum*), and the model achieved an accuracy of 95.00% (Coulibaly *et al.*, 2019).

11.2.5 Conventional mating strategies and hybrid breeding initiatives

Conventional mating designs, such as mean generation analysis, diallel crosses, line tester analysis, North Carolina designs, and triple test crosses, were made possible by the use of statistics in genetics. In order to determine the type of genetic actions (dominance, additive, and epistasis) that govern significant structural, phenological, and yield component characteristics, as well as to compute broad and narrow sense heritability and forecast the results of cross-breeding initiatives, these techniques have been applied to crops. The selection of superior combinational homozygous parental lines in F_1-hybrid breeding operations depends heavily on the prediction of parental combinations (Dezfouli *et al.*, 2019). However, when there are numerous inbred paternal lines, it becomes a difficult task with a lot of cross-combinations. Thus, the time and money needed for F_1-hybrid breeding projects may be greatly decreased by forecasting the yield performance of cross-combinations of parental lines. With mating designs, such as topcross, line tester, and diallel cross, ANNs could be used to anticipate parental combinations and determine the appropriate values of general and special combining abilities (GCA and SCA) (Khaki *et al.*, 2020).

11.3 Deep Learning Principles Enabled Crop Breeding

Deep learning operates on large volumes of data to resolve complex problems. Currently, the main data type for choosing phenotypic characteristics in plant breeding is still images (Araus and Cairns, 2014). The exceptional performance of CNN in phenotypic big data assessment has triggered fast progress in the application of deep learning to plant phenotyping image processing (Jin *et al.*, 2025). It possesses good feature modeling and extraction abilities that provide new approaches for solving data analysis problems. Deep learning-based crop breeding generally encompasses the following five steps in its process: (i) Gathering information: Accumulate phenotypic traits, environmental factors, genetic markers, and other relevant agronomic and genetic information regarding the crops to be targeted; (ii) Data preprocessing: Use methods for data augmentation, feature extraction, data normalization, and missing data handling to clean and preprocess the gathered data; (iii) Model construction: Choose the right neural network architectures based on the type of data and breeding goals to create a deep learning model that is appropriate for the particular breeding task; (iv) Model training: Develop the deep learning model with preprocessed data, adjusting model parameters to reduce discrepancies between observed and anticipated values—model weights are typically updated during training using methods, such as gradient descent and backpropagation (Zhou, 2018); and (v) Genetic optimizing parameters and selection: Forecast and assess agricultural features and crop genetic traits using the trained DNNs model.

Breeding using deep learning is based on key genes that have been found and integrates next-generation biotechnologies, multiomics, and new information technologies, such as big data and AI. Deep learning-based breeding is based on the detection of crucial genes through genomics and transgenic approaches, pioneering a new direction (Pan, 2015; Luqman *et al.*, 2025). Breeding is accelerated by emerging technologies such as big data, AI, and multiomics, which are used in conjunction with high-capacity phenotyping platforms and plant phenotypic analysis. Additionally, they speed up breeding, discover specific genes, and facilitate the evaluation of plant materials. Plant breeding is aided by the integration of multiomics data, including transcriptomics, phenomics, proteomics, metabolomics, and genomics, which helps find important genes and regulatory components and analyze biological changes and regulatory processes (Yang *et al.*, 2021). More intriguingly, telomere-to-telomere (T2T) whole genome and T2T complete genome analysis work as representative markers for precisely detecting genetic variation and boosting functional genetics and genetic advancements. Through the use of integrated genomic and environmental prediction (iGEP) based on genomes and population–environment interactions, for example, intelligent breeding strategies powered by big data and AI allow for targeted breeding. Deep learning frameworks provide automatic differentiation, which makes it possible to apply these scores effectively with only a few lines of code. They will help manage large, multifaceted genotype–phenotype–environment big data, which will enable effective selection and production of superior, disease-resistant novel kinds (Table 11.1). It is important to note that comprehensive genetic studies involving associated environmental and phenotypic data are required (Parmley *et al.*, 2019).

11.4 The Evolution of DNNs

DNNs have their roots in the first artificially generated neural network, which was suggested by McCulloch and Pitts in 1943. Deep learning has now expanded progressively and reached several noteworthy developmental milestones. Rosenblatt's 1958 "perceptron" is one of those that merits note. It proved that when the thing being learned can be represented, a perceptron will converge. Although such a model has clear limitations, and difficult tasks require multilayer perceptrons, training these models was unclear

Table 11.1. Deep learning algorithms and models used in intelligent breeding. Author's own table.

Model	Data	Application
CNN (DeepSEA)	DNA sequences	DNA and gene characteristics
Bi-GRU and LSTM (DeepCpG)	DNA sequences	DNA and gene characteristics
DBN	Gene sequences	DNA and gene characteristics
Evoformer and transformer (AlphaFold)	Protein sequences	Protein characteristics
Stacked autoencoder and DNN	Protein sequences	Protein characteristics
GCN (Decagon)	Protein–protein interaction network	Protein characteristics
Multilayer perceptron and AE	SNP	Genomics variations
GAN	RNA sequences	Genomics variations
Multilayer perceptron and AE	SNP	Genomics variations
CNN (AMBER)	Gene sequences	Genomics variations
CNN and LSTM (DeepNovo)	Peptide spectrum	Protein characteristics

CNN, convolutional neural networks; DBN, deep belief network; DNN, deep neural network; GAN, generative adversarial network; LSTM, long short-term memory; SNP, single-nucleotide polymorphism.

at the time. After that, deep learning experienced its first winter. In 2006, Hinton and coworkers employed a layer-wise pretraining architecture to address the challenge of training DNNs, greatly stimulating the field (Bahmer *et al*., 2023). Deep learning was further spurred by the concurrent suggestion of techniques for training deep autoencoders (AEs) and other deep architectures. Deep belief networks (DBNs), restricted Boltzmann machines (RBMs), CNNs, generative adversarial networks (GANs), graph neural networks (GNNs), and other deep learning methods started appearing more often after that. Two phenomenal deep learning applications amazed the world and raised a storm in the last decade. One is AlphaGo, which applied deep learning and an abundance of hardware capabilities to beat the world champion Go players. The second is AlphaFold, which resolved the difficult protein-folding issue that had persisted for 50 years. These further boosted deep learning's development in a variety of fields. The development of high-performance computing (HPC) and graphics processing units (GPUs) has made deep learning one of the best techniques available today, with exceptional results in nearly every discipline.

11.5 Core Principles of DNNs

DNNs aim to imitate the way in which numerous neurons, commonly known as nodes, send messages to each other in biological neurons. Several hidden layers, an input layer, and an output layer are among the layers of neurons that typically make up a DNN's structure (Schmidhuber, 2015) (Fig. 11.1). All neurons are connected to one another in order to transmit information. A DNN takes input in the form of numbers, words, sounds, images, and so on, and decomposes them into binary data that a computer can understand. Based on the tasks performed, the output can be binary, continuous, or categorical. Training data is necessary for a DNN to learn and gradually increase its accuracy. An algorithm modifies its weights during learning if it is unable to correctly identify a specific pattern for a task until it determines the appropriate mathematical operation to process the input in its entirety.

11.5.1 Conceptualization of neuron and activation

Each input is multiplied by its corresponding weight by a neuron, and it adds a predefined value referred to as the bias after summing the weighted inputs. Activated neurons send their output to the DNN's next layer when their output exceeds a predetermined threshold. In other words, every layer of a DNN has neurons that take input from the learned representations and the preceding layer, and then pass on the information to the next layer. The previous layer's output is used as the input for each DNN layer that comes after it. A DNN ultimately generates an output in this manner. A DNN's mechanism for activation is a crucial component. The primary

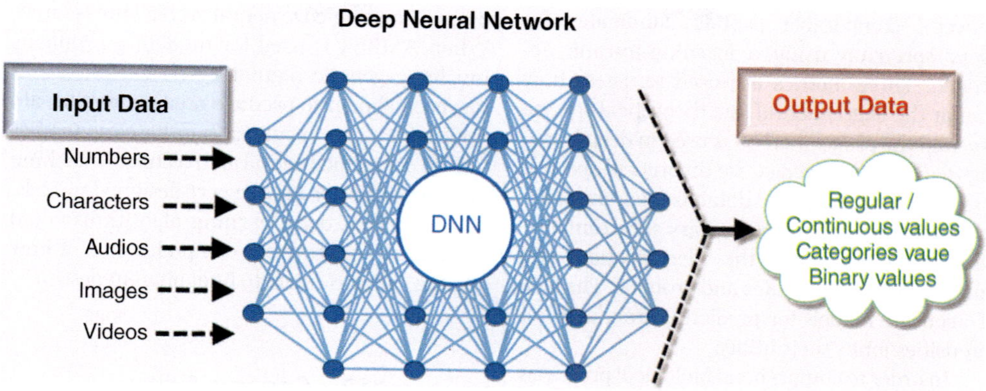

Fig. 11.1. Diagram of DNN. DNN, deep neural networks. (Author's own image.)

function of this node is to create non-linear interactions between input and output by defining the node's output and its input. ReLU activation is another very important and frequently used activation function. Its primary characteristic is that it only activates nodes with an outcome greater than 0; it fails to activate all nodes simultaneously. This type of function is therefore more computationally efficient than the tanh and sigmoid activation functions. Additionally, it helps gradient descent converge toward the loss functions. Its disadvantage is that the opposite direction of the curve, which sets the gradient value to zero, could result in potential dead nodes.

11.6 Utilization of Deep Learning in Agricultural Breeding Practices

11.6.1 Features of genes and DNA

The specificity of transcription factor (TF)–DNA binding is significantly influenced by the structure of DNA (Lai *et al.*, 2019), and deep learning models can analyze a variety of data types (Zampieri *et al.*, 2019). A thorough understanding of the sequence specificity of DNA- and RNA-binding proteins is necessary for the identification of pathogenic variations and biological system regulatory models (Wang *et al.*, 2020). TF binding characteristics can now be predicted using a number of deep learning-based techniques. DeepSEA, DeepBind, and Basset were some of the first CNNs used for genomic data analysis. While DeepSEA assembled a substantial collection of chromatin maps for non-coding variants to investigate chromatin features, DeepBind developed numerous single-task models to forecast the affinity for binding of transcription factors, and Basset anticipated DNA accessibility features. Functional non-coding variations' effects were assessed in DeFine, DFIM, and DeepSEA. It has been considered challenging to identify significant genomic regulatory regions in organisms with extensive intergenic regions and high repetitive elements. This issue was solved by using natural language processing-based techniques, including k-mer grammar, to accurately and efficiently annotate regulatory regions in corn. Regulatory region annotation, transcription factor binding site prediction, and functional non-coding variation prediction have all been significantly improved by these techniques. To improve the prediction of in vivo transcription factor binding sites, ML models have proven effective in plant biology. They can be trained on a range of sequencing data sources with additional information, for example, DNase I hypersensitivity data (Qin and Feng, 2017).

11.6.2 Features of proteins

The structure of proteins and their function are closely related. DeepNovo, which employs primary peptide sequences to extract important amino acid features, was developed using CNN. By meticulously assessing a range of protein attributes, one can find a protein's tertiary structure, which governs its activity. pNovo3 distinguishes between comparable peptide candidates for every spectrum using a learning-to-rank approach. Three metrics are used to assess how similar the experimental and theoretical spectra are, and the pDeep method is used in deep learning to accurately predict the theoretical spectra. Spectrum matching and database searching are the main methods used in mass spectrometry-based proteomics for the identification and quantification of peptides and proteins. The lack of accurate models for predicting fragment ion intensities limits their utility.

In order to comprehend biological processes and disease mechanisms, protein–protein interactions (PPIs) are crucial. Sequence-based prediction programs (Hashemifar *et al.*, 2018) and DNNs are some of the methods researchers have explored to predict PPIs. One of the methods utilizes the protein–protein interaction network to obtain distinctive protein features unsupervised. These characteristics are then utilized to forecast the functions of proteins in different tissues. Zitnik *et al.* (2018) proposed the Decagon graph CNN model, which they used to develop multimodal graphs of drug–protein target interactions, PPIs, and multidrug side effects. Some of these methods further leverage topological features of PPI networks and physicochemical characteristics of proteins to improve prediction performance via deep learning under multimodal supervision.

11.6.3 Variations in genomics

Deep learning models could be trained to predict the effect across the entire population from only a small segment of natural populations' genetic variation, despite the fact that natural populations encompass numerous genetic variances and a range of alterations (Killamsetty *et al.*, 2021). Models trained on given genes, for example, could then be applied to predict how another gene would function. Such models capture a broad spectrum of mutations, irrespective of the impact on gene function, including rare and low-frequency variants and common alleles. AMBER (Zhang *et al.*, 2021b) is an automated modeling system that is biologically informed. CNNs are easily designed and applied to genomic sequences using this fully automated methodology. AMBER creates the best models for particular biological challenges by using cutting-edge neural architecture search. When AMBER is used for modeling problems involving genetic regulatory characteristics, it has been shown to produce predictions that are noticeably more accurate than baseline models that do not include neural architectures, matching or even outperforming expert-designed models. In conclusion, deep learning algorithms could significantly improve our comprehension of how genetic variants relate to final phenotypes.

11.7 Genome Editing

11.7.1 How genome editing evolved

The science of genome editing allows for the deliberate alteration of an organism's genome, including DNA mutations such as insertion, deletion, and replacement, which change biological characteristics. Meganuclease, zinc-finger nucleases (ZFNs), transcription activator-like effector nucleases (TALENs), and clustered regularly interspaced short palindromic repeat (CRISPR)-associated protein (Cas) technology are some of the genome-editing technologies that have been developed to modify particular loci or DNA segments in organisms. Natural endodeoxyribonucleases known as meganucleases can recognize and cleave DNA sequences that are usually 20–30 base pairs long. They can also modify the genome site-directedly by re-engineering their DNA recognition site. It is vital to keep in mind that altering a protein's entire structure is a time-consuming and difficult endeavor in real life, which restricts the applications of these genome-editing approaches. A class of proteins called transcription activator-like effectors (TALEs) is released by *Xanthomonas* bacteria and has the ability to modify gene transcription in host plant cells. Repeat variable diresidues (RVDs), the main repeating component of these effectors, are composed of varying amounts of repeat units of 34 amino acids. The RVD region determines the biological specificity of each effector and its ability to identify specific DNA sequences (Li *et al.*, 2011). The extremely specific DNA-binding domains of TALEs and the FokI endonuclease domain make up TALENs, which are based on the ZFN approach and may induce double-strand breaks (DSBs) at a particular location. Even though genome editing has been successfully accomplished, the complex process of creating meganucleases, TALENs, and ZFNs has limited the applicability of these molecules. The most used genome-editing approach at the moment is CRISPR because of its ease of use and great effectiveness.

11.7.2 CRISPR/Cas editing technology is based on genetic modification

The CRISPR/Cas system is an adaptive defense mechanism in archaea and bacteria, where small RNAs are used for the specific detection and silencing of foreign nucleic acids. The system consists of a gene encoding a Cas protein that is able to cut DNA and a CRISPR array. There are individual genome-targeting sequences (spacers) with the same repetitions interspersed by them in the CRISPR array.

There are already several CRISPR/Cas systems known (Makarova *et al.*, 2011, 2020). Among them, CRISPR/Cas9 and CRISPR/Cas12a are the most popular and sophisticated techniques in genome editing. The CRISPR/Cas9 system generates new spacers by incorporating short pieces of foreign nucleic acids (protospacers) into the CRISPR locus within the host genome in response to viral and plasmid infections. The CRISPR array is the source of CRISPR RNA (crRNA), which is processed by enzymatic cleavage and forms a complex with tracrRNA. The Cas9 protein is guided by the complex to find particular foreign nucleic acid sequences so that they can be cleaved. The protospacer adjacent motif (PAM) should be present in the sequence that the CRISPR/Cas system recognizes. This is necessary for the editing system to identify binding sites and start the Cas protein's activation for cleavage. With the substitution of crRNA and tracrRNA using a specifically designed single guide RNA (sgRNA), CRISPR/Cas9 technology enhances the immune system of CRISPR/Cas9 and develops a simpler editing mechanism. Although sharing the same principles, CRISPR/Cas12a differs from CRISPR/Cas9 in sgRNA, protein, and PAM, providing special features such as improved selectivity for target sites. Several CRISPR-based genome-editing tools have emerged due to the increasing need for genome editing and research on the CRISPR system. Base editing and prime editing are two of the most notable among them. By combining a Cas9 nickase (nCas) or catalytically dead Cas (dCas) with an enzyme that catalyzes a nucleoside deamination reaction, base editing allows the direct replacement of a target base without resulting in DSBs. A guide RNA for prime editing (pegRNA) and a fusion protein composed of dCas and reverse transcriptase (RT) constitute the prime editor. The pegRNA serves as the template for reverse transcription after localizing the fusion protein to the target location, ultimately delivering the desired editing effect at the site (Anzalone *et al.*, 2019). The effective application of such cutting-edge technologies, which enable precise editing and subsequently aid in crop development, has enabled a variety of crops (Lin *et al.*, 2020b).

11.7.3 The relevance of crop breeding using genome editing

The ability to precisely modify individual genes through genome editing has transformed the study of crop gene function. Genome editing can be utilized by breeders to alter specific genes associated with desirable characteristics using techniques such as targeted knockout, insertion, and replacement (Wu *et al.*, 2022a). This creates a strong basis for crop breeding. By genome editing, breeders are able to modify cis-regulatory elements (CREs), including enhancers and promoters, to fine-tune the levels of gene expression and enhance traits. Breeders can employ genome editing to improve traits in crops in order to accomplish a range of goals, including increasing crop yields, improving nutritional content and quality, and increasing resistance to various pests and diseases (Xu *et al.*, 2022). Apart from crop improvement, genome-editing technology provides a practical way to speed up breeding. Unlike traditional breeding methods that rely on crossing or backcrossing, superior varieties are able to produce superior types with desired characteristics at a fast pace through direct genome editing. This method greatly reduces the breeding period because it does away with the time-consuming backcrossing process (Wulff and Dhugga, 2018). In addition, engineers can edit multiple gene loci in one generation at the same time due to the benefit of CRISPR's multiplex editing.

11.8 Machine Learning

11.8.1 An overview of ML

In the field of AI, ML enhances the performance of computer systems by imitating human learning processes through data and algorithms. This allows models to learn automatically from input data and make predictions or decisions when faced with unseen data. Several categories of ML exist: reinforcement learning, supervised learning, unsupervised learning, and semi-supervised learning (Fig. 11.2). A model is learned using tagged data in supervised learning, where the correct output is already known in advance. To predict outputs for new inputs, the model first needs to learn the relationship between inputs and outputs. Two other categories into which supervised learning algorithms can be divided are regression and classification (Table 11.2). Several parameters, such as the diagnosis of plant diseases, are forecast or classified based on classification algorithms. Prediction of crop production is one of the primary uses for regression algorithms, which examine and predict continuous numerical values. In unsupervised learning, a model is learned from unlabeled data without any prior information about the correct output. Its primary objective is to recognize inherent patterns and structures within data. Dimensionality reduction and clustering are essential aspects of unsupervised learning. While dimensionality reduction is focused on reducing the feature dimensions of a dataset while preserving its essential features, clustering refers to grouping items with similar characteristics into one category. A blend of supervised and unsupervised ML methods yields semi-supervised learning methods, hoping to enhance model performance by employing an abundance of labeled data

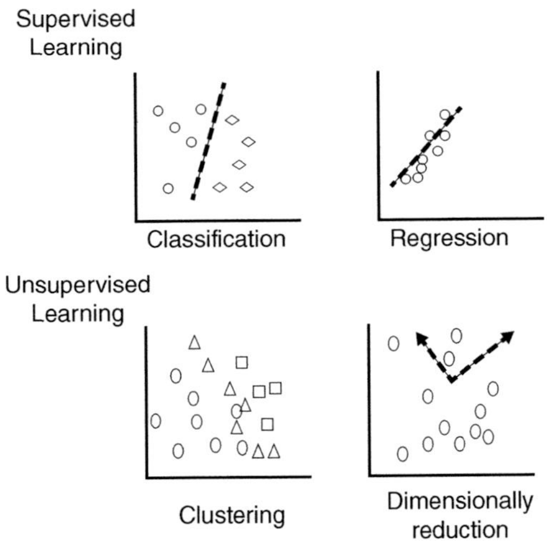

Fig. 11.2. Classification of machine learning. (Author's own image.)

Table 11.2. Typical algorithms for machine learning. Author's own table.

Type	Algorithms
Supervised learning	Classification: logistic regression, k-nearest neighbors, Naïve Bayes, linear regression, ridge regression
Semi-supervised learning	Self-training, generative models, transudative SVM
Unsupervised learning	Both regression and classification: DT, RF, Clustering: k-means clustering, hierarchical clustering, SVM
Reinforcement learning	Deep Q-networks, state-action-reward-state-action, Q-learning

DT = decision tree, RF = random forest, SVM = support vector machine.

along with a small quantity of unlabeled data. Using trial and error to train a model, reinforcement learning assigns rewards or penalties for specific behaviors.

11.9 Utilizing ML in Various Crops

There are a number of applications of ML in plant sciences and developing a model for plant species classification is one of them. ML is extensively employed in a number of other domains apart from this one, including proteomics, metabolomics, phenomics, and plant genomics. ML can be used to decipher the intricacy and diversity of plant genomes. The evolution of sequencing technology and bioinformatics methods has rendered genome assembly relatively simple. Determining the functional parts of the genome remains challenging, however. By making it feasible to mark up structural regions of the genome, with a specific focus on protein-coding genes and CREs, ML gives a solution. Numerous ML-based software tools, such as SVM and Hidden Markov Models (HMM), have been created for ab initio gene prediction utilizing data from genomic sequences. With the help of a trained annotation file, AUGUSTUS, for example, makes it possible to accurately identify and annotate genes in unknown genomes, offering crucial information for genome study. By identifying patterns and characteristics in non-coding regions, CNNs are frequently used to anticipate CREs (Umarov and Solovyev, 2017). Researchers can also find intriguing linkages throughout the genome by using ML to anticipate gene–gene interactions and build gene regulatory networks.

ML in plant proteomics is utilized to promote an extensive understanding of protein structure, function, and interactions. The creation of AlphaFold2 is undoubtedly an extremely efficient application of ML for proteomics research (Jumper *et al.*, 2021). AlphaFold2 employs an ML method to precisely predict a protein's atomic-level structure solely based on the amino acid sequence. This finding significantly reduces the effort and time required to study the structure and function of proteins. Ziyun Ding and Daisuke Kihara developed a computational method that is trained and validated with data for known PPIs in *Arabidopsis thaliana* (Ding and Kihara, 2019). In metabolomics, ML makes processes of plant growth, plasticity, and environmental responses more understandable. Applying ML to predict metabolic pathways in plants accelerates research and enables the examination of novel pathway membership and associated genes compared to classical biochemical and genetic methods (Noordiik *et al.*, 2024).

11.10 Enhancing CRISPR/Cas Editing Systems through the Use of Machine Learning

More accurate and efficient CRISPR editing is becoming more and more necessary since genome-editing technology is being used more often in breeding. Nonetheless, there are still several challenges to be addressed, including inconsistent CRISPR editing results and efficiency, as well as enduring off-target consequences. A key tactic to deal with these problems is the advancement and refinement of CRISPR/Cas editing systems. Scientists identify regularities and patterns in vast experimental data by applying ML methods to process it. These patterns help them advance editing methods, opening the door to more accurate and effective genome editing.

11.11 Enhancing the Efficiency of the Editing Process

11.11.1 Estimating how various Cas variations will affect editing

RNA secondary structure, guanine and cytosine (GC) content, nucleotide composition, and epigenetic states are a few of the variables that affect the CRISPR/Cas system's editing efficacy, which is heavily impacted by the target site.

The optimization of sgRNAs for particular genome editing is still a problem because of their complexity. In order to increase the capacity to predict efficiency and choose sgRNAs with high efficiency, ML techniques have become practical instruments for developing prediction models based on these attributes.

Various models for base editor (BE) and prime editor (PE) have been created, including

FORECasT-BE (Pallaseni *et al.*, 2022) and DeepPE (Kim *et al.*, 2021). They are developed using efficient data from high-throughput experiments and inherit features from the corresponding editing systems. However, individual performances differ markedly on the same experimental dataset owing to variations in feature algorithmic techniques, dataset usage, and selection.

11.12 Enhancing the Precision of Editing

11.12.1 Creating guide RNA with excellent specificity

Guide RNA can cause off-target effects by binding to off-target locations with few mismatches. Off-target effects can be minimized by predicting the specificity of CRISPR cleavage and designing highly specific sgRNA. Multiple software or platforms are under development for off-target prediction based on three methods: hypothesis-driven, alignment-based, and learning-based. Learning-based methods have the potential to more accurately predict guide RNA specificity through complex model building based on various features. Generally speaking, these methods include scanning the entire genome for probable off-target regions, using ML algorithms to forecast such locations, and generating only one numerical score to evaluate the specificity of guide RNA. Approaches such as Discover-seq, GUIDE-seq, and CIRCLE-seq (Wienert *et al.*, 2019) have been established to identify off-target locations, with deep learning algorithms able to process more data and derive complex features for more accurate predictive results.

11.12.2 Making the editing protein better

The discovery of highly specialized guide RNAs and editing proteins has led to the identification of new editing proteins, such as LrCas9 (Zhong *et al.*, 2023), Cas13X, and Cas13Y. In order to increase the accuracy and efficacy of known editing proteins, protein engineering techniques such as guided evolution and rational design have been applied. Time and money constraints pose challenges to traditional methods. Nevertheless, protein structure and function predictions have significantly improved with the latest advancements in ML technology, such as protein structure estimation software, such as AlphaFold, AlphaFold2, and RoseTTAFold. These systems have provided important data on protein–substrate interactions and have enhanced the refinement of Cas proteins. Regardless of the sgRNA, deaminase enzymes produce off-target editing in base editing. Deaminases with fewer off-target activities can be quickly discovered using ML techniques. By creating new opportunities for the creation of more accurate and efficient genome-editing instruments, this strategy advances and applies the method of crop breeding via genome-editing initiatives.

11.13 Forecasting the Results of Genome Editing

In CRISPR/Cas9 genome editing, microhomology-mediated healing of DSBs, end joining, and template-free non-homologous end joining are the main DNA repair mechanisms (March *et al.*, 2020). These repair outcomes are not arbitrary; they are determined by features of the DNA sequence. The results of editing in plants can be predicted using such models since mutations rely on the DNA sequences surrounding DSBs. CROTON is a deep learning model based on neural architecture search and deep multitask CNNs that is applied to automate feature and model engineering and make more accurate predictions of editing outcomes.

11.14 Machine Learning Links Breeding and Crop Genome Editing

11.14.1 Finding genes and setting priorities

The intricate regulatory linkages of trait-related genes can make their identification difficult. Using ML techniques to combine various omics data helps speed up gene discovery and prioritization (Zhao *et al.*, 2020). We identified genes and transcription factors associated with drought resistance using an ML framework and a rice

GRAiN. GWAS is more logical and links particular features to particular genotypes (Gupta *et al.*, 2021). The gene for the Green Revolution is one of 13 candidate genes that QTG-Finder has identified by prioritizing causative factors in quantitative trait loci. These genes are useful breeding options since they were edited to decrease plant height and boost yield.

11.14.2 Mining important locations and cis-regulatory components

CREs, or cis-regulatory DNA elements, and trans-acting factors control gene expression. Manipulation of the CRE allows for increased crop traits, such as improved tomato quality and increased corn yield within the *ARGOS8* promoter (Shi *et al.*, 2017) under drought stress. Using ML methods, key nucleotide residues and CREs important for genome editing can be found. Researchers have confirmed the influence of the identified crucial locations on fruit ripening commencement and predicted tomato genome-wide expression patterns using CNN models and deep learning frameworks. The precise regulation of gene expression and maximal phenotypic outcomes is facilitated through ML, which allows researchers to predict key nucleotide positions involved in gene expression. Technologies based on CRISPR, such as base and prime editing, have become possible (Mao *et al.*, 2019). The development has further been boosted through the integration of ML within genome-editing techniques.

11.14.3 Upcoming opportunities

ML, genome editing, and epigenetics

The word "epigenetics" refers to long-lasting, heritable changes in gene expression that occur apart from the DNA sequence, which include non-coding RNA control, histone modifications, and DNA methylation. CRISPR-based epigenome editing will be used to change epigenetic processes, which control gene expression and affect characteristics by chemically changing DNA bases or the chromosomal superstructure.

The target gene is either upregulated or downregulated due to a change in epigenetic signatures at specific genomic locations through the use of dCas9 along with catalytic domains of an enzyme that modifies chromatin (Nunez *et al.*, 2021). It is expected that epigenome editing will become a successful breeding approach because it is known that epigenetic control over some of the most significant agronomic characteristics of crops. *Arabidopsis* has benefited from the effective use of epigenome editing technology to produce potential breeding traits, including drought stress tolerance and an appropriate blooming time. By generating highly effective guide RNA and determining the epigenetic mechanisms governing desired traits, the application of ML is to improve the epigenetic editing system (Yang *et al.*, 2023). Even though the majority of successful cases have involved model plants, ML-assisted epigenome editing will eventually emerge as a widely utilized and successful crop breeding technique.

Genome editing, ML, and single-cell technologies

In single-cell technology, individual cells are isolated experimentally, followed by cellular-level sequencing and multiomics analysis. Single-cell CRISPR screening techniques such as Perturb-seq, CRISPR-seq, or CROP-seq enable the investigation of functional CRISPR screening at the single-cell level by fusing pooled CRISPR screening with single-cell RNA-seq. The method has been effectively used to map gene regulatory networks, investigate cellular heterogeneity, and assess the function of genes in mammalian cells. The modification has proven to be challenging for these methods to use with plants (Gaillochet *et al.*, 2021). Plant cell size variability and the need to break down cell walls have a significant impact on single-cell isolation efficiency, which means that there are not enough collected plant cells for single-cell CRISPR screening. Using an ML framework could help address this problem by enabling the downscaling of the number of cells examined in order to find key regulatory signatures brought on by genetic perturbations. Since single-cell sequencing and CRISPR screens have both been shown to work in plants, it makes sense that CRISPR screens for single cells will eventually be used in plants. Additionally, single-cell levels of plants will soon be able to undergo genome editing, screening, and functional research thanks to ongoing advancements in computational frameworks and experimental pipelines.

Improvements to crops via molecular design techniques

GWAS based on ML improves our understanding of the relationship between certain phenotypes and genetic variations by utilizing multiomics data and phenotypic data. By enhancing our understanding of the genetic basis of intricate features, this method provides a strong theoretical framework for developments in genomics (Yoosefzadeh Najafabadi *et al.*, 2022). Intelligent breeding, which is predicated on cutting-edge AI-powered language models, is another possible avenue to provide biological information and theoretical underpinnings for future breeding. In this context, we refer to the analysis of vast amounts of biology, genetics, and botany literature using ML methods and natural language processing (NLP). The aim is to comprehend plant genomics principles, predict gene interactions, define the relationship between genotypes and plant traits, and offer breeding recommendations. Breeders will use genome editing effectively and accurately to perform molecularly customized breeding based on the biological information they acquire. This will provide breeders with unprecedented independence and accuracy in developing and cultivating ideal types.

11.15 Crop Breeding by De Novo Domestication

In a new breeding method termed "de novo domestication of wild relatives," wild species are genetically edited at the gene level, retaining their beneficial traits and ability to accommodate complex environments while ensuring that their main economic characteristics are equal to or better than those of present-day cultivated varieties. By modifying gene homologs, a recent study documented the quick enhancement of agronomically significant features in wild allotetraploid rice. De novo domestication is achieved by regulating these characteristics in diploid rice (Yu *et al.*, 2021). De novo domestication will proceed more swiftly thanks to ML's ability to detect genes associated with domesticated traits more rapidly. By analyzing the genetic differences between wild and related domesticated types, ML algorithms can find key genes involved in domestication. The successful introduction of domesticated features is made possible by these identified genes, which are essential targets for genome editing. Additionally, ML can help detect the implications of diverse gene combinations through the analysis of large genomic datasets, as these genes frequently have pleiotropic effects, and the results of varied combinations of altered alleles are unpredictable. This helps scientists make more informed editing decisions to create the desired traits. By increasing our understanding of genetic interactions and opening up new avenues to increasing the precision and effectiveness of de novo domestication, ML is revolutionizing breeding in the future.

11.16 Crop Improvement Using Synthetic Biology Techniques

Biological system design, engineering, or de novo synthesis with particular functionalities using engineering techniques and life science expertise is known as synthetic biology. It aims to achieve a wide range of goals, such as creating new genes or even whole new species. In synthetic biology, genome editing is an effective technique that allows for exact control over living organisms by quickly and precisely changing their genes. For instance, a new allele called Xa23SW14, which confers resistance in rice to bacterial blight disease, was synthesized through primary editing (Gupta *et al.*, 2023). The limitations of traditional breeding will eventually be overcome when genome-editing technology is used to build, or synthesize, unique genes with specific functions that are adapted to match particular needs. In certain instances, genome editing causes several DNA DSBs at the same time, and chromosomal rearrangements may be mediated via the NHEJ and MMEJ mechanisms (Kalter *et al.*, 2025). Reproductive isolation between the designed line and a wild-type parent may result from such chromosomal rearrangements, which could entirely alter the chromosomes of edited plants (potentially lowering their number). In this situation, the transgenic line might be regarded as a new plant species.

11.17 The Influence of Deep Learning-Enhanced Crop Breeding on the Quality and Yield of Crops

Climate change is severely affecting the world's agricultural productivity, with critical impacts on crop yield and quality (Praveen and Sharma, 2019). Deep learning models for crop image analysis and identification can help in the fast identification of superior plants with desired traits, which accelerates crop breeding and selection of high-yielding, disease-resistant, and other desirable varieties. Deep learning models can also detect crop performance under stress due to adverse environmental factors, such as salinity and drought stress, promoting more resilient crop varieties (Sun et al., 2019). For example, the algorithm for corn leaf blight identification and classification attained high accuracy utilizing the CNN algorithm, which is of extreme relevance for instant crop disease detection and increasing crop yields. Diagnostic and detection solutions for crop diseases were presented through several deep learning methods. The ability of SAE neural network algorithms to extract time-series chlorophyll fluorescence features demonstrated an effective method for recognizing chlorophyll fluorescence fingerprints in *Arabidopsis thaliana* under abiotic stress, providing fresh perspectives on enhancing agricultural drought resilience.

Several photos of soybean leaves were used for deep learning classification in order to identify and classify soybean symptoms under non-biological stress. This made it possible to quickly identify non-biological stress in soybeans (Ghosal *et al.*, 2018). Early disease identification and prompt action to safeguard coffee crop yields and quality can be facilitated by training deep learning models to categorize coffee leaf images and identify whether they are affected by leaf rust disease (Shao *et al.*, 2022). Furthermore, deep learning can analyze the connection between phenotypic genomics and gene editing techniques to improve crop quality and yields, as well as enable precise genomic combination selection and optimization. For example, the deployment of superior genotypes into larger grain yield trials has advanced due to the exceptional performance of deep learning-based GS models in forecasting wheat terminal quality attributes.

11.17.1 Obstacles and opportunities

Deep learning, the application of plant phenomics, including image classification, object recognition, and semantic segmentation, is one field of ML that has advanced significantly in the field of agriculture. Plant growth, yield, maturity, and biotic/abiotic stress perception can all be predicted with this method. However, acquiring a lot of labeled data is expensive, and deep learning algorithms need a lot of it. The difficulties of gathering phenotypic data include extreme occlusion and variable lighting conditions, which lengthen the time needed for annotating. The foundation of cognitive breeding design consists of huge data related to genotype, phenotype, and environment. Nevertheless, cumulative phenotypic data remain severely lacking, and conventional manual detection techniques are becoming more and more troublesome. Developments in robotics and next-generation sensors are anticipated to speed up crop phenotyping. However, because of differences in crop genetics and environments, deep learning models might not be as flexible across different crops and climatic conditions. To achieve the full potential of deep learning in breeding, researchers in the future must collaborate to develop a large database and practice interdisciplinary collaboration and data sharing. Few-shot learning and transfer learning can enhance deep learning models' performance and reduce reliance on large datasets. Robots' routes can be planned using deep reinforcement learning, which can also help farmers with crop data collection, picking, transporting, watering, and fertilization.

11.18 Artificial Intelligence Technologies Enhancing Crop Breeding

In computer science, artificial intelligence, or AI, is a field that continues to develop and concerns itself with how to develop and manage software programs that can learn to make choices and accomplish actions independently. AI involves many technologies, some of which include SVM, RF, DNNs, and ANNs. Since AI makes decisions based on data faster and more accurately than humans, it has the potential to revolutionize agriculture entirely. The terms AI, ML, and big data are employed to explain modern-day computer operations. Big data refer to the use of enormous amounts of data with diverse forms and complex structures that cannot be easily managed when researched using conventional methods. ML, which is a subfield of AI, employs massive training datasets to discover patterns. AI is applied in breeding to assist breeders by providing continuous farm surveillance and decreasing processing and data identification time. AI methods such as high-throughput genomics and phenomics have been used to accelerate the process of developing new plant varieties. The ML method has been used in marker-assisted selection, genomic prediction, and GS. Many agribusiness companies have invested hundreds of millions of dollars in systems that are able to utilize huge volumes of data about weather, types of seeds, and types of soil to help farmers save money and boost harvests. The aim of recent research on AI tools, such as ML, deep learning, and predictive analysis is to enhance thinking ability, reasoning, action-taking, and planning capabilities.

11.18.1 Artificial intelligence technologies addressing the challenges in phenomics

Plant phenomics has seen rapid progress over the past few years, and this has facilitated high-throughput phenotyping (HTP) of complex plant traits as well as precision breeding. AI has accelerated the efficiency of data analysis and gathering in a variety of scientific disciplines, including ML, deep learning, and computer vision. Three critical components of phenomic data management include employing AI: database management to share information and resources, model building to understand genotype–phenotype interactions with environmental interactions, and algorithms and programs to convert sensory data to phenotypic information. Field-based HTP is not as developed as within phenotypic facilities; therefore, there is a need for additional work in order to construct such facilities. AI technologies require a lot of data from many sources in order to become more accurate. In order to increase the quantity and quality of data acquired, this creates the opportunity to invest more in customizing existing technologies for field data collection and utilizing AI-adaptable technology, such as smartphones. Recent research has demonstrated the successful application of AI-assisted phenotyping systems to a range of crops, including oilseed crops, wheat, and corn, disease resistance, and plant productivity.

11.18.2 Investigating the capabilities of artificial intelligence in analyzing gene function

High-throughput technology in life sciences has yielded enormous amounts of data, generally referred to as "omics." Machine intelligence, or AI, is being used to deal with huge volumes of data because it is too massive and intricate to be handled by statistics or visual inspection. AI holds promise for comprehensive exploration of the genome and is already widely utilized in plant genomics. For various types of bioinformatics analysis, including protein-coding gene prediction, CRE prediction, gene expression, subcellular localization, PPI, gene ontology, metabolic pathways, phenotypes, and genomic prediction, there are various ML tools and algorithms. AI will soon be utilized to solve many plant science genomics issues. Comparative genomic studies and information transfer from a model plant to a target crop can be solved by AI algorithms. Two models for genetic trait prediction and analysis, DeepBind and DeepSEA, have been recently proposed. For predicting gene functions or the differential effect of gene expression, one can consider various types of expression or sequencing data analysis. AI software and algorithms exist for most of the bioinformatics analyses, including gene expression, CRE discovery, protein-coding gene discovery, and others. AI algorithms can be applied to comparative genomic analysis and information transfer between model plants and target crops. Biologists can explore gene expressions in parallel with the application of AI in bioinformatics, analyzing genetic information, hypothesis testing, and research enhancement.

11.18.3 Artificial intelligence is making envirotyping data more accessible for crop breeding

The goal of AI, a quickly evolving field in computer science, is to develop and supervise software that can learn to make conclusions and perform tasks on its own without assistance from humans. A variety of technologies are included in AI, including SVM, random forests (RF), DNNs, and ANNs. AI has the ability to completely transform agriculture by making data-driven decisions more quickly and accurately than humans. The terms "big data," "machine learning," and "artificial intelligence" refer to contemporary computer operations. Utilizing vast amounts of data with a variety of forms and intricate structures that are difficult to manage when studied using traditional methods is known as big data. A subfield of AI called ML uses enormous training datasets to identify patterns. AI is utilized in breeding to support breeders' efforts by ensuring ongoing farm monitoring and reducing processing and data identification time. Breeding new plant varieties has been accelerated through the use of AI techniques, such as high-throughput genomics and phenomics.

Marker-assisted selection, genomic prediction, and genomic selection have all made use of ML techniques. To assist farmers in cutting costs and increasing yields, numerous agricultural companies have spent hundreds of millions of dollars on systems that can use vast amounts of data on weather, seed variety, and soil type. More planning, learning, reasoning, thinking, and action-taking skills are the goals of recent research on AI techniques, such as deep learning, ML, and predictive analysis.

11.19 Connecting Crop Genomes to Phenomes Using AI

Plant phenomics has developed quite a lot in the last few years, opening up the way for precision breeding. This has been made possible by the invention and accessibility of new technology that can enable HTP of sophisticated plant traits. In recent years, the use of AI in a variety of scientific fields has increased dramatically. Deep learning, ML, and computer vision are examples of successful integration of AI characteristics into non-invasive imaging methods. This incorporation is gradually increasing the effectiveness of data collection and analysis through the integration of ML for effective picture analysis. AI has also facilitated the creation of tools and software for field phenotyping data administration and collection. These comprise open-source tools and platforms that facilitate community-driven study and data exchange, offering the vast volumes of data required for trustworthy phenotyping research. Three crucial aspects of phenomic data management involve the use of AI: database management to encourage the exchange of resources and information; model development to comprehend the connections between phenotype, genotype, and ecological interactions; and software programs and algorithms to translate perceptions into phenotypic data (Nabwire *et al.*, 2021). Repeated trials under different conditions are required to screen plants for desirable properties (e.g. grain size, abiotic stress tolerance, product quality, or yield potential) while accounting for the statistical necessity for an objective evaluation. Smartphones and other AI-adaptable devices can increase the volume and caliber of data gathered. Smartphones have become popular consumer goods, and their widespread usage in agriculture is suggested by how easily the sensors they contain may be utilized. Smartphones with advanced signal processing must deal with restrictions, such as low battery life, poor processing power, and constrained bandwidth. The amount of data that can be gathered could be increased by combining professional researchers with citizen scientists (Kumar *et al.*, 2022). The primary purpose of applying these technologies and tactics is to enable data administration, analysis, and application of AI methods while providing the platform for monitoring plant development throughout the growing season.

11.19.1 Future opportunities in AI breeding

The quick advancement of technology for high-throughput in biological sciences has resulted in massive data production in recent decades. In order to represent the entire amount of DNA present in every cell of a living being, disciplines that seek to collect and examine enormous volumes of genetic information are frequently referred to as "omics," which also have the flavor of being vulnerable to serious issues. "Omics" data are now too big and complex to be examined graphically or by statistical correlations. This has resulted in the application of AI, which operates on gigantic quantities of data that are far beyond human capacity, derives conclusions that make our understanding of the system under analysis even more profound, and—most importantly—enhances itself with training data. AI has several potential uses for in-depth genome analysis in the future and is already widely employed in plant genomics. Several ML algorithms and tools exist to perform various bioinformatics analyses, including gene expression, cis-regulatory component recognition, subcellular location, gene ontology, PPI, metabolic processes, phenotypes, genomic prediction, and protein-coding gene identification (as reviewed by Mahood *et al.*, 2020). Although the successful advancements in high-throughput sequencing technology resulted in the production of a substantial amount of genomic data, the massive volume of data generated poses a considerable challenge for data storage and analysis. The measurement of the simultaneous expression of many genes, or even every gene that is found in the genome, under a variety of conditions, is made possible by the AI technology of bioinformatics (Lemmon and Lemmon, 2013). By giving them a more "relevant" representation of their data and the ability to integrate it, all of this aids biologists in analyzing their genetic data, testing and validating their ideas throughout the experimental cycle, and ultimately improving their study.

11.19.2 Integrated image processing and machine learning for effective phenotyping and precision farming

Measuring the physical characteristics of plants is made easier and more accurate by machine vision, which is essential for both traditional and *in vitro* plant breeding research. It makes it possible to measure physical characteristics non-destructively both *in vitro* and outdoors. The profiling of populations of plants and the identification of outstanding types can be accelerated and facilitated by automatic, harmless, quick scoring of plant attributes using instruments for HTP. Further developments in picture analysis are made possible by deep learning's ability to recognize plant characteristics. A potent tactic for more accurate and quicker picture-based plant phenotyping is the combination of ML with image processing. In recent years, deep learning techniques, especially CNN, have emerged as a potent instrument for image interpretation. In-field images can be analyzed using sophisticated imaging techniques, and by training DNNs with these images, desired features can be efficiently estimated and classified. Based on the spatial and temporal divergence of crop and soil elements within a field, precise farming is a method of agricultural management. Comprehensive image-based phenotyping of plant stress, from leaf and canopy to field range, can be accomplished with ML techniques.

11.20 An Approach for Using Image Processing and ML to Determine the Ploidy Level of Crops

It is essential to verify the ploidy level in chromosomal engineering experiments. Indirect techniques, such as flow cytometry, anatomical and morphological indicators, or direct techniques, such as chromosome counting, can be used for this. Although they are dependable, direct procedures are time-consuming and need experienced operators. Although they are quick, indirect methods—such as stomatal morphometric data—are not totally trustworthy. A trustworthy technique that relies on the relationship between ploidy level and nuclear DNA content is flow cytometry. Ploidy level identification can be done with ML techniques. For example, an advanced network-based object detection system for assessing stomatal density can be utilized. Remote sensing for quaking aspen and CNNs for identifying diploid and haploid corn seeds are further techniques. To detect ploidy levels in plants, a novel method uses DNNs supervised by linked image processing. Studies on genetic diversity, evolution, and species invasiveness are among the other areas of plant science that potentially benefit from this approach, which is more accurate, quick, and economical.

11.21 Platforms for Phenotyping

Plant stress phenotyping is a field that is continuously developing, with non-invasive high-throughput techniques replacing damaging, low-throughput phenotyping protocols and methods. The rapid advancement of inexpensive, non-invasive sensors and imaging methods throughout the years has changed plant phenomics. Additionally, the imaging techniques, sensors, and analytical instruments are now in harmony because of these improvements. This congruence leads to the development of one-piece, small imaging platforms for HTP research. Among these platforms are automated phenotyping tools, such as "PHENOPSIS," for studying responses of *Arabidopsis* plants to soil water stress; "GROWSCREEN FLUORO" for phenotypic analysis of chlorophyll fluorescence and leaf growth, which made it possible to identify arabidopsis' (*Arabidopsis thaliana* L.) resistance to various abiotic stressors; "HyperART" for the harmless leaf measurement characteristics, such as chlorophyll concentration and leaf disease severity across four different crop types (tomato, corn, barley, and rapeseed); "PhenoBox" for the identification of corn smut and head smut ailments on brachypodium and corn (*Zea mays* L.), respectively; "PHENOVISION" for the identification of stress caused by the drought and corn plant recovery (Asaari *et al.*, 2019); "PhénoField" for the review of different abiotic stressors in wheat (*Triticum aestivum* L.); and the "PlantScreen™ Robotic XYZ System" for the analysis of various traits related to rice (Kim *et al.*, 2020). High-throughput, automated data collection has also been done for traits relating to roots and shoots, leaves, plant height, biomass, early vigor, radiation use efficiency, and photosynthesis in various plant species, including rice, sorghum (*Sorghum bicolor* L.), wheat, corn, arabidopsis, brachypodium (*Brachypodium distachyon* L.), cotton (*Gossypium hirsutum* L.), rapeseed (*Brassica napus* L.), and barley (*Hordeum vulgare* L.), among others, with various phenotyping platforms including RootReader3D, GROWSCREEN-Rhizo, Zeppelin NT plane, Phenocart (Gill *et al.*, 2022), Phenovator, PHENOARCH (Brichet *et al.*, 2017), Field Scanalyzer, CropQuant, and MVS-Pheno (Wu *et al.*, 2020) (Table 11.3). This phenotyping platform holds promise for use in HTP of characteristics linked with stress tolerance/resistance in various crops.

11.22 Imaging Processing

In many crop species, phenomics is used to track weed presence, growth, nutrient status, drought stress, disease, pest infestations, and yield. Recent developments in technology have opened up new imaging methods for HTP. Handheld cell phones and unmanned aerial vehicle (UAV)-based drone imaging are examples of these methods. High spatial resolution is provided by UAVs; however, their battery capacity is constrained for large-scale HTP. Plant stress assessment has

Table 11.3. Information on a few chosen imaging platforms for trait phenotyping of abiotic and biotic stressors, as well as other crop morphological features. Author's own table.

Imaging Platform	Characteristics	Crop
PHENOPSIS	The reactions of plants to water stress	Arabidopsis (*Arabidopsis thaliana)*
GROWSCREENFLUORO	Chlorophyll fluorescence and leaf development, enabling the measurement of stress tolerance	Arabidopsis (*Arabidopsis thaliana)*
HyperART	Leaf characteristics, such as the amount of chlorophyll or the severity of disease	Corn (*Zea mays*), barley (*Hordeum vulgare*), tomato (*Solanum lycopersicum*), and rapeseed (*Brassica rapa*)
PhenoBox	Identification of the salt-stress response in tobacco and the effects of corn smut and head smut fungus on corn and brachypodium, respectively	Tobacco (*Nicotiana benthamiana*), corn (*Zea mays*), brachypodium (*Brachypodium distachyon)*
PHENOVISION	Identification and recovery of stress caused by drought	Corn (*Zea mays*)
PlantScreen™ robotic XYZ system	Characteristics of drought resistance	Rice (*Oryza sativa)*

been done using satellite imaging since the 1970s. Platforms for ground-based imaging are more practical and effective since they offer precise and detailed images at the level of individual plants and branches. Terabytes of data are produced every day by these methods, necessitating the application of ML as well as deep learning software for data management.

11.22.1 Satellite-derived images

The oldest technique for remote sensing is satellite imaging, which may cover whole counties or vast expanses of 1000 ha. To record ground data, earth observation satellites equipped with a range of sensors such as RGB, multispectral, hyperspectral, thermal, and time-of-flight light sensors are utilized. A spectral resolution of up to 0.25 m is possible with multispectral sensors, which target certain wavelengths, such as red, blue, and green. The ability of hyperspectral sensors to target multiple bands within a limited area makes them more useful for identifying plant diseases. RADAR and LiDAR are examples of active sensors that use radiation to collect data from the ground, whereas these passive sensors use light reflections from the Earth's surface. However, because of its high cost, satellite imaging has its limits.

11.22.2 Mobile photography/imaging devices

High-quality cameras in mobile phones are only good for simple imaging, but some manufacturers also include more sophisticated sensors, as in 3D imaging, such as LiDAR. The creation of mobile applications is increasing, and their usefulness has also been made easier by the rapid growth of smartphones with potent processors and high-resolution cameras. Additionally, elements of smartphones are being integrated into other portable instruments and devices, increasing the number of sensors and enhancing the connectivity and portability of conventional phenotyping tools. Researchers now have an advantage thanks to these developments since they can easily capture the phenotype using a portable mobile instrument. However, as it is impractical to utilize this technology to take images of every plant in the field, these advancements offer no practical utility for research. It is comparable to manual phenotyping because each plant or component would still need to be examined by researchers. Even so, mobile has no use in HTP investigations, although it can assist in the diagnosis of biotic (Hall *et al.*, 2017) along with abiotic (Naik *et al.*, 2017) stressors utilizing ML and DL. A researcher or even a farmer can use a variety of mobile applications to quickly diagnose symptoms prior to doing a more thorough examination. These mobile apps might be modified to address other significant crops and associated illnesses, which would improve integrated disease management decision-making; nevertheless, their usefulness is constrained.

11.22.3 UAV imaging

For large-scale HTP studies, UAV imaging is thought to be the most practical and cost-effective method. Multiple photos are taken by UAVs while they are in flight. Parts of the entire field are included in the photos, subsequently sewn together to create an orthomosaic, a larger picture that displays the area as a whole. These orthomosaics can be made using a variety of software programs, including Pix4D, QGIS, and OpenDroneMap. Because UAVs fly closer to the ground than satellites, their images have a higher resolution. Given their low spatial resolution, hyperspectral sensors may find this especially useful. Because of their increased payload capacity, UAVs are able to carry any sensor that can be mounted on any satellite. Their limited flight height and battery capacity, however, prevent them from covering as much ground as satellites.

11.22.4 Ground-based imaging platforms

Among all the methods, the most advanced platforms for imaging are ground-based. Imaging at a very close range for assessing plant stress is possible with ground-based devices (Mishra *et al.*, 2020). The closeness allows for more efficient hand phenotyping that is human-like. According to Vougioukas (2019), the majority of ground-based platforms are self-sufficient systems that can be combined with onboard chips to assess each phenotype's properties. These autonomous systems generate daily data volumes of up to terabytes and, when combined with test photographs, contribute a great deal more information to the project, including parameters, metrics, and phenotypic scores.

11.22.5 Spectral indicators for assessing plant stress phenotypes

After the photos taken using the previously mentioned methods are decoded, the information contained in them is evaluated using spectral indices (SIs). SIs involve performing different sets of procedures on distinct picture spectral layers. These sets of processes consist of combining spectral reflectance from two or more bands and performing certain mathematical computations. The figure that emerges through this mathematical addition represents the relative abundance of the feature in question. A number of SIs can be used to estimate different types of features that are being obtained in an image. This review focuses on vegetation indices (VI), which are spectral estimates based on different spectral bands for interpretation. Characteristics and vegetation data are recorded in a picture. Plant phenotype, plant architecture, stress level, and biomass are just a few of the many data that VIs may offer (Kokhan and Vostokov, 2020). Deciding which type of VI to use for a given application is the most crucial step in the VI usage process (Xue and Su, 2017). In HTP programs, a variety of VI types can be employed for plant assessment investigations. Abiotic and biotic stressors are mostly monitored using NIR or IR bands because plants in good health reflect more infrared light than stressed plants.

11.23 Conclusion

The study of agricultural genome editing is fast developing and holds promise for increasing crop resistance, nutritional value, and production. A potent tool for improving and automating crop genome-editing procedures is ML. In addition to optimizing editing technologies for increased speed and specificity, ML helps breeders identify important functional regions and places in the genome to meet precision breeding goals. The full capabilities of crop genetic modification for enhancing crop attributes will be unlocked by fusing the finest techniques from both domains. Future developments in genome editing, ML, biological technology, and research will all contribute to increasing the range of potential uses of genome editing and offering breeders a robust set of tools for crop improvement. When used in revolutionary breeding strategies such as de novo domestication, molecular design breeding, and synthetic biology, genome-learning and genome-editing algorithms will outperform conventional breeding techniques and transform contemporary agriculture. In many aspects of plant breeding, ML algorithms can handle massive volumes of data, which can produce more precise outcomes and greater interoperability than traditional statistical methods. Because ANNs can handle binary, continuous, categorical, and fuzzy information, they can be employed in plant tissue culture studies for classification, pattern recognition, and non-linear regression. An overview of ML applications for plant breeders can be found in this review. In subsequent research, it would be beneficial to use the appropriate data analysis methodology, which can boost study productivity.

References

Abdipour, M.; Ramazani, S.H.R.; Younessi-Hmazekhanlu, M.; Niazian, M. (2018) Modeling Oil Content of Sesame (*Sesamum indicum* L.) Using Artificial Neural Network and Multiple Linear Regression Approaches. *J. Am. Oil Chem. Soc.* 95, 283–297

Ajay, B.C.; Bera, S.K.; Singh, A.L.; Kumar, N.; Gangadhar, K.; Kona, P. (2020) Evaluation of Genotype Environment Interaction and Yield Stability Analysis in Peanut Under Phosphorus Stress Condition Using Stability Parameters of AMMI Model. *Agric. Res.*, 1–10

Ali, S.; Shafique, O.; Mahmood, T.; Hanif, M.A.; Ahmed, I.; Khan, B.A. (2018) A Review about Perspectives of Nanotechnology in Agriculture. *Pakistan J. Agric. Res.* 31, 116–121.

Anzalone AV *et al.* (2019) Search-and-replace genome editing without double-strand breaks or donor DNA. *Nature* 576:149–157. https://doi.org/10.1038/s41586-019-1711-4

Araus, J. L., and Cairns, J. E. (2014) Field high-throughput phenotyping: the new crop breeding frontier. *Trends Plant Sci.* 19, 52–61. doi: 10.1016/j.tplants.2013.09.008

Asaari MSM, Mertens S, Dhondt S *et al.* (2019) Analysis of hyper spectral images for detection of drought stress and recovery in maize plants in a high-throughput phenotyping platform. *Comput Electron Agric* 162:749–758. https://doi.org/10.1016/j.compag. 2019.05.018

Azizi, A., Abbaspour-Gilandeh, Y., Nooshyar, M. and Afkari-Sayah, A. (2016) Identifying Potato Varieties Using Machine Vision and Artificial Neural Networks. *Int. J. Food Prop.* 19, 618–635.

Bahmer, A., Gupta, D. and Effenberger, F. (2023) Modern artificial neural networks: Is evolution cleverer? *Neural Computation*, 35(5), 763–806.

Barbedo, J.G.A. (2019) Detection of nutrition deficiencies in plants using proximal images and machine learning: A review. *Comput. Electron. Agric.* 162, 482–492.

Boonsrangsom, T. (2020) Genetic diversity of 'Wan ChakMotluk' (*Curcuma comosa*Roxb.) in Thailand using morphological characteristics and random amplification of polymorphic DNA (RAPD) markers. *South Afr. J. Bot.* 130, 224–230

Coulibaly, S., Kamsu-Foguem, B., Kamissoko, D. and Traore, D. (2019) Deep neural networks with transfer learning in millet crop images. *Comput. Ind.* 108, 115–120

Dezfouli, P.M., Sedghi, M., Shariatpanahi, M.E., Niazian, M. and Alizadeh, B. (2019) Assessment of general and specific combining abilities in doubled haploid lines of rapeseed (Brassica napus L.). *Ind. Crop. Prod.* 141, 111754.

Ding, Z. and Kihara, D. (2019) Computational identification of protein–protein interactions in model plant proteomes. *Scientific Reports*, 9(1), 8740.

Emamgholizadeh, S., Parsaeian, M. and Baradaran, M. (2015) Seed yield prediction of sesame using artificial neural network. *Eur. J. Agron.* 68, 89–96.

Gaillochet C., Develtere W., Jacobs, T.B. (2021) CRISPR screens in plants: approaches, guidelines, and future prospects. *Plant Cell* 33(4):794–813. https://doi.org/10.1093/plcell/koab099

Geetharamani, G. and ArunPandian, J. (2019) Identification of plant leaf diseases using a nine-layer deep convolutional neural network. *Comput. Electr. Eng.* 76, 323–338.

Ghaffari,H., Tadayon, M.R., Razmjoo, J., Bahador, M., Soureshjani, H.K. and Yuan, T. (2020) Impact of Jasmonic Acid on Sugar Yield and Physiological Traits of Sugar Beet in Response to Water Deficit Regimes: Using Stepwise Regression Approach. *Russ. J. Plant Physiol.* 67, 482–493.

Ghosal, S., Blystone, D., Singh, A.K., Ganapathysubramanian, B., Singh, A. and Sarkar, S. (2018) An explainable deep machine vision framework for plant stress phenotyping. *Proc. Natl. Acad. Sci.* 115, 4613–4618. doi: 10.1073/pnas.1716999115

Gill, T., Gill, S.K., Saini, D.K., Chopra, Y., de Koff, J.P. and Sandhu, K.S. (2022) A comprehensive review of high throughput phenotyping and machine learning for plant stress phenotyping. *Phenomics*, 2(3), 156–183.

Gupta C., Ramegowda V., Basu S. and Pereira A. (2021) Using network based machine learning to predict transcription factors

Gupta, A., Liu, B., Chen, Q.J. and Yang, B. (2023) High-efficiency prime editing enables new strategies for broad-spectrum resistance to bacterial blight of rice. *Plant Biotechnol J* 21(7):1454–1464. https://doi.org/10.1111/pbi.14049

Haider,S.A., Naqvi, S.R., Akram, T., Umar, G.A., Shahzad, A., Sial, M.R., Khaliq, S. and Kamran, M. (2019) LSTM Neural Network Based Forecasting Model for Wheat Production in Pakistan. *Agronomy* 9, 72

Hallau, L., Neumann, M., Klatt, B. *et al.* (2017) Automated identification of sugar beet diseases using smartphones. *Plant Pathol.* 67:399–410. https://doi.org/10.1111/ppa.12741

Han, X., Hu, C., Chen, Y., Qiao, Y., Liu, D., Fan, J., Li, S. and Zhang, Z. (2020) Crop yield stability and sustainability in a rice-wheat cropping system based on 34-year field experiment. *Eur. J. Agron.* 113, 125965.

Hashemifar, S., Neyshabur, B., Khan, A.A. and Xu, J. (2018). Predicting protein protein interactions through sequence-based deep learning. *Bioinformatics* 34, i802 i810. doi: 10.1093/bioinformatics/bty573g

Jin, C., Zhou, L., Pu, Y., Zhang, C., Qi, H. and Zhao, Y. (2025) Application of deep learning for high-throughput phenotyping of seed: a review. *Artificial Intelligence Review*, 58(3), 76.

Jumper, J. *et al.* (2021) Highly accurate protein structure prediction with AlphaFold. *Nature* 596:583–589. https://doi.org/10. 1038/s41586-021-03819-2

Kalter, N., Fuster-García, C., Silva, A., Ronco-Díaz, V., Roncelli, S. *et al.* (2025) Off-target effects in CRISPR-Cas genome editing for human therapeutics: Progress and challenges. *Molecular Therapy Nucleic Acids*, 36 (3), 102636.

Khaki, S., Khalilzadeh, Z. and Wang, L. (2020) Predicting yield performance of parents in plant breeding: A neural collaborative filtering approach. *PLoS ONE* 15, e0233382

Killamsetty, K., Sivasubramanian, D., Ramakrishnan, G. and Iyer, R. (2021) "Glister: Generalization based data subset selection for efficient and robust learning," in Proceedings of the AAAI Conference on Artificial Intelligence (Menlo Park, Calif: AAAI Press) 35, 8110–8118.

Kim, H.K., Yu, G., Park, J., Min, S., Lee, S., Yoon, S. and Kim, H.H. (2021) Predicting the efficiency of prime editing guide RNAs in human cells. *Nat Biotechnol* 39:198–206. https://doi.org/10.1038/s41587-020-0677-y

Kim, J.Y. (2020) Roadmap to high throughput phenotyping for plant breeding. *J BiosystEng* 45(1):43–55

Kokhan S., Vostokov, A. (2020) Using vegetative indices to quantify agricultural crop characteristics. *J EcolEng* 21:120–127. https:// doi.org/10.12911/22998993/119808

Kumar, R.S., Keerthana, G., Murali, L., Chidambaranathan, S., Premkumar, C. and Mahaveerakannan, R. (2022) Enabling Artificial Intelligence and Cyber Security in Smart Manufacturing. In: *Artificial Intelligent Techniques for Wireless Communication and Networking*; John Wiley & Sons, Ltd.: Hoboken, NJ, USA, pp. 269–286

Lai, X., Stigliani, A., Vachon, G., Carles, C., Smaczniak, C., Zubieta, C., *et al.* (2019). Building transcription factor binding site models to understand gene regulation in plants. *Mol. Plant* 12, 743–763. doi: 10.1016/j.molp.2018.10.010

Lee, S., Jeong, Y., Son, S. and Lee, B. (2019) A Self-Predictable Crop Yield Platform (SCYP) Based on Crop Diseases Using Deep Learning. *Sustainability* 11, 3637.

Lemmon, E.M. and Lemmon, A.R. (2013) High-throughput genomic data in systematics and phylogenetics. *Annu. Rev. Ecol. Evol. Syst.* 44, 99–121

Lin, K., Gong, L., Huang, Y., Liu, C. and Pan, J. (2019) Deep Learning-Based Segmentation and Quantification of Cucumber Powdery Mildew Using Convolutional Neural Network. *Front. Plant Sci.* 10

Lin, Q.P. *et al.* (2020) Prime genome editing in rice and wheat. *Nat Biotechnol.* 38:582–585. https://doi.org/10.1038/s41587 020-0455-x

Li, T., Huang, S., Jiang, W.Z., Wright, D., Spalding, M.H., Weeks, D.P. and Yang, B. (2011) TAL nucleases (TALNs): hybrid proteins composed of TAL effectors and FokI DNA-cleavage domain. *Nucleic Acids Res.* 39(1):359–372. https://doi.org/10.1093/nar/gkq704

Luqman, T., Hussain, M., Ahmed, S.R., Ijaz, I. and Maryum, Z. *et al.* (2025) Cotton under heat stress: a comprehensive review of molecular breeding, genomics, and multi-omics strategies. *Frontiers in Genetics*, 16, 1553406.

Lv, C., Huang, Y., Sun, W., Yu, L. and Zhu, J. (2020) Response of rice yield and yield components to elevated [CO2]: Asynthesis of updated data from FACE experiments. *Eur. J. Agron.* 112, 125961.

Ma, W., Qiu, Z., Song, J., Li, J., Cheng, Q., Zhai, J. and Ma, C. (2018) A deep convolutional neural network approach for predicting phenotypes from genotypes. *Planta* 248, 1307–1318.

Mahood, E.H., Kruse, L.H. and Moghe, G.D. (2020) Machine learning: A powerful tool for gene function prediction in plants. *Appl. Plant Sci.* 8, e11376.

Makarova, K.S. *et al.* (2011) Evolution and classification of the CRISPR-Cas systems. *Nat Rev Microbiol* 9:467–477. https:// doi.org/10.1038/nrmicro2577

Makarova, K.S. *et al.* (2020) Evolutionary classification of CRISPR Cas systems: a burst of class 2 and derived variants. *Nat Rev Microbiol* 18:67–83. https://doi.org/10.1038/s41579-019 0299-x

Mao, Y.F., Botella, J.R, Liu, Y.G. and Zhu, J.K. (2019) Gene editing in plants: progress and challenges. *Natl Sci Rev* 6(3):421–437. https:// doi.org/10.1093/nsr/nwz005

March, O.P., Kocher, T. and Koller, U. (2020) Context-dependent strategies for enhanced genome editing of genodermatoses. *Cells*, 9(1), 112.

Mishra, P., Polder, G. and Vilfan, N. (2020) Close range spectral imaging for disease detection in plants using autonomous platforms: a review on recent studies. *Curr Robot Rep* 1:43–48. https:// doi. org/ 10. 1007/s43154-020-00004-7

Nabwire, S., Suh, H.K., Kim, M.S., Baek, I. and Cho, B.K. (2021) Review: Application of Artificial Intelligence in Phenomics. *Sensors* 21, 4363.

Naik, H.S., Zhang, J., Lofquist, A. *et al.* (2017) A real-time phenotyping framework using machine learning for plant stress severity rat ing in soybean. *Plant Methods* 13:23. https:// doi. org/ 10. 1186/ s13007-017-0173-7

Niazian, M., Sadat Noori, S.A., Tohidfar, M. and Mortazavian, S.M.M. (2017) Essential Oil Yield and Agro-morphological Traits in Some Iranian Ecotypes of Ajowan (*Carum copticum* L.). *J. Essent. Oil Bear. Plants* 20, 1151–1156.

Niazian, M., Sadat-Noori, S.A. and Abdipour, M. (2018) Modeling the seed yield of Ajowan (*Trachyspermumammi* L.) using artificial neural network and multiple linear regression models. *Ind. Crops Prod.*

Niedbała, G., Kurasiak-Popowska, D., Stuper-Szablewska, K. and Nawracała, J. (2020) Application of Artificial Neural Networks to Analyze the Concentration of Ferulic Acid, Deoxynivalenol, and Nivalenol in Winter Wheat Grain. *Agriculture* 10, 127

Noordijk, B., Garcia Gomez, M.L., ten Tusscher, K.H., de Ridder, D., van Dijk, A.D. and Smith, R.W. (2024) The rise of scientific machine learning: a perspective on combining mechanistic modelling with machine learning for systems biology. *Frontiers in Systems Biology*, 4, 1407994.

Nunez, J.K. *et al.* (2021) Genome-wide programmable transcriptional memory by CRISPR-based epigenome editing. *Cell* 184(9):2503–2519. https://doi.org/10.1016/j.cell.2021.03. 025

Pallaseni A *et al.* (2022) Predicting base editing outcomes using position-specific sequence determinants. *Nucleic Acids Res* 50(6):3551–3564. https://doi.org/10.1093/nar/gkac161

Pan, Y. (2015) Analysis of concepts and categories of plant phenome and phenomics. *Acta Agronom. Sin.* 41, 175–186. doi: 10.3724/SP.J.1006.2015.00175

Parmley, K. A., Higgins, R. H., Ganapathysubramanian, B., Sarkar, S. and Singh, A. K. (2019). Machine learning approach for prescriptive plant breeding. *Sci. Rep.* 9, 17132. doi: 10.1038/s41598-019-53451-4

Parsaeian, M., Shahabi, M. and Hassanpour, H. (2020) Estimating oil and protein content of sesame seeds using image processing and artificial neural network. *Journal of the American Oil Chemists' Society*, 97(7), 691–702.

Poletto,T. Poletto,I., Moraes Silva, L.M., Brião Muniz, M.F., Silveira Reiniger, L.R., Richards, N. and Stefenon, V.M. (2019) Morphological, chemical and genetic analysis of southern Brazilian pecan (*Carya illinoinensis*) accessions. *Scientia Horticulturae* 261, 108863.

Praveen, B., and Sharma, P. (2019) A review of literature on climate change and its impacts on agriculture productivity. *J. Public Affairs* 19, e1960. doi: 10.1002/pa.1960

Qin, Q. and Feng, J. (2017). Imputation for transcription factor binding predictions based on deep learning. *PloSComput. Biol.* 13, e1005403. doi: 10.1371/ journal.pcbi.1005403

Ray, A., Halder, T., Jena, S., Sahoo, A., Ghosh, B. *et al.* (2020) Application of artificial neural network (ANN) model for prediction and optimization of coronarin D content in *Hedychium coronarium*. *Ind. Crops Prod.* 146, 112186.

Schmidhuber, J. (2015) Deep learning in neural networks: An overview. *Neural Netw.* 61, 85–117.

Shi, J.R. *et al.* (2017) ARGOS8 variants generated by CRISPR-Cas9 improve maize grain yield under field drought stress conditions. *Plant Biotechnol J* 15(2):207–216. https://doi.org/10. 1111/pbi.12603

Shao, B., Hou, Y., Huang, N., Wang, W., Lu, X., and Jing, Y. (2022) "Deep learning based coffee beans quality screening," in 2022 IEEE International Conference on e Business Engineering (ICEBE). (Piscataway: IEEE), 271–275.

Srivastava, A., Gupta, S., Shanker, K., Gupta, N., Gupta, A.K. and Lal, R.K. (2020) Genetic diversity in Indian poppy (*P. somniferum* L.) germplasm using multivariate and SCoT marker analyses. *Ind. Crops Prod.* 144, 112050

Sun, D., Zhu, Y., Xu, H., He, Y. and Cen, H. (2019) Time-series chlorophyll fluorescence imaging reveals dynamic photosynthetic fingerprints of sos mutants to drought stress. *Sensors* 19, 2649. doi: 10.3390/ s19122649

Umarov, R.K. and Solovyev, V.V. (2017) Recognition of prokaryotic and eukaryotic promoters using convolutional deep learning neural networks. *PLoS ONE* 12(2):e0171410. https://doi. org/10.1371/journal.pone.0171410

Vougioukas, S.G. (2019) Agricultural robotics. *Annu Rev Control Robot Auton Syst.* 2:365–392. https:// doi. org/ 10. 1146/ annurev- contr ol-053018-023617

Wang, H., Cimen, E., Singh, N. and Buckler, E. (2020) Deep learning for plant genomics and crop improvement. *Curr. Opin. Plant Biol.* 54, 34–41. doi: 10.1016/j.pbi.2019.12.010

Wienert, B. *et al.* (2019) Unbiased detection of CRISPR off-targets in vivo using DISCOVER-Seq. *Science* 36.

Wulff, B.B.H. and Dhugga, K.S. (2018) Wheat—the cereal abandoned by GM. *Science* 361(6401):451–452. https://doi.org/10.1126/ science.aat5119

Wu, Y. *et al.* (2022a) CRISPR-BETS: a base-editing design tool for generating stop codons. *Plant Biotechnol J.* 20(3):499–510. https://doi.org/10.1111/pbi.13732

Wu, S., Wen, W., Wang, Y. *et al.* (2020) MVS-pheno: a portable and low cost phenotyping platform for maize shoots using multiview stereo 3D reconstruction. *Plant Phenomics*, 1848437. https:// doi. org/10.34133/2020/1848437

Xu, J., Zhang, N., Wang, K., Xian, Q., Dong, J. and Chen, X. (2022) Exploring new strategies in diseases resistance of horticultural crops. *Frontiers in Sustainable Food Systems*, 6, 1021350.

Xue J. and Su B. (2017) Significant remote sensing vegetation indices: a review of developments and applications. *J Sensors* 2017:1–17. https://doi.org/10.1155/2017/1353691

Yang, Y., Saand, M. A., Huang, L., Abdelaal, W. B., Zhang, J., Wu, Y., *et al.* (2021) Applications of multi-omics technologies for crop improvement. *Front. Plant Sci.* 12, 563953. doi: 10.3389/fpls.2021.563953

Yang, Q.Q., Wu, L.L., Meng, J., Ma, L., Zuo, E.R. and Sun, Y.D. (2023) Predicting sgRNA activity for CRISPR-mediated epigenome editing by deep learning. *Comput Struct Biotechnol.* 21:202–211. https://doi.org/10.1016/j.csbj.2022.11.034

Yoosefzadeh-Najafabadi, M., Eskandari, M., Torabi, S., Torkamaneh, D., Tulpan, D. and Rajcan, I. (2022) Machine-learning-based genome wide association studies for uncovering QTL underlying soybean yield and its components. *Int J Mol Sci.* 23(10):5538. https://doi.org/10.3390/ijms23105538

Yu, H. *et al.* (2021) A route to de novo domestication of wild allotetraploid rice. *Cell* 184(5):1156–1170. https://doi.org/ 10.1016/j.cell.2021.01.013

Zampieri, G., Vijayakumar, S., Yaneske, E. and Angione, C. (2019) Machine and deep learning meet genome-scale metabolic modeling. *PloSComput. Biol.* 15, e1007084. doi: 10.1371/journal.pcbi.1007084

Zhang, Z., Park, C.Y., Theesfeld, C.L., and Troyanskaya, O.G. (2021b) An automated framework for efficiently designing deep convolutional neural networks in genomics. *Nat. Mach. Intell.* 3, 392–400. doi: 10.1038/s42256-021-00316-z

Zheng, H., Li, W., Jiang, J., Liu, Y., Cheng, T., Tian, Y., Zhu, Y., Cao, W., Zhang, Y. and Yao, X.A. (2018) Comparative Assessment of Different Modeling Algorithms for Estimating Leaf Nitrogen Content in Winter Wheat Using Multispectral Images from an Unmanned Aerial Vehicle. *Remote Sens.* 10, 2026.

Zhou, X. (2018) Understanding the convolutional neural networks with gradient descent and backpropagation. *J. Phys.*: Conf. Ser. (IOP Publishing) 1004, 012028. doi: 10.1088/1742-6596/1004/1/012028

Zhao, H.N., Zhang, W.L., Zhang, T., Lin, Y., Hu, Y.D., Fang, C. and Jiang, J.M. (2020) Genome-wide MNase hypersensitivity assay unveils distinct classes of open chromatin associated with H3K27me3 and DNA methylation in Arabidopsis thaliana. *Genome Biol* 21:24. https://doi.org/10.1186/s13059-020-1927-5

Zitnik, M., Agrawal, M. and Leskovec, J. (2018) Modeling polypharmacy side effects with graph convolutional networks. *Bioinformatics* 34, i457–i466. doi: 10.1093/ bioinformatics/bty294

Zhong, Z.H. *et al.* (2023) Efficient plant genome engineering using a probiotic sourced CRISPR-Cas9 system. *Nat Commun.* 14:6102. https://doi.org/10.1038/s41467-023-41802-9

Zou, J., Hu, W., Li, Y., He, J., Zhu, H. and Zhou, Z. (2020) Screening of drought resistance indices and evaluation of drought resistance in cotton (Gossypium hirsutum L.). *J. Integr. Agric.* 19, 495–508.

12 Deep Learning for Managing Greenhouse Cultivation of Crops

Suthari Vijay Kumar[1], Charan Babu Ankela[2], Pooja Erapuram[3], Challa Dinesh Kumar Reddy[1], Akarapu Sagar[4], Shreedevi Jinnur[5], Rashadin Md Bodeul Ddoza[6], Santanu Kundu[7]*, Maksud Hasan Shah[8]*, and Akbar Hossain[9]

[1]Uttar Banga Krishi Viswavidyalaya, Cooch Behar, West Bengal, India; [2]IARI Mega University, Hyderabad Hub, ICAR–Central Research Institute for Dryland Agriculture, Hyderabad, Telangana, India; [3]Vasantrao Naik Marathwada Krishi Vidyapeeth, Parbhani, Maharashtra, India; [4]IARI Mega University, ICAR-CAZRI, Jodhpur, Rajasthan, India; [5]University of Agricultural Sciences, Bengaluru, Karnataka, India; [6]Visva-Bharati University, Shantiniketan, Birbhum, West Bengal, India; [7]ICAR-Indian Agricultural Research Institute, New Delhi, India; [8]Bidhan Chandra Krishi Viswavidyalaya, Nadia, West Bengal, India; [9]Bangladesh Wheat and Maize Research Institute, Dinajpur, Bangladesh

Abstract

This chapter examines the utilization of deep learning (DL) in greenhouse agriculture, emphasizing its capacity to augment productivity, optimizing resource allocation, and enhancing sustainability. DL utilizes data-driven algorithms to enable precise regulation of environmental factors, including temperature, humidity, and light, thereby ensuring optimal conditions for crop growth. The amalgamation of sensor networks and real-time data processing facilitates adaptive modifications to irrigation, nutrient distribution, and pest control, thereby reducing resource depletion and enhancing plant vitality. In order to monitor crop health, spot early indicators of pests or diseases, and take prompt action, advanced image recognition and anomaly detection models are essential. DL algorithms also support predictive modeling, which helps with strategic planning and climate change adaptation. DL facilitates the effective administration of greenhouse systems by automating intricate decision-making procedures, which helps to sustainably increase crop output in regulated settings.

Keywords: climate control, deep learning, greenhouse cultivation, resource management

12.1 Introduction

With the escalating global population, changing climate, and urbanization, it has become harder to imagine our future agricultural systems. According to Mok *et al.* (2020), these, in turn, have a major impact on global food security and sustainable resource use. However, a glimmer of hope in this field of worries is provided by technological developments, which have the capacity to completely transform the horticultural and agricultural sectors (Liu *et al.*, 2021). Thus, there is

*E-mail: santanukundu120@gmail.com or maksudhasanshah@gmail.com

 Deep Learning in Plant Science and Agricultural Biotechnology (Jen-Tsung Chen ed.)
DOI: 10.1079/9781836991120.0012

a targeted application of some state-of-the-art technology to build more resilient, productive, and sustainable farming ecosystems across the globe. The problem of sustainable availability of high-quality food is not limited to developing countries; wealthy countries are also facing this challenge. Present agricultural methods, which are mostly carried out in open farms, are clearly in deficit to satisfy the growing demand. According to WHO projections, food production must rise by 70% by 2050 to feed the world's expected 10 billion people (FAO, 2022), of whom 7 billion will reside in cities (FAO, 2023).

Investigating alternative production methods becomes essential to ensuring a sustainable food supply chain, not merely a choice. Controlled environment agriculture (CEA), which encompasses a range of techniques such as plant factories, high tunnels, greenhouses, vertical farms, and so on, is one viable substitute (Ojo and Zahid, 2022). Through regulated conditions, greenhouse farming has become an essential instrument for maintaining food security and stabilizing crop prices. Crops can be grown in a greenhouse to extend their growing season, protect them from weather and temperature fluctuations, and create a safe growing environment. To ensure that crops grow under the best possible conditions, modern greenhouses can also regulate environmental parameters such as humidity, temperature, radiation, and carbon dioxide (Vanthoor *et al.*, 2011; Ponce *et al.*, 2014). The use of precision agriculture in greenhouse environments has grown significantly, helping to successfully address and lessen the problems caused by natural disasters and climate change. Precision agriculture in greenhouses is becoming more feasible thanks to developments in greenhouse technology, including both hardware and software breakthroughs such as machine learning, cloud-based servers, and, most importantly, the Internet of Things (IoT) (Fig. 12.1). Cultivation in a greenhouse is affected by a number of parameters, including temperature, carbon dioxide concentrations, radiation, crop seed quality, soil and fertilization, and disease incidence (Hoogenboom *et al.*, 2000; Vanthoor *et al.*, 2011; Alhnait *et al.*, 2019). Building a suitable and efficient model that captures the link between different variables and the yield of the crop is not simple. Understanding the complexity of managing the microclimate in greenhouses, which are made up of extremely dynamic and non-linear systems, is essential.

Therefore, the development of microclimate prediction models becomes paramount for farmers.

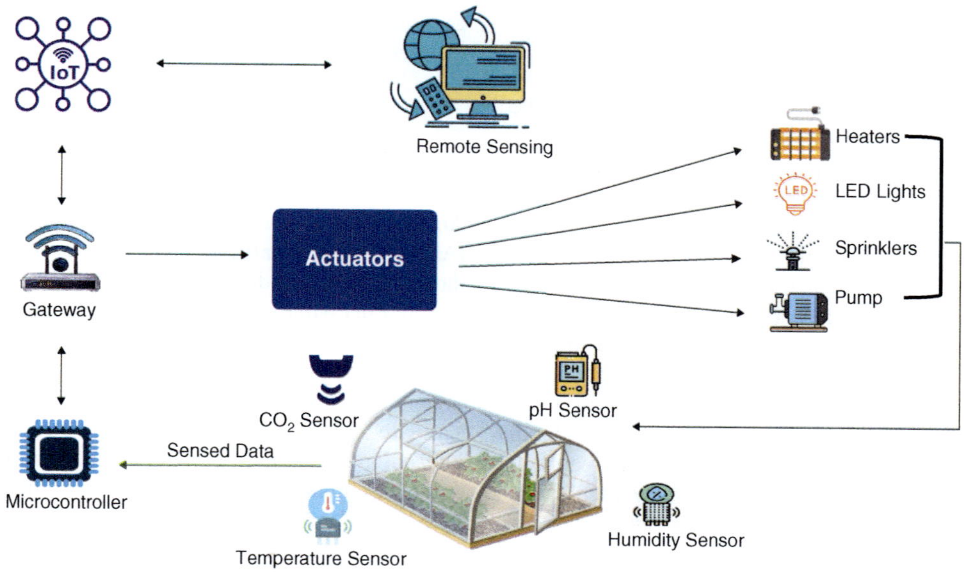

Fig. 12.1. Diagrammatic representation of an automated greenhouse cultivation using IoT. IoT, Internet of Things. (Author's own image.)

By enabling proactive greenhouse management, these models provide the best possible conditions for crops, which, in turn, increases agricultural productivity as a whole. Greenhouse farming has some obstacles, even though it has potential inside the CEA framework. The necessity for constant labor, complicated microclimate controls, high operating expenses, and frequent pest and disease monitoring makes economic sustainability a major challenge (Castro *et al.*, 2019; Rayhana *et al.*, 2020). The effectiveness and scalability of greenhouse farming may be hampered by these factors. Advanced technologies, such as deep learning (DL) and artificial intelligence (AI), must be integrated into these farming systems in order to solve these challenges. The operational features of greenhouse farming may be revolutionized by these state-of-the-art instruments. They can be used, for example, to better monitor the microenvironment and regulate the root zone, which will maximize plant growth and reduce resource waste (Ojo and Zahid, 2022). Furthermore, labor-intensive jobs can be automated with the help of AI and DL, improving overall operational efficiency. To fully utilize CEA in tackling the world's food security issues, it is imperative that AI and DL must be adopted and integrated into greenhouse farming. This is not only a strategic advantage.

DL has the capacity to streamline and improve the potential of farming practices. Using the immense data that can be collected using varied equipment, processing and analysis can be done efficiently. Such "big data" can be handled efficiently by DL algorithms, allowing for accurate and prompt crop management decision-making (Yang and Xu, 2021). Given these developments, it is important to keep in mind that large-scale farming is not the only application for these technologies. The use of these automation technologies can also be advantageous for small indoor farms, which need a lot of labor throughout the year. Long-term food security depends on this inclusive approach to agricultural technology implementation. We want to identify various DL models and their uses in greenhouse farming through this review. By highlighting these technologies, we aim to develop a secure food supply, boost horticultural output, and lead toward sustainability, while focusing prominently on their difficulties, opportunities, and prospects.

12.2 Basics of DL in Agriculture

DL falls under subsets of machine learning, which employ multilayered neural networks to handle tasks, such as regression, representation, and classification learning. DL is used to turn raw data into increasingly abstract and composite representations. It is derived from biological neuroscience and is based on arranging artificial neurons as sequential layers and "training" them to handle input data. The term "deep" indicates the application of numerous layers (LeCun *et al.*, 2015). The distributed communication nodes and information processing in biological systems, especially the human brain, inspired the design of early neural network architectures. However, existing neural networks are usually regarded as low-quality models for that purpose and are not designed to simulate the way an organism's brain functions (Trafton, 2022). The majority of contemporary DL prototypes are built on hierarchical neural networks, such as transformers and convolutional neural networks (CNNs), though they can further incorporate latent variables or propositional formulas arranged hierarchically within deep generative architectures, similar to the units in deep Boltzmann machines and deep belief networks (Bengio, 2009). In greenhouse management, recurrent neural networks (RNNs), CNNs, and generative adversarial networks (GANs) are a few well-known DL network models.

12.2.1 Recurrent neural network

An RNN is a kind of artificial neural network (ANN) created primarily for analyzing time series data. RNNs process data in a sequential, step-by-step manner while keeping an internal state that stores information from previous steps. The connections between nodes form a directed graph over a period of time, allowing the network to simulate variable temporal behavior. RNNs use internal information to grasp and analyze input sequences, making them particularly useful for time-dependent tasks. LSTM is a customized RNN architecture that is commonly used in DL. LSTM networks have feedback connections that allow them to handle a whole series of data points rather than individual ones. Because of their ability to store and exploit knowledge over time, LSTMs are especially effective at tasks involving sequential data, contributing to their reputation as one of the most influential and commercially successful advances in AI. LSTM networks are particularly well-suited for categorizing, analyzing, and predicting time series data, especially when key events are separated by unforeseen delays. They were created primarily to address the issues of exploding and vanishing gradients, which frequently impact standard RNNs. LSTMs may learn and store knowledge from events that occurred thousands or even millions of time steps ago, making them ideal for tasks that need long-term memory (Fig. 12.2).

12.2.2 Convolutional neural network

CNNs are feedforward neural networks that learn features by optimizing filters (or kernels). These DL networks have been employed to process and anticipate a variety of data types, including pictures, audio, and text (LeCun *et al.*, 2015). They are the default option in DL-based methods for computer vision (LeCun *et al.*, 1989) and visual data analysis, and have recently been superseded in certain cases by novel DL technologies, such as transformers. The regularization that results from utilizing shared weights over fewer connections prevents disappearing and exploding gradients, which were observed during backpropagation in prior neural networks (Venkatesan and Li, 2017; Balas *et al.*, 2020). For processing a 100 × 100-pixel image, each unit within the dense layer requires 10,000 weights. However, employing cascaded convolution kernels (Zhang *et al.*, 2019; Chervyakov *et al.*, 2020) requires just 25 weights for each layer to process 5 × 5-sized tiles (Homma *et al.*, 1987; Heravi, 2018). Upper-layer features are captured from more extensive frames than lower-layer traits. CNNs have a variety of applications, including video and image identification, recommender systems, image segmentation, natural language processing, image classification, medical analysis, brain imaging, monetary time streams, and computer interfaces.

12.2.3 Generative adversarial networks

A GAN is a kind of machine learning architecture that is commonly used to approach generative AI. Ian Goodfellow created the concept in 2014 (Goodfellow *et al.*, 2014). Here, two neural networks compete against each other in a zero-sum game. This approach leverages training data; it generates novel data with the same attributes, such as the source. For instance, a GAN trained on image data is capable of generating new images that appear to human viewers to be at least superficially authentic, with various natural features. GANs were first introduced as a creative framework for unsupervised methods, yet they have proven beneficial for reinforcement learning (Ho and Ermon, 2016), hybrid supervision (Salimans *et al.*, 2016), and fully supervised learning (Isola *et al.*, 2017). Training is carried out through the discriminator, which is a neural network that assesses the accuracy of the input and is dynamically changed during the training

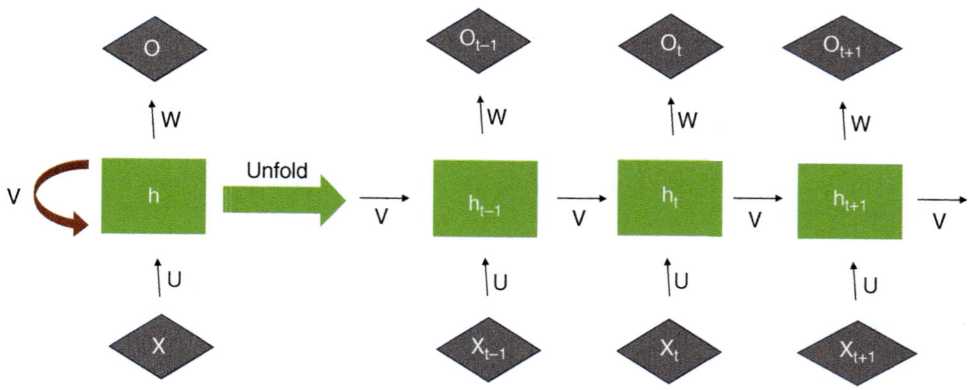

Fig. 12.2. Compressed and unfolded RNN. RNNs, recurrent neural networks. (Author's own image.)

process. This implies that the generator is not designed to decrease the gap with a particular image but instead to deceive its discriminator. This permits the model to train in an unsupervised framework (Fig. 12.3).

12.3 Applications of DL in Greenhouse Cultivation

12.3.1 Climate and environment control

To maintain optimal conditions in a greenhouse throughout every stage of plant growth, it is essential to have a deep understanding of the microclimate and its key influencing factors. By carefully managing elements, such as temperature, humidity, and carbon dioxide (CO_2) levels, the greenhouse environment can be optimized to significantly boost crop yields. For example, in colder climates, achieving higher indoor air temperatures by enhancing the greenhouse effect or implementing efficient heating systems is essential for promoting maximum plant growth.

Wireless sensor networks (WSNs) are effective in monitoring environmental conditions. In the greenhouse, WSNs are deployed in various areas to monitor temperature, humidity, soil moisture, pH, and air pressure (Table 12.1). They provide the essential communication infrastructure that allows sensors to transmit data on parameters, such as temperature, humidity, and carbon dioxide concentration. WSNs consist of numerous sensor nodes—compact devices with limited computational power, memory, and energy resources. The nodes are strategically deployed throughout the greenhouse, and each one is equipped with components for data processing, sensing, and wireless communication.

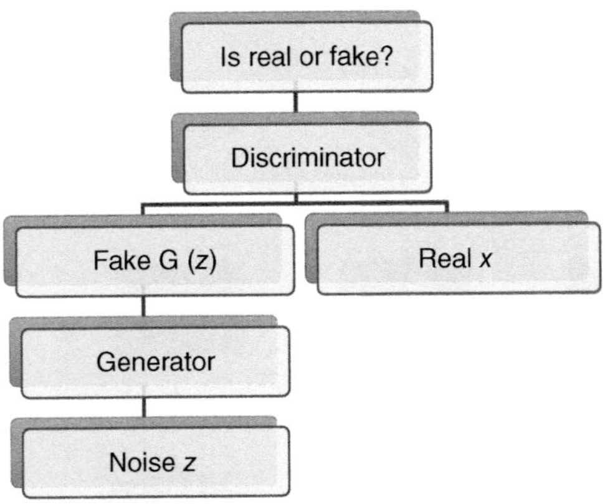

Fig. 12.3. An illustration of how a GAN works. GANs, generative adversarial networks. (Author's own image.)

The gathered data are input into an RNN that employs LSTM units to make accurate predictions. Based on these predictions, the greenhouse temperature can be adjusted to the desired level. This is achieved by activating actuators that control environmental systems, such as opening windows, turning on heaters, or shutting off fans to maintain optimal growing conditions. Yu *et al.* (2016) designed a temperature prediction framework utilizing the least squares support vector machine (LSSVM), incorporating an optimization algorithm to fine-tune the model parameters. Jiang *et al.* (2015) implemented a WSN-based system to forecast carbon dioxide concentration within a greenhouse, utilizing a backpropagation neural network (BPNN) not only to predict CO_2 levels but also to estimate the photosynthetic rate of tomato plants. Zou *et al.* (2017) employed a convex bidirectional extreme learning machine (CB-ELM) to predict humidity and temperature inside the greenhouse environment.

12.3.2 Pest and disease detection

The image monitoring system is a wireless system that consists of multiple imaging nodes that communicate over Wi-Fi. Each node includes a Raspberry Pi 3B+ embedded system coupled with a Raspberry Pi v2 camera. It is positioned in front of an A5 sticky paper trap (21.0 cm × 14.8 cm), which is held in place by an acrylic board (Rustia *et al.*, 2021. Pictures are captured at a 3280 × 2464-pixel resolution and transmitted to the server every 10 min, from morning to evening. When the system is configured, the number of wireless imaging devices is determined according to the farm's area and crop placement, based on recommendations from the farm owners.

As the wireless system captures the pictures, they are transmitted to a remote server via the Internet for processing. These images undergo analysis through an insect recognition and detection algorithm, which consists of three main

Table 12.1. Various kinds of IoT-based greenhouse sensors. Author's own table.

Sl. no.	Kind of sensors	Sensor details
1	Light sensors	1. Pyranometer sensor 2. Photosynthetically active radiation sensor 3. Light-dependent resistor sensor 4. Phototransistor sensor 5. Photodiode sensor
2	Temperature sensors	1. Silicon sensor 2. Thermocouple sensor 3. Thermistor sensor 4. Resistance temperature detector sensor
3	Soil moisture sensors	1. Capacitive sensor 2. Tensiometer sensor 3. Time-domain reflectometry sensor 4. Granular-matrix sensor
4	CO_2 sensors	1. Metal oxide-based CO_2 sensor 2. Electrochemical-based CO_2 sensor 3. Optical-based CO_2 sensor
5	Humidity sensors	1. Resistive sensor 2. Thermal-humidity sensor 3. Capacitive sensor

steps: object detection, insect-versus-non-insect classification, and multiclass classification for species identification.

Object detection

In the first stage, the object detection module identifies and outputs the bounding box coordinates of the objects present on the sticky traps. These coordinates are then used to isolate and examine each object. The subsequent classification stage determines whether each object is an insect or not. Non-insect objects are discarded, while detected insects proceed to the final classification stage, where they are identified by species. This algorithm enables the extraction of useful data, such as the temporal trends and spatial distribution of various insect species. Designed specifically for sequential and long-term monitoring, such systems can generate valuable insights over time for crop and pest management.

Based on various studies involving counting the number of pests automatically, one source of error is the effect of external factors, such as the presence of non-insect objects in the pictures and varying lighting conditions (Wen and Guyer, 2012; Zhong *et al.*, 2018). To minimize mistakes caused by variations in object scale, color, and size, this study employs a CNN-based object detector. Unlike traditional methods, such as background subtraction or color clustering, CNN-based detectors offer greater robustness and accuracy, making them less sensitive to these factors.

CNN-BASED OBJECT DETECTOR. The implementation of this algorithm begins with the pictures undergoing a series of preprocessing steps. Initially, the original image is resized (3200 × 2400 pixels) using cubic interpolation. This resizing ensures that the image can be evenly divided into smaller sections. Next, the image is tiled into 12 segments at 840 × 840 pixels. The tiling technique, used in object detection, improves the accuracy of detection by breaking down high-resolution images into smaller, more manageable parts—particularly useful for detecting small-sized objects (Huang *et al.*, 2018; Unel *et al.*, 2019).

The chosen tile with the resolution of 840 × 840 ensures that the image is downscaled by half before entering the object detector, helping to preserve important details and minimize distortion. After tiling, each picture at 840 × 840 is further reduced to 416 × 416 pixels to match the input size required by the object detection model. This final resizing to a square format ($n \times n$) helps maintain image consistency and improves the model's detection performance (Ren *et al.*, 2017).

TINY YOLO V3 OBJECT DETECTOR MODEL. This is a lightweight variant of the YOLO v3 model that uses fewer convolutional layers, which significantly enhances processing speed while maintaining reliable detection performance. This makes it well-suited for applications involving less complex images, such as those captured from sticky paper traps. Objects in the image are detected by splitting the pictures into a grid and producing bounding boxes, which are associated with class probabilities for each object.

Insect-vs-non-insect classification

The coordinates of the bounding box, which are obtained from the object detector, are used to crop objects from the real image for further classification. Each cropped image is then resized to average pixel dimensions using cubic interpolation, a size chosen based on the average dimensions of insects found in the images. This resizing method helps to minimize distortion. The cropped object pictures are used as input for the CNN-based image classifier.

CNN IMAGE CLASSIFIER. The CNN classifier is a continuous neural network designed specifically for general-purpose classification tasks. It usually comprises three layers (CL1, CL2, and CL3), each followed by a max pooling layer, which extracts low-, medium-, and high-level features from the input RGB pictures. These extracted features are then carried through a fully connected layer (FCL) with ReLU activation to capture deeper representations. Finally, a softmax layer generates the prediction probabilities for every class. The class of the image is determined by comparing these probabilities against a predefined classification threshold. The insect-vs-non-insect design assigns the output class corresponding to the highest classification probability, *P*. If the probability of the object being an insect is higher, it is passed on to a multiclass insect classifier for further identification; otherwise, it is disregarded.

Multiclass insect classification

This model categorizes insect objects into flies, gnats (Diptera: Sciaridae), fruit flies (Diptera: Drosophilidae), whiteflies (Hemiptera: Aleyrodidae), and thrips (Thysanoptera: Thripidae). These five groups represent the most prevalent pests seen in greenhouses. The multiclass insect classifier assigns an output class based on the highest predicted probability (*P*). A classification probability threshold (TH) is applied to filter the results—if the highest *P* for a classified insect falls below the threshold, it is categorized as "other" (Geng *et al.*, 2018). Xu *et al.* (2022) tackled the problems of diseases such as melon leaf detection by using a newly pruned model of YOLO v5s in combination with ShuffleNet v2.

12.3.3 Crop growth prediction

Traditionally, crop monitoring has depended largely on human observation, often resulting in delayed and inaccurate assessments. In precision agriculture, especially in greenhouses, consistent and accurate monitoring across different stages of growth is essential. Combining computer vision with the management of greenhouses enables high-precision and real-time monitoring. This technology enables early detection of subtle changes in crop health, allowing for timely interventions that promote optimal plant growth by identifying nutrient deficiencies at an earlier stage than manual inspections. Researchers have employed techniques such as YOLO, CNNs, and other object detection designs to precisely track indicators and growth stages.

CNN model for monitoring crop growth patterns

CNNs are a kind of feedforward ANN, belonging to the broader family of multilayered neural networks. They have proven particularly effective in computer vision tasks and are widely used for analyzing large-scale datasets. The CNN model is made up of four pooling layers, five convolutional layers, and an FCL. It takes digital images as input.

Zhang *et al.* (2020) utilized the strengths of CNNs for analyzing digital pictures of lettuce. The study showcased how CNNs can effectively monitor growth indicators such as leaf dry weight (LDW), leaf area, and leaf fresh weight (LFW), producing outcomes that work better than traditional methods. This model took digital pictures of lettuce with a 128 × 128 × 3 (width × height × channel) size as input. These layers used 5 × 5 kernels for extracting features, with the number of kernels in the five layers being 32, 64, 128, 216, and 512. Pooling layers used 2 × 2 kernels with a width of 2, reducing the size of the feature maps by half. The FCL contained three hidden neurons, each corresponding to one of the output traits: LFW and LDW. This configuration enabled CNN to predict all three growth traits at once. To prevent overfitting, a dropout rate of 0.5 was applied. The model optimization was done using stochastic gradient descent (SGD) with an initial learning rate of 0.001, which was decreased by a factor of 0.1 every 20 epochs. Training used a mini-batch size of 128, and the maximum number of epochs was set to 300. Lee *et al.* (2022) conducted a study utilizing IoT clip-type cameras to observe tomato growth and development. They employed DL object detection to identify both flower blooming and immature fruit development. By integrating the bloom dates with temperature forecasts, their system was able to predict harvest times with a mean margin of error of just 2 days.

12.3.4 Crop yield estimation

A deep neural network approach for the prediction of yield using RNN and TCN

A new DL approach has been developed for predicting yields in greenhouses by combining two advanced models for temporal sequence analysis: the RNN and the Temporal Convolutional Network (TCN). This hybrid architecture is specifically designed to forecast future crop yields using historical data on greenhouse parameters such as temperature, humidity, CO_2 levels, and radiation, in addition to previous yield records. Extensive testing on diverse datasets collected from multiple greenhouses over different time frames confirms the effectiveness of the RNN-TCN model. The proposed method consistently outperforms conventional machine learning techniques and other commonly used DL models, achieving notably lower Root Mean Square Error (RMSE) values in yield prediction tasks.

The methodology of the RNN-TCN hybrid model includes four parts:

1. Data normalization

To forecast crop yield over a given period, a time series dataset combining historical yield and environmental conditions is used. A temporal sequence of length N, represented by $x_{t-N}, ..., x_t$, is used as the network input. Each vector x_t in the sequence represents the recorded values at time step *t*, including crop yield (g/m^2), greenhouse temperature (°C), carbon dioxide concentration (ppm), humidity deficit (g/kg), radiation (W/m^2), and relative humidity (%). Prior to inputting the data into the network, a normalization step is performed to scale each factor to a standard range of [0, 1]. Here, x^i_t represents the i-th factor at time step *t*, and x^i_{min} and x^i_{max} denote the minimum and maximum values of that feature, respectively. Through this normalization, all input features are transformed into a uniform scale between 0 and 1, facilitating more stable and effective neural network training.

2. Recurrent neural network

The temporal data are the input source for the RNN. RNNs are widely used for handling sequential data due to their ability to model temporal dependencies and extract key features for tasks such as sequence classification and temporal prediction. The RNN is initially employed to extract meaningful features from the normalized temporal data for further processing. However, traditional RNNs suffer from issues such as gradient vanishing and explosion, which significantly hinder their performance, especially with long sequences. To address these challenges, an improved model incorporating LSTM units, referred to as LSTM–RNN, is commonly adopted. The LSTM–RNN is composed of a number of LSTM units, each performing a series of arithmetic operations.

3. Temporal convolutional network

The TCN utilizes a hierarchical structure of temporal convolutions to effectively capture representative features across various temporal scales within the input sequence. The dilated TCN architecture comprises multiple residual blocks, each containing several dilated convolution layers. Within these layers, dilated causal temporal convolution operations are applied to model long-range temporal dependencies without compromising the sequence order.

Within each residual block, an extra 1D convolution task is applied to match the input dimensions with the output dimensions of the dilated causal convolution layers, enabling their outputs to be added together. The output of each residual block serves as the input to the next, forming a sequential processing chain. The final outcome from the last block is then flattened and passed through an FCL, which produces the overall yield forecast result.

4. Fully connected layer
The outcome generated by the TCN component is initially flattened into a vector and then directed into an FCL, which produces the final productivity prediction. This FC layer consists of a single output unit and uses a ReLU activation function.

The integration of two state-of-the-art deep neural network architectures, LSTM–RNN and TCN, for temporal sequence modeling in greenhouse crops is used for yield prediction. Given a temporal input sequence that includes historical yield data and climatic parameters over a specific time period, an LSTM–RNN is first used to extract meaningful feature representations. These features are then passed through a series of residual blocks in the TCN, which further refine the temporal features for the final yield prediction task. The layered design allows for more effective feature extraction by leveraging both short- and long-range temporal dependencies, resulting in more accurate predictions. Maji *et al.* (2022) reported that wheat productivity estimation through SlypNet, a DL model that merges Mask R-CNN with U-Net, captures the morphological traits of wheat.

12.3.5 Seed quality analysis

Quality seed is a basic and important input for sustainable agriculture. Seed quality directly influences crop yield and profitability, as it affects crop emergence and population, which ultimately impact yield per unit area. Traditional seed quality analysis methods, which are done manually, are laborious, time-consuming, and not accurate. Hence, there is a need to shift to the latest technologies to assess seed quality.

Seed quality analysis has been revolutionized by DL technologies, especially CNNs, which allow for automated, precise, and bulk seed analysis. Images of seeds can be processed by these algorithms to identify diseased or non-viable seeds and even predict the germination potential of seeds. DL models assist in selecting good-quality seeds for sowing, which enhances germination rate and early plant vigor in greenhouse cultivation. Techniques such as hyperspectral imaging in combination with DL improve seed analysis by identifying internal seed characteristics (Ahmed *et al.*, 2024).

Hyper Seed is a state-of-the-art end-to-end technology specialized in providing hyperspectral data for seeds. When applied to rice seeds, the use of a 3D CNN outperformed traditional approaches such as the support vector machine model, achieving an accuracy rate of 97.5% (Goa *et al.*, 2021). To automate seed classification and quality testing, Kundu *et al.* (2021) proposed a DL-based "Mixed Cropping Seed Classifier and Quality Tester (MCSCQT)" system with the YOLO v5 algorithm. The system showed an accuracy of 99% in the classification of seeds into two groups, that is, corn and bajra. It is also effective in the segregation of diseased and healthy seeds of crops (corn). Nehoshtan (2021) proposed the first generic germination prediction model, which uses DL and RGB picture data to help in the classification of seeds based on seed usability and germinability. Nagar *et al.* (2021) proposed a novel computer vision-based automated system used for testing the quality of corn seeds. A BAL-based UI tool is built for labeling, and GANs (BigGAN) address the class imbalance by generating more image classes with the small dataset. Similarly, this tool helps in the quality testing of seeds of various vegetables by decreasing human intervention. A model by Hidayat *et al.* (2023) with Google Collaboratory determines the quality of paddy seeds based on digital image processing techniques, with image classification consisting of one rice seed object. It has an accuracy of 88%, a recall of 95%, and a precision of 93%. In greenhouse cultivation, the integration of DL techniques for seed quality analysis has shown significant ability to improve agricultural productivity and precision. The significance of DL models in greenhouse cultivation is increasing rapidly through the development of improved algorithms with larger datasets, which ultimately help promote high-efficiency and sustainable agricultural practices.

12.4 Challenges

DL has revolutionized greenhouse cultivation by enabling automatic monitoring of plant health, growth stages, pest infestation, and environmental conditions. But despite its potential, a number of challenges limit its widespread and seamless implementation.

12.4.1 Unavailability of standardized large-scale datasets

The lack of high-quality labeled data in agriculture is a significant obstacle that can be overcome by establishing large labeled datasets for DL model training and by creating novel techniques for data collection (Albahar, 2024). It takes a lot of work, effort, and specialized knowledge to create labeled datasets for purposes such as growth measurement or disease diagnosis.

12.4.2 Data variability

Variability in greenhouse conditions, including light, humidity, and background clutter, can make it challenging for computer vision models to operate consistently. It is challenging for machine learning models, particularly DL systems, to generalize efficiently and retain high accuracy across various contexts due to this high degree of unpredictability. Accurate prediction and model training are further complicated by the natural differences in plant appearance caused by various species, growth phases, or stressors.

12.4.3 Requirement for optimized and specialized models

Variations in light intensity, plant occlusion, seasonal changes, and subtle differences between healthy and stressed plants demand models that are carefully optimized for agricultural environments. Specialized models need to be lightweight enough for deployment on low-power devices yet powerful enough to maintain high accuracy. They should also be resilient to noise, capable of working with incomplete data, and adaptable to different crops and growth stages (Kamilaris and Prenafeta, 2018). Building such models requires deep domain knowledge, careful tuning, and custom architectures that balance efficiency, performance, and practical use in greenhouse systems.

12.4.4 Occlusion issues

Important components of dense plant canopies may be partially or totally masked from the camera by leaves, stems, fruits, or even supporting structures. Models find it challenging to precisely identify plant health issues, count fruits, quantify growth, or evaluate maturation phases due to this visual occlusion. Occlusion can therefore result in misleading readings, missed detections, and inconsistent monitoring results. Advanced methods such as multiview imaging, 3D reconstruction, or the use of depth cameras, which can record hidden features from several angles, are necessary to handle occlusion successfully (Charisis and Argyropoulos, 2024). Furthermore, using examples of occluded data to train models can enhance their capacity to infer missing information while maintaining high accuracy in actual greenhouse conditions.

12.4.5 Lack of multidisciplinary and specialized talent

Personnel should have proficiency in computer vision and AI, as well as knowledge of agronomy, plant physiology, environmental control systems, and sensor integration. However, professionals with extensive experience in all these fields are quite rare. While agricultural specialists may not have received training in DL model creation or data science, computer scientists might not understand the biological complexities of plant growth. Collaborative efforts, knowledge transfer, skill-building, designing user-friendly interfaces, and stronger industry–academia alliances are required to address this skills deficit and develop experts capable of successfully designing, implementing, and overseeing integrated smart agriculture systems (Araujo *et al.*, 2023).

12.5 Limitations

DL models, particularly those based on architectures such as CNNs and transformer models, require substantial computational memory, power, and energy to operate productively. However, most greenhouses depend on low-power, low-cost edge devices that are unable to support such heavy computational loads. This drawback affects the ability to perform real-time processing, leading to delays in decision-making and system responsiveness. Additionally, frequent transmission of large amounts of sensor data to remote servers for processing can lead to bandwidth problems and increase the cost of operation (Liakos *et al.*, 2018).

Overcoming these challenges requires the development of lightweight, efficient models optimized for edge computing, the use of model compression techniques such as pruning and quantization, and innovations in hardware accelerators specifically designed for agricultural IoT applications. Efficient resource management is essential to make smart greenhouse technologies affordable, scalable, and practical for farming operations. Growth Net, a lightweight CNN, dramatically lowers computational complexity and parameter counts while enabling effective real-time processing and feature extraction (Li *et al.*, 2024). The AgroTec 4.0 system, which is based on edge computing and IoT technologies, has been shown to be a useful instrument for managing environmental factors and optimizing irrigation management. Improvements in irrigation efficiency and environmental data collection were made possible by the integration of inexpensive sensors and control modules, which also helped to lower operating costs and enhance sustainable agriculture practices (Abdo-Peralta *et al.*, 2024).

There is significant work remaining to develop computer vision systems that are proficient at handling the subtleties of greenhouse environments. These problems include restricted multimodality, task-specific models, computational limits, and lack of appropriate labeled data. Although significant progress has been made, achieving the full potential of computer vision for smart greenhouse cultivation will require overcoming these obstacles through model optimization, data production, sensor fusion integration, and effective designs.

12.6 Conclusion

A significant step toward highly sustainable and precise farming methods is the application of DL. DL algorithms give ideal conditions for growth by allowing for control of critical environmental factors such as humidity, temperature, and light intensity, which increases the quantity and quality of the crop. By integrating sensor networks, adaptive management, and real-time data collection, producers may enhance plant health and minimize resource waste by making well-informed changes to pest control tactics, fertilizer distribution, and watering schedules. In order to detect biotic stressors quickly and execute timely treatments that protect production, advanced recognition of images and anomaly detection algorithms are essential. Additionally, long-term planning is made simple by DL-driven predictive modeling capabilities, which enable greenhouse operators to adjust to the effects of climate change and market volatility.

DL enhances greenhouse operations' efficiency and resilience by automating intricate decision-making processes, which reduces human error and labor reliance. DL is an important technology for the future of controlled environment agriculture due to its capacity to integrate large quantities of varied data into useful and logical conclusions. Finally, utilizing DL not only improves profitability and production but also encourages sustainable resource usage, highlighting that greenhouse farming can fulfill the rising global need for food in an ecologically responsible way.

References

Abdo-Peralta, P., García-Pumagualle, C., Carrera-Silva, K., Frey, C., Rosero-Erazo, C.R. *et al.* (2024) Implementation of an enhanced edge computing system for the optimization of strawberry crop in greenhouses: a smart agriculture approach. *Agronomy* 14(12), 3030.

Ahmed, M.T., Monjur, O. and Kamruzzaman, M. (2024) Deep learning-based hyperspectral image reconstruction for quality assessment of agro-product. *Journal of Food Engineering* 382, 112223.

Albahar, M. (2024) A survey on deep learning and its impact on agriculture: challenges and opportunities. *Agriculture* 13(3), 540.

Alhnait, B., Pearson, S., Leontidis, G., and Kollias, S. (2019) Using deep learning to predict plant growth and yield in greenhouse environments. *arXiv preprint* arXiv:1907.00624.

Araujo, S.O., Peres, R.S., Ramalho, J.C., Lidon, F. and Barata, J. (2023) Machine learning applications in agriculture: current trends, challenges, and future perspectives. *Agronomy* 13(12), 2976.

Balas, V.E., Kumar, R. and Srivastava, R. (Eds.) (2020) *Recent trends and advances in artificial intelligence and internet of things*.

Bengio, Y. (2009). Learning deep architectures for AI. *Foundations and Trends in Machine Learning* 2(1), 1–127.

Castro, A.J., Lopez-Rodriguez, M.D., Giagnocavo, C., Gimenez, M., Cespedes, L., La Calle, A., Gallardo, M., Pumares, P., Cabello, J., Rodríguez, E., and Ucles, D. (2019) Six collective challenges for sustainability of Almería greenhouse horticulture. *International Journal of Environmental Research and Public Health* 16(21), 4097.

Charisis, C. and Argyropoulos, D. (2024) Deep learning-based instance segmentation architectures in agriculture: a review of the scopes and challenges. *Smart Agricultural Technology* 8, 100448.

Chervyakov, N.I., Lyakhov, P.A., Deryabin, M.A., Nagornov, N.N., Valueva, M.V. and Valuev, G.V. (2020) Residue number system-based solution for reducing the hardware cost of a convolutional neural network. *Neurocomputing* 407, 439–453.

FAO, IFAD, UNICEF, WFP, and WHO. (2022) *The State of Food Security and Nutrition in the World*. FAO, Rome, Italy.

FAO (2023) *The State of Food Security and Nutrition in the World*. FAO, Rome, Italy.

Geng, C., Huang, S.-J. and Chen, S. (2018) Recent advances in open set recognition: a survey. arXiv preprint arXiv:1811.08581.

Gao, D., Li, M., Zhang, J., Song, D., Sun, H., Qiao, L. and Zhao, R., (2021) Improvement of chlorophyll content estimation on maize leaf by vein removal in hyperspectral image. *Computers and Electronics in Agriculture*, 184.

Goodfellow, I.J., Pouget-Abadie, J., Mirza, M., Xu, B., Warde-Farley, D., Ozair, S., Courville, A. and Bengio, Y. (2014) Generative adversarial nets. *Advances in Neural Information Processing Systems* 27.

Heravi, E.J. (2018) *Guide to convolutional neural networks: A practical application to traffic-sign detection and classification*.

Hidayat, S.S., Rahmawati, D., Prabowo, M.C.A., Triyono, L., and Putri, F.T. (2023) Determining the rice seeds quality using convolutional neural network. *JOIV: International Journal on Informatics Visualization* 7(2), 527–534.

Ho, J. and Ermon, S. (2016). Generative adversarial imitation learning. *Advances in Neural Information Processing Systems* 29.

Homma, T., Atlas, L.E. and Marks, R.J. (1987) An artificial neural network for spatiotemporal: application to phoneme classification. In *Proceedings of the 1st International Conference on Neural Information Processing Systems*, 31–40.

Hoogenboom, G. (2000) Contribution of agrometeorology to the simulation of crop production and its applications. *Agricultural and Forest Meteorology* 103, 137–157.

Huang, B., Reichman, D., Collins, L.M., Bradbury, K. and Malof, J.M. (2018) Tiling and stitching segmentation output for remote sensing: Basic challenges and recommendations. arXiv preprint arXiv:1805.12219.

Isola, P., Zhu, J.Y., Zhou, T., and Efros, A.A. (2017). Image-to-image translation with conditional adversarial networks. In *Proceedings of the IEEE Conference on Computer Vision and Pattern Recognition*, 1125–1134.

Jain, L.P., Scheirer, W.J. and Boult, T.E. (2014) Multi-class open set recognition using probability of inclusion. In: European Conference on Computer Vision. Springer, Zurich, Switzerland.

Jiang, Y., Li, T., Zhang, M., Sha, S. and Ji, Y. (2015) WSN-based control system of CO_2 concentration in greenhouse. *Intelligent Automation and Soft Computing* 21, 285–294.

Kamilaris, A. and Prenafeta-Boldu, F.X. (2018) Deep learning in agriculture: a survey. *Computers and Electronics in Agriculture* 147, 70–90.

Kundu, N., Rani, G. and Dhaka, V.S. (2021) Seeds classification and quality testing using deep learning and YOLO v5. In *Proceedings of the International Conference on Data Science, Machine Learning and Artificial Intelligence*, 153–160.

LeCun, Y., Bengio, Y. and Hinton, G. (2015) Deep learning. *Nature* 521(7553), 436–444.

LeCun, Y., Boser, B., Denker, J.S., Henderson, D., Howard, R.E., Hubbard, W. and Jackel, L.D. (1989) Back propagation applied to handwritten zip code recognition. *Neural Computation* 1(4), 541–551.

Lee, U., Islam, M.P., Kochi, N., Tokuda, K., Nakano, Y., Naito, H., Kawasaki, Y., Ota, T., Sugiyama, T. and Ahn, D.H. (2022) An automated, clip-type, small internet of things camera-based tomato flower and fruit monitoring and harvest prediction system. *Sensors* 22(7), 2456.

Li, Z., Wang, J., Gao, G., Lei, Y., Zhao, C., Wang, Y. and Li, Q. (2024) SGSNet: a lightweight deep learning model for strawberry growth stage detection. *Frontiers in Plant Science* 15, 1491706.

Liakos, K.G., Busato, P., Moshou, D., Pearson, S. and Bochtis, D. (2018) Machine learning in agriculture: a review. *Sensors* 18(8), 2674.

Liu, Y., Ma, X., Shu, L., Hancke, G.P., and Abu-Mahfouz, A.M. (2020) From industry 4.0 to agriculture 4.0: Current status, enabling technologies, and research challenges. *IEEE Transactions on Industrial Informatics* 17(6), 4322–4334.

Maji, A.K., Marwaha, S., Kumar, S., Arora, A., Chinnusamy, V. and Islam, S. (2022) SlypNet: spikelet-based yield prediction of wheat using advanced plant phenotyping and computer vision techniques. *Frontiers in Plant Science* 13. DOI: 10.3389/fpls.2022.954128.

Mok, W.K., Tan, Y.X. and Chen, W.N. (2020) Technology innovations for food security in Singapore: a case study of future food systems for an increasingly natural resource-scarce world. *Trends in Food Science and Technology* 102, 155–168.

Nagar, S., Pani, P., Nair, R. and Varma, G. (2021) Automated Seed Quality Testing System Using GAN and Active Learning. In *International Conference on Pattern Recognition and Machine Intelligence*, 509–519.

Nehoshtan, Y., Carmon, E., Yaniv, O., Ayal, S. and Rotem, O. (2021). Robust seed germination prediction using deep learning and RGB image data. *Scientific Reports* 11(1), 22030.

Ojo, M.O. and Zahid, A. (2022) Deep learning in controlled environment agriculture: A review of recent advancements, challenges and prospects. *Sensors* 22(20), 7965.

Ponce, P., Molina, A., Cepeda, P., Lugo, E. and MacCleery, B. (2014) *Greenhouse Design and Control*. CRC Press: Boca Raton, FL, USA.

Rayhana, R., Xiao, G. and Liu, Z. (2020) Internet of things empowered smart greenhouse farming. *IEEE Journal of Radio Frequency Identification* 4(3), 195–211.

Ren, S., He, K., Girshick, R. and Sun, J. (2017) Faster R-CNN: towards real-time object detection with region proposal networks. *IEEE Transactions on Pattern Analysis and Machine Intelligence* 39(6), 1137–1149. DOI: 10.1109/TPAMI.2016.2577031.

Rustia, D.J.A., Chao, J.J., Chiu, L.Y., Wu, Y.F., Chung, J.Y., Hsu, J.C. and Lin, T.T. (2021) Automatic greenhouse insect pest detection and recognition based on a cascaded deep learning classification method. *Journal of Applied Entomology* 145(3), 206–222.

Salimans, T., Goodfellow, I., Zaremba, W., Cheung, V., Radford, A. and Chen, X. (2016) Improved techniques for training GANs. *Advances in Neural Information Processing Systems* 29.

Trafton, A. (2022) Study urges caution when comparing neural networks to the brain. *MIT News*. Massachusetts Institute of Technology.

Unel, F.O., Ozkalayci, B.O. and Cigla, C. (2019) The power of tiling for small object detection. In: Conference on Computer Vision and Pattern Recognition. IEEE, Long Beach, CA, USA.

Vanthoor, B., De Visser, P., Stanghellini, C. and Van Henten, E. (2011) A methodology for model-based greenhouse design: description and validation of a tomato yield model. *Biosystems Engineering* 110, 378–395.

Vanthoor, B., Stanghellini, C., Van Henten, E., and De Visser, P. (2011) A methodology for model-based greenhouse design: description and validation of a tomato yield model. *Biosystems Engineering* 110, 363–377.

Venkatesan, R., and Li, B. (2017) *Convolutional neural networks in visual computing: a concise guide*. CRC Press.

Wen, C. and Guyer, D. (2012) Image-based orchard insect automated identification and classification method. *Computers and Electronics in Agriculture* 89, 110–115. DOI: 10.1016/j.compag.2012.08.008.

Xu, Y., Chen, Q., Kong, S., Xing, L., Wang, Q and Cong, X. (2022) Real-time object detection method of melon leaf diseases under complex background in greenhouse. *Journal of Real-Time Image Processing* 19(5), 985–995.

Yang, B., and Xu, Y. (2021) Applications of deep-learning approaches in horticultural research: a review. *Horticulture Research* 8.

Yu, H., Chen, Y., Hassan, S.G. and Li, D. (2016) Prediction of the temperature in a Chinese solar greenhouse based on LSSVM optimized by improved PSO. *Computers and Electronics in Agriculture* 122, 94–102.

Zhang, L., Xu, Z., Xu, D., Ma, J., Chen, Y. and Fu, Z. (2020) Growth monitoring of greenhouse lettuce based on a convolutional neural network. *Horticulture Research* 7, 124. DOI: 10.1038/s41438-020-00345-6.

Zhang, Y., Soon, H.G., Ye, D., Fuh, J.Y.H. and Zhu, K. (2019) Powder-bed fusion process monitoring by machine vision with hybrid convolutional neural networks. *IEEE Transactions on Industrial Informatics* 16(9), 5769–5779.

Zhong, Y., Gao, J., Lei, Q. and Zhou, Y. (2018) A vision-based counting and recognition system for flying insects in intelligent agriculture. *Sensors* 18(5), 1489. DOI: 10.3390/s18051489.

Zou, W., Yao, F., Zhang, B., He, C. and Guan, Z. (2017) Verification and predicting temperature and humidity in a solar greenhouse based on convex bidirectional extreme learning machine algorithm. *Neurocomputing* 249, 72–85.

13 Deep Learning for Real-Time Plant Monitoring Systems

Mani Manoj[1], Esakkimuthu Balaji[2], Anandan Lokesh[3], Vadivelu Bharathi[4], Thamaraiselvan Nandhini Devi[5], Ramasamy Manikandan[6], Jeyabal Philomenathan Antony Prabhu[1*], and Arumugam Vijaya Anand[1*]

[1]Department of Human Genetics and Molecular Biology, Bharathiar University, Coimbatore, Tamil Nadu, India; [2]Prashant Fertility and Research Centre, Chennai, Tamil Nadu, India; [3]Department of Genetics, University of Madras, Chennai, Tamil Nadu, India; [4]Department of Biochemistry, Thanthai Periyar Government College, Tiruchirappalli, Tamil Nadu, India; [5]Department of Biomedical Engineering, Selvam College of Technology, Namakkal, Tamil Nadu, India; [6]Department of Biochemistry, Shrimati Indira Gandhi College, Tiruchirappalli, Tamil Nadu, India

Abstract

This chapter explores the integration of deep learning algorithms in real-time plant monitoring systems, a departure from traditional observational methods to intelligent, automatic monitoring. Using artificial intelligence (AI) models such as Vision Transformers, 3D convolutional neural networks, neural ordinary differential equations, and graph neural networks, advanced systems can interpret complex plant behaviors, stress responses, and growth patterns at high accuracy rates. Multimodal sensor fusion of RGB, hyperspectral, light detection and ranging (LiDAR), and bio-signals enable robust biotic and abiotic stress detection in diverse conditions. Advances in Edge AI, TinyML, and federated learning for scalable and field-level deployment are also explored. Synthetic plant simulators, digital twins, and explainable AI frameworks enhance the transparency and predictive ability of the systems. Ethical concerns, biodiversity protection, and sustainable AI are emphasized throughout. Real-world test cases and benchmarking provide some insight into real-world applications as well as future directions, such as AI-based closed-loop systems in space agriculture. The chapter formulates a revolutionary platform for intelligent, resilient, and sustainable plant ecosystem monitoring.

Keywords: deep learning, phenotyping, real-time plant monitoring, Vision Transformers

13.1 Next-Generation Real-Time Plant Monitoring: Concepts and Framework

The transition from traditional methodologies to artificial intelligence (AI)-enabled monitoring of crops in agriculture is a major change in agricultural practices, prompted by the demands of increased efficiency and accuracy. Traditional methods, which heavily depend on human intervention to carry out tasks such as measuring soil moisture and plant health, are time-consuming and less effective in large-scale agricultural production. The convergence of AI and Internet of

*E-mail: avahgmb@buc.edu.in or jtonybio@gmail.com

DOI: 10.1079/9781836991120.0013

Things (IoT) technologies has transformed this sector by providing automated, real-time monitoring systems. For example, IoT device-based systems can handle and monitor vital variables such as temperature, humidity, and soil moisture, thereby minimizing human intervention and maximizing efficiency (Ahmad *et al.*, 2021). The historical and technological transition of plant monitoring is explained in Fig. 13.1. AI-based solutions, especially computer vision and machine learning (ML)-based solutions, enable computerized measurement of plant growth and health through the analysis of images to extract features such as leaf area and plant height (Bhamare *et al.*, 2023). Moreover, these systems help detect disease in plants at an early stage using deep learning (DL) models, which are capable of processing aerial photos taken using drones, presenting a scalable and efficient alternative to human monitoring (Sornalakshmi *et al.*, 2022). Apart from disease detection, AI technologies bring strength to monitoring frameworks with real-time alert correlation and pattern detection, drastically enhancing incident detection, response time of the system, and operational effectiveness in large-scale agricultural monitoring (Dutta *et al.*, 2024). In short, the convergence of AI and ML in agriculture not only revolutionizes plant health monitoring but also enforces the principle of precision agriculture,

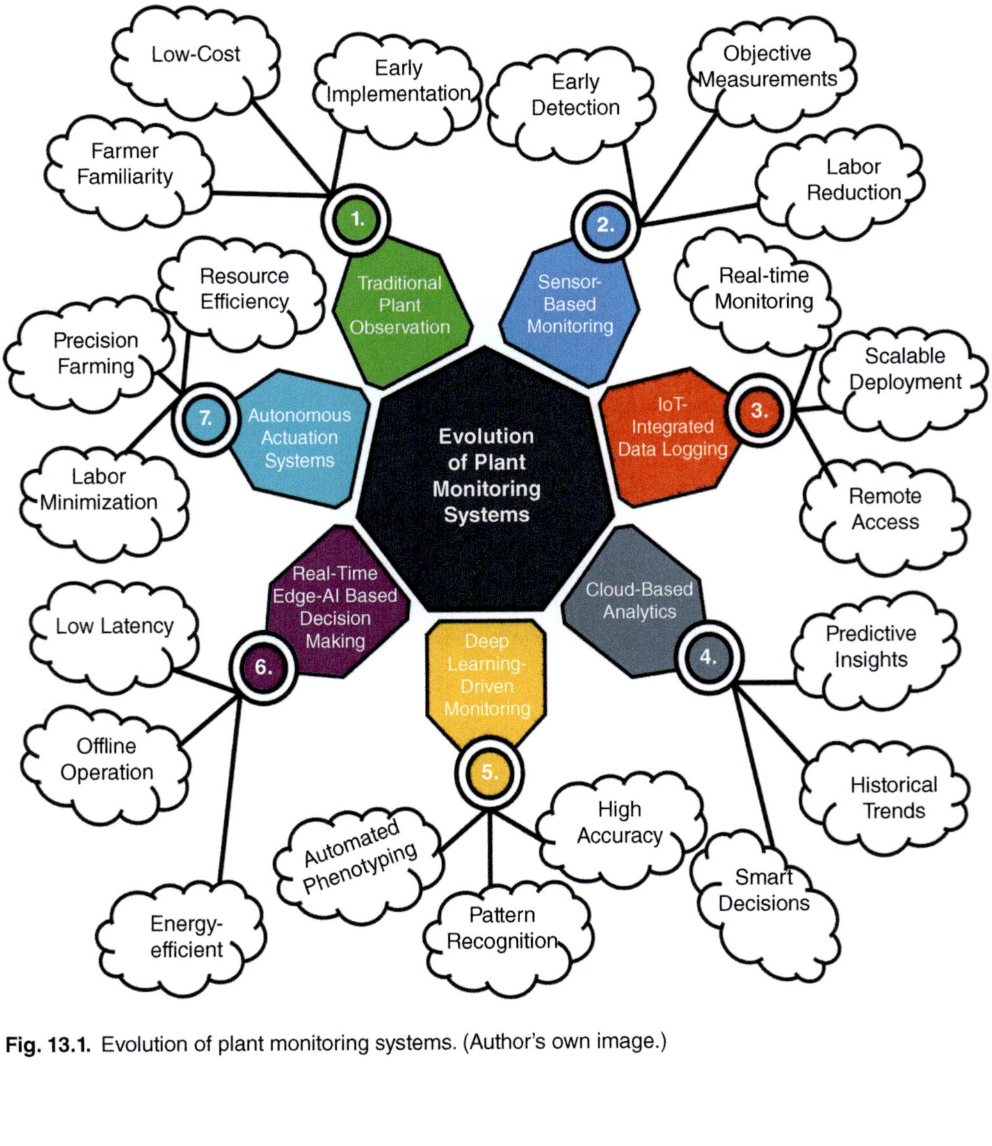

Fig. 13.1. Evolution of plant monitoring systems. (Author's own image.)

ensuring resource efficiency and sustainable food production practices (Meenakshi *et al.*, 2023).

13.1.1 Role of IoT, edge computing, and real-time sensing networks

IoT is the centre of today's real-time plant monitoring systems, as it combines advanced sensor technologies with cloud computing to transform agriculture. Such systems utilize a sensor network that constantly collects information on significant parameters such as soil moisture, temperature, humidity, and light intensity, all of which are critical to determining plant health (Sharma *et al.*, 2025). Such data are then sent to centralized servers, where it is analyzed in real time, hence allowing timely interventions such as automation of irrigation or pest management. Additionally, IoT-based systems enable remote monitoring and control, maximizing crop yield, reducing the need for manual intervention, and increasing resource usage while at the same time solving critical issues such as scalability of deployment and data security (Rane *et al.*, 2022). A critical technological innovation within IoT-based systems is the use of edge computing, which enables localized processing of data in close proximity to the sensor nodes. Such an approach minimizes latency, increases processing speed, and ensures real-time responsiveness to environmental dynamics. For example, real-time data analysis performed on edge devices enables direct processing of data from field sensors, allowing real-time responses to operational anomalies irrespective of cloud-based processing (Li *et al.*, 2023a). Integrated monitoring systems that support customizable algorithms also expand this framework by allowing real-time adjustment of monitoring parameters in line with particular field conditions (Xiao *et al.*, 2023). A combination of deep reinforcement learning (DRL) and edge computing introduces sophisticated control mechanisms that maximize resource allocation and operational parameters dynamically, ultimately contributing to reduced failure rates, increased sustainability, and minimized maintenance expenses (Xu *et al.*, 2024b). Such sophisticated systems interact synergistically with real-time sensing networks, which are critical to the modern plant monitoring practices. Through the utilization of a decentralized network of sensor technologies (e.g. soil moisture, temperature, and light sensors), these systems collect vital information on environmental and plant physiology conditions on a continuous basis. Wireless protocol communication, for example, ZigBee, enables dynamic data transfer from fields to centralized or edge-based processing systems, thereby enabling rapid and accurate control over agricultural inputs. This integration of advanced learning algorithms and sensor networks not only ensures effective resource management but also creates real-time responses to the requirements of plants, thereby enhancing the objectives of precision agriculture and sustainable agricultural practices. More sophisticated systems employ ML algorithms to analyze these data to detect patterns that are indicative of plant stress, nutrient deficiencies, or the development of diseases, thereby enabling timely and informed interventions based on data analysis. Through the deployment of such integrated technologies, contemporary sensing networks considerably enhance resource efficiency, crop yields, and environmental sustainability in the agricultural industry (Venkatesh *et al.*, 2024).

13.1.2 Architecture of autonomous plant surveillance systems

The design of autonomous plant monitoring systems is defined by the combination of various emerging technologies aimed at optimizing operational efficiency while minimizing reliance on human interventions. Unmanned aerial vehicles (UAV)-based systems, as elaborated by Mondal *et al.* (2020), utilize automated task allocation and mission planning techniques to perform agricultural surveillance, thereby enabling real-time monitoring of crop health and optimizing the accuracy of treatment applications, which ultimately optimizes productivity while minimizing input costs. In parallel, Manjah *et al.* (2024) advocate for a multisensory architecture founded on a model that is agent-based, providing scalability and flexibility in terms of the capacity to add new sensors and decentralized decision units that autonomously respond to varying field conditions. Claro *et al.* (2014) concentrate on the use of robot collaboration in monitoring systems, emphasizing the development of communication protocols and interoperable architecture that allow autonomous robots to communicate with one another and with agricultural infrastructure already in place, maximizing efficiency and reducing labor dependency. Likewise, Klose *et al.* (2010) present a modular system for autonomous field-based plant phenotyping, which involves the use of a range of sensors alongside real-time data storage facilities that enable high-resolution plant analysis and flexible system setups. Expanding on such developments, Samy *et al.* (2022) present an intelligent monitoring system for plant growth using image sensors and actuators that adaptively modify environmental conditions based on sensed plant requirements, demonstrating the potential of real-time responsive monitoring systems for optimizing plant development.

13.2 Advanced DL Architectures in Plant Intelligence

13.2.1 Transformers and Vision Transformers (ViTs) for dynamic plant vision

The advent of ViTs has revolutionized dynamic plant vision, offering a more sophisticated alternative to traditional image analysis methods and convolutional neural networks (CNNs). Through the application of the self-attention mechanism, ViTs effectively capture long-range dependencies and global contextual relationships in high-dimensional plant images, which is particularly useful for the identification of subtle phenotypic variations in tasks such as disease diagnosis, growth monitoring, and interspecies classification (Ali *et al.*, 2025). Unlike CNNs, which are localized convolutional computation-based, ViTs involve holistic representations, hence enhancing model generalizability across heterogeneous environmental conditions and plant morphologies. Recent innovations, such as GreenViT and LeafViT, are prime examples of the domain-specific application of ViTs in the plant sciences. These models involve domain-specific modifications to solve problems such as high interclass similarity, imbalance, and the curse of dimensionality. For example, GreenViT utilizes hierarchical tokenization to preserve spatial hierarchies, while LeafViT involves attention regularization methods to focus on diagnostically important leaf areas. Both models have shown exceptional performance, registering classification accuracies over 97% on benchmark plant pathology datasets (Parez *et al.*, 2023), hence showcasing the potential of transformer-based models in keystone agricultural diagnostics. The advent of Dynamic Vision Transformers (DViTs) has further extended the application of ViTs to include dynamic and spatiotemporally varying environments. DViTs enhance feature extraction from aerial and remote sensing imagery through the fusion of visual inputs with auxiliary geolocation metadata, enabling the accuracy of species recognition even under the presence of occlusions, scale variations, and spectral noise (Pan *et al.*, 2022). This innovation has paved the way to real-time ecological monitoring, biodiversity surveys, and precision agriculture, especially when integrated within drone-based or satellite imaging platforms.

Apart from this, ViTs have shown immense promise in seedling classification, which is critical to crop monitoring at the early stage and yield outcome prediction. A comparative study found that ViTs surpassed traditional CNN architectures with a classification accuracy of 95.2% across different plant species, thus cementing their viability in low-resolution datasets and morphological ambiguity (Patel, 2024). The attention maps generated by ViTs also improve interpretability, allowing for post hoc analysis of morphological traits relevant to agronomic decision-making. In spite of these developments, the practicality of applying ViTs in edge computing environments, such as IoT-based agricultural systems, entails substantial computational loads. ViTs are resource-intensive in nature due to their high parameter numbers and quadratic attention complexity. However, ongoing research efforts are actively engaged in overcoming these drawbacks through model compression, knowledge distillation, and designing efficient attention mechanisms (Kempelis *et al.*, 2024). Early prototype combinations of lightweight versions of ViTs with Edge AI processors have shown promise in delivering near real-time inference for plant health monitoring in field applications. Overall, the integration of ViTs in plant sciences is a critical move toward the construction of autonomous, scalable, and interpretable systems for plant phenotyping. As research continues to advance both algorithmic frameworks and hardware acceleration strategies, ViTs are set to become an essential component of the digital revolution in agricultural practices.

13.2.2 3D convolutional neural networks (3D-CNNs) and spatiotemporal models for structural growth pattern analysis

Spatiotemporal models for the analysis of structural growth patterns in plants are a critical juncture of computational modeling, environmental systems analysis, and high-resolution imaging technologies. The models seek to capture temporal dynamics as well as spatial heterogeneity in plant development and, therefore, provide more advanced insights into morphological and physiological responses to environmental stimuli. As emphasized by Debbagh *et al.*, the integration of deterministic, probabilistic, and generative modeling frameworks within high-throughput phenotyping (HTP) platforms facilitates predictive pattern recognition under varying climatic and soil conditions. These approaches are pivotal in propelling precision agriculture and optimizing genotype–environment interactions to enhance crops (Debbagh *et al.*, 2024). A landmark in this context is the work of Basu and Pal, who developed a 3D spatiotemporal reconstruction methodology for root growth analysis. The technique enables visualization and quantification of root architecture and localized growth zones as a function of time, which is crucial to derive insights into subterranean dynamics that are underemphasized in traditional phenotyping approaches. Their model takes anisotropic growth patterns as well as mechanical interactions in the rhizosphere and, therefore, derives deeper insights into root–soil biophysics and nutrient acquisition strategies (Basu and Pal, 2012).

The modulation of environmental parameters on plant growth has been successfully modeled using spatiotemporal models. Walter highlights the importance of dynamic light conditions in regulating photosynthetic activity, illustrating that spatiotemporal modeling can model interactions at the canopy scale and evaluate photosynthetic efficiency during diurnal and seasonal fluctuations. Such models enable the optimization of controlled-environment agriculture by informing the design of artificial lighting that aligns with species-specific light response traits (Walter, 2008). To address the complexity involved with non-rigid biological forms, Zhang *et al.* present a novel image registration framework that supports morphological changes during the development of plants. Their model, applying deformable image registration algorithms in a time sequence, enhances the accuracy of monitoring growth at the organ scale, particularly in phenotyping applications with irregular developmental trajectories or responses to mechanical stress. The method greatly boosts longitudinal plant studies by ensuring geometric consistency among time points (Zhang *et al.*, 2023). In practical forestry applications, the utilization of spatiotemporal models also supports silvicultural decision-making processes. O'Rourke and Kelly use hierarchical spatiotemporal covariance models to quantify the impacts of thinning regimes on tree growth patterns. Their method enables forest managers to predict yield and carbon sequestration capacity more accurately, demonstrating the use of spatiotemporal models in natural and managed ecosystems (O'Rourke and Kelly, 2015).

The use of 3D-CNNs has transformed the structural analysis of plant growth by allowing volumetric feature extraction and spatially consistent phenotype mapping. Research by Boogaard *et al.* emphasizes the advantage of 3D point cloud analysis in the accurate capture of complex morphological features such as internode curvature and leaf orientation in crops, that is, cucumber. In comparison to 2D image-based approaches, 3D-CNNs have better performance in delineating non-linear growth patterns and can efficiently work in situations involving occluded or overlapping structures (Boogaard *et al.*, 2023). In addition to voxel-based approaches, point-based DL architectures, including PointNet++ and Point-Voxel convolutional neural network (PVCNN), have emerged as state-of-the-art tools for plant structural segmentation. Turgut *et al.* (2022) demonstrated that PointNet++ is particularly effective in segmenting complex structures in rosebush models, even with small, labeled datasets, owing to its hierarchical feature aggregation and local region sampling abilities. Likewise, Saeed *et al.* (2023) demonstrated the use of PVCNN in accurate segmentation of cotton plant organs with high accuracy and computational efficiency, a breakthrough for the enhancement of real-time field phenotyping and the acceleration of breeding programs. In general, these advances mark a giant leap in our ability to model, monitor, and manipulate plant growth in natural and engineered environments.

13.2.3 Neural ordinary differential equations (ODEs) and physics-informed neural networks (PINNs) for modeling physiological processes

Neural ordinary differential equations (Neural ODEs) are a profound paradigm shift in dynamic modeling of biological processes, offering a continuous-time, differentiable model that elegantly combines classical mechanistic modeling with state-of-the-art DL. Neural ODEs provide a strong mathematical basis for the depiction of complex, non-linear relationships between fluctuating external factors and internal biological control in plant physiology. Neural ODEs are particularly well suited to handle irregular time-series data, a ubiquitous aspect of many plants phenotyping and environmental monitoring experiments and, therefore, enhance the temporal resolution and interpretability of physiological processes (De Brouwer and Krishnan, 2023). Recent methodological advances have significantly increased the expressive power of Neural ODE models. For instance, the use of neural event functions enables the description of hybrid systems, enabling precise modeling of sudden physiological transitions, such as stomatal closure during drought or photoperiodic regulation of flowering, within a continuous framework (Chen *et al.*, 2020). This hybrid modeling ability is critical in capturing the discrete-continuous dynamics of biological regulation accurately. Furthermore, the incorporation of symmetrical regularization during Neural ODE training ensures that the resulting models are not only predictive but also physically consistent. Even in situations with limited data, these models show improved generalizability and resilience when they incorporate well-known conservation principles and structural invariants (such as mass balance and energy conservation) (Hao, 2023). This ability is useful in plant systems biology, where detailed mechanistic understanding is often lacking or experimentally inaccessible.

To demonstrate the combination of knowledge-based and data-driven modeling, hybrid frameworks such as PHOENIX (Prior-informed Hybrid ODE Network for Interpretable expression) modeling are developed. PHOENIX leverages gene expression data combined with preexisting regulatory network topologies to learn interpretable models that describe gene regulatory dynamics and thus further advance knowledge of developmental processes and stress response in plants (Hossain *et al.*, 2024). Such frameworks open the door to explainable AI (XAI) in systems biology by bridging flexibility with biological significance. Neural ODEs, in turn, have shown considerable promise toward modeling metabolic kinetics, enabling efficient simulation and parameter estimation within complex biochemical networks. By replacing traditional numerical integration mechanisms with neural approximators learned from kinetic data, such models have been shown to drastically reduce computational loads while maintaining accuracy compared with established reaction dynamics and, thus, further extend the exploration of metabolic flux in the context of fluctuating environmental and genetic perturbations (van Lent *et al.*, 2024). Coupled with Neural ODEs, PINNs have developed as a flexible tool for modeling systems described by partial differential equations or differential-algebraic formulations. In contrast to opaque black-box models, PINNs impose physical constraints such as transport laws, energy conservation relations, or diffusion kinetics into the loss function to enable learning accordant with first principles (Velioglu *et al.*, 2025). Such an approach is of specific relevance in plant physiology, where empirical evidence is typically poor or afflicted with noise.

One of the strongest aspects of PINNs is their ability to infer unknown states and unknown parameters, a requirement for reconstructing hidden physiological variables, such as internal nutrient fluxes and hormonal gradients. Empirical evidence from applications to bioprocess systems, such as continuously stirred tank reactors (CSTRs), indicates that PINNs are superior to data-driven models in terms of generalization and robustness, even in cases of partial observability. Enhancements such as the Karush–Kuhn–Tucker-constrained hard PINNs (KKT-hPINNs) ensure that optimization under equality constraints, common in stoichiometric balances or resource allocation models, is retained strictly during training, thereby ensuring greater model fidelity and usability across domains (Chen *et al.*, 2024). Furthermore, the extension to physics-informed neural control (PINC) allows these capabilities to be extended to the domain of dynamic system regulation. The combination of PINNs with control-theoretic paradigms, researchers are able to design optimal interventions within plant systems, such as irrigation or nutrient delivery schedule regulation based on real-time physiological predictions, thereby uniting experimental agronomy with novel smart farming paradigms (Antonelo *et al.*, 2021). Overall, the confluence of Neural ODEs and PINNs in plant physiology offers a principled and robust framework for interpreting and controlling dynamic biological systems. These models overcome the limitations of static or empirical approaches by harnessing mechanistic reasoning in flexible neural structures, thereby not only allowing for high-fidelity predictions but also offering biological insight, interpretability, and controllability pillars of the next generation of plant systems modeling.

13.2.4 Equivariant neural networks (E3NNs) for morphology recognition

E3NNs have emerged as a game-changing tool for morphological evaluation of biological organisms due to their inbuilt capability of being equivariant to a large variety of group transformations, including scaling, rotation, reflection, and translation. The property is essential in the morphology of plants, whose phenotypic variation is multiscale and orientation-dependent in nature. Unlike the CNNs, which suffer from generalizability across such transformations in the absence of excessive data augmentation, E3NNs leverage representation theory of groups in their formulation, allowing for efficient and robust feature extraction from geometrically varying data. Embedding morphological scale-spaces into E3NN architectures, as proposed by Sangalli *et al.*, further improves the scalability and adaptability of such models. With multiresolution representations of morphological arrangements, E3NNs extract efficiently and effectively global and local geometric features across varying spatial scales, a requirement that is central to the investigation of plant organs such as leaves, roots, and inflorescences, which show broad structural heterogeneity by species and growth stage. Such a scale-invariant approach leads to greatly enhanced efficiency in critical applications such as supervised classification, semantic segmentation, and morphological phenotyping, where traditional CNNs are hindered by issues in overfitting or loss of spatial specificity with scale variability (Sangalli *et al.*, 2021). The efficacy of E3NNs further improves with utilization in combination with ViTs, a comparatively new DL model, that has a strength in handling long-range dependencies through attention-based mechanisms. Recent research by Chen *et al.* highlights the benefits of the use of ViTs over CNN-based models in high-precision applications such as the detection of plant material, especially in those cases involving complex or subtle morphological attributes. Application of equivariant hybridization in ViTs promises a lot in the application areas of fine-grained phenotypic discrimination tasks, including species classification, disease symptom recognition, and developmental stage evaluation (Chen *et al.*, 2024).

Concurrent with this, Duits *et al.*'s work on partial differential equation-based group convolutional neural networks (PDE-G-CNNs) offers a geometrically motivated extension of the equivariant paradigm by incorporating continuous symmetry transformations into the network's convolutional kernels through the use of PDE formulations. This mathematically formalized approach not only renders the model more interpretable but also allows for the construction of biologically meaningful representations of morphological transformations critical in the understanding of plant growth dynamics, organ patterning, and tissue-level symmetry (Duits *et al.*, 2021). Additionally, the use of DL models in the exploration of cryptic morphological variations, as in the case of *Marchantia polymorpha*, demonstrates the potential of neural networks to outperform traditional morphometric methods by quantifying subtle phenotypic features that may go unnoticed to human observers. Tomizawa *et al.* demonstrated that computerized morphological classification pipelines have the potential to reveal hidden phenotypic dimensions corresponding to genetic or environmental variations, thus enabling the discovery of new genotype–phenotype correlations and expanding the horizons of developmental plant biology. The synergy of combining E3NNs with attention-based models (e.g. ViTs), PDE-informed convolutional architectures, and deep unsupervised learning frameworks represents a revolutionary breakthrough in plant phenotyping. Such an integrated architecture enhances not only the accuracy of morphological identification in terms of scale and orientation but also the interpretability of model outputs, which is a critical property in scientific applications where explainability is essential in biological interpretations. As these models continue to evolve, they are set to revolutionize the field of computational botany by enabling the possibility of performing scalable, accurate, and biologically contextualized analyses of morphology in controlled and natural environments (Tomizawa *et al.*, 2023).

13.2.5 Self-supervised and contrastive learning in plant health diagnostics

Self-supervised learning (SSL) is now an emerging state-of-the-art method in plant health analysis, particularly in plant disease detection. The conventional ML models applied to plant disease detection are typically dependent on large amounts of well-labeled data, which are time- and cost-consuming to obtain. Use of labeled datasets has been a prime challenge, leading to slow disease detection and thereby huge losses in agricultural production (Al Mamun *et al.*, 2024). SSL methods, which learn strong feature representations using unlabeled data with no manual labeling, provide a pragmatic solution to the challenge by enabling efficient model training with minimal human intervention and considerably lower costs. Recent developments in SSL, including the use of Siamese Networks and Masked Autoencoders, have been extremely efficient in generating discriminative feature embeddings from unlabeled data. For instance, a model that applied a Masked Autoencoder strategy achieved a phenomenal accuracy of 99.61% in a company-owned plant disease database, demonstrating its capacity to learn latent representations and accurately detect plant diseases regardless of labeled training datasets (Wang *et al.*, 2024). In contrast, a study that applied momentum contrastive learning resulted in an accuracy of 94.31%, further demonstrating the capability of contrastive learning-based methods under SSL for plant health analysis. Such methods, by concentrating on relationships and similarities between unlabeled instances, provide a strong basis for comprehending patterns of plant diseases even without vast amounts of labeled datasets.

The integration of instance-aware detection techniques, including visual saliency maps and positional encoding, has continued to advance the frontiers of SSL in plant disease evaluation. The techniques have been effective in addressing the class imbalance issue, which is common in agricultural datasets where some conditions or diseases may be underrepresented. Highlighting the most informative parts within an image and introducing spatial context, these SSL techniques have attained state-of-the-art results in a range of plant disease classification tasks, indicating their promise in circumventing the challenges of unstructured real-world agricultural data (Kim *et al.*, 2022). Besides, contrastive learning, a leading SSL technique, has gained significant attention for its promise to address some of the biggest challenges in plant disease identification, particularly those involving dataset biases and class imbalance. Chung's study is a proof of the promise of contrastive learning by focusing on the disease condition rather than relying on specific plant species. This crop-agnostic model, when evaluated on the PlantDoc dataset, attained a high accuracy of 87.42%, indicating the applicability of contrastive learning in addressing the generalizable aspect of plant disease detection (Chung, 2024). Choi and Devakirubai (2023) also demonstrated the promise of unsupervised contrastive learning techniques in learning discriminative features from diverse plant species, attaining a high accuracy of 85.88% and addressing common dataset biases, that is, the tendency to overfit to species-specific features.

Another development in this area is the use of few-shot learning, a method where the models are trained on a few labeled samples. Mu *et al.* introduced a new supervised contrastive learning method combined with meta-learning principles and attained an accuracy rate of 79.51% despite the constraint of limited availability of labeled samples. This ability to attain optimal learning results from smaller datasets is extremely crucial in situations where the resources are scarce, since the collection of enormous labeled datasets may not be feasible (Mu *et al.*, 2024). Uzhinskiy demonstrated the potential of contrastive learning for application in few-shot setups by comparing various few-shot learning methods and confirming that contrastive methods far outperformed other methods in terms of accuracy and generalization on a range of plant disease datasets. Collectively, these developments in contrastive learning and SSL have tremendous potential for the future of automating plant health monitoring, enabling extremely efficient and accurate plant disease detection in real-world agricultural environments. The ability of SSL models to generalize across various crops and disease types and to learn from limited amounts of labeled samples makes them crucial parts of advanced plant diagnostic systems. By minimizing dependence on expert-labeled datasets and mitigating the problems associated with unstructured, imbalanced data, these approaches have the potential to revolutionize the face of agriculture, leading to timely disease detection and, in the long term, more sustainable crop management practices (Uzhinskiy, 2025).

13.3 Sensor Fusion and Multimodal DL Systems

Cross-modal fusion of heterogeneous data sources of RGB, hyperspectral, light detection and ranging (LiDAR), infrared (IR), chlorophyll fluorescence, and acoustic data is an advanced task in remote sensing tasks. The primary intention of such fusion is to utilize the complementary information present in each data type to enhance the classification, analysis, and interpretation of environmental datasets more accurately and robustly. Combination of such heterogeneous data sources provides richer knowledge of the phenomenon being analyzed but involves serious technical hurdles, particularly feature alignment, maintaining modality-specific features, and designing optimum data fusion algorithms. Therefore, several advanced methodologies have been proposed to overcome such difficulties and capitalize on the synergy advantage of multimodal information. One of the latest leading techniques is the Modality Fusion Vision Transformer (MF-ViT) technique, which makes use of multimodal cross-attention and spectral self-attention mechanisms to fuse hyperspectral and LiDAR data efficiently. By mitigating the need for explicit spatial alignment between feature spaces of the two modalities, the model retains critical spatial features while taking benefits of the rich spectral information embedded in hyperspectral data. Use of attention mechanisms enables the selective combination of appropriate features and improves the ability of the model to classify intricate scenes accurately (Yang *et al.*, 2024). The proposed methodology not only improves the performance of classification but also preserves intrinsic spatial topologies in the data regardless of the differences in resolution or coordinate systems between the modalities. Similarly, the cross-modal hierarchical frequency fusion network (HFNet) bridges the inherent differences between hyperspectral and LiDAR datasets by focusing on cross-modal frequency features. This approach entails a frequency consistency loss during training such that it becomes feasible to combine spectral information from hyperspectral data with topographic and geometric information from LiDAR. By matching the frequency components of these two datasets, HFNet offers improved classification accuracy, particularly in harsh environmental conditions where the interaction of multiple physical processes is of significant complexity (Zeng *et al.*, 2024).

The multiple information collaborative fusion network (MICF-Net) introduces a novel dual-branch fusion mechanism based on a cross-modal attention fusion transformer. The model bifurcation enables it to preserve modality-specific information while facilitating cross-guided learning through semantic-guided fusion of features. The importance of this solution is that each modality brings its strengths to the integrated model without compromising the modalities' coherence. Inter-modal collaborative fusion allows the model to adaptively fuse complementary information to attain high performance in classification tasks of various types, from vegetation type classification to land cover mapping (Tang *et al.*, 2024). Interconnected fusion, however, takes a holistic approach by investigating intra- and inter-modality features. The use of sophisticated attention mechanisms, including self-attention and cross-attention, allows the framework to effectively produce a composite multimodal representation with the ability to classify complex environmental datasets. The self-attention mechanism allows the modalities to attend their respective corresponding features in isolation, and the cross-attention module allows the model to detect and combine primary relationships among modalities. This solution is particularly beneficial in remote sensing applications, where interactions of the environment with sensors are varied (Huo *et al.*, 2023).

In addition, the dual-channel cross-transformer network possesses high-level fusion of self-attention mechanisms within every modality and cross-modal attention to fuse features. This process allows the model to fuse and enhance features from every modality, thus enhancing classification effectiveness in dynamic and complex environments. Using the process of applying self-attention in analyzing modality-specific features and cross-attention in fusing them, this method demonstrates better performance in classification tasks, especially in complex environments where multiple environmental factors need to be treated simultaneously. Such methods highlight the potential of cross-modal fusion to drive innovation in remote sensing technologies, especially in applications such as environmental monitoring, precision agriculture, and land-use classification. Nevertheless, despite the tremendous advancement in hyperspectral and LiDAR data fusion, the fusion of other modalities, such as RGB, IR, chlorophyll fluorescence, and acoustic data, remains a vibrant area of research. The distinctive nature and complementary information found in these other data types pose tremendous opportunities for enhancing the robustness and effectiveness of remote sensing models. Fusing these modalities can provide models that are more sophisticated and accurate, with the ability to differentiate complex environmental dynamics, thus facilitating more effective decision-making in applications such as ecosystem management, crop health assessment, and disaster relief (Zhou *et al.*, 2023).

Innovation in deep fusion transformers and multistream CNNs has revolutionized the fusion of information obtained from UAVs, ground stations, and environmental sensors in plant phenotyping and monitoring and brought these technologies to the forefront of precision agriculture. By combining information from diverse sources such as hyperspectral, thermal, and LiDAR sensors mounted on UAVs, soil, root zone, and climatic information collected from ground stations, these systems enable multimodal trait predictions with unprecedented spatial resolution and accuracy. For example, Nguyen *et al.* established that a multitask deep CNN, trained on UAV and ground-based data sources, could predict key phenotypic traits in corn with high accuracy, outperforming conventional ML approaches. This multitask architecture not only ensured efficient resource exploitation but also enabled simultaneous prediction of multiple traits, improving throughput in HTP experiments (Nguyen *et al.*, 2023). Building upon these functions, ConViTX, a hybrid model combining CNNs and ViTs, offers improved generalizability and explainability in disease identification-related tasks. As explained by Thakur *et al.* (2025), this architecture showed strong performance on both field-collected and UAV-acquired image datasets and showed the potential to bridge gaps in laboratory-to-field translation.

Second, ViTs have performed extremely well in crop and weed classification on UAV imagery. Their ability to capture long-range dependencies and contextual relationships allows them to outperform conventional CNNs, especially where training sets are limited or subject to class imbalance. Reedha *et al.* (2021) reported that ViTs not only improved classification but also reduced the need for elaborate manual feature engineering, thus making them acceptable to deploy in real-world agricultural settings. The advancement of ground–air interaction systems, reported by Jingzhao and Mingming, also favors this convergence. These systems use UAVs as dynamic nodes for data acquisition, processing, and transmission from fixed ground sensors, thus making near real-time crop condition monitoring and dynamic decision-making possible in large-scale farming operations (Jingzhao and Mingming, 2019). A critical next step in this development is the use of multistream CNNs to combine aerial and subterranean plant data, capturing the root–shoot continuum. Biological research has long established that root and shoot system interaction is modulated by complex molecular, hydraulic, and electrophysiological signals in response to environmental stimuli (Shabala *et al.*, 2015). Aerial platforms, through DL processing of UAV imagery, offer accurate geolocation and structural evaluation of crops, as reported in corn and citrus orchard research (Osco *et al.*, 2021). Combined with this, ground-penetrating radar (GPR) with deep CNNs provides non-invasive root architecture mapping, delivering informative information on belowground phenotypes under varied soil and moisture regimes (Lu and Lu, 2022). Despite these developments, most root models, such as ArchiSimple, today remain detached from shoot phenotyping platforms and more general agroecological models (Pagès *et al.*, 2012). With the application of multistream DL architectures, in which different CNN branches handle aerial, soil, and root sensor data before being combined within a single, unified feature fusion module, researchers can develop holistic, end-to-end models capable of capturing the complex architectural and physiological characteristics of plants. Such fusion has the potential to revolutionize crop modeling, stress detection, and genotype-by-environment interaction studies through the delivery of dynamic, systems-level understanding of phenotypic traits.

13.4 DL for Biotic and Abiotic Stress Detection

The YOLOv8 architecture, a cutting-edge DL model, has been very successful in early disease detection in plants, indispensable for reducing crop losses and promoting agricultural production. Praised for its incredible speed and accuracy, YOLOv8's real-time object detection allows for rapid and accurate diagnostics under on-site conditions. Empirical tests report impressive performance metrics, such as mean Average Precision (mAP) of 0.79, precision of 0.61, and recall of 0.79 on the PlantDoc dataset, hence outperforming previous YOLO variants in detection speed and accuracy (Shen *et al.*, 2025; Cheng, 2024). Moreover, the optimized YOLOv8 design also cuts training time and GPU resource consumption by up to 47% while maintaining mAP levels, enhancing computational efficiency (Hussain, 2023). Practical validation in situations using unseen images further supports its strength and adaptability in varied agricultural environments (Khan *et al.*, 2025). Enhancements in YOLOv8, including the incorporation of GhostNet and coordinate attention modules, significantly improve its accuracy and computational efficiency, achieving an effectiveness of 72.2% when combined with transfer learning, which is highly beneficial for small datasets (Bui and Do Le, 2025). Comparative studies report YOLOv8 and YOLOv9-C exhibit better F1-scores and recall rates, proving their effectiveness in precision agriculture. Nevertheless, sustaining consistent real-time performance amid variable and dynamic field conditions remains a constant challenge. Continued breakthroughs in architectural design and module combinations further boost YOLOv8's robustness and operating reliability (Wang and Liu, 2024).

In addition to disease detection, DL is also transforming plant stress detection and modeling rare diseases. State-of-the-art architectures allow for consistent classification of biotic as well as abiotic stressors. Generative models, such as generative adversarial networks (GANs) and variational autoencoders (VAEs), are used to model rare disease phenotypes, enabling a deeper understanding of plant pathogenesis and support in the development of personalized interventions (Zafar *et al.*, 2023). Gradient-weighted class activation mapping (Grad-CAM) has enabled the detection and quantification of foliar stress, with success in cross-species transfer learning demonstrated in crops such as soybean. Moreover, hybrid architectures such as CNN-LSTM networks also efficiently extract spatiotemporal dynamics to identify water stress in crops such as chickpea earlier compared to conventional classification methods (Bounoua *et al.*, 2024). Stress-driven evolutionary synthesis frameworks have also been proposed to design compact high-speed neural networks for plant stress detection, which optimize inference time and resource usage (Shafiee *et al.*, 2019). In harmony, DL has also been applied to stress assessment in human beings, with remote photoplethysmography (rPPG) from facial video input integrated in hybrid DL models to identify stress at a high level of accuracy using remote monitoring (Fontes *et al.*, 2024).

The integration of epigenetic data with zero-shot and few-shot learning models further enhances new plant pathogen and environmental stressor detection and control. Such epigenetically guided, data-efficient solutions extend DL's predictive capabilities in crop protection (Gou *et al.*, 2024). CNNs, for example, architectures including GoogLeNet, ResNet50, and VGG16, have been popularly used in the classification of biotic stress, notably fungal diseases of vegetables and fruits (Ibrahim *et al.*, 2024). Though applied less often in abiotic stress detection, DL models can be useful for stress phenotyping by enabling swift, precise detection and quantification of environmental stresses. Integration of genomic and epigenomic datasets into DL pipelines further enables prediction of plant responses to stress by explaining complex gene regulatory networks underlying resilience and productivity (Zhang *et al.*, 2024). Additionally, immune-related gene network construction using DL assists in detecting key genes involved in host–pathogen interactions and yields valuable information on plant defense mechanisms.

13.5 Smart Phenotyping and Morphological Tracking

Semantic and instance segmentation of plant organs with DL architectures such as U-Net, DeepLabv3+, and Mask R-CNN is the basis of automated plant phenotyping and morphological analysis. These architectures allow accurate delimitation of organs such as leaves and stems, thus allowing for accurate measurement of growth and breeding optimization. Recent research has explored methodological developments and tested their performance on various species and environmental conditions. For example, a segmentation network based on a double-neighborhood feature extraction block and an attention module attained high semantic and instance accuracy in 3D plant point clouds, surpassing benchmark models such as PointNet++ and PlantNet (Li *et al.*, 2022). Likewise, Mirande *et al.* (2022) introduced a hierarchical graph-based approach that maintains the structural coherence of plants by extracting local geometric and spectral descriptors, with segmentation refinement through a quotient graph representation. Other developments include the plant segmentation transformer (PST) architecture, specifically tailored for high-resolution point clouds, which uses a dynamic voxel feature encoder and attention mechanisms to learn contextual dependencies. This approach has shown higher performance in segmenting complex plant architectures, surpassing state-of-the-art models (Du *et al.*, 2023). Other developments include AC-UNet, an enhanced U-Net variant using VGG16 and multiscale feature extraction that minimizes feature edge loss considerably and attains accurate segmentation of *Betula luminifera* stems and leaves, surpassing traditional models such as DeepLabV3 and PSPNet (Yi *et al.*, 2023).

Despite such advances, challenges persist, specifically segmenting extremely complex structures and handling high-resolution plant data. New modules and attention mechanisms show promise; however, more work is needed to improve scalability and responsiveness across species and environmental regimes. DL has significantly changed automatic phenotyping through the high-throughput and accurate analysis of agronomic characteristics such as leaf area, flowering, tillering, and senescence. Image-based methods have also allowed close tracking of growth, making targeted agricultural intervention. AlexNet-based models trained on datasets such as LCC CVPPP 2017 have been shown to deliver improved performance in leaf segmentation and counting, besting conventional techniques in terms of difference in count and mean squared error (Hati and Singh, 2020). Semantic segmentation of grapevine leaf structures has enabled structural and functional analyses, allowing life cycle monitoring and precision agrochemical applications (Tamvakis *et al.*, 2022). While not specifically developed for flowering and tillering, DL models can be trained to handle these phenological stages through analyzing temporal image data, making stage detection and crop monitoring. Models such as DeepPheno have successfully detected stages of plant development, including senescence, providing real-time feedback on plant life cycles and making possible optimized crop production (Yalcin, 2018).

However, automated phenotyping still encounters setbacks in the form of plant structural complexity, variability in the environment, and demands for large datasets with annotations. Nonetheless, developments in DL architectures and methods of data collection are continually advancing the robustness and usability of these methods within agriculture. One such revolutionary improvement in this space is the intersection of neural radiance field (NeRF) with depth estimation methods for reconstructing 3D plants. By transcending the shortcomings of 2D imaging, NeRF allows precise, non-invasive phenotypic data capture, which is crucial for today's agricultural industries. NeRF models produce dense point clouds using RGB images, which allow precise 3D morphology reconstruction with high correlations with conventional field measurements, recording R^2 values in excess of 0.95 for measurements such as node length, leaf area, and fruit volume in tomato (Choi *et al.*, 2024). In container plants, NeRF, using path analysis and density clustering, facilitates effective segmentation of organs and accurate trait extraction such as plant height and leaf size (Zhu *et al.*, 2024). Single-RGB-image depth estimation through the P3ES-Net DL model is supported and outperforms commercial depth cameras, facilitating inexpensive monitoring of canopy size and plant height (Zhao *et al.*, 2022). Techniques based on NeRF have shown resilience across conditions, recording 74.65% F1-score in outdoor environments (Arshad *et al.*, 2024) and mean distance errors as low as 0.865 mm, further signifying their relevance for scalable high-precision modeling in greenhouse systems. However, issues with regard to computational needs and the requirements for large datasets continue to hold, requiring the continued development to increase the efficiency and generalizability of such approaches.

13.6 Plant–Environment Interaction Modeling

Graph neural networks (GNNs) provide a robust platform for complex plant–environment interaction analysis utilizing graph-structured data to model multiscale environmental factor and plant property relationships. Performance is derived from the capacity to represent relational interdependencies, which increases predictive efficiency and modeling parsimony, especially under changing environmental conditions. The inclusion of new tools such as the Walk Index Sparsification (WIS) algorithm enhances the functional capacity of GNNs through preservation of dominant interaction pathways and retention of the core modeling ability, reducing redundancy in computational resources (Razin *et al.*, 2023). GNNs provide for detailed modeling of phenotypic plasticity, acclimation processes, and the feedback impacts of plant responses on ecosystem processes, making them essential in ecological prediction and adaptive agriculture (Reichstein *et al.*, 2014). Demonstration of successful application in traffic prediction demonstrates the high potential of GNNs for such use in plant sciences (Diehl *et al.*, 2019). Complete resolution of the complex dynamics of plant–environment interactions remains challenging. Coupling of GNNs with supporting models and empirical datasets, such as remote sensing or omics, may improve scalability and ecological accuracy (Raj *et al.*, 2023).

In parallel to GNNs, DRL provides a data-driven platform for adaptive decision-making for agricultural resource management. DRL employs high-dimensional sensor inputs to optimize irrigation and fertilizer application schedules to maximize yield at minimum water use. For example, DRL-based irrigation scheduling for wheat production in Australia achieved profit margins of up to 17.3% in dry seasons compared to conventional methods with similar water savings (Saikai *et al.*, 2023; Zhao *et al.*, 2023). By integrating key variables such as leaf area index, evapotranspiration, and soil moisture into decision rules, DRL systems learn to dynamically respond to environmental inputs, maximizing real-time use of resources. Although under-explored in nutrient management, DRL integration with sensor technologies has promise in integrated crop-environment management (Yang *et al.*, 2022).

One of the most valuable lines of research entails the combination of spatiotemporal long short-term memory (LSTM) networks and GNNs. LSTMs are especially effective for the extraction of temporal patterns, for example, daily and seasonal plant activity, while GNNs are effective for capturing spatial relationships, for example, the effect of microclimatic conditions on transpiration and photosynthesis (Jones, 1985; Jones, 2014). The combined framework supports a deeper understanding of plant–environment interaction, especially under heterogeneous field conditions. Moreover, ecological feedback processes, where vegetation modifies and reacts to its microclimate, are necessary for simulating local adaptation and resilience (Pałubicki *et al.*, 2022). Although the auspicious potential of the combination of LSTM and GNN is great, issues such as the demand for high-quality data and high computing resources need to be solved to enable wider application in agricultural and environmental modeling.

13.7 Edge AI, TinyML, and Federated Learning in the Field

TinyML optimization for microcontrollers and edge inference is transforming real-time plant monitoring by enabling intelligent data processing on low-power, resource-constrained devices. TinyML allows ML models to be deployed on microcontrollers to enable low-latency and energy-efficient inference, which is important for agriculture applications (Soro, 2021). Software such as Edge Impulse and TensorFlow Lite for microcontrollers have become a go-to tools for model development and deployment. Edge Impulse enables user-friendly interfaces and robust deployment pipelines, while TensorFlow allows sophisticated training flexibility. Integration of both software allows practitioners to take advantage of TensorFlow's model optimization features and Edge Impulse's efficient deployment software (Arthur *et al.*, 2024). Power efficiency and computational frugality are central strategies in TinyML optimization. Cutting the data acquisition rate by 75% has, for instance, been shown to cut energy consumption by up to 70%, a critical benefit for battery-powered IoT systems in agriculture (Samanta *et al.*, 2024). In smart irrigation systems, TinyML models are utilized to process crop and environmental data to enable precision irrigation scheduling, thereby mitigating water scarcity issues (Hidalgo *et al.*, 2023).

Alongside TinyML, federated plant learning is a decentralized ML platform that is formulated for agricultural environments, as illustrated in Table 13.1. By utilizing the application of local model training on several farms and then centrally aggregating the learned parameters, federated learning (FL) protects data privacy while encouraging collaborative intelligence. The approach is found to be especially beneficial for sensitive agricultural applications such as disease detection, crop classification, and yield prediction (Dwarampudi and Yogi, 2024). FL efficiently manages diversity in agricultural environments by allowing the construction of location-based models, thereby improving generalization across geographical locations. However, FL is affected by a number of challenges such as communication overhead and non-independent and identically distributed (non-IID) data-related challenges. To combat such issues, pruning methods have been

Table 13.1. AI and signal-based technologies for plant monitoring. Author's own table.

Technology/ approach	Functionality	Applications	References
Electrophysiological sensors	Capture bioelectrical signals	• Real-time stress and health monitoring in plants	Cai and Qi (2017); Tran *et al.* (2019)
Machine learning	Classify plant responses to stress and environmental stimuli	• Disease detection • Growth modeling	Choi *et al.* (2016); Fichman and Mittler (2021)
Biohybrid AI systems	Use plants as biosensors interfaced with AI systems	• Environmental monitoring • Robotic feedback	Buss *et al.* (2022); Levi *et al.* (2023)
Deep learning	Analyze plant images to detect early signs of disease	• Precision agriculture • Yield forecasting	Priyadharshini *et al.* (2023); Shao and Devaraj (2024)
TinyML	Deploy lightweight ML models on microcontrollers for real-time edge inference	• Irrigation optimization • Stress detection	Soro (2021); Arthur *et al.* (2024)
Federated learning	Collaborative, decentralized model training across farms	• Crop classification • Yield prediction	Idoje *et al.* (2023); Dwarampudi and Yogi (2024)
Model compression and pruning	Reduce model size and latency for edge deployment	• In-field AI systems for agriculture	Fountsop *et al.* (2020); Kwak *et al.* (2024)

proposed for models to accommodate local models to local data distributions, thereby reducing both communication requirements and inference time. For yield prediction tasks, inference accuracy improvement by 15.5–20%, besides considerable communication cost reduction, has been recorded with pruned federated models (Li *et al.*, 2024). Besides, edge-based FL provides extra fault tolerance and allows for real-time plant disease detection without the necessity for cloud infrastructure (Thonglek and Rattanatamrong, 2023). Importantly, decentralized FL methods have experienced accelerated convergence and increased classification accuracy in comparison to centralized models in large-scale crop analysis (Idoje *et al.*, 2023).

Despite these advancements, in-field deployment of DL models remains constrained by the edge hardware's constrained resources. Therefore, model compression techniques such as pruning, quantization, and knowledge distillation are necessary. These techniques greatly reduce model size and computation requirements without compromising predictive accuracy. For example, compression of conventional architectures, such as LeNet5, VGG16, and AlexNet, can compress model size by a factor of 38 and reduce VGG16's floating-point operations (FLOPs) by a factor of 99, enabling deployment in seedling classification tasks (Fountsop *et al.*, 2020). Additionally, neural architecture search (NAS) and latency-aware optimization algorithms play a crucial role in model adaptation for specific hardware. The use of $\frac{\ell_1}{\ell_2}$ latency surrogate approximations enables end-to-end optimization for inference speed and energy efficiency (Nasery *et al.*, 2024). Additionally, integer linear programming (ILP)-based pruning techniques can optimize convolutional filter structures, trading accuracy and latency in resource-constrained deployment environments (Kwak *et al.*, 2024). These advancements, collectively, enable scalable, autonomous, and sustainable agricultural systems capable of operating in real-world farming environments.

13.8 Digital Twins and Synthetic Plant Simulations

The synergy of generative models, including StyleGAN3, diffusion models, and sim2real approaches, has resulted in a revolutionary shift in the field of plant science, with a prime focus on HTP, conservation of biodiversity, and autonomous agriculture. All these developments cater to major issues such as domain shift, limited labeled datasets, and phenotypic heterogeneity observed across controlled and natural environments. Synthetic-to-real image translation advancements, which employ architectures such as stable diffusion, low-rank adapters (LoRA), and ControlNets, have resulted in improved efficiency compared to conventional approaches such as CycleGAN, particularly for intricate tasks such as instance segmentation of leaves (Hartley *et al.*, 2024). Such advancements are particularly effective in cases where it is not feasible to generate large, annotated datasets, thus providing robust alternatives for further phenotypic analysis. Furthermore, the generation of synthetic images has enabled scalable classification of rare species, thus facilitating zero-shot and few-shot learning processes that demonstrate performance improvements of 29% and 31%, respectively (Dasgupta *et al.*, 2024). Such functionality is essential in biodiversity research, especially in under-sampled or ecologically sensitive habitats where *in situ* data acquisition is limited.

Diffusion models have come out as effective tools for the knowledge distillation process. Li *et al.* (2023b) showed that low-fidelity synthetic images generated by diffusion models can be employed to effectively train high-performing networks, illustrating the instructional capability of generative data. Sim2real methods improve such capabilities further by utilizing physically based structured light simulations to create photorealistic RGB and depth images, allowing the sim2real gap to be closed. Such a method improves the robustness and transferability of computer vision models when applied in real-world agricultural environments (Bai *et al.*, 2024). At the same time, the generation of photorealistic digital twins via the utilization of Unity3D, PyTorch3D, and NeRF is a landmark in plant modeling. The application of NeRFs allows for view-consistent, high-fidelity three-dimensional reconstructions of plant morphology, allowing for accurate phenotypic assessment, that is, fruit detection and yield estimation under varying lighting and sensor conditions (Chopra *et al.*, 2024). The Fast Plant Digital Twin Network (FPDTN) improves this further with scalable and affordable solutions specifically tailored for IoT-enabled greenhouse systems, providing better morphological simulation accuracy compared to conventional approaches (Xu *et al.*, 2024a). In real-world deployments, NeRF-based models have been found to achieve F1-scores of up to 74.65%, confirming their resilience in varying environmental conditions (Arbab *et al.*, 2024). Architectural improvements, such as Instant-NGP and Instant-NSR, have further improved image quality and geometric convergence; however, computational costs and scene complexity remain as challenges (Hu *et al.*, 2024). The utilization of NeRFs in virtual reality systems is a promising field of research and education in plant science, with current systems attaining real-time rendering rates of 30 frames per second and a resolution of 1280 × 720 per eye (Li *et al.*, 2022).

In conjunction with these technological advancements, the development of closed-loop feedback simulation paradigms is being explored to enhance autonomous plant care. A significant contribution by Jiang *et al.* (2021) introduced a hybrid control system incorporating pixel-level domain adaptation and real-time policy transfer, thereby enabling robotic agents that have been trained in simulation to perform optimally in real-world contexts. Additionally, Jordan *et al.* (2019) introduced a biologically inspired closed-loop neural toolchain enabling constant interaction between neural agents and their environment, supporting reinforcement learning (RL) and the evolution of adaptive behavior. The foundation of these systems is model-free non-linear feedback optimization techniques, which dynamically interleave policy iteration with physical system responses to ensure both efficiency and stability (He *et al.*, 2023). For greenhouse management, model-based reinforcement learning (MBRL) presents a safe and sample-efficient alternative to standard policy training techniques. Zhang *et al.* (2021) illustrated the utilization of ensemble environment models and sample dropout modules to ensure robustness during extreme conditions, thereby enhancing the reliability of autonomous control in dynamic horticultural settings. Importantly, the integration of a secondary feedback loop providing proprioceptive environmental signals has been found to significantly augment the traditional sensorimotor loop, leading to enhanced self-monitoring and the evolution of adaptive behavior in autonomous systems (Toutounji and Pasemann, 2013).

13.9 Explainable AI and Causal Inference in Plant Monitoring

In this fast-paced era of precision agriculture, the need for XAI has grown exponentially as a method of establishing trust in plant monitoring automated systems. Interpretable models derived from methods, such as SHAP (SHapley Additive exPlanations), Grad-CAM, and LIME (Local Interpretable Model-agnostic Explanations), have been shown to be extremely potent in elucidating the underlying reasoning behind advanced ML and DL models. These methods improve global and local interpretability, thus enabling informed decision-making and permitting users, especially agronomists and farmers, to understand model outputs in the context of their own specific domain knowledge. For example, SHAP and LIME have been effectively used in applications such as gas well productivity appraisal and smart agriculture, illustrating how different input features influence predictive outcomes (Ma *et al.*, 2023). In plant disease diagnosis, deep CNNs such as Inception V3 and ResNet-9 are increasingly being used for plant pathogen classification based on image processing. Methods such as Grad-CAM enable visual interpretation of these models by generating saliency heatmaps that highlight the areas of an image that contribute most to the model's decision-making process, thus providing valuable insights into the focus of the model and its potential biases (Rakesh and Indiramma, 2022). However, one major limitation of perturbation-based explainers such as SHAP and LIME is their reliance on artificially perturbed input samples, which may not be representative of potential actual scenarios. To mitigate this, spectral zone-based SHAP/LIME methods have been proposed. These methods group features into semantically consistent groups, thus providing more realistic and intuitive explanations that are more in accordance with biological and ecological attributes (Contreras *et al.*, 2024). Despite ongoing debates over their validity and computational requirements, SHAP and LIME remain pillars of the field of XAI, especially due to their flexibility and versatility across different applications (Cartolano *et al.*, 2024).

Based on the idea of interpretability, causal DL has been a revolutionary technique for estimating the treatment effect in plant management systems. Causal models such as CURE (causal inference with unsupervised representations) demonstrate the application of large pretrained models in deriving causal relationships from observational data, thereby avoiding the requirement of expensive large-scale randomized controlled trials. This technology finds special relevance in plant science, where experimentation might be costly or pose logistical issues (Li, 2024). Similarly, the causal-deep neural network (CDNN) uses a two-stage training process, which first encodes pretreatment covariates and then predicts the treatment effects. Local Neyman orthogonality employed in CDNN makes the model robust to nuisance parameter variation, a commendable characteristic for an agronomic intervention that functions under varying environmental conditions (Nair *et al.*, 2022). Further, to make causal estimations even more transparent, the Match2 framework combines self-organizing maps (SOMs) with neural networks, allowing visually interpretable treatment matching. This attribute can be especially beneficial in plant management contexts, where visual tools can easily fill the gap between data scientists and field experts (Shou *et al.*, 2021). The Deep Treatment-Adaptive Network (DTANet) solves another significant problem in observational learning, that is, treatment-induced covariate shift, by building treatment-invariant feature representations. This method improves the reliability of causal estimates in longitudinal plant care applications where variables tend to vary after the intervention (Li *et al.*, 2022).

In support of causal inference and interpretability, counterfactual explanations (CEs) are becoming increasingly popular as powerful instruments for efficient and action-guided intervention planning in integrated pest management (IPM) and other agricultural approaches. The premise behind CEs is to propose small but optimal changes in input features that would have transformed an undesirable prediction, such as an outbreak of pests, into a preferable one. However, without robustness, the interventions might collapse when confronted with environmental fluctuation or noise in measurements. According to Virgolin and Fracaros, robust CEs are vital to ensure proposed strategies remain feasible and cost-effective under perturbations of real-world settings (Virgolin and Fracaros, 2023). New breakthroughs, such as the iterative CE generation technique by Maragno *et al.* (2024), improve the robustness and plausibility of CEs by approximating explanation regions stable against small variations in inputs. In IPM, Tsoumas *et al.* (2023) have proposed an integrated framework combining CEs with interpretable pest prediction models. This facilitates field-specific and adaptive treatment recommendations, averting farmer mistrust and the adoption of climate-smart and economically feasible farming practices.

13.10 Robustness, Continual Learning, and Concept Drift

To provide the trustworthiness of AI-based plant monitoring systems under real-world environments, it is crucial to take a multiperspective approach toward addressing adversarial attacks, data distribution variations, and sensor aging. Previous research has emphasized the need to incorporate adversarial robustness techniques, continuous learning techniques, and drift detection techniques into system design to improve the performance and robustness of systems under different agricultural and industrial environments. Adversarial robustness has emerged as a primary concern in the case of vision-based monitoring systems under uncontrolled environmental conditions. Adversarial training (AT) techniques have been found to be effective in improving object detection models such as YOLOv8 by boosting generalization abilities. With the inclusion of adversarial examples during training, AT enhances immunity to variations in environmental weather conditions, occlusion, and variation in light intensity, leading to an upsurge in mAP up to 5.6% and a decrease in false alarms by 10% under particular conditions (de Lange, 2024). However, the benefit of AT has the tendency to flatten under training across an extensive range of environmental variability, reflecting the need for more advanced generalization methodologies.

In multisensor fusion systems, especially those combining LiDAR and RGB image streams, as in autonomous driving technologies, adversarial robustness is limited by its weaknesses in 3D–2D alignment. Although the addition of feature denoising in AT can prevent some attacks, physical adversarial perturbations continue to undermine system integrity (Tu *et al.*, 2021). To overcome these weaknesses, Test-Time Transformation Ensembling (TTE) offers a model-agnostic improvement method, applying uniform transformations both during training and inference times. TTE enhances adversarial robustness against adversarial examples without damaging clean-sample performance and has been promising even under certified robustness constraints (Pérez *et al.*, 2021). Aside from adversarial defense, continual learning is critical in ensuring model adaptability in dynamic, constantly changing environments. Methods such as elastic weight consolidation (EWC) and online fine-tuning allow plant monitoring systems to maintain previously acquired knowledge and incorporate new knowledge. The Self-Refining Task-Free Diagnosis (SRTFD) framework is just one of the scalable methods by combining online continual learning (OCL) with Retrospective Coreset Selection and confidence-driven pseudo-labeling. This combination of the two removes data redundancy and class imbalance, enabling high-precision fault diagnosis in real-time industrial processes (Zhao *et al.*, 2024). Likewise, online RL has been crucial in adaptive optimization and predictive maintenance, enabling systems to refine decisions iteratively through real-time interaction with complex production environments (Doskenov and Okuyelu, 2025).

The assessment of OCL frameworks often struggles with over-reliance on aggregate accuracy metrics. Conventional accuracy metrics can mask model degradation through deceptive label correlations. To address this problem, Al Kader Hammoud *et al.* (2023) propose an alternative metric that places a high value on accuracy in near-future samples and thus a more precise measure of generalization and model stability. Besides, the Online-LoRA framework employs lightweight online regularization and adaptation of pretrained models to counteract catastrophic forgetting, particularly critical when task boundaries are vague or data streams are heavily non-stationary (Wei *et al.*, 2024). Drift detection and self-healing AI processes are also critical in continuous monitoring of plant phenotypes, particularly in the presence of genotypic evolution and environmental changes. Processes of genetic drift and adaptive response, that is, changes in flowering time, necessitate models that can dynamically recalculate in order to remain biologically and agronomically sound (Haag, 2021). The detection methods of concept drift are thus increasingly incorporated into predictive systems. Systems by Cerquitelli *et al.* incorporate self-assessment features for identifying when new data statistically differ from training distributions, thus triggering timely model updates (Cerquitelli *et al.*, 2019). In high-velocity data streams, the Fast Hoeffding Drift Detection Method (FHDDM) provides a statistically robust and efficient mechanism for detecting distributional changes with low false alarm rates and rapid adaptation (Pesaranghader and Viktor, 2016).

Industrial process application, especially in the production of synthetic natural gas, relies heavily on the reliability of sensors. Auto-associative kernel regression has been used with enormous success to detect sensor drift and thus maintain data integrity during environmental condition variations (Lee *et al.*, 2017). In plant phenotyping systems, deployment of such sensor health diagnostics guarantees the reliability of ensuing decisions, such as yield estimation, disease detection, and stress response classification. Overall, the use of adversarial robustness techniques, continual learning architecture, and drift detection algorithms forms a general framework for advanced and robust systems of plant monitoring. These innovations enable real-time adaptation, adaptive resistance, and maintain consistent effectiveness in dynamic and complex environments, ushering in the beginning of a new era of autonomous and self-optimizing agricultural intelligence.

13.11 Real-Time Bio-Signal Integration and Plant Neurobiology

Integration of electrophysiological signals such as action potentials and calcium waves into real-time plant monitoring systems is a fast-developing field. This process combines sensor technologies with ML to assess plant health and stress response with high precision, as explained in Fig. 13.2. Sophisticated sensor configurations have been designed to non-invasively detect minute electrical signals from plants even in uncontrolled environments. For example, an amplification device based on integrated operational amplifiers in combination with patch electrodes has been able to detect electrical signals from stressed aquatic *Scindapsus* and *Guaiacum* species, exhibiting characteristic patterns of signals in response to environmental stressors such as mechanical pressure and flooding (Cai & Qi, 2017). Emerging technologies in the pipeline are sensors capable of being deployed outside of Faraday cages, enabling real-time measurement of bioelectrical activity and automatic classification of plant physiological conditions using supervised ML. Such systems are able to detect changes due to water stress or circadian rhythms and thus provide key information for agronomic decision-making (Tran *et al.*, 2019). Plants also employ sophisticated long-distance signaling networks, including electrical, calcium, hydraulic, and reactive oxygen species (ROS) signals, to coordinate systemic responses to environmental stimuli. These signals control gene expression and developmental processes, where receptor-like proteins play key roles in signal transmission (Choi *et al.*, 2016; Fichman and Mittler, 2021). For further research in this direction,

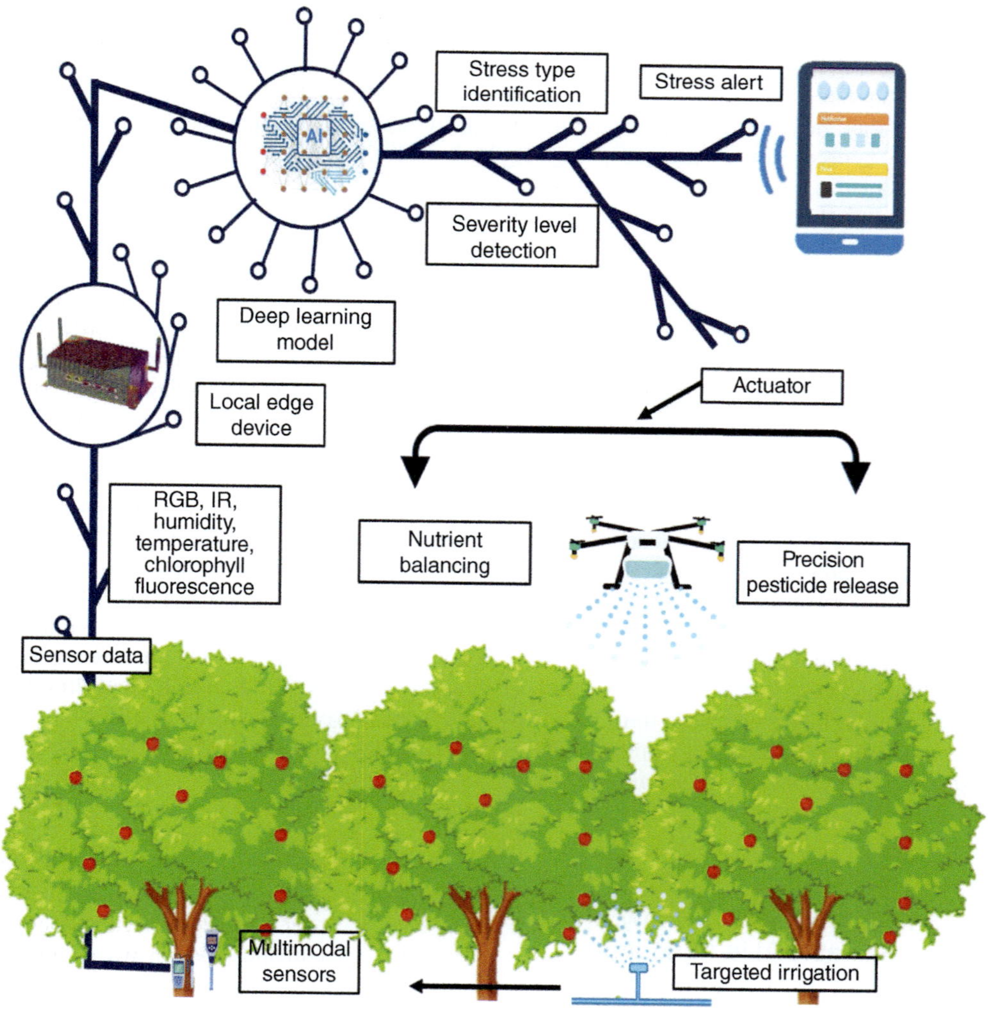

Fig. 13.2. Real-time plant stress detection and autonomous response system. (Author's own image.)

new phytosensing systems are being developed that autonomously convert electrophysiological signals to actionable information, potentially probing into plant cognition and interfacing with AI-enabled platforms for continuous monitoring (Sun *et al.*, 2023).

At the interface of biological science and AI, biohybrid AI models are being created as cutting-edge tools for the cause of environmental monitoring and management. The PhytoNodes system is an exemplar of this interfacing by making use of living plants, *Zamioculcas zamiifolia*, as biological sensors. The plants are capable of perceiving external stimuli, including light and temperature, and their electrophysiological responses are processed with artificial neural networks on energy-efficient microcontrollers to acquire accurate classification (Buss *et al.*, 2022). Moreover, the BiœmuS system is an exemplar of real-time simulation of plant neural activity, paving the way for biohybrid experimentation and future neuroprosthetic use (Levi *et al.*, 2023). Interactive robotics research has also demonstrated that ML can imitate plant behavior, allowing robots to modulate and enhance plant growth through intelligent modulation of environmental parameters (Wahby *et al.*, 2018). In parallel, DL has emerged as an essential element in the early warning systems of agriculture, particularly with respect to disease detection. Methods such as CNNs and DenseNet architectures have been used to identify early indications of plant disease from images of leaves, allowing timely intervention to avoid crop loss (Priyadharshini *et al.*, 2023). These models use data augmentation and are increasingly being used with autonomous drones for efficient monitoring of vast agricultural fields. However, challenges such as environmental variability, scarce availability of labeled datasets, and generalizability across various crop species must be overcome to guarantee the scalability and reliability of DL-based methodologies (Shao and Devaraj, 2024).

13.12 Ethics, Biodiversity, and Sustainable AI in Agriculture

The moral application of AI in agriculture calls for a high-tech combination of boosting crop yields and preserving biodiversity. On the one hand, AI technologies such as predictive analytics, ML, and computer vision have revolutionized yield enhancement by facilitating precision agriculture, optimizing resource use, and improving food security, particularly through advances in plant monitoring and disease diagnosis (Chavan *et al.*, 2024). But these advances create infrastructural and economic costs, particularly in the developing world. On the other hand, AI holds a revolutionary capability for biodiversity protection. Initiatives such as CAPTAIN use AI for spatial conservation prioritization, which surpasses traditional methods by integrating complicated ecological datasets and citizen science inputs to conserve a greater number of species from extinction (Antonelli *et al.*, 2021). Additionally, models with continuous land protection optimization highlight the necessity for localized ecological issues and policy mechanisms to deliver sustainable results (Armsworth *et al.*, 2020), while international compacts such as the Convention on Biological Diversity highlight the imperative of conserving microbial diversity and greater ecological networks (Arya and Shinde, 2022).

The abovementioned advances highlight the need for a robust data governance structure, particularly for AI-driven systems processing sensitive, cross-modal plant-related data. The General Data Protection Regulation (GDPR) offers a balanced legal basis in the form of principles of legality, transparency, data minimization, and confidentiality (Schwartmann *et al.*, 2024), while the EU AI Act supplements these with algorithmic accountability and moral deployment of AI (Musch *et al.*, 2024). However, there are challenges in mapping the growing autonomy of AI systems to existing regulatory regimes (Jędrzejczak, 2024), which highlights the need for adaptive technical frameworks integrating privacy protection over multimodal data (Lei *et al.*, 2020). Globally, variations in data governance, such as Japan's APPI and China's PIPL, highlight the need for harmonized, contextual regulatory frameworks to facilitate cross-border AI innovation while instilling trust and conformance (Laksito *et al.*, 2024). Simultaneously, AI-facilitated technologies are revolutionizing conservation agriculture and tracking rare species, as seen through platforms such as Conservation AI, which utilize CNNs and transformer models to automate species identification and poaching detection from visual and thermal images, thereby providing immediate response as well as long-term ecological monitoring (Fergus *et al.*, 2024). These advances consolidate multimodal environmental data for proactive interventions, strategic habitat decision-making, and transition toward sustainable farming (Dwivedi *et al.*, 2024), thus highlighting AI's status as a double agent for productivity and ecological conservation.

13.13 Global Case Studies and Benchmarking

Plantix, PEAT, and Taranis case studies demonstrate the intersection of agricultural applications in real-world settings using technology-driven solutions, bridging research and field applicability. Plantix, an intelligent farming application, leverages smart information and communication technology to aid farmers in identifying crop diseases and pests using mobile phones. Analysis of user reviews on Google Play Store using latent Dirichlet allocation (LDA) topic modeling identified positive comments, along with suggestions for enhancing plant-specific functionalities, emphasizing the utility of the application in generating early warnings on diseases and providing actionable interventions (Siow *et al.*, 2023). Specific technical features of PEAT and Taranis were not mentioned in the reviewed literature but are indicative of the broader interest in applied technologies in educational and agricultural contexts. The initiative of Open Case Studies provides a pedagogical model for the incorporation of real-world agricultural issues into data science courses with peer-reviewed, standalone data guides, promoting active learning. Similarly, experiential learning models in engineering education have been shown to enhance conceptual understanding and contextual applicability, which are key to the development of data-literate professionals who can solve real-world problems (Wright *et al.*, 2024).

Open datasets such as CVPPP, PlantCLEF, Pheno4D, and TerraByte are essential HTP resources that enable scientists to plan and test sophisticated tools for yield prediction, climate-resilient crop selection, and pest detection from multimodal sensor data, for example, RGB and hyperspectral imaging (Danilevicz *et al.*, 2021). The datasets combat common computer vision-based plant phenotyping challenges through enabling quantitative analysis and geometric trait analysis (Meraj *et al.*, 2024), while online resources such as plant-image-analysis.org enable researchers to discover and utilize more than 90 plant image analysis software packages for specific experimental purposes (Lobet *et al.*, 2013). Metrics such as the F1-score, Intersection over Union (IoU), and Fréchet inception distance (FID) are essential to determining the quality and usability of synthetic plant image data. The F1-score, balancing precision and recall perfectly, has been found useful in applications such as Hayai-Annotation Plants for gene functional annotations assessment (Ghelfi *et al.*, 2023), while IoU is a common measure in image segmentation used to measure accuracy in leaf counting and plant localization (Miao *et al.*, 2019). FID is a key benchmark used to determine the authenticity of generated plant images; for instance, in synthetic weed image generation, diffusion models have been proven superior to GANs by attaining lower FID scores, hence ensuring higher fidelity and diversity of synthetic samples (Chen *et al.*, 2024). Overall, these metrics provide a solid framework for the assessment of synthetic data in plant recognition and classification processes, hence ensuring improvement in automated agriculture, rare species identification, and weed detection (Dasgupta *et al.*, 2024).

13.14 Emerging Directions in Space and Controlled Agriculture

DL plays a vital role in the design of closed-loop systems in space agriculture, especially in missions such as NASA's Veggie and ESA's MELiSSA, aimed at facilitating sustainable food production under microgravity conditions during long-duration space missions. The use of machine vision and multisensor fusion in autonomous plant growth systems enables real-time monitoring and management of crops, thereby reducing reliance on human intervention while inducing the growth of nutrient-rich crops such as Sugar Ann Snap Peas and Red Robin Tomatoes (Morales-Badajoz *et al.*, 2023). Among various soilless cultivation techniques, fogponics is especially useful for agricultural operations in space because it can provide nutrient-rich mist directly to the plant roots, providing tremendous benefits in cases where gravity-based systems are not feasible. However, optimization of these systems is a major challenge, which requires the use of AI-enhanced closed-loop control techniques to improve operational efficiency and responsiveness to the environment (Suganob *et al.*, 2024). The history of developing space plant growth modules, from simple experiments to sophisticated bio-regenerative systems such as Veggie, highlights the importance of technological progress in controlled-environment agriculture, which is the backbone of life support systems for extraterrestrial environments (Zabel *et al.*, 2016).

Within this paradigm, neural network controllers demonstrate considerable potential for system stabilization under uncertain extraterrestrial conditions, particularly when trained by reinforcement learning or model predictive control techniques (Lopez *et al.*, 2019). The combination of AI and biospheric systems is a revolutionary advancement in space agriculture, especially the combination of AI with fogponics, which allows continuous monitoring of the environment and adaptive adjustment of major parameters such as humidity, light intensity, and aspects of nutrient delivery required to optimize plant growth in microgravity (Kaya, 2025). Hydroponics, another gravity-free method of cultivation, can also be employed in extraterrestrial agriculture through misting technologies and temperature gradients to manage fluid dynamics and nutrient uptake (Hasenstein and Miklave, 2024). Moreover, the introduction of liquid bio-stimulant media, made up of precisely balanced nutrients and growth accelerators, adds the hardiness of plants to the extraterrestrial stressors imposed by microgravity (Katiyar, 2024). Taken together, these technologies not only affirm the viability of sustainable food and oxygen generation in prolonged space missions but also facilitate waste recycling and regeneration of the life support system. The inclusion of AI in these systems allows predictive maintenance and anticipatory decision-making, a crucial step toward the development of autonomous, self-sustaining agricultural systems for deep space exploration (Kaya, 2025).

13.15 Conclusion

As agriculture is facing mounting challenges due to climate change, food security needs, and sustainability issues, the combination of DL and real-time plant monitoring is a compelling argument. Throughout this chapter, the manner in which cutting-Edge AI architectures, multimodal sensor fusion, and edge-deployable intelligence can revolutionize the way we observe, understand, and respond to plant behavior in diverse environments, from traditional farmlands to extraterrestrial biospheres, has been illustrated. Following the developments in continuous learning, explainability, and biohybrid modeling, plant monitoring systems become more resilient, adaptive, and ethics-compliant. The future direction implies the design of scalable, privacy-aware, and sustainable AI ecosystems that possess the potential to make plant intelligence a fundamental part of future agricultural practices. Merging computational sciences with biology does not merely optimize crop growth but also enhances the broader objective of sustaining ecological stability and preserving biodiversity.

Acknowledgment

All the authors are thankful to their respective universities and institutions for their valuable support.

Data Availability

The authors confirm that the data supporting the findings of this study are available within the book/chapter.

References

Ahmad, I., Shariffudin, S.E., Ramli, A.F., Maharum, S.M.M., Mansor, Z. *et al.* (2021) Intelligent plant monitoring system via IoT and fuzzy system. In: *2021 IEEE 7th International Conference on Smart Instrumentation, Measurement and Applications (ICSIMA)*, IEEE, pp. 123–127.

Al Kader Hammoud, H.A., Prabhu, A., Lim, S.N., Torr, P.H., Bibi, A. *et al.* (2023) Rapid adaptation in online continual learning: Are we evaluating it right? In: *Proceedings of the IEEE/CVF International Conference on Computer Vision*, pp. 18852–18861.

Al Mamun, A., Ahmedt-Aristizabal, D., Zhang, M., Hossen, M.I., Hayder, Z. *et al.* (2024) *Plant Disease Detection Using Self-supervised Learning: A Systematic Review*. IEEE Access.

Ali, M., Salma, M., El Haji, M. and Jamal, B. (2025) Plant disease detection using vision transformers. *International Journal of Electrical and Computer Engineering (*IJECE)15(2), 2334–2344.

Antonelli, A., Goria, S., Sterner, T. and Silvestro, D. (2021) Optimising biodiversity protection through artificial intelligence. *bioRxiv*.

Antonelo, E.A., Camponogara, E., Seman, L.O., de Souza, E.R., Jordanou, J.P. *et al.* (2021) Physics-informed neural nets-based control. arXiv. *arXiv Preprint arXiv:2104.02556*.

Armsworth, P.R., Benefield, A.E., Dilkina, B., Fovargue, R., Jackson, H.B. *et al.* (2020) Allocating resources for land protection using continuous optimization: Biodiversity conservation in the United States. *Ecological Applications* 30(6), e02118.

Arshad Arbab, M., Jubery, T., Afful, J., Jignasu, A., Balu, A. *et al.* (2024) Evaluating nerfs for 3d plant geometry reconstruction in field conditions.

Arthur, E.A.E., Wulnye, F.A., Gookyi, D.A.N., Agyekum, K.O.B.O., Danquah, P. *et al.* (2024) Edge impulse vs tensorFlow: A comparative analysis of TinyML platforms for maize leaf disease identification. In: *2024 Conference on Information Communications Technology and Society (ICTAS)*, IEEE, pp. 1–6.

Arya, R. and Shinde, S. (2022) Legal protection of microbial biodiversity. In: *Microbial Diversity in Hotspots*. Academic Press, pp. 349–364.

Bai, K., Zhang, L., Chen, Z., Wan, F. and Zhang, J. (2024) Close the sim2real gap via physically-based structured light synthetic data simulation. In: *2024 IEEE International Conference on Robotics and Automation (ICRA)*, IEEE, pp. 17035–17041.

Basu, P. and Pal, A. (2012) A new tool for analysis of root growth in the spatio-temporal continuum. *New Phytologist* 195(1), 264–274.

Bhamare, A., Upadhyay, V. and Bansal, P. (2023) AI based plant growth monitoring system using computer vision. In: *2023 IEEE Technology & Engineering Management Conference-Asia Pacific (TEMSCON-ASPAC)*, IEEE, pp. 1–5.

Boogaard, F.P., van Henten, E.J. and Kootstra, G. (2023) The added value of 3D point clouds for digital plant phenotyping–A case study on internode length measurements in cucumber. *Biosystems Engineering* 234, 1–12.

Bounoua, I., Saidi, Y., Yaagoubi, R. and Bouziani, M. (2024) Deep learning approaches for water stress forecasting in arboriculture using time series of remote sensing images: Comparative study between convlstm and cnn-lstm models. *Technologies* 12(6), 77.

Bui, T.D. and Do Le, T.M. (2025) Ghost-Attention-YOLOv8: Enhancing rice leaf disease detection with lightweight feature extraction and advanced attention mechanisms. *AgriEngineering* 7(4), 93.

Buss, E., Rabbel, T.L., Horvat, V., Krizmancic, M., Bogdan, S. *et al.* (2022) Phytonodes for environmental monitoring: Stimulus classification based on natural plant signals in an interactive energy-efficient bio-hybrid system. In: *Proceedings of the 2022 ACM Conference on Information Technology for Social Good*, pp. 258–264.

Cai, W. and Qi, Q. (2017) Study on electrophysiological signal monitoring of plant under stress based on integrated Op-Amps and patch electrode. *Journal of Electrical and Computer Engineering* 2017(1), 4182546.

Cartolano, A., Cuzzocrea, A. and Pilato, G. (2024) Analyzing and assessing explainable AI models for smart agriculture environments. *Multimedia Tools and Applications* 83(12), 37225–37246.

Cerquitelli, T., Proto, S., Ventura, F., Apiletti, D. and Baralis, E. (2019) Automating concept-drift detection by self-evaluating predictive model degradation. arXiv. *arXiv Preprint arXiv:*1907.*08120*.

Chavan, L., Kumbhalkar, R., Agrawal, P., Rakesh, N., Parvathi, R. *et al.* (2024) AI in agriculture: yield optimization and food security. In: *2024 4th International Conference on Technological Advancements in Computational Sciences (ICTACS)*, IEEE, pp. 536–541.

Chen, D., Qi, X., Zheng, Y., Lu, Y., Huang, Y. *et al.* (2024) Synthetic data augmentation by diffusion probabilistic models to enhance weed recognition. *Computers and Electronics in Agriculture* 216, 108517.

Chen, H., Flores, G.E.C. and Li, C. (2024) Physics-informed neural networks with hard linear equality constraints. *Computers & Chemical Engineering* 189, 108764.

Chen, R., Zhang, Y., Song, W.J., Zhao, T.T., Wang, J.N. *et al.* (2024) High-precision identification of highly similar *Pinelliae rhizoma* and adulterated *Rhizoma pinelliae* pedatisectae through deep neural networks based on vision transformers. *Journal of Food Science* 89(11), 7372–7379.

Chen, R.T., Amos, B. and Nickel, M. (2020) Learning neural event functions for ordinary differential equations. arXiv. *arXiv Preprint arXiv: 2011.03902*.

Cheng, J.Y. (2024) Plant disease detection using deep learning. Doctoral dissertation, UTAR.

Choi, D. and Devakirubai, B. (2023) Towards generalizable crop-agnostic plant disease recognition: An unsupervised-learning approach. *Journal of Student Research* 12.

Choi, H.B., Park, J.K., Park, S.H. and Lee, T.S. (2024) NeRF-based 3D reconstruction pipeline for acquisition and analysis of tomato crop morphology. *Frontiers in Plant Science* 15, 1439086.

Choi, W.G., Hilleary, R., Swanson, S.J., Kim, S.H. and Gilroy, S. (2016) Rapid, long-distance electrical and calcium signaling in plants. *Annual Review of Plant Biology* 67(1), 287–307.

Chopra, S., Cladera, F., Murali, V. and Kumar, V. (2024) AgriNeRF: Neural radiance fields for agriculture in challenging lighting conditions *arXiv preprint arXiv: 2409.15487*.

Chung, B. (2024) Addressing data imbalance in plant disease recognition through contrastive learning. In: *2024 IEEE 3rd International Conference on AI in Cybersecurity (ICAIC)*, IEEE, pp. 1–6.

Claro, J., Dias, B., Rodrigues, B., Pimentão, J.P., Sousa, P. *et al.* (2014) Autonomous robot integration in surveillance system: Architecture and communication protocol for systems cooperation. In: *2014 16th International Power Electronics and Motion Control Conference and Exposition*, IEEE, pp. 713–719.

Contreras, J., Winterfeld, A., Popp, J. and Bocklitz, T. (2024) Spectral zones-based SHAP/LIME: Enhancing interpretability in spectral deep learning models through grouped feature analysis. *Analytical Chemistry* 96(39), 15588–15597.

Danilevicz, M.F., Bayer, P.E., Nestor, B.J., Bennamoun, M. and Edwards, D. (2021) Resources for image-based high-throughput phenotyping in crops and data sharing challenges. *Plant Physiology* 187(2), 699–715.

Dasgupta, D., Mondal, A. and Chakrabarti, P.P. (2024) Can synthetic plant images from generative models facilitate rare species identification and classification? In: *Proceedings of the IEEE/CVF Conference on Computer Vision and Pattern Recognition*, pp. 7530–7540.

De Brouwer, E. and Krishnan, R.G. (2023) Anamnesic neural differential equations with orthogonal polynomial projections. arXiv. *arXiv Preprint arXiv: 2303.01841*.

de Lange, T. (2024) Improving Real-Life Robustness by Increasing Adversarial Robustness.

Debbagh, M., Sun, S. and Lefsrud, M. (2024) Predictive pattern recognition techniques towards spatiotemporal representation of plant growth in simulated and controlled environments: A comprehensive review. arXiv. *arXiv Preprint arXiv: 2412.10538*.

Diehl, F., Brunner, T., Le, M.T. and Knoll, A. (2019) Graph neural networks for modelling traffic participant interaction. In: *2019 IEEE Intelligent Vehicles Symposium (IV)*, IEEE, pp. 695–701.

Doskenov, B. and Okuyelu, O. (2025) Advancing production systems with online reinforcement learning: Real-time monitoring, control, and optimization. *Current Journal of Applied Science and Technology* 44(2), 1–22.

Du, R., Ma, Z., Xie, P., He, Y. and Cen, H. (2023) PST: Plant segmentation transformer for 3D point clouds of rapeseed plants at the podding stage. *ISPRS Journal of Photogrammetry and Remote Sensing* 195, 380–392.

Duits, R., Smets, B., Bekkers, E. and Portegies, J. (2021) Equivariant deep learning via morphological and linear scale space PDEs on the space of positions and orientations. In: *International Conference on Scale Space and Variational Methods in Computer Vision*, Springer International Publishing, pp. 27–39.

Dutta, P.K., Raj, P., Sundaravadivazhagan, B. and Selvan, C.P. (eds) (2024) *Artificial Intelligence Solutions for Cyber-Physical Systems*. Taylor & Francis Limited.

Dwarampudi, A. and Yogi, M.K. (2024) Application of federated learning for smart agriculture system. *International Journal of Information Technology and Computer Science (IJITCS) ISSN* 2455–5290.

Dwivedi, D.N., Mahanty, G. and nath Dwivedi, V. (2024) Intelligent conservation: A comprehensive study on AI-enhanced environmental monitoring and preservation. In: *The Convergence of Self-Sustaining Systems with AI and IoT*. IGI Global, pp. 215–226.

Fergus, P., Chalmers, C., Longmore, S. and Wich, S. (2024) Harnessing artificial intelligence for wildlife conservation. *Conservation* 4(4), 685–702.

Fichman, Y. and Mittler, R. (2021) Integration of electric, calcium, reactive oxygen species and hydraulic signals during rapid systemic signaling in plants. *The Plant Journal* 107(1), 7–20.

Fontes, L., Machado, P., Vinkemeier, D., Yahaya, S., Bird, J.J. *et al.* (2024) Enhancing stress detection: A comprehensive approach through rPPG analysis and deep learning techniques. *Sensors* 24(4), 1096.

Fountsop, A.N., Ebongue Kedieng Fendji, J.L. and Atemkeng, M. (2020) Deep learning models compression for agricultural plants. *Applied Sciences* 10(19), 6866.

Ghelfi, A., Shirasawa, K. and Isobe, S. (2023) Benchmarking hayai-annotation plants: A re-evaluation using standard evaluation metrics. *bioRxiv* 2023–09.

Gou, C., Zafar, S., Fatima, N., Hasnain, Z., Aslam, N. *et al.* (2024) Machine and deep learning: Artificial intelligence application in biotic and abiotic stress management in plants. *Frontiers in Bioscience-Landmark* 29(1).

Haag, C. (2021) Separating adaptation from drift: A cautionary tale from a self-fertilizing plant. *Peer Community in Evolutionary Biology* 100128.

Hao, W. (2023) Symmetry-regularized neural ordinary differential equations. arXiv. *arXiv Preprint arXiv: 2311.16628*.

Hartley, Z.K., Lind, R.J., Pound, M.P. and French, A.P. (2024) Domain targeted synthetic plant style transfer using stable diffusion LoRA and ControlNet. In: *Proceedings of the IEEE/CVF Conference on Computer Vision and Pattern Recognition*, pp. 5375–5383.

Hasenstein, K.H. and Miklave, N.M. (2024) Hydroponics for plant cultivation in space–A white paper. *Life Sciences in Space Research* 43, 13–21.

Hati, A.J. and Singh, R.R. (2020) Towards smart agriculture: A deep learning based phenotyping scheme for leaf counting. In: *2020 International Conference on Smart Technologies in Computing, Electrical and Electronics (ICSTCEE)*, IEEE, pp. 510–514.

He, Z., Bolognani, S., He, J., Dörfler, F. and Guan, X. (2023) Model-free nonlinear feedback optimization. *IEEE Transactions on Automatic Control* 69(7), 4554–4569.

Hidalgo, C.H., González-Vidal, A. and Skarmeta, A.F. (2023) Edge computing in smart agriculture scenario based on TinyML for irrigation control. In: *2023 IEEE 9th World Forum on Internet of Things (WF-IoT)*. IEEE, pp. 01–08.

Hossain, I., Fanfani, V., Fischer, J., Quackenbush, J. and Burkholz, R. (2024) Biologically informed neural-odes for genome-wide regulatory dynamics. *Genome Biology* 25(1), 127.

Hu, K., Ying, W., Pan, Y., Kang, H. and Chen, C. (2024) High-fidelity 3D reconstruction of plants using neural radiance fields. *Computers and Electronics in Agriculture* 220, 108848.

Huo, L., Xia, J., Zhang, L., Zhang, H. and Xu, M. (2023) Multimodal hyperspectral image classification via interconnected fusion. arXiv. *arXiv Preprint arXiv: 2304.00495*.

Hussain, M. (2023) YOLO-v1 to YOLO-v8, the rise of YOLO and its complementary nature toward digital manufacturing and industrial defect detection. *Machines* 11(7), 677.

Ibrahim, S.S., Chaladi, P., Galam, B.S., Maridu, B.V. and Matta, D. (2024) Stress detection using facial emotion recognition: Comparative analysis of VGG-16, resnet-50, and a custom CNN architecture. In: *2024 5th International Conference on Data Intelligence and Cognitive Informatics (ICDICI)*, IEEE, pp. 1247–1253.

Idoje, G., Dagiuklas, T. and Iqbal, M. (2023) Federated learning: Crop classification in a smart farm decentralised network. *Smart Agricultural Technology* 5, 100277.

Jędrzejczak, M. (2024) Protection of personal data processed in artificial intelligence systems. *Gdańskie Studia Prawnicze* 3(64), 86–98.

Jiang, R., He, B., Wang, Z., Zhou, Y., Xu, S. *et al.* (2021) A novel simulation-reality closed-loop learning framework for autonomous robot skill learning. *IEEE Transactions on Cognitive and Developmental Systems* 14(4), 1520–1531.

Jingzhao, L. and Mingming, G. (2019) A monitoring system of ground-air interaction multi-sensory fusion plant growth condition. In: *International Conference on Applications and Techniques in Cyber Security and Intelligence ATCI 2018: Applications and Techniques in Cyber Security and Intelligence*, Springer International Publishing, pp. 325–332.

Jones, H.G. (2014) *Plants and microclimate: A quantitative approach to environmental plant physiology*. Cambridge university press.

Jones, M.B. (1985) Plant microclimate. In: *Techniques in Bioproductivity and Photosynthesis*. Pergamon, pp. 26–40.

Jordan, J., Weidel, P. and Morrison, A. (2019) A closed-loop toolchain for neural network simulations of learning autonomous agents. *Frontiers in Computational Neuroscience* 13, 46.

Katiyar, R. (2024) Optimizing plant growth in microgravity: A biostimulant growth medium for agriculture in deep space environments.

Kaya, C. (2025) Intelligent environmental control in plant factories: Integrating sensors, automation, and AI for optimal crop production. *Food and Energy Security* 14(1), e70026.

Kempelis, A., Romanovs, A. and Patlins, A. (2024) Review on application of vision transformers in IoT edge devices for plant sensor measurement forecasting. In: *2024 IEEE 11th Workshop on Advances in Information, Electronic and Electrical Engineering (AIEEE)*. IEEE, pp. 1–7.

Khan, A.T., Jensen, S.M. and Khan, A.R. (2025) Advancing precision agriculture: A comparative analysis of YOLOv8 for multi-class weed detection in cotton cultivation. *Artificial Intelligence in Agriculture*.

Kim, T., Kim, H., Baik, K. and Choi, Y. (2022) Instance-aware plant disease detection by utilizing saliency map and self-supervised pre-training. *Agriculture* 12(8), 1084.

Klose, R., Möller, K., Vielstädte, C. and Ruckelshausen, A. (2010) Modular system architecture for individual plant phentotyping with an autonomous field robot. In: *Proceedings of the 2nd International Conference of Machine Control & Guidance*, pp. 299–307.

Kwak, J., Jang, S.J., Shin, H. and Park, J. (2024) Latency-aware pruning of neural networks via structure search with integer programming. In: *2024 24th International Conference on Control, Automation and Systems (ICCAS)*, IEEE, pp. 15–16.

Laksito, J., Pratiwi, B. and Ariani, W. (2024) Harmonizing data privacy frameworks in artificial intelligence: Comparative insights from Asia and Europe. *Perkara: Jurnal Ilmu Hukum Dan Politik* 2(4), 579–588.

Lee, T., Kim, J.Y., Shin, J., Kim, J. and Yeom, C. (2017) Sensor drift detection in SNG plant using auto-associative kernel regression. In: *2017 IEEE International Systems Engineering Symposium (ISSE)*, IEEE, pp. 1–4.

Lei, Y., Duan, Y. and Song, M. (2020) Technical implementation framework of AI governance policies for cross-modal privacy protection. In: *International Conference on Collaborative Computing: Networking, Applications and Worksharing*, Springer International Publishing, pp. 431–443.

Levi, T., Beaubois, R., Cheslet, J., Duenki, T., Khoyratee, F. *et al.* (2023) BiœmuS: A new tool for neurological disorders studies through real-time emulation and hybridization using biomimetic spiking neural network.

Li, S. (2024) Large pre-trained models for treatment effect estimation: Are we there yet? *Patterns* 5(6).

Li, A., Markovic, M., Edwards, P. and Leontidis, G. (2024) Model pruning enables localized and efficient federated learning for yield forecasting and data sharing. *Expert Systems with Applications* 242, 122847.

Li, C., Li, Y., Zhang, Z., Wan, J., Cao, Y. *et al.* (2023a) A data-driven edge computing solution for real-time monitoring of industrial load conditions. In: *2023 IEEE International Conference on Energy Technologies for Future Grids (ETFG)*, IEEE, pp. 1–5.

Li, Z., Li, Y., Zhao, P., Song, R., Li, X. *et al.* (2023b) Is synthetic data from diffusion models ready for knowledge distillation? *arXiv Preprint arXiv: 2305.12954*.

Li, D., Li, J., Xiang, S. and Pan, A. (2022) PSegNet: Simultaneous semantic and instance segmentation for point clouds of plants. *Plant Phenomics* 2022. DOI: 10.34133/2022/9787643

Li, K., Rolff, T., Schmidt, S., Bacher, R., Frintrop, S. *et al.* (2022) Immersive neural graphics primitives. *arXiv Preprint arXiv: 2211.13494*.

Li, Q., Wang, Z., Liu, S., Li, G. and Xu, G. (2022) Deep treatment-Adaptive network for causal inference. *The VLDB Journal* 31(5), 1127–1142.

Lobet, G., Draye, X. and Périlleux, C. (2013) An online database for plant image analysis software tools. *Plant Methods* 9, 1–8.

Lopez, D.M., Musau, P., Tran, H.D. and Johnson, T.T. (2019) Verification of closed-loop systems with neural network controllers. In: *ARCH@ CPSIoTWeek*. pp. 201–210.

Lu, Y. and Lu, G. (2022) 3d modelling beneath ground: Plant root detection and reconstruction based on ground-penetrating radar. In: *Proceedings of the IEEE/CVF Winter Conference on Applications of Computer Vision*, pp. 68–77.

Ma, X., Hou, M., Zhan, J. and Liu, Z. (2023) Interpretable predictive modelling of tight gas well productivity with SHAP and LIME techniques. *Energies* 16(9), 3653.

Manjah, D., Galland, S., De Vleeschouwer, C. and Macq, B. (2024) Autonomous methods in multisensor architecture for smart surveillance. In: *ICAART (3)*. pp. 823–831.

Maragno, D., Kurtz, J., Röber, T.E., Goedhart, R., Birbil, Ş.I. *et al.* (2024) Finding regions of counterfactual explanations via robust optimization. *INFORMS Journal on Computing* 36(5), 1316–1334.

Meenakshi, B., Sumathy, B., Haariharan, N.C., Krishnakanth, L. and Soumya, M. (2023) Artificial intelligence enhanced-plant and soil monitoring systems. In: *2023 Intelligent Computing and Control for Engineering and Business Systems (ICCEBS)*. IEEE, pp. 1–6.

Meraj, T., Sharif, M.I., Raza, M., Alabrah, A., Kadry, S. *et al.* (2024) Computer vision-based plants phenotyping: A comprehensive survey. *Iscience* 27(1).

Miao, C., Hoban, T.P., Pages, A., Xu, Z., Rodene, E. *et al.* (2019) Simulated plant images improve maize leaf counting accuracy. *BioRxiv,* 706994.

Mirande, K., Godin, C., Tisserand, M., Charlaix, J., Besnard, F. *et al.* (2022) A graph-based approach for simultaneous semantic and instance segmentation of plant 3D point clouds. *Frontiers in Plant Science* 13, 1012669.

Mondal, S., Williamson, A., Xu, Z. and Tsourdos, A. (2020) Autonomous architecture for uav-based agricultural survey. In: *In AIAA Scitech 2020 Forum*. p. 2298.

Morales-Badajoz, A.M., Elieh, N., Diederich, A., Sadler, E., Glover, J. *et al.* (2023) Astro cultivators: Autonomous growth system for space farming based on machine vision and multi-sensor fusion. In: *Proceedings of Cyber-Physical Systems and Internet of Things Week 2023*, pp. 385–390.

Mu, J., Feng, Q., Yang, J., Zhang, J. and Yang, S. (2024) Few-shot disease recognition algorithm based on supervised contrastive learning. *Frontiers in Plant Science* 15, 1341831.

Musch, S., Borrelli, M.C. and Kerrigan, C. (2024) Bridging compliance and innovation: A comparative analysis of the EU AI act and GDPR for enhanced organisational strategy. *Journal of Data Protection & Privacy* 7(1), 14–40.

Nair, N., Gurumoorthy, K.S. and Mandalapu, D. (2022) Individual treatment effect estimation through controlled neural network training in two stages. *arXiv Preprint arXiv: 2201.08559*.

Nasery, A., Shah, H., Suggala, A.S. and Jain, P. (2024) End-to-end neural network compression via l1/l2 regularized latency surrogates. In: *Proceedings of the IEEE/CVF Conference on Computer Vision and Pattern Recognition*, pp. 5866–5877.

Nguyen, C., Sagan, V., Bhadra, S. and Moose, S. (2023) UAV multisensory data fusion and multi-task deep learning for high-throughput maize phenotyping. *Sensors* 23(4), 1827.

O'Rourke, S. and Kelly, G.E. (2015) Spatio-temporal modelling of forest growth spanning 50 years–the effects of different thinning strategies. *Procedia Environmental Sciences* 26, 101–104.

Osco, L.P., de Arruda, M.D.S., Gonçalves, D.N., Dias, A., Batistoti, J. *et al.* (2021) A CNN approach to simultaneously count plants and detect plantation-rows from UAV imagery. *ISPRS Journal of Photogrammetry and Remote Sensing* 174, 1–17.

Pagès, L., Moreau, D., Sarlikioti, V., Boukcim, H. and Nguyen, C. (2012) ArchiSimple: A parsimonious model of the root system architecture. In: *2012 IEEE 4th International Symposium on Plant Growth Modelling, Simulation, Visualization and Applications*, IEEE, pp. 297–303.

Pałubicki, W., Makowski, M., Gajda, W., Hädrich, T., Michels, D.L. *et al.* (2022) Ecoclimates: Climate-response modelling of vegetation. *ACM Transactions on Graphics (TOG)* 41(4), 1–19.

Pan, H., Xie, L. and Wang, Z. (2022) Plant and animal species recognition based on dynamic vision transformer architecture. *Remote Sensing* 14(20), 5242.

Parez, S., Dilshad, N., Alghamdi, N.S., Alanazi, T.M. and Lee, J.W. (2023) Visual intelligence in precision agriculture: Exploring plant disease detection via efficient vision transformers. *Sensors* 23(15), 6949.

Patel, S. (2024) Vision transformer with self-attention for automated plant seedling classification. In: *2024 2nd International Conference on Advancement in Computation & Computer Technologies (InCACCT)*, IEEE, pp. 707–710.

Pérez, J.C., Alfarra, M., Jeanneret, G., Rueda, L., Thabet, A. *et al.* (2021) Enhancing adversarial robustness via test-time transformation ensembling. In: *Proceedings of the IEEE/CVF International Conference on Computer Vision*, pp. 81–91.

Pesaranghader, A. and Viktor, H.L. (2016) Fast hoeffding drift detection method for evolving data streams. In: *Joint European conference on machine learning and knowledge discovery in databases*, Springer International Publishing, Cham, pp. 96–111.

Priyadharshini, S.I., Sree, T.R. and Balasaraswathi, V.R. (2023) Enhancing plant health: Deep learning approaches for precise disease detection. In: *2023 Second International Conference on Advances in Computational Intelligence and Communication (ICACIC)*, IEEE, pp. 1–6.

Raj, P., Susila, N. and Usha, S. (2023) Illustrating graph neural networks (GNNs) and the distinct applications.

Rakesh, S. and Indiramma, M. (2022) Explainable AI for crop disease detection. In: *2022 4th International Conference on Advances in Computing, Communication Control and Networking (ICAC3N)*, IEEE, pp. 1601–1608.

Rane, A., Vidhale, B., Kale, P.H. and Khekare, G. (2022) Design of an IoT based smart plant monitoring system. In: *2022 10th International conference on emerging trends in engineering and technology-signal and information processing (ICETET-SIP-22)*, IEEE, pp. 1–5.

Razin, N., Verbin, T. and Cohen, N. (2023) On the ability of graph neural networks to model interactions between vertices. *Advances in Neural Information Processing Systems* 36, 26501–26545.

Reedha, R., Dericquebourg, E., Canals, R. and Haane, A. (2021) Vision transformers for weeds and crops classification of high resolution UAV images.

Reichstein, M., Richardson, A.D., Migliavacca, M. and Carvalhais, N. (2014) Plant–Environment interactions across multiple scales. In: *Ecology and the Environment*. Springer, pp. 1–27.

Saeed, F., Sun, S., Rodriguez-Sanchez, J., Snider, J., Liu, T. *et al.* (2023) Cotton plant part 3D segmentation and architectural trait extraction using point voxel convolutional neural networks. *Plant Methods* 19(1), 33.

Saikai, Y., Peake, A. and Chenu, K. (2023) Deep reinforcement learning for irrigation scheduling using high-dimensional sensor feedback. *PLoS Water* 2(9), e0000169.

Samanta, R., Saha, B., Ghosh, S.K. and Roy, R.B. (2024) Optimizing TinyML: The impact of reduced data acquisition. *arXiv Preprint arXiv: 2409.10942*.

Samy, M.K.G., Deixler, P., Vangalapat, T. and Jaehan, K.O.H. (2022) *U.S. Patent Application No.17/777,406*.

Sangalli, M., Blusseau, S., Velasco-Forero, S. and Angulo, J. (2021) Scale equivariant neural networks with morphological scale-spaces. In: *International Conference on Discrete Geometry and Mathematical Morphology*, Springer International Publishing, pp. 483–495.

Schwartmann, R., Keber, T., Zenner, K. and Kurth, S. (2024) Data protection aspects of the use of artificial intelligence—Initial overview of the intersection between GDPR and AI act. *Computer Law Review International* 25(5), 145–150.

Shabala, S., White, R.G., Djordjevic, M.A., Ruan, Y.L. and Mathesius, U. (2015) Root-to-shoot signaling: Integration of diverse molecules, pathways and functions. *Functional Plant Biology* 43(2), 87–104.

Shafiee, M.J., Chwyl, B., Li, F., Chen, R., Karg, M. *et al.* (2019) StressedNets: Efficient feature representations via stress-induced evolutionary synthesis of deep neural networks. *Neurocomputing* 352, 93–105.

Shao, Y. and Devaraj, M. (2024) Application of plant disease identification and detection based on deep learning. In: *2024 3rd International Conference on Robotics, Artificial Intelligence and Intelligent Control (RAIIC)*, IEEE, pp. 437–441.

Sharma, A., Dharrao, D.S., Joshi, K., Tiwari, V., Kumar, S. *et al.* (2025) IoT-enabled monitoring system for plant health growth. *Journal of Intelligent Systems & Internet of Things* 14(2).

Shen, Y., Yang, Z., Khan, Z., Liu, H., Chen, W. *et al.* (2025) Optimization of improved YOLOv8 for precision tomato leaf disease detection in sustainable agriculture. *Sensors* 25(5), 1398.

Shou, X., Gao, T., Subramanian, D. and Bennett, K.P. (2021) Match2: Hybrid self-organizing map and deep learning strategies for treatment effect estimation. In: *Proceedings of the 12th ACM International Conference on Bioinformatics, Computational Biology, and Health Informatics*, pp. 1–10.

Siow, M.Q., Han, M.M.C., Lee, Y.N., Yu, S.Y., Noh, M.J. *et al.* (2023) Empowering agriculture: Exploring user sentiments and suggestions for plantix, a smart farming application. *Smart Media Journal* 12(10), 38–46.

Sornalakshmi, K., Sujatha, G., Sindhu, S. and Hemavathi, D. (2022) A technical survey on deep learning and AI solutions for plant quality and health indicators monitoring in agriculture. In: *2022 3rd International Conference on Smart Electronics and Communication (ICOSEC)*, IEEE, pp. 984–988.

Soro, S. (2021) TinyML for ubiquitous edge AI. *arXiv Preprint arXiv: 2102.01255*.

Suganob, N.J., Arroyo, C.L. and Concepcion, R. (2024) Interplay of fogponics and artificial intelligence for potential application in controlled space farming. *AgriEngineering* 6(3), 2144–2166.

Sun, C.C., Chan, F.C., Ahamad, A. and Yao, Y.H. (2023) Improved natural plant electrophysiology sensor design for phytosensing system. *IEEE Transactions on Instrumentation and Measurement* 72, 1–10.

Tamvakis, P.N., Kiourt, C., Solomou, A.D., Ioannakis, G. and Tsirliganis, N.C. (2022) Semantic image segmentation with deep learning for vine leaf phenotyping. *Ifac-Papersonline* 55(32), 83–88.

Tang, X., Zou, Y., Ma, J., Zhang, X., Liu, F. *et al.* (2024) Multiple information collaborative fusion network for joint classification of hyperspectral and LiDAR data. *IEEE Transactions on Geoscience and Remote Sensing.*

Thakur, P.S., Chaturvedi, S., Seal, A., Khanna, P., Sheorey, T. *et al.* (2025) An ultra lightweight interpretable convolution-vision transformer fusion model for plant disease identification: Convitx. *IEEE Transactions on Computational Biology and Bioinformatics.*

Thonglek, K. and Rattanatamrong, P. (2023) Decentralized federated learning for agricultural plant diseases identification on edge devices. In: *2023 18th International Joint Symposium on Artificial Intelligence and Natural Language Processing (iSAI-NLP)*, IEEE, pp. 1–6.

Tomizawa, Y., Minamino, N., Shimokawa, E., Kawamura, S., Komatsu, A. *et al.* (2023) Harnessing deep learning to analyze cryptic morphological variability of *Marchantia polymorpha*. *Plant and Cell Physiology* 64(11), 1343–1355.

Toutounji, H. and Pasemann, F. (2013) Autonomous learning needs a second environmental feedback loop. In: *International Joint Conference on Computational Intelligence*, Springer International Publishing, pp. 455–472.

Tran, D., Dutoit, F., Najdenovska, E., Wallbridge, N., Plummer, C. *et al.* (2019) Electrophysiological assessment of plant status outside a faraday cage using supervised machine learning. *Scientific Reports* 9(1), 17073.

Tsoumas, I., Sitokonstantinou, V., Giannarakis, G., Lampiri, E., Athanassiou, C. *et al.* (2023) Causality and explainability for trustworthy integrated pest management. *arXiv Preprint arXiv: 2312.04343*.

Tu, J., Li, H., Yan, X., Ren, M., Chen, Y. *et al.* (2021) Exploring adversarial robustness of multi-sensor perception systems in self-driving. *arXiv Preprint arXiv: 2101.06784*.

Turgut, K., Dutagaci, H., Galopin, G. and Rousseau, D. (2022) Segmentation of structural parts of rosebush plants with 3D point-based deep learning methods. *Plant Methods* 18(1), 20.

Uzhinskiy, A. (2025) Evaluation of different few-shot learning methods in the plant disease classification domain. *Biology* 14(1), 99.

van Lent, P., Bunkova, O., Planken, L., Schmitz, J. and Abeel, T. (2024) Neural ordinary differential equations inspired parameterization of kinetic models. *bioRxiv* 2014–12.

Velioglu, M., Zhai, S., Rupprecht, S., Mitsos, A., Jupke, A. *et al.* (2025) Physics-informed neural networks for dynamic process operations with limited physical knowledge and data. *Computers & Chemical Engineering* 192, 108899.

Venkatesh, N., Maheswari, S. and Triveni, P. (2024) Harnessing IoT for real-time plant health monitoring: Challenges and opportunities.

Virgolin, M. and Fracaros, S. (2023) On the robustness of sparse counterfactual explanations to adverse perturbations. *Artificial Intelligence* 316, 103840.

Wahby, M., Heinrich, M.K., Hofstadler, D.N., Zahadat, P., Risi, S. *et al.* (2018) A robot to shape your natural plant: The machine learning approach to model and control bio-hybrid systems. In: *Proceedings of the Genetic and Evolutionary Computation Conference*, pp. 165–172.

Walter, A. (2008) Plant growth dynamics: Analysis of basic spatial and temporal growth patterns on the background of photosynthetic energy gain and interactions with the environment. In: *Plant Growth Signaling*. Springer, Berlin, Heidelberg, pp. 63–88.

Wang, X. and Liu, J. (2024) Vegetable disease detection using an improved YOLOv8 algorithm in the greenhouse plant environment. *Scientific Reports* 14(1), 4261.

Wang, Y., Yin, Y., Li, Y., Qu, T., Guo, Z. *et al.* (2024) Classification of plant leaf disease recognition based on self-supervised learning. *Agronomy* 14(3), 500.

Wei, X., Li, G. and Marculescu, R. (2025) Online-lora: Task-free online continual learning via low rank adaptation. In: *2025 IEEE/CVF Winter Conference on Applications of Computer Vision (WACV)*, IEEE, pp. 6634–6645.

Wright, C., Meng, Q., Breshock, M.R., Atta, L., Taub, M.A. *et al.* (2024) Open case studies: Statistics and data science education through real-world applications. *Journal of Statistics and Data Science Education* 32(4), 331–344.

Xiao, H., Hu, W., Liu, G. and Zhou, H. (2023) Edge computing-based unified condition monitoring system for process manufacturing. *Computers & Industrial Engineering* 177, 109032.

Xu, F., He, J. and Huang, H. (2024a) FPDTN: Fast plant digital twin model for IoT automated greenhouse based on NeRF. In: *2024 International Conference on Intelligent Computing and Robotics (ICICR)*, IEEE, pp. 13–21.

Xu, J., Wan, W., Pan, L., Sun, W. and Liu, Y. (2024b) The fusion of deep reinforcement learning and edge computing for real-time monitoring and control optimization in IoT environments. In: *2024 3rd International Conference on Energy and Power Engineering, Control Engineering (EPECE)*, IEEE, pp. 193–196.

Yalcin, H. (2018) Phenology recognition using deep learning: DeepPheno. In: *2018 26th Signal Processing and Communications Applications Conference (SIU)*, IEEE, pp. 1–4.

Yang, B., Wang, X., Xing, Y., Cheng, C., Jiang, W. *et al.* (2024) Modality fusion vision transformer for hyperspectral and LiDAR data collaborative classification. *IEEE Journal of Selected Topics in Applied Earth Observations and Remote Sensing*.

Yang, Y., Hongxin, K.O.N.G., Hu, J., Porter, D.O., Marek, T.H. *et al.* (2022) *U.S. Patent Application No.17/624,594*.

Yi, X., Wang, J., Wu, P., Wang, G., Mo, L. *et al.* (2023) AC-UNet: An improved UNet-based method for stem and leaf segmentation in *Betula luminifera*. *Frontiers in Plant Science* 14, 1268098.

Zabel, P., Bamsey, M., Schubert, D. and Tajmar, M. (2016) Review and analysis of over 40 years of space plant growth systems. *Life Sciences in Space Research* 10, 1–16.

Zafar, I., Anwar, S., Yousaf, W., Nisa, F.U., Kausar, T. *et al.* (2023) Reviewing methods of deep learning for intelligent healthcare systems in genomics and biomedicine. *Biomedical Signal Processing and Control* 86, 105263.

Zeng, Z., Song, T., Ma, X., Jiu, Y. and Sun, H. (2024) Joint classification of hyperspectral and LiDAR data using cross-modal hierarchical frequency fusion network. In: *ICASSP 2024-2024 IEEE International Conference on Acoustics, Speech and Signal Processing (ICASSP)*, IEEE, pp. 5390–5394.

Zhang, T., Elnashef, B. and Filin, S. (2023) Spatio-temporal registration of plants non-rigid 3-D structure. *ISPRS Journal of Photogrammetry and Remote Sensing* 205, 263–283.

Zhang, W., Cao, X., Yao, Y., An, Z., Xiao, X. *et al.* (2021) Robust model-based reinforcement learning for autonomous greenhouse control. In: *Asian Conference on Machine Learning*, PMLR, pp. 1208–1223.

Zhang, X., Ibrahim, Z., Khaskheli, M.B., Raza, H., Zhou, F. *et al.* (2024) Integrative approaches to abiotic stress management in crops: Combining bioinformatics educational tools and artificial intelligence applications. *Sustainability* 16(17), 7651.

Zhao, D., Sharma, K., Yin, H., Qi, Y. and Zhang, S. (2024) SRTFD: Scalable real-time fault diagnosis through online continual learning. *arXiv Preprint arXiv: 2408.05681*.

Zhao, G., Cai, W., Wang, Z., Wu, H., Peng, Y. *et al.* (2022) Phenotypic parameters estimation of plants using deep learning-based 3-d reconstruction from single RGB image. *IEEE Geoscience and Remote Sensing Letters* 19, 1–5.

Zhao, H., Di, L., Guo, L., Li, L., Zhang, C. *et al.* (2023) Optimizing irrigation scheduling using deep reinforcement learning. In: *2023 11th International Conference on Agro-Geoinformatics (Agro-Geoinformatics)*, IEEE, pp. 1–4.

Zhou, B., Wang, Q., Zhang, J. and Wang, Y. (2023) Combined classification of hyperspectral and lidar data based on dual-channel cross-transformer. In: *Proceedings of the 2023 5th International Conference on Video, Signal and Image Processing*, pp. 43–49.

Zhu, X., Huang, Z. and Li, B. (2024) Three-dimensional phenotyping pipeline of potted plants based on neural radiation fields and path segmentation. *Plants* 13(23), 3368.

Appendix—Abbreviations

3D-CNNs: 3D convolutional neural networks
AI: artificial intelligence
AT: adversarial training
CEs: counterfactual explanations
CNNs: convolutional neural networks
CSTRs: continuously stirred tank reactors
CURE: causal inference with unsupervised representations
DL: deep learning
DRL: deep reinforcement learning
DTANet: deep treatment-adaptive network
DViTs: Dynamic Vision Transformers
E3NNs: equivariant neural networks
EWC: elastic weight consolidation
FHDDM: Fast Hoeffding Drift Detection Method
FID: Fréchet inception distance
FL: federated learning
FLOPs: floating-point operations
FPDTN: Fast Plant Digital Twin Network
GDPR: General Data Protection Regulation
GNNs: graph neural networks
HFNet: Hierarchical Frequency Fusion Network
HTP: high-throughput phenotyping

ILP: integer linear programming
IoT: Internet of Things
IoU: Intersection over Union
IPM: integrated pest management
KKT-hPINN: Karush–Kuhn–Tucker-constrained hard PINN
LDA: latent Dirichlet allocation
LSTM: spatiotemporal long short-term memory
mAP: mean Average Precision
MBRL: model-based reinforcement learning
MF-ViT: Modality Fusion Vision Transformer
MICF-Net: multiple information collaborative fusion network
ML: machine learning
NAS: neural architecture search
NeRF: neural radiance field
non-IID: non-independent and identically distributed
OCL: online continual learning
ODEs: ordinary differential equations
PDE-G-CNNs: partial differential equation-based group convolutional neural networks
PHOENIX: prior-informed hybrid ODE network for interpretable expression modeling
PINC: physics-informed neural control
PINNs: physics-informed neural networks
PVCNN: Point-Voxel convolutional neural network
ROS: reactive oxygen species
rPPG: remote photoplethysmography
SRTFD: Self-Refining Task-Free Diagnosis
SSL: self-supervised learning
TTE: Test-Time Transformation Ensembling
ViTs: Vision Transformers
WIS: Walk Index Sparsification
XAI: explainable artificial intelligence

14 Deep Learning for Advancing Weed Management

Shreedevi Jinnur[1], Charan Babu Ankela[2], D.J. Kotresh[1], Chappali Harendra[3], Jhonson Raju Sankati[3], T.M. Tanjina Tarannum[4], Subrata Goswami[5], Maksud Hasan Shah[6]*, Santanu Kundu[3]*, and Md Mazharul Islam[7]

[1]University of Agricultural Sciences Bangalore, Karnataka, India; [2]IARI Mega University, Hyderabad hub, ICAR–CRIDA, Hyderabad, Telangana, India; [3]ICAR-Indian Agricultural Research Institute, New Delhi, India; [4]Department of Agronomy, Bangladesh Agricultural University, Maimansingh, Bangladesh; [5]Institute of Agricultural Sciences, Banaras Hindu University, Varanasi, Uttar Pradesh, India; [6]Bidhan Chandra Krishi Viswavidyalaya, Mohanpur, Nadia, West Bengal, India; [7]School of Agriculture, University of the Virgin Islands, St Croix, Virgin Islands

Abstract

Traditional methods of weed management often lead to herbicide overuse, environmental harm, and the emergence of herbicide-resistant weed species. To address these challenges, smart spraying systems equipped with sensor technology and advanced AI models have emerged, offering sustainable, site-specific weed control solutions, which are collectively referred to as deep learning (DL). Despite the similarities between crops and weeds, DL techniques show promising results in extracting and distinguishing traits. This chapter emphasizes DL-driven weed management, highlighting the possibility of real-time weed detection, high-accuracy classification, and control of weeds.

Keywords: convolutional neural networks, deep learning, weed classification, weed detection

14.1 Introduction

As the global population continues to grow and agricultural productivity declines, fulfilling the food demands of both humans and livestock is becoming increasingly challenging. Among all the challenges faced by farmers, one of the major challenges is pest management. Weeds, in particular, pose a major competition to crop growth, leading to a severe decline in crop yields. The annual food loss due to weeds is approximately 13.2% (Liu *et al.*, 2021), so effective weed management is crucial for enhancing crop yields and agricultural productivity.

Manual weeding could be very effective and economical for small areas, but for managing weeds in large farmlands, this method is extremely slow and expensive (Naveed *et al.*, 2023). The challenge is further intensified during periods of continuous rainfall and labor shortages, making manual weed management particularly difficult. Hence, to overcome these shortcomings, farmers

*E-mail: maksudhasanshah@gmail.com or santanukundu120@gmail.com

DOI: 10.1079/9781836991120.0014

have resorted to chemical weed management. However, manual identification of weeds and blanket application of herbicides may lead to an excess quantity of chemical application, posing risks to both the environment and human health (MacLaren *et al.*, 2020; Roslim *et al.*, 2021). Also, it increases the overall cost of agrochemicals for the farmer. Hence, there is a pressing need for precise application of herbicides.

Site-specific weed management involves identifying areas within the crop field that are infested with weeds and applying herbicides exclusively to those targeted patches, ensuring more efficient and localized treatment, which delays the development of resistance in weeds and also prevents crop injury. Satellites and unmanned aerial vehicles (UAVs) equipped with appropriate sensors capture high-resolution aerial images of the field. While satellite images provide the advantage of wider coverage, UAVs offer flexibility and rapid collection of detailed field data at very high spatial resolution and low cost (Bah *et al.*, 2018; Oghaz *et al.*, 2019).

As accurate weed detection and classification form the crux of efficient weed management, the acquired imagery is processed and classified to identify the weeds by machine learning (ML) tools. Traditionally, weed detection from aerial images was based on ML algorithms such as support vector machines (SVMs), random forests (RF), and extreme gradient boosting (XGB) (Gašparovi *et al.*, 2020; Amarasingam *et al.*, 2023). The abovementioned technologies have also been used to classify weeds from crops by manual feature extraction technique. This process involves selecting specific characters such as shape, size, and color, based on expert knowledge, which are then used as input data for classification. This requires considerable expertise in the domain.

As these classifiers are trained with limited data, they cannot perform efficiently in complex environments containing unknown objects, weed look-alikes, and shadows. Hence, using them for real-time weed classification in fields is not effective. Therefore, deep learning (DL) techniques with enhanced capability of classification in real time were introduced in weed management.

DL has been developed to imitate the processing of the human brain and utilizes neural networks, where individual nodes (or neurons) are connected through various channels. A DL model is typically composed of three types of layers: (i) the input layer, where data is introduced into the system; (ii) hidden layers, where data is processed; and (iii) the output layer, where the final result is generated.

DL models include numerous hidden layers, which is why they are referred to as "deep" networks. The large volumes of data fed into the classifier are passed through the hidden layers, where data is transformed by forming connections between the relevant neurons. After the data is processed, the output is generated. During training, if an error is detected in the output, the system adjusts the weights of the neurons by back-propagating the error, allowing the model to learn and improve its accuracy. As large volumes of data are used for training, DL can avoid confusion between visually similar crops and weeds, thereby showing effectiveness in field deployment (Fan and Su, 2022; Qu and Su, 2024).

Furthermore, by incorporating hierarchical functions and adding depth to data, DL increases the complexity of ML. Because of its intricate models, which enable large parallelization, it is very good at addressing complicated issues (Wang *et al.*, 2020). When extensive datasets are available, DL can improve classification accuracy or decrease errors in regression studies. Depending on which network architecture is being utilized, DL might consist of different components (unsupervised pretrained networks, convolutional neural networks (CNNs), recurrent neural networks (RNNs), and recursive neural networks). It can handle a wide range of complicated data analysis problems due to its substantial learning capacity and hierarchical structure (Wang *et al.*, 2023). Although natural language processing (DL) is widely used in systems that work with raster-based data, it may be used with any type of data, including speech, audio, natural language, weather data, and soil chemistry. Figure 14.1 depicts the CNN architecture. CNNs are deep neural networks (DNNs), that use learnable weights and biases to classify and identify items such as weeds by assigning these weights to different characteristics and objects of input photos. In order to uncover buried information, CNNs automatically learn key features from training data, eliminating the need for human feature selection. A wide range of objects with varying scales, orientations, and degrees of occlusion may be reliably classified by CNNs. With a significantly smaller number of estimable

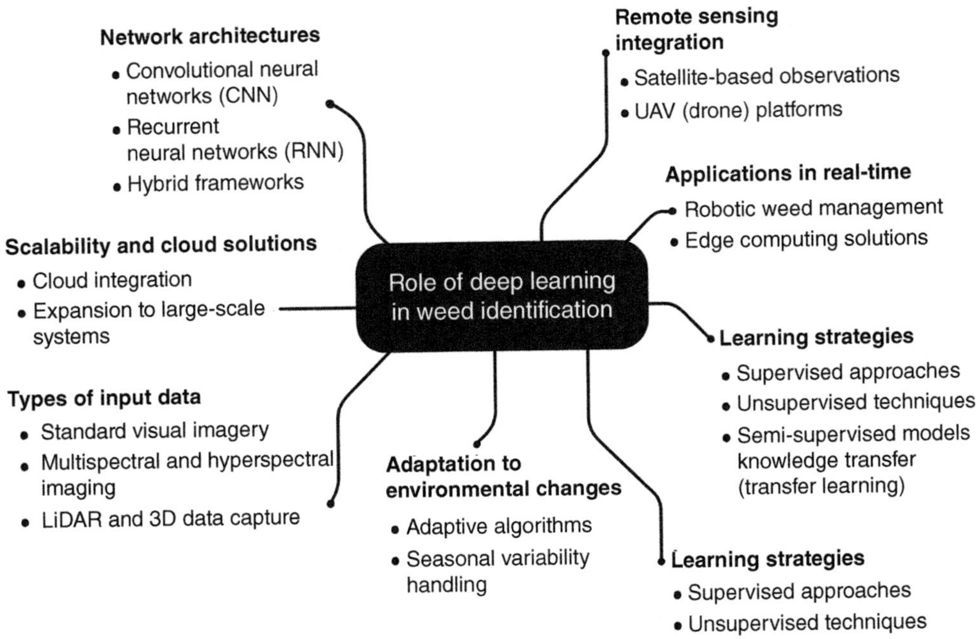

Fig. 14.1. CNN architecture. (Author's own image.) CNN, convolutional neural network.

parameters and processing time, CNNs offer superior and more effective image processing by autonomously capturing the spatial and temporal dependencies of the input image through pertinent filters.

The revolutionary potential of a DL-based weed management system is examined in this chapter. Experimental results and comparative field investigations regarding the effectiveness and environmental benefits of DL techniques over traditional methods form the basis of the discussion. This will support the data that DL is a dependable and long-lasting weed control method.

14.2 DL in Weed Management

The integration of DL with modern technologies paved the way for its wider application in weed management. From precise weed identification using trained algorithms to supporting integrated weed management, DL has emerged as a transformative tool in modern weed management (Rai *et al.*, 2023).

The processing and meaningful interpretation of the images taken from aerial or ground-based technologies—by trained models such as CNN, you only look once (YOLO), and k-nearest neighbor (KNN)—will help in the precise identification of weeds and weed species, by analyzing several physical features of the vegetation for accurate discrimination from crops and their surroundings. The inclusion of well-trained modules under DL capable of generating temporal data on weed emergence and their growth patterns can serve as a tool in planning timely interventions with optimized herbicide application rates, thereby enhancing weed management efficiency. DL enhances the efficiency of automatic robots by improving their weed-detecting abilities, which not only removes the weeds by mechanical or laser techniques but also reduces the reliability of herbicides, thus paving the way for sustainable approaches. By creating a field map of weed infestation, DL generates weed distribution maps, which give precise knowledge about site-specific weed management and implement variable-rate herbicide applications, supporting ecosystem sustainability and farm income. The simultaneous collection of temporal and spatial data on weeds in the field will shed light on weed dynamics and a shift in their population, if any. Thus, it can indicate the development of resistant weeds against a particular weed management strategy, which helps in planning appropriate alternatives to control and overcome resistance development. The continuous monitoring of fields using DL techniques against weeds and other factors such as soil fertility and environmental conditions may help reveal correlations between these factors and identify the favorable conditions for specific weed emergence.

The location- and time-specific detection of weeds boosts the implementation of site-specific weed management strategies and also helps in adopting integrated approaches such as hand and mechanical weeding at lower weed density sites and herbicide applications at sites with higher weed density. Overall, DL offers a holistic approach to weed management, ensuring more sustainable, timely, and economically sound agricultural practices (Rai *et al.*, 2023). Weed management using DL involves a streamlined process that includes

1. weed detection;
2. weed mapping; and
3. implementation of targeted control strategies.

14.2.1 DL in weed detection

The process of weed detection using DL consists of four primary steps: image data acquisition, preprocessing of collected data, feature extraction, and weed classification using algorithms.

Image data acquisition

Weed detection relies on remote sensing technologies, including satellites, aircraft, drones (UAVs), and handheld devices, integrated with hyperspectral, multispectral, and RGB (red, green, blue) sensors to capture real-time imagery of target areas (Zhang *et al.*, 2024).

The hyperspectral imaging (HSI) technique works by merging the spectral characteristics with imaging technology to capture continuous, narrow-band data with high spectral resolution (Zhang *et al.*, 2024). Hyperspectral spectroscopy operates based on molecular vibrations in the infrared region, enabling material classification and quality assessment through wavelength-specific absorbance. Each weed species exhibits a unique spectral signature, which HSI can detect and model. In a study, three crucial spectral regions (550–700 nm, 1000–1200 nm, and 1300–1500 nm) revealed distinct reflectance differences among four weed species, that is, annual winged thistle, annual yellow bristle grass, perennial giant buttercup, and wind grass (Li *et al.*, 2021).

Multispectral imaging (MSI) has emerged as a cost-effective alternative to HSI, as MSI captures images across multiple spectral bands while offering faster processing and requiring lighter hardware. This technology enables data collection across different spectrums, such as near-UV, visible, near-IR, mid-IR, and far-IR wavelengths, combining spatial and spectral information for efficient analysis (Fan and Su, 2022).

RGB sensors detect and digitize an object's color, offering advantages such as high precision, speed, low power consumption, and affordability. However, collecting a large volume of image data is crucial for early training and later model validation. Proper image selection and processing significantly influence model accuracy (Qu and Su, 2024). A case study using DJI Phantom 4 drone equipped with an RGB camera produced a vegetation map of giant black grass, revealing a strong correlation (0.91) with actual ground cover (Rozenberg *et al.*, 2023).

COMPARISON OF SENSORS FOR WEED DETECTION IN DL. HSI provides detailed spectral information, capturing the full spectrum at each pixel. MSI is more cost-effective and practical, distinguishing weeds from crops with high-resolution images. RGB cameras have affordability and ease of use but lack advanced spectral differentiation capabilities. Multispectral cameras are preferred for their balanced cost and spectral diversity (Fig. 14.2). Unlike RGB cameras, they include additional spectral bands (such as red-edge and near-infrared) and typically operate across four to six bands. The integration of a reflectance calibration panel also minimizes environmental variations, making multispectral cameras valuable for smart agricultural applications (Amarasingam *et al.*, 2023).

EVOLUTION OF DL IN WEED DETECTION. According to reference (Zhang *et al.*, 2022), sensing technologies dictate the collection of weed

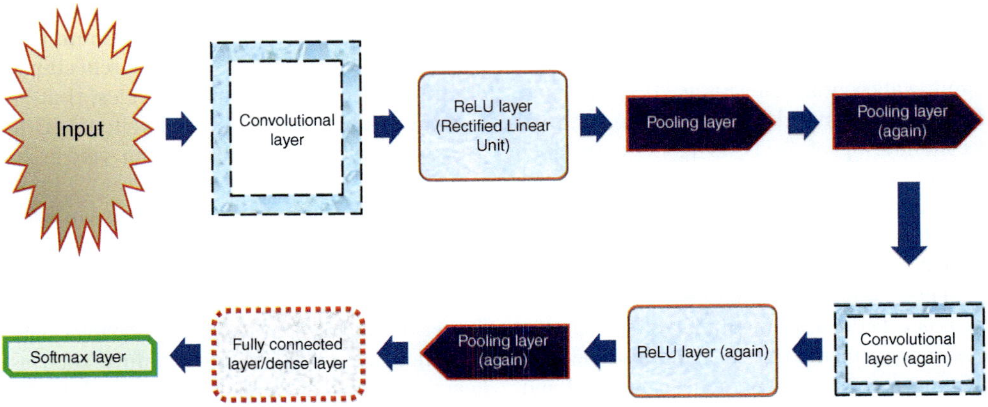

Fig. 14.2. A comparison between the outputs of multispectral and hyperspectral images. (Author's own image.)

data and weed management tactics. To develop and compare weed identification methods, weed data is necessary. Interest in image-based weed detection has grown as a result of advancements in imaging techniques, including depth imaging, MSI, and near-infrared imaging. The availability of large public datasets in the field facilitates the development of new algorithms for weed identification tasks (Zhang *et al.*, 2019; Zhang *et al.*, 2024). The public datasets offer helpful annotation information and benchmarking photographs; however, they differ in terms of contextual information and metadata reporting restrictions. Developing effective weed management tactics requires an understanding of the many types of weeds. Annotated dataset creation takes a lot of effort and can lead to overfitting and insufficient variability in weed control scenarios. To address this, a number of data augmentation methods have been employed to increase the number and caliber of training sets, including rotation, random cropping, and generative approaches.

The efficiency of weed identification algorithms is assessed using several criteria, depending on the identification method. Four distinct results for binary image classification can be obtained based on the categorization of an input sample: true positive (TP), false positive (FP), true negative (TN), and false negative (FN). There are four types of research on weed identification: weed picture classification, weed object detection, weed object segmentation, and weed instance segmentation.

Preprocessing of collected image data

Optimizing image pixel size reduces computational complexity and speeds up DL model processing, ensuring optimal performance. Preprocessing enhances visual interpretation and improves algorithm selection through several techniques such as histogram equalization for contrast enhancement, image quality enhancement and denoising to reduce noise interference, and contrast adjustments for better feature detection. Reedha *et al.* (2022) recorded recognition accuracy of more than 99% using the ViT B-16 model (a self-attention model), only after enhancing the images using Keras ImageDataGenerator. Image segmentation is used to remove background elements, such as unwanted vegetation and soil surfaces, ensuring a more focused analysis (Qu and Su, 2024). Optimized soil-adjusted vegetation index (OSAVI) was employed to extract the spectra of the vegetation canopy by improving the contrast between the soil background and the vegetation canopy (Lou *et al.*, 2022)

Extraction of weed features from images

Feature extraction plays a critical role in accurately distinguishing weeds from crops. Qu and Su (2024) identified three major feature categories: (i) visual texture features: textural variations such as soft, rough, and horizontal corrugations versus hard, fine, and vertical corrugations help differentiate plants. For instance, these features successfully identified tobacco plants from surrounding vegetation (Tufail *et al.*, 2021). (ii) Spatial context features: since most crops follow a predetermined planting pattern, spatial context and position data improve classification accuracy. (iii) Morphological features: shape, structure, size, pattern, and color help differentiate crop species and weeds (Qu and Su, 2024).

Weed classification using deep learning algorithms

DL algorithms automatically extract features by learning from errors, unlike traditional methods that require manual feature selection (Qu and Su, 2024). Various neural network architectures support weed classification: CNNs, RNNs, generative adversarial networks (GANs), and graph neural networks (GNNs) (Zhang *et al.*, 2024). When training a deep CNN model, input images pass through convolutional layers with filters, followed by flattening, pooling, and fully connected layers. CNNs effectively capture spatial and temporal dependencies, leading to enhanced image processing (Qu and Su, 2024). In a comparative study on weed identification in pea and strawberry fields between shallow and DL algorithms, R-CNN demonstrated the highest accuracy and a perfect kappa coefficient (Khan *et al.*, 2021). Additionally, the combination of shallow learning models such as SVMs and KNNs, with DL algorithms such as Visual Geometry Group Network (VGGNet), has been applied for weed detection in a winter rape field (Tao *et al.*, 2022). Various DL algorithms used for weed classification in crop fields are presented in Table 14.1.

Among DL techniques, CNNs are especially prominent in image-based weed detection due to their ability to capture spatial hierarchies from raw images. CNNs have been used to classify weed species with high precision; for instance, Olsen *et al.* (2019) developed a model to identify 22 weed species in Danish fields, achieving accuracy rates above 95%. Architectures such as AlexNet, VGG16, ResNet, and MobileNet are commonly used for such classification tasks (Luo *et al.*, 2023). To improve performance on smaller agricultural datasets, transfer learning is often employed. It allows models pretrained on large datasets to be fine-tuned for specific agricultural tasks (Fig. 14.3).

14.2.2 Weed mapping

Weed mapping involves understanding weed spatial distribution, which is assessing how weeds are spread across an area. This is vital for precision weed management (Somerville *et al.*, 2020). Traditionally, this has been done through manual observation and sampling, which is labor-intensive and often imprecise (Zhang *et al.*, 2019).

Advancements in remote sensing and ML have enabled more efficient and accurate methods of weed mapping. Technologies such as satellite imagery and UAVs provide high-resolution data, which can be analyzed using ML algorithms to detect and monitor weed infestations over large areas (Corceiro *et al.*, 2023). For example, Nyamekye *et al.* (2021) demonstrated the utility of combining remote sensing with ML to track spatial and temporal variations of aquatic weeds such as water hyacinths in the Lower Volta River.

Geographic information systems (GIS) further enhance this capability by integrating diverse data sources into a unified framework. When paired with ML, GIS can generate detailed maps that highlight weed distribution patterns, aiding researchers and practitioners in identifying infestation hotspots and applying control measures accordingly (Andujar and Martinez-Gaunter, 2022). This integrated approach improves both strategic planning and real-time weed control efforts.

Predictive models for weed spatial distribution

Predictive modeling leverages environmental factors, historical data, and other parameters to forecast where weeds are likely to appear. ML-based models—such as RFs, SVMs, and neural networks—analyze these variables to identify correlations and generate accurate predictions (Sarker, 2021). These models are instrumental in determining weed-prone zones, thereby enabling resource optimization and proactive intervention strategies.

To improve prediction robustness, ensemble methods such as gradient boosting, bagging, and stacking are increasingly being used (Ryo

Table 14.1. Classification of weeds and crops for weed detection through deep learning. From Murad *et al.*, 2023, CC BY.

Weed type	Crop type	DL algorithms	Best accuracy
Vegetables	N/M	Faster R-CNN, CenterNet, YOLO-v3	97% [YOLO-v3]
Cocklebur	Peanut, corn, wheat	AlexNet	99% [AlexNet]
Japanese hop	Peanut, wheat, corn	AlexNet	99% [AlexNet]
Velvetleaf	Tomato, cotton	DenseNet, SVM, ResNet, Xception, VGGNet, MobileNet	99% [SVM, DenseNet]
Chickweed	Cranesbill, wheat, corn, charlock, sugar beet	CNN, FCN, Mask R-CNN, VGGNet	95% [Mask R-CNN]
Cleaver	Sugar beet, wheat, corn	FCN, Mask R-CNN	95% [Mask R-CNN]
Cereal crops	Monocotyledon and dicotyledon	YOLO	83% [YOLO]
Lantana camara	N/M	CNN	98% [CNN]
Foxtail, redroot pigweed, giant ragweed, and cocklebur	Soybean, corn	VGG16, YOLOv3, ResNet50, Inception30	98% [VGG16]
Shepherd's purse	Wheat, sugar beet, corn	Mask R-CNN, SegNet, FCN	95% [Mask R-CNN]
Crowfoot	Corn, rice	GoogleNet	98% [GoogleNet]
N/M	Spinach, bean, beet	CNN	93% [CNN]
N/M	N/M	ResNet50, DenseNet, VGG16	91% [ensemble]
Black nightshade	Cotton, tomato	Xception, ResNet, DenseNet, MobileNet, VGGNet, SVM	99% [SVM, DenseNet]
Dandelion, ground ivy, spotted spurge	N/M	DCNN	92% [DCNN]
Turnip weed	Sugar beet	ANN, SVM	95% [SVM]
Fat-hen	Sugar beet, corn, wheat, peanut	AlexNet, Mask R-CNN, GCN, FCN, YOLO, VGGNet, ResNet	99% [YOLO]
Crowfoot	Corn, rice	GoogleNet	98% [GoogleNet]
Blackgrass	Wheat, sugar beet	CNN, Mask R-CNN, FCN	97% [CNN]
N/M	Soybean	CNN, SSD, Faster RCNN	65% [RCNN]
Dicots	Sugar beet, soybean	FCN	87% [FCN]
Chinee apple	N/M	CNN	98% [CNN]
Meadow grass	Wheat	VGGNet, CNN	84% [CNN, VGGNet]
Pigweed	Wheat, sugar beet	ANN, VGGNet, SVM, CNN	95% [SVM]
Sugar beet	N/M	ResNet, U-Net	96% [U-net, ResNet]
Nutsedge	Chrysanthemum, corn	ResNet, AlexNet, VGGNet, CNN, GCN	98% [CNN, GCN]
Indian jointvetch	Rice, corn	SSD	86% [SSD]
Crowfoot	Corn, rice	GoogleNet	98% [GoogleNet]
Jungle rice	Corn, rice	SSD, GoogleNet	98% [GoogleNet]
Dockleaf	Pasture, bahiagrass, soybean, turfgrass	AlexNet, GoogleNet, CaffeNet, DetectNet, VGGNet, CNN	99% [AlexNet, CaffeNet]
Grass weed	Sugar beet, corn, wheat	FCN	87% [FCN]
Hedge bindweed	Sugar beet	YOLO	89% [YOLO]

Continued

Table 14.1. Continued.

Weed type	Crop type	DL algorithms	Best accuracy
Canadian thistle	Cereal crops, corn	AlexNet, VGGNet, DT, ResNet, GCN, YOLO, ANN, KNN, SVM	99% [YOLO]
Little seed	Rice, corn	GoogleNet, SSD	98% [GoogleNet]
Paragrass	Chrysanthemum	VGGNet, CNN, AlexNet, GCN	98% [CNN, GCN]
Bluegrass	Corn	YOLO, ResNet	99% [ResNet, YOLO]
Field pansy	Wheat	VGGNet, CNN	84% [CNN, VGGNet]
Snakeweed	N/M	CNN	98% [CNN]
Mayweed	Wheat, carrot, corn, sugar beet	Mask R-CNN, FCN, CNN, VGGNet	95% [Mask R-CNN]
Bengal dayflower	Rice	SSD	86% [SSD]

CNN, convolutional neural networks; KNN, k-nearest neighbor; DL, deep learning; SVM, support vector machine; YOLO, you only look once.

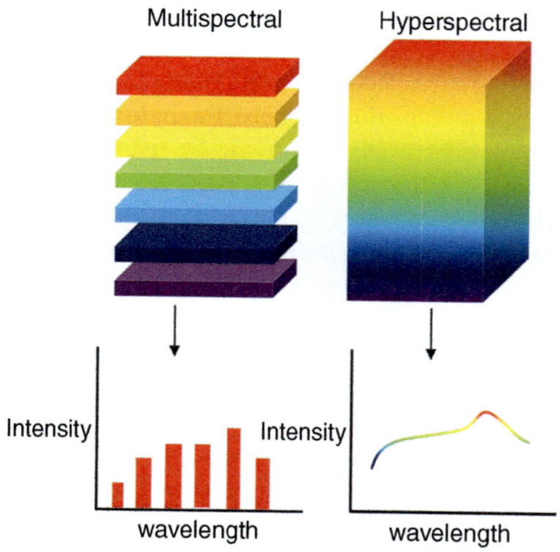

Fig. 14.3. Steps involved in the processing of acquired data. (Author's own image.)

et al., 2022). These methods combine multiple models to enhance accuracy and reduce overfitting, making them well suited for complex ecological forecasting.

Weed spatial distribution analysis

Spatial analysis using DL focuses on uncovering patterns, relationships, and processes within geographical datasets. These tools allow rapid identification of weed growth hotspots, facilitating targeted interventions and efficient herbicide application. DL models also support forecasting the potential spread of weeds, enabling preemptive action.

In practice, DL has been used in real-time weed detection. For instance, Zhang *et al.* (2022) applied machine vision in peanut fields, where the EM-YOLOv4-Tiny network demonstrated high efficiency in detecting individual weeds from imagery. Furthermore, DL systems can be integrated into precision agriculture tools such as automated sprayers and robotic weeders, minimizing chemical usage and environmental impact.

Integration of classification and mapping

Weed classification and mapping are interlinked in precision weed management. Classification identifies and labels weed species based on visual or spectral data, while mapping determines their spatial distribution and density across a field. Together, these processes underpin site-specific weed management (SSWM) by enabling selective and localized weed control (Christensen *et al.*, 2009).

14.2.3 DL for weed control

Weeds can be controlled mechanically or chemically, depending on the intensity of infestation. Traditional approaches have much greater recurring costs, especially from labor and wasteful use of chemicals. Farmers tend to apply herbicides uniformly across fields, resulting in resource waste and decreased return on yield (Lottes *et al.*, 2018; Azghadi *et al.*, 2024). Equipping the sensors with integrated DL algorithms to the machinery, such as tractors and drones, for data capture, near real-time processing, weed mapping, informed decision-making, and weed control action could significantly enhance weed management.

The inclusion of DL impacts weed management by enhancing the weed control indices, offering accuracy in identification and additionally making it cost-effective and ecologically sustainable.

Impact on weed control efficiency

Multiple research studies have demonstrated that sprayers powered by a direct line can destroy weeds effectively. Independent research on AutoWeed robotic sprayers revealed 97% better performance than broadcast spraying (Azghadi *et al.*, 2024). Research involving 25 ha of sugarcane land produced findings that proved both effective weed control and reduced herbicide runoff.

Accuracy in weed management

DL is far superior to conventional approaches in sensing, pinpointing, and responding to a weed in real-time. More advanced DL models, such as YOLOv4, RetinaNet, and VGG-16, achieve more than 95% precision and can identify several weed species at different growth stages (Saqib *et al.*, 2023; Upadhyay *et al.*, 2024) using ResNet50 with transfer learning on patch regional datasets and achieved 95.1% accuracy. Another example is Islam *et al.* (2024), who trained RetinaNet and DETR to classify 16 weed species at 11 growth stages, demonstrating the scalability and robustness of DL in real-world scenarios.

Cost and functional effectiveness

DL-based sprayers cut labor needs and can be operated in real time. Due to higher precision and efficacy, it minimizes the need for multiple passes in the field and on-the-ground scouting. Even though these systems have high initial costs, they save enough on herbicide use and labor to pay for themselves within two to three cropping seasons (Azghadi *et al.*, 2024). Integrating a spray volume and rate sensor (SVRS) device with YOLOv3-Tiny yielded more than 40% economic savings through liquid pesticide use along with precise target delivery compared to farmer practice (Hussain *et al.*, 2020). Several other researchers have also reported the functional effectiveness of DL-based sprayers, as presented in Table 14.2.

Effect on the quantity of herbicides and the environment

Evidence from real-world implementation shows that DL-based systems have quantifiably lesser damaging effects on the environment. Research has shown that conventional methods produce 39% higher herbicide contamination rates and 54% greater chemical load in irrigation-induced runoff compared to AutoWeed-equipped machinery, which is a system that uses computer vision and DL to identify and target weeds in real time during farming operations (Azghadi *et al.*, 2024).

Systems that can manage real-time spray parameters based on weed density measurements have been developed through enhanced DeepLab V3+ segmentation models. Such techniques exhibit effective spot spraying by reducing the droplet size and densifying droplet distribution, which collectively saves herbicide consumption. He *et al.* (2024) experimented with one such advanced precision spraying system and recorded a decrease in droplet size to 138.13 μm, densified at 38.87 droplets/cm^2, thus saving 46.3% of herbicide. Smart sprayers conducted an evaluation of potato crops, which led to a 47% and 51% decrease in agrochemical utilization for detecting both weeds and diseases. The sprayer implemented automatic adjustments to lighting differences, velocity changes, and distinct plant forms to deliver stable outcomes throughout (Farooque *et al.*, 2023).

14.2.4 Important DL models in weed control

1. CNNs:
The most common DL algorithms used to detect and classify weeds in agricultural images are CNNs (Murad *et al.*, 2023). These models, based on RGB and multispectral images, can learn to distinguish crops from weeds very well. Dyrmann *et al.* (2016) reported more than 95% accuracy in crop and weeds at pixel level, and Milioto *et al.* (2018) achieved 94% accuracy for real-time semantic segmentation of sugar beet fields.

Table 14.2. Functional effectiveness offered by DL-based sprayers. Author's own table.

Crop species	Task	Model	F1 score	Precision	Accuracy	Recall	References
Sesame/ weed	Object detection	YOLO-sesame	92	90.30	–	94.53	Chen *et al.* (2022)
Wheat/ weed	Object detection	YOLOv4+Tiny	92	89	–	96	Haq *et al.* (2023)
Crop/weed	Image segmentation	VGG-UNet		99		94.85	Fawakherji *et al.* (2020)
Carrot/ weed	Image segmentation	U-Net++	89.20	88.84	94.02	89.56	Fathipoor *et al.* (2023)
Crop/weed	Image classification	ConvNeXt	90.37	–	92.33	–	Benchallal *et al.* (2024)

2. RNNs and long short-term memory models (LSTMs):
RNNs and LSTMs, in particular, have been used on time-series data to predict and monitor weed infestations. Kamilaris and Prenafeta-Boldú (2018) noted that LSTMs could be useful for monitoring the weed–crop relationship over time.
3. GANs:
The channel attention GAN (CA-GAN) variation of GANs, in particular, has demonstrated great promise in improving weed detection by artificially producing realistic and varied weed images. When there is a lack of real-world annotated data, these artificial datasets are essential for training DL models (Li *et al.*, 2024).

When tested with a ResNet-based auxiliary classifier, CA-GAN-generated synthetic images obtained a classification accuracy of 82.63% on the public segmented plant seedling dataset (sPSD) and 93.46% on the Institute for Sustainable Agro-ecosystem Services (ISAS) dataset. CA-GAN may produce class-distinguishable weed images that are appropriate for DL training pipelines, as demonstrated by the results, which surpass those of current GAN models.

14.3 Challenges of DL in Weed Management

1. Adversarial pictures:
These elements in an image can create noise or even a little pixel-level disturbance, which can deceive a DL network into incorrectly categorizing an item (Bai *et al.* 2021).
2. Data labeling for retraining supervised classification models:
The quantity of labeled data needed for retraining these models to adjust to a new area makes the process challenging (Milioto *et al.*, 2018).
3. Environmental impact:
As DL models are trained on specific data from certain geographies or lighting conditions, shifting environmental variables have an impact on the effectiveness of DL-based weed management (Ahmed *et al.*, 2024; Gryshova *et al.*, 2024; Upadhyay *et al.*, 2024).
4. Expensive:
The high cost of robotic platforms, cameras, and processors for DL hinders adoption, particularly in developing nations (Upadhyay *et al.*, 2024).
5. Internet connectivity challenges in remote locations:
In remote locations, access to steady and reliable internet for real-time data gathering and processing might be difficult (Pandey *et al.*, 2024).

14.4 Future Directions on Applying Novel Techniques for Weed Detection Using DL

Over the past 10 years, there has been a noticeable advancement in the field of DL to support SSWM. Significant efforts should be made to address some of the difficulties in using DL for weed detection in light of the current findings. These difficulties generally relate to fine-tuning models by testing them on an entirely unknown dataset with a different distribution after training on a small and varied dataset. Future research in this domain may explore the scope for breakthroughs across five efficient approaches. These innovative strategies could include the following:

1. Training a neural network from scratch. This essentially entails creating a new CNN model from scratch and training it. As previously stated, most weed identification methods have mostly depended on using a CNN model that has already been trained. In the agricultural world, this has emerged as one of the most popular strategies. The basic layers of the pretrained models contain weight information that could lead to bias on the new and complex dataset, even if these models have demonstrated an acceptable level of accuracy in both still picture context and real-time scenarios. More research should concentrate on developing neural networks that are based on targeted learning of representations from intricate training datasets in order to address this difficulty. For the network to have a strong learning mechanism, a large number of photos must be fed into it in order to complete this step. However, gathering, preprocessing, and annotating data will take a significant amount of time when training models on a variety of datasets. If high-end computational resources are available, training could occasionally take weeks as well. In order to alleviate this requirement, a few additional strategies are discussed in the following sections that may be used to enhance the trained models' capacity for weed identification generalization.

2. Adopting domain adaptation (DA) approach (Dev *et al.* 2022). A unique kind of TL technique called the DA technique helps a model generalize more effectively on datasets with different distributions than the training dataset (test images with novel context). For instance, the distribution of a dataset of weed images taken with a handheld camera will differ significantly from that of a dataset obtained with an unmanned aerial system (UAS). Thus, by modifying and refining the trained model on the new test dataset with an entirely different distribution, the DA technique can be applied in this situation.

3. Training using an adversarial approach (Bai *et al.*, 2021). The use of adversarial pictures is one of the main problems DL algorithms encounter. These adversaries can appear in an image as noise or even as a tiny pixel-level disturbance, which can deceive a DL model into incorrectly categorizing an object. In order to address this issue, researchers have recently started developing an adversarial training method known as GANs (Espejo-Garcia *et al.*, 2021).

4. Model- and module-level ensemble approach (Ganaief *et al.*, 2022). The ensemble approach is a method that forecasts a result for an unknown dataset using prediction results from the foundation of several models (Ofori *et al.* 2022). In contrast, a more sophisticated method called module re-parameterization divides the convolutional blocks while the inferencing process is underway (Wang *et al.*, 2022a).

5. Integrating attention module for DL (Zhu *et al.*, 2021). The attention mechanism approach has gained popularity in the DL community in recent years. The decoder that makes use of the greatest and most extensive feature information from the provided input sequences is typically made possible by the attention mechanism technique. In a recent study by Wang *et al.* (2022), for instance, the incorporation of an attention module into the YOLOv5 model led to a notable improvement in the detection accuracy of dual seedlings. Additionally, the agricultural field is falling behind in investigating model interpretability-based research studies (Li *et al.*, 2021). The fact that DL is merely a black box-based technique limits a widely held belief about its implementation. In order to tackle this idea, scientists who are trying to identify or categorize weeds from crop plants in a lab or industrial environment need to create methods that will assist in identifying a pattern that a trained CNN model has learned. Published technical articles in reputable journals should also heavily emphasize this. By doing this, a deeper comprehension of the DL model feature learning technique or tuned models to learn targeted features for weed classification from crop plants will be developed (Lee *et al.*, 2017).

14.5 Conclusion

This chapter offers a concise introduction to DL, highlighting its growing relevance in modern agriculture. In the context of weed management, DL involves detection of weeds, mapping of their distribution, and implementation of the necessary strategies. Weed control systems powered by DL provide a flexible, sustainable, and scalable replacement for the blanket application of chemical herbicides. Making such approaches a part of agriculture is imperative for the future of sustainable farming. CNNs, RNNs, GANs, and GNNs have potential applications in weed management. Of all the algorithms, CNNs stand out as the most reliable and studied network for precise weed detection. Advanced DL models, such as YOLOv4, RetinaNet, and VGG-16, offer accurate mapping. DL reduces the cost of labor and herbicides, making it cost-effective. Hence, making such approaches a part of agriculture is imperative for the future of sustainable farming.

References

Ahmed, S., Basu, N., Nicholson, C.E., Rutter, S.R., Marshall, J.R., Perry, J.J. and Dean, J.R. (2024) Use of machine learning for monitoring the growth stages of an agricultural crop. *Sustainable Food Technology,* 2, 104–125. https://doi.org/10.1002/sft2.121.

Amarasingam, N., Hamilton, M., Kelly, J.E., Zheng, L., Sandino, J., Gonzalez, F., Dehaan, R.L. and Cherry, H. 2023 Autonomous detection of mouse-ear hawkweed using drones, multispectral imagery and supervised machine learning. *Remote Sensing*, 15, 1633. https://doi.org/10.3390/rs15061633

Andújar, D. and Martínez-Guanter, J. 2022 An overview of precision weed mapping and management based on remote sensing. *Remote Sensing*, 14, 3621. https://doi.org/10.3390/rs14123621

Azghadi, M.R., Olsen, A., Wood, J., Saleh, A., Calvert, B., Granshaw, T., Fillols, E. and Philippa, B. (2024) Precise robotic weed spot-spraying for reduced herbicide usage and improved environmental outcomes--a real-world case study. agrirXiv, preprint https://doi.org/10.48550/arXiv.2401.13931.

Bah, M. D., Hafiane, A. & Canals, R. (2018) Deep learning with unsupervised data labelling for weed detection in line crops in UAV images. *Remote sensing*, 10(11), 1690. https://doi.org/10.3390/rs10101690

Bai, T., Luo, J., Zhao, J., Wen, B. and Wang, Q. (2021) Recent advances in adversarial training for adversarial robustness. agriRXiv preprint. https://doi.org/10.48550/arXiv.2102.01356.

Bai, T., Luo, J., Zhao, J., Wen, B. and Wang, Q. (2021) Recent advances in adversarial training for adversarial robustness. arXiv preprint arXiv:2102.01356.

Baozhou, Z., Hofstee, P., Lee, J. and Al-Ars, Z. (2021) An attention module for convolutional neural networks. arXiv preprint arXiv:2108. 08205.

Benchallal, F., Hafiane, A., Ragot, N. and Canals, R. (2024). ConvNeXt based semi-supervised approach with consistency regularization for weeds classification. *Expert Systems with Applications*, 239: 122222. https://doi.org/10.1016/j.eswa.2023.122222

Brilhador, A., Gutoski, M., Hattori, L. T., de Souza Inácio, A., Lazzaretti, A. E. and Lopes, H. S. (2019). Classification of weeds and crops at the pixel-level using convolutional neural networks and data augmentation. In 2019 IEEE Latin American conference on computational intelligence (LA-CCI). *IEEE*, pp. 1-6.

Chen, J., Wang, H., Zhang, H., Luo, T. Wei, D. Long, T. and Wang, Z. (2022). Weed detection in sesame fields using a YOLO model with an enhanced attention mechanism and feature fusion. *Computers and Electronics in Agriculture*, 202, 107412. https://doi.org/10.1016/j.compag.2022.107412

Corceiro, A., Alibabaei, K., Assunção, E., Gaspar, P. D. and Pereira, N. (2023) Methods for detecting and classifying weeds, diseases and fruits using AI to improve the sustainability of agricultural crops: a review. *Processes*, 11(4), 1263. https://doi.org/10.3390/pr11041263 MDPI

Dev, K., Ashraf, Z., Muhuri, P. K. and Kumar, S. (2022) Deep auto encoder based domain adaptation for transfer learning. *Multimedia tools and applications*, 81(16), 22379-22405. https://doi.org/10.1007/s11042-022-12226-2.

Espejo-Garcia, B., Mylonas, N., Athanasakos, L., Vali, E. and Fountas, S. (2021). Combining generative adversarial networks and agricultural transfer learning for weeds identification. *Biosystems Engineering*, 204, 79-89. https://doi.org/10.1016/j. biosystemseng.2021.01.014

Fan, K. J. and Su, W. H. (2022) Applications of fluorescence spectroscopy, RGB-and multispectral imaging for quality determinations of white meat: a review. *Biosensors*, 12(2), 76. https://doi.org/10.3390/bios12020076

Farooque, A.A., Hussain, N., Schumann, A.W., Abbas, F., Afzaal, H., McKenzie-Gopsill, A., Esau, T., Zaman, Q. and Wang, X. (2023) Field evaluation of a deep learning-based smart variable-rate sprayer for targeted application of agrochemicals. *Smart Agricultural Technology*, 3, 73. https://doi.org/10.1016/j.atech.2022.100073.

Farooq, A., Hu, J. and Jia, X. (2018) Analysis of spectral bands and spatial resolutions for weed classification via deep convolutional neural network. *IEEE Geoscience and Remote Sensing Letters*, 16(2), 183-187.

Fathipoor, H., Shah-Hosseini, R. and Arefi, H. (2023). Crop and weed segmentation on ground-based images using deep convolutional neural network. *ISPRS Annals of the Photogrammetry, Remote Sensing and Spatial Information Sciences*, X-4/W1-202: 195–200

Fawakherji, M., Youssef, A., Bloisi, D.D., Pretto, A. and Nardi, D. (2020). Crop and weed classification using pixel-wise segmentation on ground and aerial images. *International Journal of Robotics and Computing*, 2, 39–57. https://doi.org/10.1504/IJRC.2020.108345.

Ganaie, M.A., Hu, M., Malik, A.K., Tanveer, M. and Suganthan, P.N. (2020). Ensemble deep learning: A review. *Engineering Applications of Artificial Intelligence*, 115, 105151. https://doi.org/10.1016/j.engappai.2022.105151.

Gašparović, M., Zrinjski, M., Barković, Đ. and Radočaj, D., (2022) An automatic method for weed mapping in oat fields based on UAV imagery. *Computers and electronics in agriculture*, 173, 105385. https://doi.org/10.1016/j.compag.2020.105385

Gryshova, I., Balian, A., Antonik, I., Miniailo, V., Nehodenko, V. and Nyzhnychenko, Y. (2024) Artificial intelligence in climate smart in agricultural: toward a sustainable farming future. *Access Journal*, 5(1), 125-40.

Haq, S.I.U., Tahir, M.N. and Lan, Y. (2023). Weed detection in wheat crops using image analysis and artificial intelligence (AI). *Applied Sciences*, 13: 8840. https://doi.org/10.3390/app13158840

He, Z., Ding, L., Ji, J., Jin, X., Feng, Z. and Hao, M. (2024) Design and Experiment of Variable-Spray System Based on Deep Learning. *Applied Sciences*, 14(8), 3330. https://doi.org/10.3390/app14083330.

Hussain, N., Farooque, A.A., Schumann, A.W., McKenzie-Gopsill, A., Esau, T., Abbas, F., Acharya, B. and Zaman, Q., (2020) Design and development of a smart variable rate sprayer using deep learning. *Remote Sensing*, 12(24), 4091. https://doi.org/10.3390/rs12244091.

Islam, T., Sarker, T. T., Ahmed, K. R., Rankrape, C. B. and Gage, K. (2024) Weed Vision: Multi-Stage Growth and Classification of Weeds using DETR and RetinaNet for Precision Agriculture. In 2024 International Conference on Machine Learning and Applications (ICMLA) (pp.1374-1381). IEEE. https://doi.org/10.48550/arXiv.2502.14890.

Kamilaris, A. and Prenafeta-Boldú, F. X. (2018) Deep learning in agriculture: A survey. *Computers and electronics in agriculture*, 147, 70-90.

Khan, S., Tufail, M., Khan, M. T., Khan, Z. A. and Anwar, S. (2021) Deep learning-based identification system of weeds and crops in strawberry and pea fields for a precision agriculture sprayer. *Precision Agriculture*, 22(6), 1711-1727. https://doi.org/10.1007/s11119-021-09887-2

Lee, S. H., Chan, C. S., Mayo, S. J. and Remagnino, P. (2017) How deep learning extracts and learns leaf features for plant classification. *Pattern recognition*, 71, 1-13. https://doi. org/10.1016/j.patcog.2017.05.015.

LeCun, Y., Bengio, Y. and Hinton, G. (2015) Deep learning. *Nature*, 521(7553), 436-444.

Li, T., Asai, M., Kato, Y., Fukano, Y. and Guo, W. (2024). Channel Attention GAN-Based Synthetic Weed Generation for Precise Weed Identification. *Plant Phenomics*, 6, 0122. https://doi.org/10.34133/plant-phenomics.0122.

Li, Y., Al-Sarayreh, M., Irie, K., Hackell, D., Bourdot, G., Reis, M. M. and Ghamkhar, K. (2021) Identification of weeds based on hyperspectral imaging and machine learning. *Frontiers in Plant Science*, 11, 611622. https://doi.org/10.3389/fpls.2020.611622

Liu, Y., Pu, H. and Sun, D. W. (2021) Efficient extraction of deep image features using convolutional neural network (CNN) for applications in detecting and analysing complex food matrices. *Trends in Food Science & Technology*, 113, 193-204. https://doi.org/10.1016/j.tifs.2021.04.010

Li, Y., Al-Sarayreh, M., Irie, K., Hackell, D., Bourdot, G., Reis, M. M. and Ghamkhar, K. (2021) Identification of weeds based on hyperspectral imaging and machine learning. *Frontiers in Plant Science*, 11, 611622. https://doi.org/10.3389/fpls.2020.611622.

Lottes, P., Behley, J., Milioto, A. and Stachniss, C. (2018) Fully convolutional networks with sequential information for robust crop and weed detection in precision farming. *IEEE Robotics and Automation Letters*, 3(4), 2870-2877. https://doi.org/10.1109/LRA.2017.2699782.

Lou, Z., Quan, L., Sun, D., Li, H. and Xia, F. (2022) Hyperspectral remote sensing to assess weed competitiveness in maize farmland ecosystems. *Science of The Total Environment*, 844, 157071. https://doi.org/10.1016/j.scitotenv.2022.157071

Luo, T., Zhao, J., Gu, Y., Zhang, S., Qiao, X., Tian, W. and Han, Y. (2023). Classification of weed seeds based on visual images and deep learning. *Information Processing in Agriculture*, 10(1), 40-51. https://doi.org/10.1016/j.inpa.2021.10.002 aeeisp.com

Machleb, J., Peteinatos, G. G., Kollenda, B. L., Andújar, D. and Gerhards, R. (2020) Sensor-based mechanical weed control: Present state and prospects. *Computers and electronics in agriculture*, 176, 105638.

MacLaren, C., Storkey, J., Menegat, A., Metcalfe, H. and Dehnen-Schmutz, K. (2020) An ecological future for weed science to sustain crop production and the environment. A review. *Agronomy for Sustainable Development*, 40, 1-29. https://doi.org/10.1007/s13593-020-00631-3

Milioto, A., Lottes, P. and Stachniss, C. (2018) Real-time semantic segmentation of crop and weed for precision agriculture robots leveraging background knowledge in CNNs. In 2018 IEEE international conference on robotics and automation (ICRA). IEEE, pp. 2229-2235. https://doi.org/10.1109/ICRA.2018.8461052.

Murad, N. Y., Mahmood, T., Forkan, A. R. M., Morshed, A., Jayaraman, P. P. and Siddiqui, M. S. (2023) Weed detection using deep learning: A systematic literature review. *Sensors*, 23(7), 3670. https://doi.org/10.3390/s23073670.

Naveed, A., Muhammad, W., Irshad, M. J., Aslam, M. J., Manzoor, S. M., Kausar, T. and Lu, Y. (2023) Saliency-based semantic weeds detection and classification using UAV multispectral imaging. *IEEE Access*, 11, 11991-12003. https://doi.org/10.1109/ACCESS.2023.3278901

Nyamekye, C., Ofosu, S.A., Arthur, R., Osei, G., Appiah, L.B., Kwofie, S., Ghansah, B. and Bryniok, D. (2021) Evaluating the spatial and temporal variations of aquatic weeds (Biomass) on Lower Volta River using multi-sensor Landsat Images and machine learning. *Heliyon*, 7(5), 07080. https://doi.org/10.1016/j.heliyon.2021.e07080 PMC

Ofori, M., El-Gayar, O., O'Brien, A. and Noteboom, C. (2022) A deep learning model compression and ensemble approach for weed detection. http://hdl.handle.net/10125/79470 (Accessed: February 21, 2025).

Oghaz, M. M. D., Razaak, M., Kerdegari, H., Argyriou, V. and Remagnino, P. (2019) Scene and environment monitoring using aerial imagery and deep learning. In 2019 15th International Conference on Distributed Computing in Sensor Systems (DCOSS). IEEE, pp. 362-369. https://doi.org/10.1109/DCOSS.2019.00069

Olsen, A., Konovalov, D.A., Philippa, B., Ridd, P., Wood, J.C., Johns, J., Banks, W., Girgenti, B., Kenny, O., Whinney, J. and Calvert, B., (2019) Deep Weeds: A multiclass weed species image dataset for deep learning. *Scientific reports*, 9(1), 2058. https://doi.org/10.1038/s41598-019-38526-9.

Pandey, D. R. and Mishra, N. (2024) IoT integration for enhanced turmeric cultivation: a case study in smart agriculture. In BIO web of conferences. EDP Sciences, 82, 05008. https://doi.org/10.1051/bioconf/20248205008.

Pan, S. Y. (2010) Q.: A survey on transfer learning. *IEEE Transactions on Knowledge and Data Engineering*, 22(10), 1345-1359.

Dyrmann, M., Mortensen, A. K., Midtiby, H. S. and Jørgensen, R. N. (2016) Pixel-wise classification of weeds and crops in images by using a fully convolutional neural network. In Proceedings of the International Conference on Agricultural Engineering, Aarhus, Denmark (pp. 26-29).

Qu, H. R. and Su, W. H. (2024) Deep learning-based weed–crop recognition for smart agricultural equipment: A review. *Agronomy*, 14(2), 363. https://doi.org/10.3390/agronomy14020363

Rai, N., Zhang, Y., Ram, B. G., Schumacher, L., Yellavajjala, R. K., Bajwa, S. and Sun, X. (2023) Applications of deep learning in precision weed management: A review. *Computers and Electronics in Agriculture*, 206, 107698. https://doi.org/10.1016/j.compag.2023.107698

Reedha, R., Dericquebourg, E., Canals, R. and Hafiane, A. (2022) Transformer neural network for weed and crop classification of high resolution UAV images. *Remote sensing*, 14(3), 592. https://doi.org/10.3390/rs14030592

Roslim, M. H. M., Juraimi, A. S., Che'Ya, N. N., Sulaiman, N., Manaf, M. N. H. A., Ramli, Z. and Motmainna, M. (2021) Using remote sensing and an unmanned aerial system for weed management in agricultural crops: A review. *Agronomy*, 11(9), 1809. https://doi.org/10.3390/agronomy11091809

Rozenberg, G., Dias, J. L. C., Anderson, W. M., Sellers, B. A., Boughton, R. K., Piccolo, M. B. and Blank, L. (2023) Using a low-cost unmanned aerial vehicle for mapping giant smut grass in bahia grass pastures. *Precision Agriculture*, 24(3), 971-985. https://doi.org/10.1007/s11119-022-09915-1

Ryo, M. (2022) Explainable artificial intelligence and interpretable machine learning for agricultural data analysis. *Artificial Intelligence in Agriculture*, 6, 257-265. https://doi.org/10.1016/j.aiia.2022.11.003 Science Direct

Saqib, M. A., Aqib, M., Tahir, M. N. and Hafeez, Y. (2023) Towards deep learning based smart farming for intelligent weeds management in crops. *Frontiers in Plant Science*, 14, 1211235. https://doi.org/10.3389/fpls.2023.1211235.

Sarker, I. H. (2021) Machine learning: Algorithms, real-world applications and research directions. *SN computer science*, 2(3), 160. https://doi.org/10.1007/s42979-021-00661-0

Somerville, G. J., Sønderskov, M., Mathiassen, S. K. and Metcalfe, H. (2020) Spatial modelling of within-field weed populations; a review. *Agronomy*, 10(7), 1044. https://doi.org/10.3390/agronomy10071044

Tao, T. and Wei, X. (2022) A hybrid CNN–SVM classifier for weed recognition in winter rape field. *Plant Methods*, 18(1), 29. https://doi.org/10.1186/s13007-022-00898-1

Tufail, M., Iqbal, J., Tiwana, M. I., Alam, M. S., Khan, Z. A. and Khan, M. T. (2021) Identification of tobacco crop based on machine learning for a precision agricultural sprayer. *IEEE access*, 9, 23814-23825. https://doi.org/10.1109/ACCESS.2021.3059459

Upadhyay, A., Sunil, G. C., Zhang, Y., Koparan, C. and Sun, X. (2024) Development and evaluation of a machine vision and deep learning-based smart sprayer system for site-specific weed management in row crops: An edge computing approach. *Journal of Agriculture and Food Research*, 18, 101331. https://doi.org/10.1016/j.jafr.2024.101331.

Wang, A., Xu, Y., Wei, X. and Cui, B. (2020) Semantic segmentation of crop and weed using an encoder-decoder network and image enhancement method under uncontrolled outdoor illumination. *IEEE Access*, 8, 81724-81734.

Wang, C. Y., Bochkovskiy, A. and Liao, H. Y. M. (2023) YOLOv7: Trainable bag-of-freebies sets new state-of-the-art for real-time object detectors. In Proceedings of the IEEE/CVF conference on computer vision and pattern recognition (pp. 7464-7475). 10.48550/arXiv.2207.02696.

Wang, H., Xu, J.W., Sun, C., Yan, R.Q. and Chen, X.F. (2022a) Intelligent Fault diagnosis for planetary gearbox using time-frequency representation and deep reinforcement learning. *Ieee-Asme Transactions on Mechatronics,* 27, 985–998.

Wang, Q., Cheng, M., Huang, S., Cai, Z., Zhang, J. and Yuan, H. (2022b) A deep learning approach incorporating YOLO v5 and attention mechanisms for field real-time detection of the invasive weed *Solanum rostratum* Dunal seedlings. *Computers and Electronics in Agriculture*, 199, 107194. https://doi.org/10.1016/j.compag.2022.107194.

Zhang, S., Guo, J. and Wang, Z. (2019) Combing K-means clustering and local weighted maximum discriminant projections for weed species recognition. *Frontiers in Computer Science*, 1, 4. https://doi.org/10.3389/fcomp.2019.00004

Zhang, H., Wang, Z., Guo, Y., Ma, Y., Cao, W., Chen, D., Yang, S. and Gao, R. (2022) Weed detection in peanut fields based on machine vision. *Agriculture*, 12(10), 1541. https://doi.org/10.3390/agriculture12101541 MDPI

Zhang, J., Yu, F., Zhang, Q., Wang, M., Yu, J. and Tan, Y. (2024) Advancements of UAV and deep learning technologies for weed management in Farmland. *Agronomy*, 14(3), 494. https://doi.org/10.3390/agronomy14030494 MDPI

Zhu, W., Sun, L., Huang, J., Han, L. and Zhang, D. (2021) Dual Attention Multi-Instance Deep Learning for Alzheimer's Disease Diagnosis with Structural MRI. *IEEE Transactions on Medical Imaging*, 40 (9), 2354-2366. ISSN 0278-0062

15 Deep Learning in Studying Regulatory Networks in Plants

Sanam Ashraf[1†], Zubaria Tul Ain[1], Javed Iqbal[2*†], Banzeer Ahsan Abbasi[3], Ghulam Murtaza[4], Rashid Iqbal[5,6], Muhammad Anas[1,7], Tariq Mahmood[1], Sana Basharat[8,9,10], and Muhammad Waseem[8,9,10*]

[1]Department of Plant Sciences, Faculty of Biological Sciences, Quaid-i-Azam University Islamabad, Pakistan; [2]Department of Botany, Bacha Khan University, Charsadda, Khyber Pakhtunkhwa, Pakistan; [3]Department of Botany, Rawalpindi Women University, Rawalpindi, Pakistan; [4]School of Agriculture, Yunnan University, Kunming, China; [5]Department of Agronomy, Faculty of Agriculture and Environment, The Islamia University of Bahawalpur, Bahawalpur, Pakistan; [6]Department of Life Sciences, Western Caspian University, Baku, Azerbaijan; [7]Department of Soil Science and Plant Nutrition, University of Selcuk, Konya, Turkiye; [8]School of Breeding and Multiplication (Sanya Institute of Breeding and Multiplication), Sanya, China; [9]College of Tropical Agriculture and Forestry, Hainan University, Sanya, China; [10]Fang Zhiyuan Academician Team Innovation Center, Sanya, China

Abstract

Research into the detailed control systems responsible for plant genetic expression along with development and adaptative responses is fundamental for developing better crops and plant technology. Heavy demand exists today for interpreting large-scale omics data produced by the accelerated growth of high-throughput technologies in plant biology. Deep learning (DL) is a useful subgroup of machine learning that reveals concealed patterns and forecasts biological activities within dense datasets. This chapter offers detailed insights into the DL applications that help study plant regulatory networks through research on gene regulatory networks (GRNs), transcription factor binding prediction, and epigenetic regulation. It explains DL architectures consisting of convolutional neural networks (CNNs) along with recurrent neural networks (RNNs) and graph neural networks (GNNs), because these networks work optimally with spatial and temporal connections of plant systems and topological dynamics. Plant sciences benefit from DL applications, as demonstrated by the studies showing the practical use of DL in the areas such as Ayurvedic medicinal plant taxonomy, mobile-based rice disease detection, and nitrogen pathway and sorghum phenotyping analysis using aerial images. The practical implementation of DL techniques for plant regulatory network study faces several obstacles despite its prospective uses. Model application faces three main obstacles: restricted labeled dataset availability, unexplained model workings and susceptibility to small sample overfitting, and requirement of extensive computational power. The experimental verification of learned attributes encounters biological challenges that delay the process of merging computational forecast data with laboratory testing. Model generalizability across different species encounters obstacles because of specific genomic complexities and genomic variations that exist between species. A successful continuation of DL

[†]Equal contribution

[*]E-mail: javed89qau@gmail.com or m.waseem.botanist@gmail.com

DOI: 10.1079/9781836991120.0015

applications in plant regulatory network research demands collaborative efforts between different fields while improving explainable AI systems and data collaboration protocols. The written guide directs researchers on how to utilize computationally advanced methods to explore biological complexities and foster breakthroughs in plant science.

Keywords: AI systems, data collaboration, deep learning, plant regulatory networks

15.1 Introduction

15.1.1 Background on gene regulatory networks (GRNs) in plants

Researchers in plant biology study the effects that gene interactions throughout different cell regions, tissue types, and organ systems have on plant biological growth and developmental stages, including germination, flowering, regeneration, and stress adaptation. A detailed understanding of this developmental subject will contribute to enhanced plant resilience, sustainable agricultural practices, and improved food security (Li *et al.*, 2015). Organisms use GRNs, including numerous molecular interaction networks, to control their genetic expression. The system consists of dynamic biological interactions among transcription factors (TFs), regulatory proteins, RNA molecules, and other biomolecules, which determine whether particular genes are activated or remain inactive, thereby guiding developmental processes, environmental responses, and cellular operations. Studies on GRNs are essential for plant biology because they control the critical functions that regulate plant growth and development, stress response, and pathogenic responses. Research on plant GRNs has progressed rapidly because of advancements in molecular biology, new sequencing technologies, and computational methods that reveal complex molecular plant relationships (Ko and Brandizzi, 2020; Mercatelli *et al.*, 2020; Khan *et al.*, 2023).

The regulation of plant gene expression in response to internal and external signals can only be understood properly through GRNs. TFs regulate essential biological processes by interacting with target genes at the intersection points, enabling development, stress response, and environmental adaptation functions in the biological networks. The interpretation of plant GRNs' structural features and dynamic behavior remains vital for advancing crop agriculture and enhancing plant growth, plant breeding, and ecological responses (Kulkarni and Vandepoele, 2020; Mercatelli *et al.*, 2020). Plants being sessile organisms are subject to rapid environmental changes that lead them to swift adaptations in gene expression due to their complex GRNs according to Artur and Kajala (2021) and Carfora *et al.* (2025). The main purpose of plant systems biology is to forecast both practical and scientific outcomes from regulatory networks when exposed to new environmental situations. Researchers have developed models for prokaryotic GRNs to predict both untested environmental condition effects on gene expression levels and gene-to-gene relationship networks. Researchers have achieved little success in creating predictive network models for multicellular organisms, including plant models, as shown in Fig. 15.1 (Bonneau, 2008). New RNA sequencing methods, known as RNA-seq, revolutionized GRN investigation because they provide a comprehensive analysis of gene expression data (Park, 2009; Junker *et al.*, 2010). Support vector machines (SVMs), together with sequencing techniques, have demonstrated considerable success in detecting regulatory connections. The computational methods provide researchers with better ways to find gene regulatory relationships in plants, because they offer higher efficiency in addition to feasibility for handling big datasets or tricky networks.

15.1.2 Deep learning: basics and applications in regulatory networks

The development of "deep learning" (DL) by Hinton *et al.* (2006) originated from the artificial neural network (ANN) principle. Neural network research experienced a resurgence after DL emerged as an important cutting-edge topic, because experts refer to these modern networks as "new-generation neural networks." The proper training of deep networks leads to substantial achievement in multiple classification and

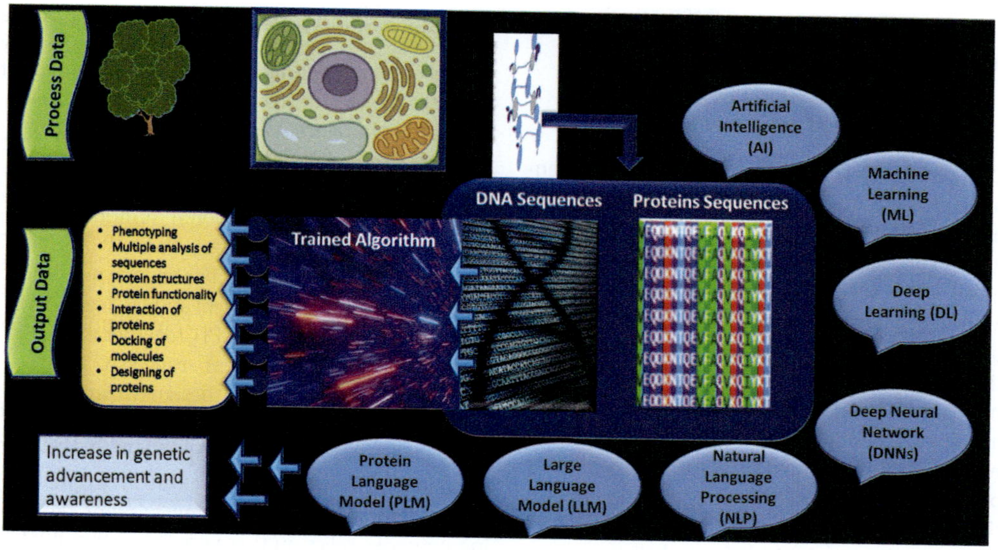

Fig. 15.1. An illustration of how predictive network models are used in plant biology to speed up research and genetic advancement. (Author's own image.)

regression challenges (Karhunen *et al.*, 2015). DL serves as a core technology of the Fourth Industrial Revolution and belongs to the field of artificial intelligence and machine learning (ML). DL technology emerged from ANN technology to become a popular topic in computing because of its ability to learn from data and its wide applications across healthcare, visual recognition, text analytics, cybersecurity, and numerous other domains. The creation of suitable DL models presents intense obstacles because of the changing characteristics found in real-world data and problems (Sarker, 2021).

Modern researchers dedicate themselves to studying DL because its plant science applications remain at a basic developmental phase. Plant phenotyping has achieved improved performance through the application of DL, because it replaced traditional hand-engineered computer vision pipelines. Researchers use DL for leaf counting, age estimation, mutant classification, and plant disease detection and diagnosis from leaf images, as well as classification of fruits and other organs and pixel-wise localization of root and shoot tips and ears (Mohanty *et al.*, 2016; Ubbens and Stavness, 2017). DL belongs to the ML field through its approach of using ANNs based on brain neural organization. The networks consist of multiple layers with interconnected units called "neurons," which process the input data to generate meaningful outputs (Zhang *et al.*, 2020). DL, advanced artificial intelligence system, functions as a flexible instrument that processes large-scale and diverse datasets and generates precise predictions even in complicated and ambiguous situations. The tools are becoming increasingly prevalent among plant scientists for analyzing the growing datasets derived from high-throughput phenotyping and genotyping systems (Singh *et al.*, 2018). This method helps researchers detect the complex TF–target gene interactions within plant GRNs. DL models provide sequential data transformation across their layers, enabling the network to recognize gene regulatory patterns that help determine how genes respond to biological signals (Kulkarni and Vandepoele, 2020; Muley, 2023).

15.1.3 Application of DL to processes governed by the central dogma of molecular biology

Tools developed through DL provide researchers with the ability to study biological complexity by analyzing vast biological datasets across the DNA, RNA, and protein stages of the central dogma. These modern tools provide substantial advantages for scientists, helping them understand the genomic data, map TF binding sites (TFBSs), and predict protein structures that control gene regulation processes.

Genomic and epigenomic insights

DL techniques serve as primary tools for genomic element investigations, enabling the study of cis-regulatory elements (CREs), which control gene expression in plants. DL models that integrate DNA accessibility data from DNase-seq, MNase-seq, and formaldehyde-assisted isolation of regulatory elements (FAIRE) with TF information from ChIP-seq use convolutional neural networks (CNNs) to discover actual protein binding sites. DeepBind and TFImpute showed effective results in human DNA- and RNA-binding protein research, and their plant system applications have successfully contributed to regulatory code identification in *Arabidopsis thaliana* and *Zea mays*. DeepSEA and Deep Feature Interaction Maps (DFIM) enable researchers to analyze non-coding variant effects on biological processes in a manner that assists environmental stress studies of plant genetic changes. The research initially focused on human datasets due to the abundance of available data; however, scientists apply similar frameworks to analyze complex plant genomic data (Yue *et al.*, 2018; Wang *et al.*, 2020).

Transcriptomic applications

The DL method excels in evaluating RNA-seq and identifying transcriptomic networks (TRNs) through their analysis of data. Scientists apply k-mer grammars together with CNNs to annotate gene regulatory regions in the corn genome, especially when it has large repetitive content. The analysis enables efficient data interpretation and has proven to be useful for plant systems lacking extensive functional annotation data according to Wang *et al.* (2025). The ability to extract features belongs to CNNs, although they have a black box system, which produces results that k-mer-based models easily understand. The sequence segmentation approach of k-mer models provides biologically straightforward and fast reference-free sequence classification through their ability to divide DNA fragments into nucleotide strings. Research communities have combined their approaches in order to maximize the benefits of each technology according to Shen *et al.* (2018).

Proteomics and protein function prediction

DL techniques have transformed the way humans study structural protein information and discern its functional characteristics. AlphaFold, from DeepMind, marked a breakthrough in tertiary protein structure prediction, achieving a high level of accuracy that enables functional annotation in non-model plant species (Tran *et al.*, 2017). The prediction tools sequence based enzyme function prediction by DL (DEEPre) and DL-based protein–protein interaction (DPPI), together with raw multiple sequence alignment (rawMSA), employ CNNs and other architectural designs to predict enzyme functions, protein–protein interaction (PPIs), transmembrane topology, and solvent accessibility. Understandable posttranscriptional and posttranslational regulation in plant signaling networks relies on these prediction results (Li *et al.*, 2018). Various issues need to be resolved before implementing these prediction models in plant research. Plants present restricted and variable dataset resources, and many regulatory components lack proper description. The functional understanding of deep models is difficult because their interpretation lacks clarity, particularly during validation experiments. The application of DL will revolutionize plant gene regulation research, since the collection of genomic resources continues to expand for models and crop species. The application of DL models to plant research enables their use of gene expression data to forecast GRNs during seed development or stress responses and growth regulation processes. Network optimization in large-scale plant genomic studies can be achieved through the combination of three techniques, stochastic gradient descent, adaptive moment estimation (ADAM), and alternating direction method of multipliers (ADMM), according to Haque *et al.* (2019) and Devarapu (2021).

15.1.4 ML foundations for GRN in plants

The process of tracing GRNs has been improved through ML, since these tools provide enhanced capabilities for analyzing complex multidimensional plant research transcriptome data, as shown in Fig. 15.2. By employing Bayesian networks, dynamic factor graph models, and various other ML-based algorithms, GENIE3 constructs models of gene regulatory networks in relation to growth and environmental factors across developmental

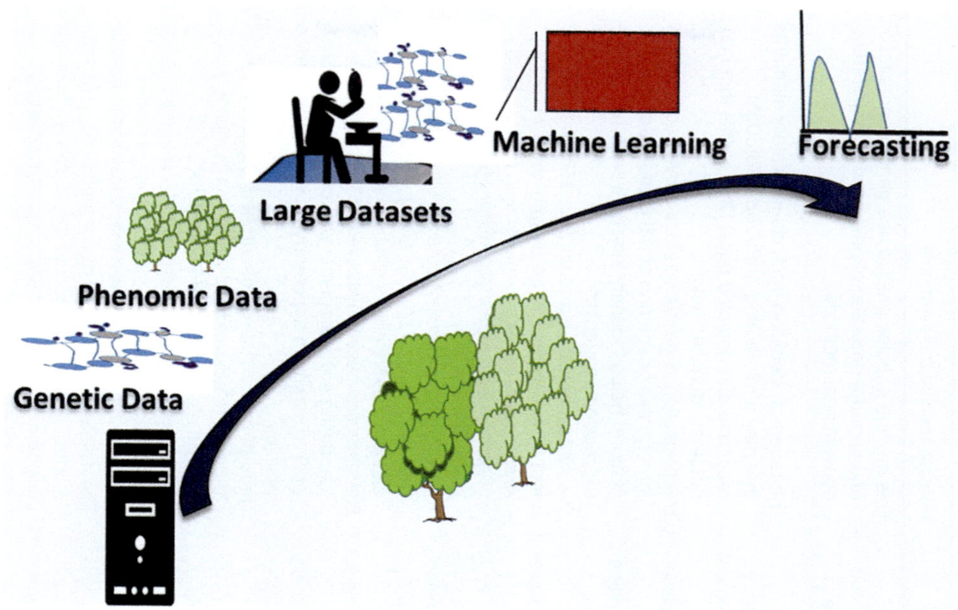

Fig. 15.2. The use of machine learning in plants for big data analysis. (Author's own image.)

and ecological contexts. GENIE3 serves as a tree-based ML method that has proven its effectiveness in constructing tissue-focused GRNs through extensive RNA-seq analysis of *Arabidopsis* and corn (Guo *et al.*, 2016; Banf and Rhee, 2017). The derivative implementation of Bayesian networks, known as MinReg, managed to reconstruct complete regulatory networks of *Fusarium graminearum* by combining transcriptomic information with PPI data. Research applying these methods successfully predicted important regulatory factors KN1, FEA4, and CRF4, which were subsequently confirmed through ChIP-seq analysis and the TARGET experimental system. The strong role of ML becomes evident through its techniques, which unravel complex regulatory systems. With the growing quantity of temporal and multi-omics data, the need to manage complex non-linear interactions alongside the plant regulatory network architecture drives interest in furthering DL models (Mochida *et al.*, 2018). The key DL models used in plant biology along with their primary characteristics and important accomplishments have been summarized in Table 15.1. It provides a clear overview of all major DL models used in plant biology. The contrast between these models creates a reference system that helps readers grasp the various applications covered within the following sections, as stated by Krouk *et al.* (2013).

15.1.5 Case studies and applications of DL in plants

Advancements in plant science rely on DL because this technology allows scientists to analyze complex omics data to uncover molecular gene control and stress reaction processes. Private research labs employ DL to study plant–pathogen dynamics to identify genes involved in stress response to develop survival strategies in crops. The self-organizing maps (SOMs) DL approach helps researchers identify how gene expression patterns change in response to stress testing as demonstrated by diverse research projects, particularly in cold defense studies on grapevine plants. CNNs maintain high precision in analyzing the morphological characteristics of somatic embryos, thus enabling researchers to explore developmental gene regulatory mechanisms. DL advancements enable scientists to find vital regulatory genes while automatically analyzing phenotypes, which results in the development of climate-resistant, high-yielding crop varieties through innovative plant molecular breeding approaches (Konecny *et al.*, 2023; Eftekhari *et al.*, 2024).

Table 15.1. Overview of all major deep learning models used in plant studies. Author's own table.

Model name	Type	Primary application	Key features	Notable results/achievements
DeepGene	Hybrid CNN–RNN	Genome annotation and gene prediction	The software tool identifies coding sequences, predicts exon–intron boundaries, and detects sites where transcription begins.	This method enables precise rice genome annotation and superior detection of new genes compared to conventional methods.
DanQ	Hybrid CNN–LSTM	Regulatory element detection	The tool detects DNA sequence characteristics and identifies regulatory elements in addition to promoter element analysis.	The research identified fresh cis-regulatory elements that play a role during *Arabidopsis* drought response and detected novel DNA motifs that had not been identified before.
ExPecto	CNN	Gene expression prediction	The system enables gene expression level prediction while performing DNA sequence evaluation and multi-condition analysis.	The prediction of genes in maize proved successful through this adaptation of the platform, and the platform helped identify essential patterns within expression sequences.
DeepPheno	CNN	High-throughput phenotyping	Physical plant features are analyzed through the system, which also monitors plant growth patterns by studying standard color images.	The system achieved greater than 95% precision in trait measurement, together with effective *Arabidopsis* development tracking and discovery of new genetic areas that influence physical attributes.
PlantVillage model	CNN	Disease detection	This system combines disease recognition, leaf picture assessment, and on-site diagnosis capability.	The technology demonstrates 99% success rate in detecting 14 crop species along with 26 different diseases and is now available as mobile software.
3D CNN (hyperspectral)	3D CNN	Stress detection	The system examines hyperspectral images while determining stress response processes and detecting conditions early.	The detection system achieved 95% accuracy in identifying soybean plants with charcoal rot 2 days prior to symptom appearance, using its diagnostic abilities.
DeepGWAS	Deep neural network	GWAS	This system captures genetic marker interactions, models environmental factors, and predicts agronomic traits.	Through their initiative, researchers detected new drought-related genetic variations in maize, which proved superior to GWAS-based studies.
DeepOmix	Variational autoencoder	Multi-omics integration	The system processes multiple data types to analyze regulatory relationships in order to predict phenotypic end results.	Researchers used multi-omics data analysis techniques to study tomato ripening in order to identify newly discovered regulatory frameworks.
GNN-based model	GNN	Multi-omics integration	This system models biological entity interactions, integrates gene expression and metabolomics, and maps complex relationships.	The research discovered major metabolic genes that control *Arabidopsis* stress responses while bringing together primary and secondary metabolic processes.

DL in plant genomic prediction

From 2008 to 2016, approximately 2247 locations received corn yield forecasts from the authors Saeed and Wang (2019), using deep neural networks, which integrated genetic and environmental factors. DL research obtained better results than random forests and fully connected neural networks in forecasting agricultural traits due to its exceptional predictive capability. DL provides a method to bring together varied genotypic information with environmental data, which enhances both agricultural prediction models and precision farming targets. Wang *et al.* (2023) developed DeepGS, through which CNNs serve to enhance genomic selection (GS). Multi-omics data—consisting of genomics, transcriptomics, metabolomics, and proteomics—enabled the successful prediction of complex phenotypic traits in plants. The Deep Neural Network Genomic Prediction (DNNGP) demonstrated better predictive accuracy than genomic best linear unbiased prediction (GBLUP) and LightGBM and several other popular prediction methods when tested on wheat, corn, and tomato datasets according to this approach. The DL technology demonstrates promising potential to enhance genomic predictions and improve breeding efficiencies by analyzing diverse genomic information.

DL for Ayurvedic medicinal plant classification

Dileep and Pournami (2019) conducted research on DL technology and demonstrated exceptional potential for identifying and categorizing medicinal plants even though expert botanical knowledge is traditionally required for this process. The development of AyurLeaf marks the researchers' achievement in creating an automatic medicinal plant classifier through DL for Ayurvedic medicine. The system operated using a specialized dataset containing 40 native medicinal plants that originate in Kerala, India, because of the region's strong ties with Ayurvedic practices. An AlexNet-inspired CNN architecture formed the core of the system for detection purposes by processing leaf features that encompassed shape, size, color, and texture points. The Softmax and SVM classifiers conducted classification through a five-fold cross-validation process, resulting in a high accuracy rate of 96.76%. The research demonstrates CNN performance in plant identification while developing a digital platform that protects native botanical knowledge systems. DL innovations have proven their ability to connect traditional botanical wisdom with present-day computer systems and increase educational access as well as plant conservation initiatives.

DL-based rice disease detection

The study conducted by Andrianto *et al.* (2020) analyzed how increasing global population drives the modification of plants to achieve higher yields. Plant diseases are a major obstacle, preventing both crop yield improvement and production quality enhancement. The standard practice of disease symptom detection among farmers depends on sight-based observations. Subjective ways to detect diseases in crops may produce unreliable results and imprecise measurements when infections start. Researchers resolved this issue by creating computer-based disease detection systems through DL development. A smartphone-based rice disease detection system uses CNNs alongside cloud computing for its operation. The system enables farmers to take pictures of the rice leaf through its mobile application, followed by cloud server processing with a VGG16-based DL method to identify the disease type. Through this method, a farmer can conduct immediate plant health surveillance from any location, including distant agricultural sites. All predictions made on the training dataset were 100% accurate by the model, which indicates strong performance of this dataset. The test accuracy of the model reached 60%, although this result suggests possible data quality limitations as well as overfitting tendencies. The researchers reported that raising the accuracy levels would be possible through the addition of a wider and varied dataset. DL has shown its capability to identify plant diseases using smartphone applications in this study. The implementation of novel detection technologies provides a financially viable method to handle agricultural diseases across all resource availability levels.

Transcriptional logic of nitrogen response via DL

According to Varala *et al.* (2018), a significant research advancement occurred in plant GRN study through DL-based network modeling, which included time as a fundamental component. The study investigated plant nitrogen signaling dynamics by applying time-resolved ML techniques to a dataset comprising 2174 nitrogen-responsive genes derived from time-series transcriptional profiles. Through "just-in-time" analytical approaches, the researchers identified temporally regulated transcriptional elements, elucidating the sequential activation patterns underlying the nitrogen response. Using experimentally proven TFs, including CRF4, SNZ, and CDF1, the researchers perfected a high-confidence GRN, which consisted of 155 TFs that regulated 608 target genes. The binding data from TGA1 and HHO5/6 along with PHL1 led scientists to develop an N-specificity index, which provided exact measures of TF-target bindings. The validated TFs, identified as regulators, controlled 50% of the nitrogen uptake and assimilation pathway genes. The root developmental patterns, together with biomass accumulation and nitrate absorption under low-nitrogen conditions, underwent substantial changes when CRF4 dosage increased. The work uncovers beneficial knowledge on the integration of time-series DL models with transcriptional cascade analysis for achieving better agricultural nitrogen use efficiency. Using this approach, scientists can evaluate dynamic biological processes in plant science along with agriculture and other fields because of its high potential.

DL in sorghum phenotyping

Sorghum yield (*Sorghum bicolor* L. Moench) depends on head distribution among different branching structures, thus necessitating precise head number measurements for plant breeding purposes according to Ghosal *et al.* (2019). The conventional process of measuring phenotypic traits through manual methods requires significant labor and allows both inefficiencies and human error to appear primarily during large breeding experiments. The solution to this challenge comes from DL approaches through deep CNNs, which provide effective phenotyping methods. A research team developed an active learning device that operates through weak supervision to detect and count sorghum heads within images obtained from UAV-based systems. A semi-trained CNN model uses restricted labeled data for its initial training until it produces synthetic annotations that minimize human annotation requirements. The method produced exceptional results through automatic annotation generation, which ended up in very similar head counts compared to human readings (0.88 correlation). Model interpretability was enhanced by visualizing the key features learned during training, thereby improving user trust through increased transparency and facilitating a clearer understanding of the underlying decision-making processes. The method both minimizes human labor requirement and establishes a scalable solution for precision agriculture that enables its prospective use for various crops, thus improving effective and accurate plant breeding and crop monitoring.

Broader impacts of DL in plant sciences

The paper by Gao *et al.* (2017) carried out a comprehensive analysis and revealed that DL excels beyond gene regulatory decoding in various plant science fields. The automated recognition, together with the classification of flowers and plants, utilizes computer vision technology as a notable application. Recently, a DL-based mobile system in the scientific field used videos instead of standard photographs to increase plant recognition precision. The solution delivered higher extraction accuracy by incorporating diverse viewpoints and ambient illumination, thus achieving 95% confident results in classifying 122 plant species from 125 taxonomic entities and 47 genera members from 50 detected groups. Cloud-based resources enhanced the system's processing speed, which reduced the system response time. The study proves that DL models have various uses in plant science, such as biodiversity analysis, visible trait detection, and outbreak diagnostics, which help create advanced agricultural tools.

15.2 Challenges and Limitations

Several barriers restrict DL from reaching its maximum potential despite its increased use in plant regulatory network research. The barriers to using DL technology consist of the following

Fig. 15.3. The challenges and limitations of deep learning technology. (Author's own image.)

major factors, including data challenges, computational requirements, and biological interpretability needs. Figure 15.3 presents various challenges and limitations in DL research, as found in recent investigations.

15.2.1 Data quality and availability issues

Inadequate plant data resources form the main obstacle to deploying DL in plant systems biology research. The effective pattern learning of DL models, especially CNNs and recurrent neural networks (RNNs), requires extensive quantities of high-quality data that includes proper annotations. The number of extensive plant genomic databases remains very low when compared to human genomic resources (Wang *et al.*, 2023). Plants need to collect multi-omics data by combining genomics, transcriptomics, proteomics, and metabolomics in order to build a complete understanding of their GRNs according to Badia *et al.* (2023). The wide differences in format, resolution, and coverage among these datasets make their integration into one system very challenging (Armengaud *et al.*, 2014). The use of DL approaches faces challenges because most plant species, including non-model organisms, feature genomes that are either poorly documented or missing annotations (Carpentier *et al.*, 2008). The use of DL for corn yield prediction proved challenging because of unstable environmental and genetic dataset information (Akande, 2024; Sitienei, 2024). The training and generalization performance of DL models is significantly hindered by challenges such as missing data, experimental noise, and batch effects. Furthermore, RNNs and long short-term memory networks (LSTMs) cannot be effectively trained due to the limited availability of high-resolution time-series gene expression data in publicly accessible plant datasets (Debbagh *et al.*, 2024).

15.2.2 Model interpretability and biological validation

DL models are criticized due to the lack of clear internal representation reasoning frameworks, unlike simpler statistical methods and ML tools such as random forests and SVMs. Plant regulatory network interpretations need essential interpretability because they allow both prediction explanation and generation of meaningful biological understanding and hypothesizing (Starke, 2022). The prediction capabilities of CNNs in both gene expression forecasting and image-based phenotypic trait identification remain unclear to users because of their complex hidden-layer structures. DLGWAS functions as a phenotype prediction method that utilizes dual-stream CNN technology yet faces interpretation obstacles regarding biological causation between neural features (Zini and Awad, 2022; Lu *et al.*, 2025). The absence of transparency in computational methods makes their wet-lab experimental confirmation more complicated. Scientists need two essential capabilities from prediction models: both precise forecasting capabilities and interpretable explanations to create testable scientific inquiries about genetic control systems as well as signaling pathways and TF management mechanisms. The plant systems use SHapley Additive exPlanations (SHAP) and integrated gradients techniques for DL model interpretation; yet, the applications remain scarce. The lack of improved interpretability tools makes it challenging to confirm that the model's decision-making processes match biological concepts, thus generating potentially incorrect results (Mkhatshwa, 2023; SSL *et al.*, 2024).

15.2.3 Computational resource constraints

An area of DL application that poses the greatest challenge for regulatory network analysis is the high cost of computation. Plant science operations in developing nations face limited access to training DL models due to lack of hardware facilities, such as processing power, memory, and storage capacity (Singh *et al.*, 2018; Zhu and Shen, 2025). The internal structure of Transformers or CNNs with extensive parameters necessitates graphics processing unit (GPU) or tensor processing unit (TPU) equipment for effective acceleration during training processes. DeepGS model needed intensive GPU processing to train on plant genotypes because of its demand for extensive GPU computation. This significantly increases research costs and energy consumption, raising concern about large resources, which in turn creates sustainability issues. To achieve correct parameter settings, researchers must perform multiple training runs with successive experiments, which cause further wear on their existing computer systems. Researching trends in DL model size and complexity have resulted in growing requirements for specialized and proprietary tools and frameworks that remain hard to access according to Wang *et al.* (2022), Chen and Skold (2023), and Ma *et al.* (2018). New data architectures and cloud-based methods are being introduced, but these strategies raise new issues regarding privacy, ownership of the data/information, and associated costs.

15.2.4 Lack of standardized benchmark datasets

Research progress in plants faces challenges because there are no commonly accepted benchmark datasets. Benchmark datasets are set for common use, such as ImageNet for image classification and GLUE for natural language processing. These improve model performance and help unify performance metrics across different tasks. Plant genomics, together with regulatory network analysis, faces challenges because the availability of commonly accepted datasets is limited. Each independent research team selects individually held or small specialized plant datasets that reduce the authenticity and wider applicability of their results (Bassel *et al.*, 2012; Liu *et al.*, 2025). New models created to process *Arabidopsis* data will typically show poor performance when used for wheat or rice species, since their genomic sequences differ. These problems become worse because there is no standardization of data preprocessing as well as annotation and labeling procedures. Genomic prediction benefits from source integration, but such methods highlight inconsistencies that appear during preprocessing of data (Hamid *et al.*, 2009).

15.2.5 Biological complexity and multi-omics data integration

Biological entities contain complicated structures because they include many multidimensional relationships and recurrent loops as well as disorder within their operational patterns. Within plant regulatory networks, numerous genes and various TFs interact in non-linear ways with environmental stimuli. DL models excel in handling complicated systems despite the problem that noise and random patterns in data (Lones, 2024; Scorzato, 2024) might mislead their operations. Overfitting stands out as a main challenge because models achieve good results on training data yet struggle to predict new conditions. While the risk of overfitting becomes more pronounced during work with small noisy datasets, these issues often arise in plant regulatory network investigations. Models attempting to predict crop yield and stress responses to different conditions on the field show reduced performance when compared to lab settings. The process of uniting information across multiple omics layers presents significant challenges to researchers (Kang *et al.*, 2022). The application of autoencoders and graph neural networks (GNNs) in plant studies is in its early stages even though researchers try using these DL architectures. The majority of current research faces challenges when it comes to generating observations across varying time spans linked with spatial variations between different datasets such as root proteomes and leaf transcriptomes during drought stress conditions. The process of data integration becomes more complex because of inconsistent data formats, missing information, and differences between data batches. The current implementations of DL systems remain negatively affected by such inconsistencies because they produce inaccurate predictions that lose biological signal value (Wang *et al.*, 2016; Ngara *et al.*, 2021; Munikoti *et al.*, 2023).

15.3 Future Directions

DL applications within plant GRN research experience quick development along with creating various promising solutions for future operational applications. Developments in biological data generation and agricultural requirements create a positive environment for enhancing DL techniques. According to Wang *et al.* (2020), the advancements in DL architectures bring significant changes to the field of plant GRN modeling. The continuous utilization of attention-based models with Transformers along with GNNs, RNNs, and CNNs showcases their ability to best process spatial and temporal gene expression data (Mienye *et al.*, 2024; Ponzi and Napoli, 2025). The natural ability of GNN to represent GRNs lies in its ability to express gene–gene interactions through graph nodes and edges. Research studies validate GNNs for their ability to enhance GRN predictions through analysis of graph topology characteristics with network hierarchical structure information. The natural language processing program Transformers serves biological sequence data, using genomic sequences to obtain efficient long-range associations beyond what RNNs (Huang *et al.*, 2023; Pham *et al.*, 2025) achieve. These developing architectures will form part of multimodal systems that process genomic and environmental data along with transcriptomic information to obtain more precise and comprehensive GRN models. Using present transcriptomic methodology in combination with DL methods allows scientists to monitor GRN changes with increasing precision levels (Ge *et al.*, 2025). Developing DL models to process streaming biological data allows them to identify transient regulatory states as well as feedback loops and oscillatory behaviors that occur in plant cells during development and under stress conditions (Goshisht, 2024). Integrating state-space models and dynamic Bayesian networks with DL algorithms has enabled inference of time-evolving GRNs of *Arabidopsis* under circadian rhythms and nitrate treatments (Shanks *et al.*, 2024). The implementation of online learning approaches enables the real-time prediction of GRNs by processing continuous data streams until their availability for living plant tissue dynamic analysis. Understanding plant adaptive mechanisms in gene control programs becomes enhanced by studying their reaction to environmental cues, pathogen attacks, and nutrient fluctuations. Real-time water management programs for agricultural systems integrate DL algorithms with biosensors. Alongside smart agriculture tools, they convert gene expression regulatory signals into irrigation and fertilizer application instructions, as well as plans for stress adaptation. DL techniques, when applied to general regulatory network studies, will establish fundamental progress that accelerates crop breeding approaches even as it enhances sustainability in farming. The use of DL models to represent TF control mechanisms of complex traits results in discoveries of key genetic determinants while making predictions about environmental performance, yield performance, drought survival, and nutrient acquisition (Montesinos-López *et al.*, 2018; Sagar and Birje, 2025). The use of ML-based GRN prediction models has enabled researchers to study TFs that mediate stress responses in rice and corn, including OsbHLH148 in drought-tolerant rice, where knockout studies proved its validity (Gupta *et al.*, 2021). The prediction accuracy of DL technology for trait prediction significantly increases because it integrates different omics layers while uncovering complex linear and non-linear associations that standard techniques typically overlook. Plants require these capabilities to develop climate resilience in agriculture since they must withstand rising abiotic stress caused by heat waves, salt exposure, and limited water availability. Through the CRISPR-Cas9 genome editing system, scientists can now target strong crop cultivar development with DL-predicted regulatory elements and transcriptional hubs. Furthermore, the expanding availability of cloud computing and AI systems means that even resource-poor research environments will soon be able to apply DL technologies to GRN research. This technology's widespread availability could enhance crop quality in developing nations, aiding global food security and progress toward sustainable development goals (SDGs).

15.4 Conclusion

Currently, plants are benefiting from DL technologies that analyze GRNs, enabling scientists to study previously elusive regulatory intricacies. This chapter examined GRNs' essential function in directing plant development alongside explanations about modern technologies, which assist researchers in studying intricate networks. Research groups can now decipher biological patterns more precisely because DL operates across the DNA-, RNA-, and protein-level sequences of the central dogma. CNNs and RNNs, alongside other advanced hybrid deep architectures, effectively identify TFBSs, model CREs, and define the functional features of non-coding regions. Plant science benefits greatly from DL techniques because these methods succeed in Ayurvedic plant classification, rice disease detection, nitrogen signaling network modeling, and sorghum phenotypic trait estimation tasks. The achievements made in this field encounter significant obstacles. The majority of obstacles in DL for plant science include restricted annotated datasets, demanding computational requirements, complex model interpretation, and biological data uncertainty. To obtain complete knowledge of plant systems, researchers must develop new methods to unite data types that are diverse in nature within multi-omics research frameworks. DL research leads to complete systemic integration because it enables the continuous tracking of GRNs, their ability to deliver personalized crop solutions, and sustainable agricultural frameworks. The integration of DL into plant research labs with agricultural sciences appears promising because of continuous progress in data collection, algorithm design, and biological experiments. The mutual connection between plant biology and DL creates a transformative change for regulatory network study and establishes efficient, sustainable approaches in plant science.

References

Akande, O. (2024) Challenges and Opportunities in Machine Learning for Bioenergy Crop Yield Prediction: A Review. *Available at SSRN 4898518*.

Andrianto, H., Faizal, A., & Armandika, F. (2020, October) Smartphone application for deep learning-based rice plant disease detection. *In 2020 international conference on information technology systems and innovation (ICITSI)* (pp. 387-392). IEEE.

Armengaud, J., Trapp, J., Pible, O., Geffard, O., Chaumot, A., & Hartmann, E. M. (2014) Non-model organisms, a species endangered by proteogenomics. *Journal of proteomics*, 105, 5-18.

Artur, M.A., & Kajala, K. (2021) Convergent evolution of gene regulatory networks underlying plant adaptations to dry environments. *Plant, cell & environment*, 44(10), 3211-3222.

Badia-i-Mompel, P., Wessels, L., Müller-Dott, S., Trimbour, R., Ramirez Flores, R. O., Argelaguet, R., & Saez-Rodriguez, J. (2023) Gene regulatory network inference in the era of single-cell multi-omics. *Nature Reviews Genetics*, 24(11), 739-754.

Banf, M., & Rhee, S. Y. (2017) Enhancing gene regulatory network inference through data integration with markov random fields. *Scientific reports*, 7(1), 41174.

Bassel, G. W., Gaudinier, A., Brady, S. M., Hennig, L., Rhee, S. Y., & De Smet, I. (2012) Systems analysis of plant functional, transcriptional, physical interaction, and metabolic networks. *The Plant Cell*, 24(10), 3859-3875.

Bonneau, R. (2008) Learning biological networks: from modules to dynamics. *Nature chemical biology*, 4(11), 658-664.

Carfora, A., Lucibelli, F., Di Lillo, P., Mazzucchiello, S.M., Saccone, G. *et al.* (2025) Genetic responses of plants to urban environmental challenges. *Planta*, 261(5), 1-21.

Carpentier, S.C., Coemans, B., Podevin, N., Laukens, K., Witters, E. *et al.* (2008) Functional genomics in a non-model crop: transcriptomics or proteomics? *Physiologia plantarum* 133, no. 2, 117–130.

Carpentier, S. C., Panis, B., Vertommen, A., Swennen, R., Sergeant, K., Renaut, J., ... & Devreese, B. (2008) Proteome analysis of non-model plants: a challenging but powerful approach. *Mass spectrometry reviews*, 27(4), 354-377.

Chen, F., & Sköld, W. (2023) Hardware Acceleration of Machine Learning.

Debbagh, M., Sun, S., & Lefsrud, M. (2024) Predictive Pattern Recognition Techniques Towards Spatiotemporal Representation of Plant Growth in Simulated and Controlled Environments: A Comprehensive Review. *arXiv preprint arXiv:2412.10538*.

Devarapu, K. (2021) Advancing Deep Neural Networks: Optimization Techniques for Large-Scale Data Processing. *NEXG AI Review of America*, 2(1), 47-61.

Dileep, M. R., & Pournami, P. N. (2019, October) AyurLeaf: a deep learning approach for classification of medicinal plants. In *TENCON 2019-2019 IEEE Region 10 Conference (TENCON)* (pp. 321-325). IEEE.

Eftekhari, M., Ma, C., & Orlov, Y. L. (2024) Applications of artificial intelligence, machine learning, and deep learning in plant breeding. *Frontiers in Plant Science*, 15, 1420938.

Gao, M., Lin, L., & Sinnott, R. O. (2017, October) A mobile application for plant recognition through deep learning. In *2017 IEEE 13th International Conference on e-Science (e-Science)* (pp. 29-38). IEEE.

Ge, S., Sun, S., Xu, H., Cheng, Q., & Ren, Z. (2025) Deep learning in single-cell and spatial transcriptomics data analysis: advances and challenges from a data science perspective. *Briefings in Bioinformatics*, 26(2), bbaf136.

Ghosal, S., Zheng, B., Chapman, S. C., Potgieter, A. B., Jordan, D. R., Wang, X., ... & Guo, W. (2019) A weakly supervised deep learning framework for sorghum head detection and counting. *Plant Phenomics*.

Goshisht, M. K. (2024) Machine learning and deep learning in synthetic biology: Key architectures, applications, and challenges. *ACS omega*, 9(9), 9921-9945.

Guo, L., Zhao, G., Xu, J. R., Kistler, H. C., Gao, L., & Ma, L. J. (2016) Compartmentalized gene regulatory network of the pathogenic fungus Fusarium graminearum. *The New Phytologist*, 211(2), 527.

Gupta, C., Ramegowda, V., Basu, S., & Pereira, A. (2021) Using network-based machine learning to predict transcription factors involved in drought resistance. *Frontiers in Genetics*, 12, 652189.

Haque, S., Ahmad, J. S., Clark, N. M., Williams, C. M., & Sozzani, R. (2019) Computational prediction of gene regulatory networks in plant growth and development. *Current opinion in plant biology*, 47, 96-105.

Hamid, J. S., Hu, P., Roslin, N. M., Ling, V., Greenwood, C. M., & Beyene, J. (2009) Data integration in genetics and genomics: methods and challenges. *Human genomics and proteomics: HGP*, 2009, 869093.

Hinton, G.E., Osindero, S., & Teh, Y.W. (2006) A fast learning algorithm for deep belief nets. *Neural computation*, 18(7), 1527-1554.

Huang, Y., Yu, G., & Yang, Y. (2023) MIGGRI: A multi-instance graph neural network model for inferring gene regulatory networks for Drosophila from spatial expression images. *PLOS Computational Biology*, 19(11), e1011623.

Junker, A., Hartmann, A., Schreiber, F., & Bäumlein, H. (2010) An engineer's view on regulation of seed development. *Trends in plant science*, 15(6), 303-307.

Kang, M., Ko, E., & Mersha, T. B. (2022) A roadmap for multi-omics data integration using deep learning. *Briefings in Bioinformatics*, 23(1), bbab454.

Karhunen J., Raiko T., Cho K.H. Unsupervised deep learning: a short review. In: Advances in independent component analysis and learning machines. 2015; p. 125–42.

Khaki, S., Wang, L., & Archontoulis, S. V. (2020) A CNN-RNN framework for crop yield prediction. *Frontiers in Plant Science*, 10, 1750.

Khan, M., Ali, S., Al Azzawi, T.N.I., Saqib, S., Ullah, F. *et al.* (2023) The key roles of ROS and RNS as a signaling molecule in plant–microbe interactions. *Antioxidants*, 12(2), 268.

Ko, D.K., & Brandizzi, F. (2020) Network-based approaches for understanding gene regulation and function in plants. *The Plant Journal*, 104(2), 302-317.

Konecny, T., Nikoghosyan, M., & Binder, H. (2023) Machine learning extracts marks of thiamine's role in cold acclimation in the transcriptome of Vitis vinifera. *Frontiers in Plant Science*, 14, 1303542.

Krouk, G., Lingeman, J., Colon, A. M., Coruzzi, G., & Shasha, D. (2013) Gene regulatory networks in plants: learning causality from time and perturbation. *Genome biology*, 14, 1-7.

Kulkarni, S.R., & Vandepoele, K. (2020) Inference of plant gene regulatory networks using data-driven methods: A practical overview. *Biochimica et Biophysica Acta (BBA)-Gene Regulatory Mechanisms*, 1863(6), 194447.

Kulkarni, S. R., & Vandepoele, K. (2020) Inference of plant gene regulatory networks using data-driven methods: A practical overview. *Biochimica et Biophysica Acta (BBA)-Gene Regulatory Mechanisms*, 1863(6), 194447.

Li, Y., Pearl, S.A., & Jackson, S.A. (2015) Gene networks in plant biology: approaches in reconstruction and analysis. *Trends in plant science*, 20(10), 664-675.

Li, Y., Wang, S., Umarov, R., Xie, B., Fan, M., Li, L., & Gao, X. (2018) DEEPre: sequence-based enzyme EC number prediction by deep learning. *Bioinformatics*, 34(5), 760-769.

Liu, C., Dong, Y., Xiang, W., Yang, X., Su, H., Zhu, J., ... & Zheng, S. (2025) A comprehensive study on robustness of image classification models: Benchmarking and rethinking. *International Journal of Computer Vision*, 133(2), 567-589.

Lones, M. A. (2024) Avoiding common machine learning pitfalls. *Patterns*, 5(10).

Lu, Q., Xu, J., Zhang, R., Liu, H., Wang, M. *et al.* (2025) RiceSNP-ABST: a deep learning approach to identify abiotic stress-associated single nucleotide polymorphisms in rice. *Briefings in Bioinformatics*, 26(1), bbae702.

Ma, W., Qiu, Z., Song, J., Li, J., Cheng, Q., Zhai, J., & Ma, C. (2018) A deep convolutional neural network approach for predicting phenotypes from genotypes. *Planta*, 248, 1307-1318.

Mercatelli, D., Scalambra, L., Triboli, L., Ray, F., & Giorgi, F.M. (2020) Gene regulatory network inference resources: A practical overview. *Biochimica et Biophysica Acta (BBA)-Gene Regulatory Mechanisms*, 1863(6), 194430.

Mercatelli, D., Scalambra, L., Triboli, L., Ray, F., & Giorgi, F.M. (2020) Gene regulatory network inference resources: A practical overview. *Biochimica et Biophysica Acta (BBA)-Gene Regulatory Mechanisms*, 1863(6), 194430.

Mienye, I. D., Swart, T. G., & Obaido, G. (2024) Recurrent neural networks: A comprehensive review of architectures, variants, and applications. *Information*, 15(9), 517.

Mkhatshwa, J. (2023) *Comparative analysis of explainable deep learning models for identification of plant nutrient deficiencies* (Doctoral dissertation, Cape Peninsula University of Technology).

Mochida, K., Koda, S., Inoue, K., & Nishii, R. (2018) Statistical and machine learning approaches to predict gene regulatory networks from transcriptome datasets. *Frontiers in Plant Science*, 9, 1770.

Mohanty, S. P., Hughes, D. P., & Salathé, M. (2016) Using deep learning for image-based plant disease detection. *Frontiers in plant science*, 7, 215232.

Montesinos-López, O. A., Montesinos-López, A., Crossa, J., Gianola, D., Hernández-Suárez, C. M., & Martín-Vallejo, J. (2018) Multi-trait, multi-environment deep learning modeling for genomic-enabled prediction of plant traits. *G3: Genes, genomes, genetics*, 8(12), 3829-3840.

Muley, V. Y. (2023) Deep Learning for Predicting Gene Regulatory Networks: A Step-by-Step Protocol in R. In *Reverse Engineering of Regulatory Networks* (pp. 265-294) New York, NY: Springer US.

Munikoti, S., Agarwal, D., Das, L., Halappanavar, M., & Natarajan, B. (2023) Challenges and opportunities in deep reinforcement learning with graph neural networks: A comprehensive review of algorithms and applications. *IEEE transactions on neural networks and learning systems.*

Ngara, R., Goche, T., Swanevelder, D. Z., & Chivasa, S. (2021) Sorghum's whole-plant transcriptome and proteome responses to drought stress: A review. *Life*, 11(7), 704.

Park, P.J. (2009) ChIP–seq: advantages and challenges of a maturing technology. *Nature reviews genetics*, 10(10), 669-680.

Pham, P., Bui, Q. T., Nguyen, N. T., Kozma, R., Yu, P. S., & Vo, B. (2025) Topological data analysis in graph neural networks: Surveys and perspectives. *IEEE Transactions on Neural Networks and Learning Systems.*

Ponzi, V., & Napoli, C. (2025) Graph Neural Networks: Architectures, Applications, and Future Directions. *IEEE Access.*

Saeed, K. and Wang, L. (2019) Crop yield prediction using deep neural networks. *Frontiers in plant science* 10, 621.

Sagar, S., & N Birje, M. (2025) Precision Agriculture Using Internet of Things and Cloud Computing: A Review. *SN Computer Science*, 6(4), 1-18.

Sarker, I.H. (2021) Deep learning: a comprehensive overview on techniques, taxonomy, applications and research directions. *SN computer science*, 2(6), 1-20.

Scorzato, L. (2024) Reliability and Interpretability in Science and Deep Learning. *Minds and Machines*, 34(3), 27.

Shanks, C. M., Rothkegel, K., Brooks, M. D., Cheng, C. Y., Alvarez, J. M., Ruffel, S., ... & Coruzzi, G. M. (2024) Nitrogen sensing and regulatory networks: it's about time and space. *The Plant Cell*, 36(5), 1482-1503.

Shen, Z., Bao, W., & Huang, D. S. (2018) Recurrent neural network for predicting transcription factor binding sites. *Scientific reports*, 8(1), 15270.

Singh, A. K., Ganapathysubramanian, B., Sarkar, S., & Singh, A. (2018) Deep learning for plant stress phenotyping: trends and future perspectives. *Trends in plant science*, 23(10), 883-898.

Singh, A. K., Ganapathysubramanian, B., Sarkar, S., & Singh, A. (2018) Deep learning for plant stress phenotyping: trends and future perspectives. *Trends in plant science*, 23(10), 883-898.

Sitienei, M.C. (2024) *machine learning in agriculture with application in maize (zea mays) yield prediction modeling in uasin gishu county, Kenya* (doctoral dissertation, university of eldoret).

SSL, D. A., & Kumar, P. (2024, May) Predicting Crop Yield using Long Short-Term Memory, Integrated Gradients and Shapley Additive Explanations. *In 2024 4th International Conference on Pervasive Computing and Social Networking (ICPCSN)* (pp. 975-983). IEEE.

Starke, G. M. J. (2022) *Trusting Black-Box Algorithms? Ethical Challenges for Biomedical Machine Learning* (Doctoral dissertation, University_of_Basel).

Tran, N. H., Zhang, X., Xin, L., Shan, B., & Li, M. (2017) De novo peptide sequencing by deep learning. *Proceedings of the National Academy of Sciences*, 114(31), 8247-8252.

Ubbens, J.R. & Stavness, I. (2017) Deep plant phenomics: a deep learning platform for complex plant phenotyping tasks. *Frontiers in plant science*, 8, 1190.

Varala, K., Marshall-Colón, A., Cirrone, J., Brooks, M. D., Pasquino, A. V., Léran, S., ... & Coruzzi, G. M. (2018) Temporal transcriptional logic of dynamic regulatory networks underlying nitrogen signaling and use in plants. *Proceedings of the National Academy of Sciences*, 115(25), 6494-6499.

Wang, H., Cimen, E., Singh, N., & Buckler, E. (2020) Deep learning for plant genomics and crop improvement. *Current opinion in plant biology*, 54, 34-41.

Wang, H., Cimen, E., Singh, N., & Buckler, E. (2020) Deep learning for plant genomics and crop improvement. *Current opinion in plant biology*, 54, 34-41.

Wang, K., Abid, M. A., Rasheed, A., Crossa, J., Hearne, S., & Li, H. (2023) DNNGP, a deep neural network-based method for genomic prediction using multi-omics data in plants. *Molecular Plant*, 16(1), 279-293.

Wang, Y., Liu, L., & Wang, C. (2023) Trends in using deep learning algorithms in biomedical prediction systems. *Frontiers in Neuroscience*, 17, 1256351.

Wang, X., Wei, Y., Xiong, Y., Huang, G., Qian, X., Ding, Y., ... & Li, L. (2022, November) Lightseq2: Accelerated training for transformer-based models on gpus. In *SC22: international conference for high performance computing, networking, storage and analysis* (pp. 1-14). IEEE.

Wang, X., Cai, X., Xu, C., Wang, Q., & Dai, S. (2016) Drought-responsive mechanisms in plant leaves revealed by proteomics. *International journal of molecular sciences*, 17(10), 1706.

Wang, Z., Yuan, H., Yan, J., & Liu, J. (2025) Identification, characterization, and design of plant genome sequences using deep learning. *The Plant Journal*, 121(1), e17190.

Yue, T., Wang, Y., Zhang, L., Gu, C., Xue, H., *et al.* (2018) Deep learning for genomics: A concise overview. *arXiv preprint arXiv:1802.00810*.

Zini, J. E., & Awad, M. (2022) On the explainability of natural language processing deep models. *ACM Computing Surveys*, 55(5), 1-31.

Zhu, Y., & Shen, T. (2025, January) The role of machine learning in enhancing computer vision processing. In *The 2nd International scientific and practical conference "Innovative technologies for training and educating young people"(January 14–17, 2025) Boston, USA. International Science Group. 2025. 309 p.* (p. 115).

Zhang Z, Hu X, Zhang Y, Jiang Z, Peng H. Integrating physiological and multi-omics methods to elucidate heat stress tolerance for sustainable rice production. *Physiol Mol Biol Plants.* 2020;26(6):1165-77.

16 Deep Learning for Studying Plant Transcriptional Factors

Rooma Waqar[1†], Javed Iqbal[2*†], Banzeer Ahsan Abbasi[3], Abdul Samad Mumtaz[4], Akhtar Munir[5], Muhammad Kaleem[4], Muhammad Anas[4], Muhammad Muzamil Sattar[4], Wadie Chalgham[6], Shumaila Ijaz[7], Zakir Ullah[4], Ghulam Murtaza[8], Rashid Iqbal[9,10], Sana Basharat[11,12,13], and Muhammad Waseem[11,12,13]

[1]Department of Botany, Faculty of Sciences, Mohi-ud-Din Islamic University, Azad Kashmir, Pakistan; [2]Department of Botany, Bacha Khan University, Khyber Pakhtunkhwa, Pakistan; [3]Department of Botany, Rawalpindi Women University, Rawalpindi, Pakistan; [4]Department of Plant Sciences, Faculty of Biological Sciences, Quaid-i-Azam University Islamabad, Pakistan; [5]Department of Chemistry, Quaid-i-Azam University, Islamabad, Pakistan; [6]Department of Mechanical and Aerospace Engineering, University of California, Los Angeles, California; [7]Department of Burn and Plastic Surgery, Shenzhen Institute of Translational Medicine, The First Affiliated Hospital of Shenzhen University, School of Biomedical Engineering, Shenzhen University Medical School, Shenzhen University, Shenzhen, China; [8]School of Agriculture, Yunnan University, Kunming, China; [9]Department of Agronomy, Faculty of Agriculture and Environment, The Islamia University of Bahawalpur, Bahawalpur, Pakistan; [10]Department of Life Sciences, Western Caspian University, Baku, Azerbaijan; [11]School of Breeding and Multiplication (Sanya Institute of Breeding and Multiplication), Sanya, China; [12]College of Tropical Agriculture and Forestry, Hainan University, Sanya 572025, China; [13]Fang Zhiyuan Academician Team Innovation Center, Hainan, China

Abstract

Transcription factors (TFs) are proteins that identify and attach to specific DNA sequences, which influence the structure of chromatin and regulate gene transcription. These factors play a critical role in controlling gene expression and ensuring that the genes are either repressed or activated in reaction to different environmental conditions and cellular signals. By facilitating gene expression under a complex regulatory network, TFs contribute to essential biological processes such as development, differentiation, and stress response. How TFs affect gene expression is a major concern for contemporary genomic researchers. However, identifying the fastening (binding) sites of TF (TFBS) and determining the precise role of TFs in the regulation of transcription remain significant challenges due to the complexity and variability of DNA–protein interactions. This chapter provides an in-depth overview of research on TFBS prediction. First, it reviews experimental methods for identifying TFBS, highlighting various laboratory techniques and available databases that store TF–DNA interaction data. These methods help

*E-mail: javed89qau@gmail.com

†Equal contribution

 Deep Learning in Plant Science and Agricultural Biotechnology (Jen-Tsung Chen ed.)
DOI: 10.1079/9781836991120.0016

in mapping TF binding patterns across the genome, providing valuable insights into gene regulation. Second, it explores computational approaches for TFBS prediction, with a particular emphasis on deep learning models. These advanced techniques leverage large-scale genomic data to improve the efficiency and accuracy of TFBS identification. Finally, this chapter discusses the major issues in the prediction of TFBS, including data limitations, sequence variability, and the influence of chromatin accessibility.

Keywords: deep learning, genomic research, transcription factor binding site

16.1 Introduction

Regulation of gene expression at the transcriptional stage takes place with the help of proteins known as transcription factors (TFs) (Lambert *et al.*, 2018). Among their roles are the initiation and control of transcription, which is contingent upon the type of cell, stage of development, and state of disease. TFs use their DNA binding domain (DBD) to directly interact with the DNA in a sequence-specific way. The combination of TF and its DBD is termed a TF binding site (TFBS), and these sites often have different sequences and range in length from 6 to 20 bp. Finding TFBSs throughout the genome is essential to comprehending transcriptional control. A mathematical model based on determining the TF binding specificity aids in the prediction of TFBSs that have not yet been identified by testing, as it is not possible to experimentally predict all TFBSs for every cell type. These models can be used to predict the impacts of TFBSs in a group of sequences (Imrichová *et al.*, 2015), the effect of changes on TF binding (Fornes *et al.*, 2018), and the exact site of TF interaction throughout the genome. Even though there has been remarkable development in understanding TFBSs, new techniques for accurately identifying TFBSs have revealed that TF binding is far more intricate and involves several structural and regulatory modifications than previously thought. The foundational step for good data prediction using machine learning for data analysis is extraction of features. The two categories of complete genome sequencing techniques used for TFBS discovery are motif search-based techniques and genome-based comparison techniques.

The foundation of the comparative genomics approach is the idea that, in contrast to non-functional components, functional elements (such as motifs) are inherited from a common ancestor. Pairwise and multiple sequence alignment methods can be used to identify these conserved functional components. GenomeVISTA (Poliakov, *et al.*, 2014), limited area global alignment of nucleotides (LAGAN)/multi-LAGAN (MLAGAN) (Brudno *et al.*, 2003), AVID, and multiple-sequence local alignment (MULAN) (Al Ait *et al.*, 2013) are some of the frequently used alignment methods for conservation studies among orthologous sequences. Nevertheless, when it comes to process speed and multiple sequences, alignment techniques are ineffective.

Heuristic methods, including anchoring, threaded block sets, and greedy search algorithms, are employed to speed up the genome alignment process (Al Ait *et al.*, 2013). Comparative genomic techniques are useful for identifying conserved motifs; however, they frequently miss certain conserved functional motifs. However, the motif search-based approach finds related TFs in the input dataset using the annotated topic profile database. Typically, topics are presented as variations (Stormo and Zhao, 2010) or position weight matrices (PWMs) (Narasimhan *et al.*, 2015). Next-generation sequencing techniques—including chromatin immunoprecipitation sequencing (ChIP-seq) (Johnson *et al.*, 2007), ChIP-exo (Venters, 2018), and ChIP-nexus (He *et al.*, 2015)—have been used in a few studies that are involved in protein–DNA complexes to demonstrate the specificity of sequence in TFs' identification of TFBS.

Numerous computing tools, particularly for deep learning techniques, have been presented. Researchers from a variety of fields have made extensive use of deep learning because it can automatically extract features from the input data. The creation of a graphics processing unit, the availability of vast volumes of data, and the creation of a learning algorithm are all responsible for the extraordinary success of deep learning. Deep learning is capable of learning about positive traits within feature representations by analyzing data, and it has been linked to breakthrough accomplishments. Molecular evolution analysis (Zhang *et al.*, 2019), molecular immunology (Xiang *et al.*, 2018), gene and splicing regulation (Smith *et al.*, 2017), protein grouping (Vidaki *et al.*, 2017), genetic recombination (Srivastava 2019), analysis of nucleic acid sequence (Subedi *et al.*, 2018), gene cloning (Ma *et al.*, 2018), genomic diagnosis (Rama *et al.*, 2018), gene network construction (Wang *et al.*, 2019), and TFBS prediction (Alipanahi *et al.*, 2015) are among the known uses of deep learning. The deep learning approach works very well when there are a lot of samples. Less than 1000 sequencing samples are available for use in genetic engineering and medical applications, nevertheless.

Consequently, the scarcity of instruction databases to create an intensive learning replica without the effect of overfitting is one of the primary obstacles to the use of deep learning in genetic engineering and medicine. Over the years, researchers have developed a number of strategies to address the problem of a limited number of samples. These strategies include: (i) in addition to sequence information for consideration in structural information that expands the feature information and increases the input dimension and (ii) manually generating some samples through data that enhance the dataset. The latest developments in TFBS research have been outlined in this review, with an emphasis on deep learning-based techniques. We have started by outlining the fundamental ideas of TFBS and the experimental techniques utilized to identify them. The fundamental theory underlying the various deep learning models—including convolutional neural network (CNN), long short-term memory (LSTM), generative adversarial network (GAN), and embedding—is then examined after reviewing a few TFBS databases. We then present the use of deep learning and machine learning for TFBS projection. We also outline the fundamental steps involved in using deep learning to forecast TFBSs. Furthermore, we go over the elements that influence TF–DNA binding. Finally, the challenges and potential limitations are discussed. A systematic workflow highlighting the key steps in TFBS identification and analysis—integrating experimental methods, computational modeling, and deep-learning approaches—is presented in Fig. 16.1.

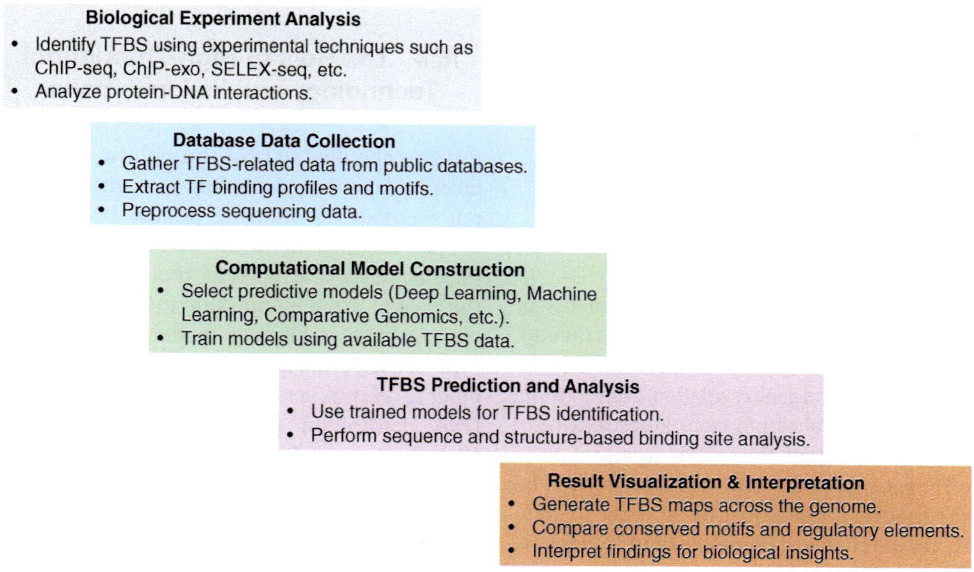

Fig. 16.1. The structured approach outlines the key steps in TFBS identification and analysis, integrating experimental, computational, and deep learning techniques. (Author's own image.) TFBS, transcription factor binding site.

16.2 Formulating Problems and Research Technologies

Identification of TFBS has now become more feasible due to the development of high-throughput methods for measuring the protein–DNA attachment and the creation of intricate DNA binding models for TF using machine learning. Although the experimental dataset that is used for complex modeling of TFBS contains inherent bias and noise, these problems can be easily overfit by complex models when compared to the simpler PWM model, that exhibits greater robustness to such imperfections. Developing models specific to TFBSs using high-throughput data has been the subject of some recent research (Weirauch *et al.*, 2013). On independent *in vivo* data, however, even the most effective models with particular *in vitro* datasets do not always function well, which improves our comprehension of the specificity of binding of protein–DNA. For instance, numerous prediction techniques that depend on *in vitro* sequence data have been created, including the SELEX and protein binding microarray (PBM) techniques. PBM and SELEX (Slattery *et al.*, 2014) techniques are quicker and less expensive than ChIP-seq. However, when dealing with highly specific antibodies, they did not appear to be independent. TFBSs that exhibit low binding to DNA, not accounted for by conventional DNA binding models, were made easier to characterize due to the PBM data, which examined the TF binding specifications of all conceivable 8-bp sequences.

16.3 Chlp Technology Development

By selectively concentrating the fragments of DNA that interact with proteins in cells (living), ChIP is a useful technique for researching the mechanisms behind gene regulation. Conventional protein–DNA localization techniques lack resolution to differentiate the spatial organization of the TFs because of the similarities between TFBS and enhancers. Consequently, in order to comprehend the chemical process behind TF binding, ChIP-exo resolution around a single base is crucial (Slattery *et al.*, 2014). At the same time, ChIP-exo can significantly increase resolution and decrease noise (Matteau and Rodrigue, 2015). Despite being more difficult to master than ChIP-seq, the ChIP-exo technique has gained widespread use in an effort to achieve a unique ultrahigh resolution by utilizing various living systems. Furthermore, ChIP-nexus evolved as a variation of the ChIP-exo technique, aiming to add complexity to the ChIP-exo library by substituting a circular ligation step for the conventional double-stranded DNA linear ligation (He *et al.*, 2015).

Furthermore, in the ChIP-based techniques, the protein that is bounded with the chromatin capture (PAtCh-Cap) method includes a processing stage that involves the bead attachment. To increase DNA accessibility for subsequent chemical treatments—such as chromatin immunoprecipitation of target protein–associated complexes via carboxylate groups—and to reduce non-specific capture of chromatin-bound proteins. When using the PAtCh-Cap method, a CCCTC-binding factor, known as CTCF, was also found (Terooatea *et al.*, 2016). For the ChIP-seq methodology, Wallerman *et al.* (2015) developed a bead library (lobChIP) to address the issue of limited throughput of traditional protocols for ChIP and preparation of the library. The suggested input technique has several uses, such as improved artifact removal from ChIP-exo data and precise identification of peaks (Gustafsson *et al.*, 2019).

16.4 The Role of High-Throughput Technology and Its Limitations

Proteins can bind to DNA through two or more different methods, according to high-throughput methods for studying protein–DNA interaction. To understand the precise mechanism (biochemical) that governs the connection of the TF at TFBSs, it is important to identify the TFs with versatile binding patterns. A single ChIP-seq experiment contains multiple types of binding events generated by diverse protein–DNA connection patterns (Venters, 2018). High-flux *in vivo* DNA analysis is presented in Table 16.1.

16.5 TFBS Database

Our ability to simultaneously produce DNA motifs and binding sites (genomic) has improved

Table 16.1. High-flux *in vivo* DNA analysis. Author's own table.

Experiment name	Experimental description	Year	References
ChIP-seq	Following chromatin immunoprecipitation, high-throughput sequencing	2007	(Johnson *et al.*, 2007)
ChIP-exo	High-throughput sequencing after chromatin immunoprecipitation and exonuclease digestion	2017	(Venters, 2018)
DNAase SEQ	Cutting with DNAase and high-throughput sequencing	2009	(Hesselberth *et al.*, 2009)
ATAC seq	High-throughput assessment of the chromatin that transpose can access	2013	(Buenrostro *et al.*, 2013)
ChIP mentation	This technique creates a sequencing library by combining chromatin immunoprecipitation	2015	(Schmidl *et al.*, 2015)
ChIP-nexus	For a high nucleotide resolution chip experiment, use a single connection and nucleic acid exonuclease	2015	(He *et al.*, 2015)
HT-ChIP mentation	ChIP-seq data creation from a relatively small cell population in a single day	2019	(Gustafsson *et al.*, 2019)

over the past few decades, yielding hitherto unheard-of volumes of information on TF–DNA interactions. The motif was initially found via experimental techniques. To put it another way, motif analysis was not limited to ChIP-seq. Pribnow identified the TATAAT box in 1975 (Pribnow, 1975), and the sequence (specific) binding site of RNA polymerase is the upstream TTGACA motif. Furthermore, not every binding site matched the motif exactly; in fact, the majority only matched seven to nine out of the 12 bases. The strength of the protein–DNA binding is frequently correlated with the degree of motif and binding site matching. These days, a growing number of motifs are acknowledged as being discovered. Numerous databases compile the TFBSs that have been verified through biological experiments. The *Homo sapiens* comprehensive model (HOCOMOCO) (Kulakovskiy *et al.*, 2018), for instance, is a compilation of chosen entries from several sources devoted to human TFs. HOCOMOCOV11 has been updated with some significant changes.

The most recent HOCOMOCO includes binding models for 680 human and 453 mouse TFs, including 576 dinucleotide PWMs and 1302 single-nucleotide PWMs, which characterize each TF's primary binding preference and trustworthy alternative binding specificity. The FlyFactorSurvey (Zhu *et al.*, 2011) and OnTheFly (Shazman *et al.*, 2014) have a lot of patterns related to fruition. ScerTF is a database specifically designed for yeast. Although there were relatively few known TF motifs in *Caenorhabditis elegans* until recently, around 40% of the PWM has already been identified or inferred, and this information is available in the CisBP database. Plant TFs and the corresponding binding sites are listed in the AGRIS (Jolma *et al.*, 2013). Heat maps illustrating DNA structural characteristics—such as minor groove width (MGW) (Dutta and Bandyopadhyay) and helix twist (HelT)—are generated from TF datasets across 23 distinct species, utilizing the TFBSshape database (Yang *et al.*, 2014). Among 1900 TFs from 39 species (different), the data information has expanded to 2428 structureless in the most recent TFBS shaped at a base (Chiu *et al.*, 2020). Each TFBS entry now includes structural profiles comprising four shape features of methylated DNA, thirteen shape features of conventional DNA, and the electrostatic potential of the minor groove. In addition, for introducing the new tool that compares the non-methylated and methylated structures of TFs, TFBShape has increased the recision of shape-based TFs that bind and figured out how nucleotide mutations in DNA shape affect the binding of TFs. Table 16.2 presents a database of binding sites for TFs.

16.6 Introducing Deep Learning Techniques

A node is the fundamental unit in neural networks, formed by biological neurons in the mammalian brain. A neural network with many layers and parameters is referred to as deep learning. It can be broadly divided into (i) CNNs, (ii) LSTM, (iii) Word2vec, (iv) attention mechanisms, and

Table 16.2. Database of binding sites for transcription factors. Author's own table.

Database	Description	References
TFBSshape	It is a motif database that analyzes the structural profile of TFBSs.	(Chiu *et al.*, 2020)
OnTheFLY	It includes the systematic collection of TFs in *Drosophila*.	(Shazman *et al.*, 2014)
FlyFactorSurvey	It consists of a DNA binding-specific database of TF in *Drosophila*.	(Zhu *et al.*, 2011)
UniPROBE	It is a collection of experimental data from universal protein binding microarray experiments.	(Hume *et al.*, 2015)
JASPAR	It is a set of transcriptional factor DNA binding preferences that is modeled as a matrix.	(Fornes *et al.*, 2020)

TFBSs, transcription factor binding sites.

(v) graph convolutional networks. Each of them is briefly introduced in the paragraphs that follow.

16.6.1 Convolutional neural network (CNN)

Since it is widely popular, CNN is used in a variety of artificial intelligence research fields, including image recognition (Ng *et al.*, 2015), audio recognition (Ganapathy and Peddinti, 2018), computer vision (Karpathy *et al.*, 2014), and others (Chen *et al.*, 2016). Wang *et al.* (2015) looked at the crucial connections between fuzzy classifiers' fuzziness and generalization abilities. The study offers convincing evidence that a fuzzy classifier's higher fuzziness might be a sign of its better generalization skills. A CNN is composed of pooling, convolutional, and fully connected layers. For the extraction and representation of local information from the original features, several feature maps and kernels should be used in the convolutional layer.

16.6.2 Long short-term memory (LSTM)

Recurrent neural networks (RNNs) recur temporally. Compared with other neural networks, RNNs include one or more memory units. The input to an RNN comprises two components: the current input xt, which updates the real-time state, and the previous hidden state ht-1. While the network architecture may vary in specific features, this fundamental structure remains consistent. Nevertheless, in RNN optimization, the back-propagation computation procedure is dependent on time. There is a multiplicative coefficient that is continuous while updating the gradient of parameter W because the derivative at time t will propagate to $t1$, $t2$, and tn time. Different issues have always been raised with multiplication: gradient explosion and vanishing. In addition, the (long) dependence problem occurs as the influence of the initial input on the subsequent nodes decreases in the course of the forward process. This results in the loss of the ability to memorize.

16.7 Utilizing Conventional Machine Learning Techniques

16.7.1 Display of the consensus sequence

A consensus indicates the predominant ring nucleotide at each position inside the binding site of the TF. In practice, we discovered that some binding sites have comparatively similar, or even exactly equal, frequencies of two or three specific nucleotides. In this instance, the degree of conservation cannot be completely reflected by merely employing a single nucleotide that represents the site. Thus, the explanation of binding locations of TFs also uses the expression degenerate consensuses. In the degenerate consensus sequence, multiple nucleotides that occupy the same location—rather than simply one—are indicated by symbols. Patterns of degenerate consensus sequences, using the International Union of Pure and Applied Chemistry (IUPAC) codes (Cornish-Bowden, 1985), are displayed in Table 16.3. Although the consensus sequence-based representation method is straightforward to comprehend, it is unable to reflect the likelihood of several bases at each point.

This mode of communication sacrifices sensitivity and specificity. The issue of random base replacement at each motif site during pattern matching was resolved by WEEDER and selective microfluidics-based ligand enrichment (SMiLE) (Isakova *et al.*, 2017). This type of approach builds the sequence's beginning index using a suf × tree structure, and then it thoroughly searches for all consistent candidate sequences. The SMiLE algorithm looks at the number of subsequences found in the input sequence set. The WEEDER algorithm compares the likelihood that a subsequence will occur in each promoter area in the same set with the actual number of occurrences (Pavesi *et al.*, 2001).

16.8 PWM Representation

The position-specific frequency matrix may reflect the likelihood of various bases at each place in comparison to the consensus sequence technique. This model's premise is that the likelihood of bases occurring at each location is unrelated to one another. Position frequency matrix (PFM) is derived from the position count matrix (PCM). In order to account for the bias of base composition in the DNA sequence, the PFM is often simultaneously changed into a PWM. PWM (Stormo and Zhao, 2010) is an expression matrix that serves to represent TF and is also referred to as a weighting matrix or a position-specific scoring matrix. There are a few things to think about while predicting the TFBS with position weights: (i) how to use the search technique in conjunction with the PWM to forecast, (ii) how to get the PWM, and (iii) the drawbacks of predicting TFBS with PWM. Initially, PWM assigns a score to every base that could be present at every location within the binding site. Every unique TF has a PWM that describes its properties. The frequency of each base at each position is simply recorded in a counting matrix that is derived from the aligned positions. The statistical measurements of these subsequences in relation to the background sequence are computed using the created probability matrix in order to assess conservatism and specificity. The motif component that is linked is the probability matrix that has the highest score. Second, the motif finding issue transforms into a combination enhancement challenge after the PWM is formed. In this process, every subsequent combination that could possibly exist is investigated to identify the combination yielding the maximum measurement value. Making motif recognition algorithms, including the greedy algorithm, random search, local search, and genetic algorithm, requires extensive use of a variety of combinatorial optimization approaches.

Training iterations continue until the maximum measurement value is reached. For the motif discovery challenge, Lawrence and Reilly (1990) initially employed a possible approach based on iteration training. The expectation maximization is a well-known algorithm. Also, it is believed that when background components are split, the phantom recognition signal in the PWM probability model will be largely affected. This assumption could appear unrealistic, though, because not every nucleotide in the input sequence is impacted by the neighboring nucleotides (Weirauch *et al.*, 2013).

Table 16.3. Patterns of nucleotide sequences are represented using IUPAC degenerate codes. Author's own table.

IUPAC code	Nucleotide
M	C or A
Y	T or C
S	G OR C
K	T or G
R	G or A

IUPAC, International Union of Pure and Applied Chemistry.

16.9 Statistical Learning Method

Commonly utilized PWM-based computational techniques for TFBS prediction typically have a high false positive rate. Furthermore, due to the vast number of TFs involved, computational investigations of the regulation of transcription in eukaryotes often demand multiple PWM models of TFBSs. Khamis *et al.* (2018) created DRAF models, a revolutionary technique for TFBS prediction, to address these issues. DRAF models use more features than PWM models because they incorporate information from TFBS sequences and the physical properties of TF DBDs into machine learning models. Phantom signal recognition is greatly impacted by PWM's assumption of an autonomous distribution of nucleotide bases. Gao and Ruan (2017) built a multiple-instance learning-based strategy to tackle the analysis of large-scale documents.

16.10. Utilizing Deep Learning in TFBS Applications

16.10.1 A range of deep learning-based applications

Deep learning has been used in many different fields in recent years. A few benchmark deep learning models are used in TFBS prediction-based research. Though studies on natural language processing that provide useful insights into DNA analysis, such as DeepBind and iDeep (Pan and Shen, 2017), currently developed applications for TF–DNA binding prediction, it is instinctive to conceptualize the DNA sequence as an input phrase composed of the four nucleotides (A, C, G, and T), rather than as a picture. A conceptual deep learning model for TFBS prediction is shown in Table 16.4. The neural network is trained independently using the sequence of DNA as input data, allowing it to modify parameters and outcomes. Additionally, Table 16.4 summarizes the deep learning-based TFBS prediction techniques.

16.10.2 Application of CNN

DeepBind (Alipanahi *et al.*, 2015) demonstrates how sequence specificity can be determined from experimental data using deep learning technologies. It offers a unified, flexible, and scalable mathematical method for finding patterns (Alipanahi *et al.*, 2015). Furthermore, the requirement for precise depiction of protein target binding motifs is addressed for the first time by DeepBind. The growing interest in CNN is driven by the deep convolution approach. To overcome the drawbacks of traditional CNN, Zhang *et al.* (2018) suggested a high-order CNN (HOCNN) architecture. HOCNN uses a structure that first recognizes changes in binding lengths (differing) and then ties the coding strongly to the characters (nucleotides). Blum and Kollmann (2019) introduced a circular filter that uses circular permutation versions of the same filter to convolute the sequences.

Previous literature introduced a hybrid strategy that combined the CNN-seq method with the kernel method (Chen *et al.*, 2019). It denotes a Hilbert space with distinct properties to express prediction functions while retaining the neural network to give adequate representation for learning challenges. Figure 16.2 represents a basic flowchart for TFBS prediction using intelligent computing.

16.10.3 Application of LSTM

According to certain research, RNNs and their variations perform better when processing time-series data. This is especially true for learning systems that process biological data. Most of the proposed TFBS prediction methods only use low-order dependencies for prediction due to the lack of efficient approaches for extracting higher-order dependencies. The author of this paper suggested a novel way to use CNN on histone modification features in order to extract high-order dependence. Next, a brand-new TFBS prediction technique known as CNN_TF, which

Table 16.4. Summary of the deep learning-based TFBS prediction approach. Author's own table.

Method	Time	Technology	References
DESSO	2019	BD + CNN	(Yang *et al.*, 2019)
DLBSS	2019	CNN	(Zhang, Shen, & Huang, 2019)
WSCNN	2018	MIL + CNN	(Zhang, Zhu, Bao *et al.*, 2018)
gkm-DNN	2018	DNN	(Zeng *et al.*, 2020)
TFImpute	2017	CNN	(Qin & Feng, 2017)
DeepBind	2016	CNN	(Hassanzadeh & Wang, 2016)
DeepBind	2015	CNN	(Alipanahi *et al.*, 2015)

CNN, convolutional neural network; TFBS, transcription factor binding site.

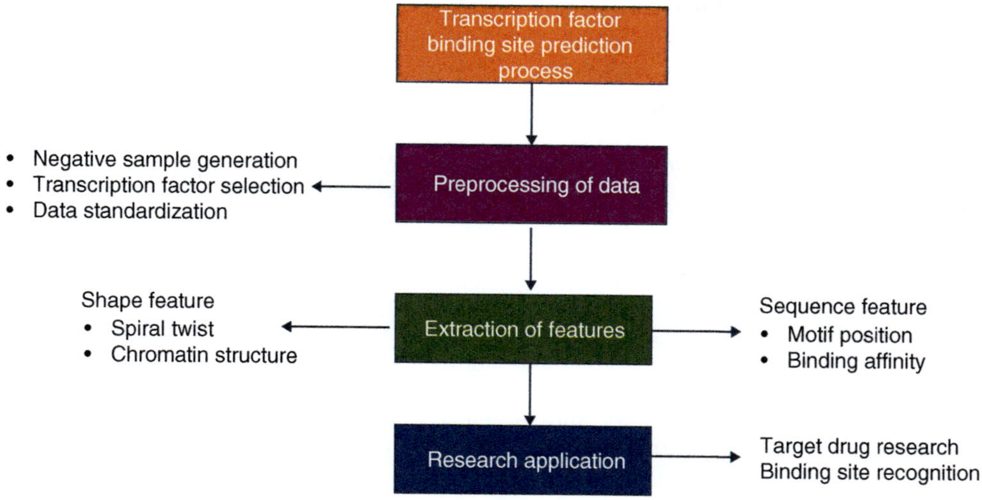

Fig. 16.2. Basic flowchart for transcription factor binding site prediction using intelligent computing. (Author's own image.)

integrates high- and low-order dependencies, is presented (Chen *et al.*, 2019).

Each k-mer, as represented in the order-preserving model, is processed using the Word-2Vec algorithm.. Third, a deep bidirectional gated recurrent unit model is constructed for classification and feature learning. The stride window, embedded vector size, and k-mer length variations are responsible for the resilience of k-mer GRU (Shen *et al.*, 2018). By implicitly assuming that each probe has no more than one motif, DeepBind is unable to capture the dynamism of the probe sequence, despite its ingenious design. The training process may be misguided by this preexisting concept. In various instances, effective motifs in the probe are known to have a high affinity for molecules, owing to their distinct characteristics. Due to the strong affinity of the entire probe, the TFBS prediction model seeks to modify the weight of the motif detector (kernel) in this particular case, punishing all but one motif while overweighing the others (Hassanzadeh and Wang, 2016).

16.11 TFBS Prediction Process by Deep Learning

16.11.1 Data preprocessing

The objective of data preprocessing in the current study is to analyze DNA shape, protein sequence, and DNA sequence. High-throughput data selection, data standardization, and TF selection are some of the methods that fall under the category of sequence processing. Each component has been thoroughly discussed in the following sections.

High-throughput data selection

High-throughput sequencing offers high-resolution TF binding datasets both *in vitro* and *in vivo*, as was previously discussed. Our capacity to forecast the placement of TFBS on genomic DNA remains inadequate, even with the growing volume of such datasets. Contextual information serves as the foundation that is selected when predicting the TFBS in a particular species.

Data standardization

The approach we utilize to generate the background (negative) sequence is crucial since each DNA strand has a large number of redundant TFBSs in both the forward and reverse orientations. It is well known that the resulting motif may be erroneous if the background sequence is not chosen to fit the statistical characteristics of the foreground set.

Sequence truncation

It is difficult to identify likely TFBSs in long DNA unless you know the tissue or the species, because many of the computational techniques now in use are either species-specific, tissue-specific, or restricted to DNA sequences (short). These drawbacks make the model less reliable when used to predict lengthy series. Based on the input, the current approaches are classified into two groups.

Feature extraction

The prediction of TFBSs greatly benefits from thorough feature extraction. Since different features have varying degrees of relevance, the accuracy of TFBS prediction will be directly impacted by its ability to fully extract all of the features of the DNA. The shape feature extraction is directed at protein and DNA. The features were separated into form features and sequence features. Nucleotide dependence, binding affinity, and phantom position are examples of sequence features, while MGW, rolling, ProT, and HelT are examples of form features. We will not go into further detail about the description because shape features are not considered in the current literature.

NUCLEOTIDE DEPENDENCE. Predicting the binding sites first requires determining how DNA and regulatory proteins interact, particularly between TFs and the associated binding sites. K-mers, such as dinucleotides, can clearly show how nucleotide locations in TFBSs depend on one another.

MOTIF POSITION. Although TFBSs are often unique, this may not hold true if they do not exhibit a one-to-one correlation. The binding site sequence and the non-coding region sequences on either side of the binding site are often included in the model's input sequence. The model can identify the border position in high-resolution data (Salekin *et al.*, 2018).

BINDING AFFINITY. Apart from the binding motifs, other sequence features that influence the TFs' binding affinity include the flanking DNA and particular targets for the flanking symmetry of repetition sequences. The length of the attaching site sequence should be considered for improved TF–DNA binding specificity.

16.12 Factors Affecting the Interaction of Proteins with DNA

According to recent research, TFs contribute to TF–DNA binding specificity by recognizing a subset of putative DNA binding sites that are outside of the core binding site. The nucleotide sequence, the structure (3D) of the attaching site, the attachment of TF–DNA with cofactors, chromatin accessibility and nucleosome, indirect cooperation with nucleosomes, DNA methylation, and so on are some of the characteristics that influence TF–DNA readout on several levels. Furthermore, there are interactions between each of these elements that could change the TFBS in a particular cell type. Additionally, TFs are able to identify the structural characteristics of their binding sites (DNA), which include the sequence-dependent DNA unwinding and bending (Slattery *et al.*, 2014).

16.12.1 Nucleosomes

DNA and histone proteins make up nucleosomes, which are the fundamental structural components of chromatin. A histone octamer is encircled 1.75 times by 146 bp of DNA in each nucleosome. TF binding is sterically hindered by nucleosome–DNA interactions, which also accelerate TF–DNA dissociation rates. Most of the TFBSs found by the ENCODE consortium are in parts of the genome that are easy for proteins to access, which is consistent with this idea.

16.12.2 Histone proteins and chromatin status

In nucleosomes, histones and their variants exchange DNA with labeled promoter regions (Xu *et al.*, 2019). This demonstrates the significance of histones in the transcription process (Ernst and Kellis, 2013). The combination of chromatin state and TF binding data indicates that many TFs have unique histone posttranslational modifications preferences that are consistent across different cell types. It is frequently unclear, nevertheless, whether a certain chromatin state is the outcome of TF binding, actively guides TF binding, or is merely permissive to TF binding. Our model that targets TF–DNA binding interactions is likely to be influenced by additional mechanistic clarification of the connections between TFs and histone PTMs. 3D-shape DNA is another technique for inadvertently recording nucleotide dependence is the DNA shape feature.

The intricate dependency between the several TFBS sites is integrated into the DNA shape. Without being aware of personal dependencies explicitly, this integration is carried out implicitly. The process by which nucleotides (di, tri) might increase the accuracy of motif descriptions is convincingly explained by the combination of DNA forms, which also lowers the number of parameters required (Zhou *et al.*, 2015). According to recent data, the structure of the DNA at the chosen target site may account for important features of TF binding.

16.12.3 Enhancer and single-nucleotide polymorphism (SNP)

SNP enhancer plays a crucial role in determining cell identity and, in conjunction with tissue-specific TFs, preserves the patterns of gene expression specific to a certain cell type. An intricate high-rise structure comprising several TFs that are closely connected to one another to control gene expression is called an enhancer. By changing the TF–DNA binding, SNPs have an impact on gene regulation. According to a recent study, only a small number of regulatory SNPs can be regulated by TFs. SNP inheritance leads to human genetic variation and, in certain situations, the emergence of illnesses. Nonetheless, the majority of SNPs linked to disease are found in non-coding areas of the genome (Pan and Shen, 2017), which also include a large number of enhancers and TFBSs (Chen *et al.*, 2019). Nevertheless, little is known about how much SNPs change TF binding. Figure 16.3 presents deep learning applications in plant TF research, aiming for improved crop resilience, precision breeding, and sustainable agriculture.

16.13 Conclusion

TFBS prediction is a crucial area of genomic research, as it provides insights into the regulatory mechanisms that govern gene expression. While experimental methods offer direct validation of TFBS, they are often time-consuming and resource-intensive. Computational approaches, particularly those based on machine learning and deep learning, have emerged as powerful tools for predicting TFBS with greater efficiency and scalability. However, challenges such as sequence variability, chromatin accessibility, and data limitations continue to hinder precise TFBS identification. Advancements in high-throughput sequencing technologies, coupled with improved computational models, are expected to enhance the accuracy of TFBS prediction in future. Integrating

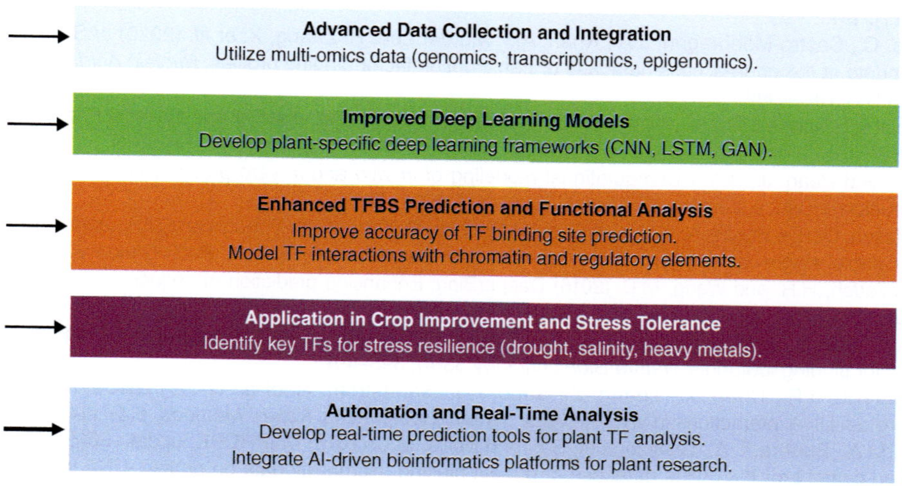

Fig. 16.3. This flowchart outlines the roadmap for advancing deep learning applications in plant transcription factor research, aiming for improved crop resilience, precision breeding, and sustainable agriculture. CNN, convolutional neural network; GAN, generative adversarial networks; LSTM, long short-term memory; TF, transcription factor; TFBS, transcription factor binding site (Author's own image.)

multi-omics data, refining predictive algorithms, and leveraging artificial intelligence will further strengthen our ability to decode complex gene regulatory networks. By addressing these challenges, researchers can achieve a deeper understanding of transcriptional regulation, paving the way for breakthroughs in genetics, biotechnology, and personalized medicine.

References

Al Ait, L., Yamak, Z. and Morgenstern, B. (2013) DIALIGN at GOBICS—multiple sequence alignment using various sources of external information. *Nucleic Acids Research* 41(W1), W3–W7.

Alipanahi, B., Delong, A., Weirauch, M.T. and Frey, B.J. (2015) Predicting the sequence specificities of DNA-and RNA-binding proteins by deep learning. *Nature Biotechnology* 33(8), 831–838.

Blum, C.F. and Kollmann, M. (2019) Neural networks with circular filters enable data efficient inference of sequence motifs. *Bioinformatics* 35(20), 3937–3943.

Brudno, M., Do, C.B., Cooper, G.M., Kim, M.F., Davydov, E. *et al.* (2003) LAGAN and multi-LAGAN: Efficient tools for large-scale multiple alignment of genomic DNA. *Genome Research* 13(4), 721–731.

Buenrostro, J.D., Giresi, P.G., Zaba, L.C., Chang, H.Y. and Greenleaf, W.J. (2013) Transposition of native chromatin for fast and sensitive epigenomic profiling of open chromatin, DNA-binding proteins and nucleosome position. *Nature Methods* 10(12), 1213–1218.

Chen, D., Jacob, L. and Mairal, J. (2019) Biological sequence modeling with convolutional kernel networks. *Bioinformatics* 35(18), 3294–3302.

Chen, Y., Jiang, H., Li, C., Jia, X. and Ghamisi, P. (2016) Deep feature extraction and classification of hyperspectral images based on convolutional neural networks. *IEEE Transactions on Geoscience and Remote Sensing* 54(10), 6232–6251.

Chiu, T.P., Xin, B., Markarian, N., Wang, Y. and Rohs, R. (2020) TFBSshape: An expanded motif database for DNA shape features of transcription factor binding sites. *Nucleic Acids Research* 48(D1), D246–D255.

Cornish-Bowden, A. (1985) Nomenclature for incompletely specified bases in nucleic acid sequences: Recommendations 1984. *Nucleic Acids Research* 13(9), 3021.

Dutta, D. and Bandyopadhyay, P. (2009) Performance of chickpea (*Cicer arietinum* l.) to application of phosphorus and bio-fertilizer in laterite soil. *Archives of Agronomy and Soil Science* 55(2), 147–155.

Ernst, J. and Kellis, M. (2013) Interplay between chromatin state, regulator binding, and regulatory motifs in six human cell types. *Genome Research* 23(7), 1142–1154.

Fornes, O., Gheorghe, M., Richmond, P.A., Arenillas, D.J., Wasserman, W.W. *et al.* (2018) MANTA2, update of the mongo database for the analysis of transcription factor binding site alterations. *Scientific Data* 5(1), 1–7.

Fornes, O., Castro-Mondragon, J.A., Khan, A., Van der Lee, R. Zhang, X. *et al.* (2020) JASPAR 2020: update of the open-access database of transcription factor binding profiles. *Nucleic Acids Research* 48(D1), D87–D92.

Ganapathy, S. and Peddinti, V. (2018) 3-D CNN models for far-field multi-channel speech recognition. In: *2018 IEEE international conference on acoustics, speech and signal processing (ICASSP)*.

Gao, Z. and Ruan, J. (2017) Computational modeling of *in vivo* and *in vitro* protein-DNA interactions by multiple instance learning. *Bioinformatics* 33(14), 2097–2105.

Gustafsson, C., De Paepe, A., Schmidl, C. and Månsson, R. (2019) High-throughput ChIPmentation: Freely scalable, single day ChIPseq data generation from very low cell-numbers. *BMC Genomics* 20, 1–6.

Hassanzadeh, H.R. and Wang, M.D. (2016) DeeperBind: Enhancing prediction of sequence specificities of DNA binding proteins. In: *2016 IEEE International conference on bioinformatics and biomedicine (BIBM)*.

He, Q., Johnston, J. and Zeitlinger, J. (2015) ChIP-nexus enables improved detection of *in vivo* transcription factor binding footprints. *Nature Biotechnology* 33(4), 395–401.

Hesselberth, J.R., Chen, X., Zhang, Z., Sabo, P.J., Sandstrom, R. *et al.* (2009) Global mapping of protein-DNA interactions in vivo by digital genomic footprinting. *Nature Methods,* 6(4), 283–289.

Hume, M.A., Barrera, L.A., Gisselbrecht, S.S. and Bulyk, M.L. (2015) UniPROBE, update 2015: new tools and content for the online database of protein-binding microarray data on protein–DNA interactions. *Nucleic acids research,* 43(D1), D117-D122.

Imrichová, H., Hulselmans, G., Kalender Atak, Z., Potier, D. and Aerts, S. (2015) I-cistarget 2015 update: Generalized cis-regulatory enrichment analysis in human, mouse and fly. *Nucleic Acids Research* 43(W1), W57–W64.

Isakova, A., Groux, R., Imbeault, M., Rainer, P., Alpern, D. *et al.* (2017) SMiLE-seq identifies binding motifs of single and dimeric transcription factors. *Nature Methods* 14(3), 316–322.

Johnson, D.S., Mortazavi, A., Myers, R.M. and Wold, B. (2007) Genome-wide mapping of *in vivo* protein-DNA interactions. *Science* 316(5830), 1497–1502.

Jolma, A., Yan, J., Whitington, T., Toivonen, J., Nitta, K.R. *et al.* (2013) DNA-binding specificities of human transcription factors. *Cell* 152(1), 327–339.

Karpathy, A., Toderici, G., Shetty, S., Leung, T., Sukthankar, R. *et al.* (2014) Large-scale video classification with convolutional neural networks. In: *Proceedings of the IEEE conference on Computer Vision and Pattern Recognition*.

Khamis, A.M., Motwalli, O., Oliva, R., Jankovic, B.R., Medvedeva, Y.A. *et al.* (2018) A novel method for improved accuracy of transcription factor binding site prediction. *Nucleic Acids Research* 46(12), e72–e72.

Kulakovskiy, I.V., Vorontsov, I.E., Yevshin, I.S., Sharipov, R.N., Fedorova, A.D. *et al.* (2018) HOCOMOCO: Towards a complete collection of transcription factor binding models for human and mouse via large-scale ChIP-Seq analysis. *Nucleic Acids Research* 46(D1), D252–D259.

Lambert, S.A., Jolma, A., Campitelli, L.F., Das, P.K., Yin, Y. *et al.* (2018) The human transcription factors. *Cell* 172(4), 650–665.

Lawrence, C.E. and Reilly, A.A. (1990) An expectation maximization (EM) algorithm for the identification and characterization of common sites in unaligned biopolymer sequences. *Proteins: Structure, Function, and Bioinformatics* 7(1), 41–51.

Ma, K.Y., Zhang, S.F., Wang, S.S. and Qiu, G.F. (2018) Molecular cloning and characterization of a gonadotropin-releasing hormone receptor homolog in the Chinese mitten crab, *Eriocheir sinensis*. *Gene* 665, 111–118.

Matteau, D. and Rodrigue, S. (2015) Precise identification of DNA-binding proteins genomic location by exonuclease coupled chromatin immunoprecipitation (ChIP-exo). *DNA-Protein Interactions: Principles and Protocols* 173–193.

Narasimhan, K., Lambert, S.A., Yang, A.W., Riddell, J., Mnaimneh, S. *et al.* (2015) Mapping and analysis of *Caenorhabditis elegans* transcription factor sequence specificities. *Elife* 4, e06967.

Pan, X. and Shen, H.B. (2017) RNA-protein binding motifs mining with a new hybrid deep learning based cross-domain knowledge integration approach. *BMC Bioinformatics* 18, 1–14.

Pavesi, G., Mauri, G., & Pesole, G. (2001). An algorithm for finding signals of unknown length in DNA sequences. *Bioinformatics* 17(suppl_1), S207-S214.

Poliakov, A., Foong, J., Brudno, M. and Dubchak, I. (2014) GenomeVISTA—an integrated software package for whole-genome alignment and visualization. *Bioinformatics* 30(18), 2654–2655.

Pribnow, D. (1975) Nucleotide sequence of an RNA polymerase binding site at an early T7 promoter. *Proceedings of the National Academy of Sciences* 72(3), 784–788.

Qin, Q. and Feng, J. (2017) Imputation for transcription factor binding predictions based on deep learning. *PLoS computational biology* 13(2), e1005403.

Rama, M., Duflos, C., Melki, I., Bessis, D., Bonhomme, A. *et al.* (2018) A decision tree for the genetic diagnosis of deficiency of adenosine deaminase 2 (DADA2): A French reference centres experience. *European Journal of Human Genetics* 26(7), 960–971.

Salekin, S., Zhang, J.M. and Huang, Y. (2018) Base-pair resolution detection of transcription factor binding site by deep deconvolutional network. *Bioinformatics* 34(20), 3446–3453.

Schmidl, C., Rendeiro, A.F., Sheffield, N.C. and Bock, C. (2015) ChIPmentation: fast, robust, low-input ChIP-seq for histones and transcription factors. *Nature Methods* 12(10), 963–965.

Shazman, S., Lee, H., Socol, Y., Mann, R.S. and Honig, B. (2014) OnTheFly: A database of Drosophila melanogaster transcription factors and their binding sites. *Nucleic Acids Research* 42(D1), D167–D171.

Shen, Z., Bao, W. and Huang, D.S. (2018) Recurrent neural network for predicting transcription factor binding sites. *Scientific Reports* 8(1), 15270.

Slattery, M., Zhou, T., Yang, L., Machado, A.C.D., Gordân, R. *et al.* (2014) Absence of a simple code: How transcription factors read the genome. *Trends in Biochemical Sciences* 39(9), 381–399.

Smith, S.J., Nemr, C.R. and Kelley, S.O. (2017) Chemistry-driven approaches for ultrasensitive nucleic acid detection. *Journal of the American Chemical Society* 139(3), 1020–1028.

Srivastava, V. (2019) Gene stacking in plants through the application of site-specific recombination and nuclease activity. *Transgenic Plants: Methods and Protocols* 267–277.

Stormo, G.D. and Zhao, Y. (2010) Determining the specificity of protein–DNA interactions. *Nature Reviews Genetics* 11(11), 751–760.

Subedi, D., Vijay, A.K., Kohli, G.S., Rice, S.A. and Willcox, M. (2018) Nucleotide sequence analysis of NPS-1 β-lactamase and a novel integron (In 1427)-carrying transposon in an MDR *Pseudomonas aeruginosa* keratitis strain. *Journal of Antimicrobial Chemotherapy* 73(6), 1724–1726.

Terooatea, T.W., Pozner, A. and Buck-Koehntop, B.A. (2016) PAtCh-cap: Input strategy for improving analysis of ChIP-exo data sets and beyond. *Nucleic Acids Research* 44(21), e159–e159.

Venters, B.J. (2018) Insights from resolving protein–DNA interactions at near base-pair resolution. *Briefings in Functional Genomics* 17(2), 80–88.

Vidaki, A., López, C.D., Carnero-Montoro, E., Ralf, A., Ward, K. *et al.* (2017) Epigenetic discrimination of identical twins from blood under the forensic scenario. *Forensic Science International: Genetics* 31, 67–80.

Wallerman, O., Nord, H., Bysani, M., Borghini, L. and Wadelius, C. (2015) LobChIP: From cells to sequencing ready chip libraries in a single day. *Epigenetics & Chromatin* 8, 1–9.

Wang, H., Sham, P., Tong, T. and Pang, H. (2019) Pathway-based single-cell RNA-seq classification, clustering, and construction of gene-gene interactions networks using random forests. *IEEE Journal of Biomedical and Health Informatics* 24(6), 1814–1822.

Wang, X.Z., Ashfaq, R.A.R. and Fu, A.M. (2015) Fuzziness based sample categorization for classifier performance improvement. *Journal of Intelligent & Fuzzy Systems* 29(3), 1185–1196.

Weirauch, M.T., Cote, A., Norel, R., Annala, M., Zhao, Y. *et al.* (2013) Evaluation of methods for modeling transcription factor sequence specificity. *Nature Biotechnology* 31(2), 126–134.

Xiang, B., Xiao, C., Shen, T. and Li, X. (2018) Anti-inflammatory effects of anisalcohol on lipopolysaccharide-stimulated BV2 microglia via selective modulation of microglia polarization and down-regulation of NF-κb p65 and JNK activation. *Molecular Immunology* 95, 39–46.

Xu, W., Zhu, L. and Huang, D.S. (2019) DCDE: An efficient deep convolutional divergence encoding method for human promoter recognition. *IEEE Transactions on Nanobioscience* 18(2), 136–145.

Yang, L., Zhou, T., Dror, I., Mathelier, A., Wasserman, W.W. *et al.* (2014) TFBSshape: A motif database for DNA shape features of transcription factor binding sites. *Nucleic Acids Research* 42(D1), D148–D155.

Yang, J., Ma, A., Hoppe, A.D., Wang, C., Li, Y. *et al.* (2019) Prediction of regulatory motifs from human Chip-sequencing data using a deep learning framework. *Nucleic acids research* 47(15), 7809–7824.

Yue-Hei Ng, J., Hausknecht, M., Vijayanarasimhan, S., Vinyals, O., Monga, R. *et al.* (2015) Beyond short snippets: Deep networks for video classification. In: *Proceedings of the IEEE conference on computer vision and pattern recognition*.

Zeng, Y., Gong, M., Lin, M., Gao, D. and Zhang, Y. (2020) A review about transcription factor binding sites prediction based on deep learning. *IEEE Access* 8, 219256–219274.

Zhang, Q., Zhu, L., Bao, W. and Huang, D.-S. (2018) Weakly-supervised convolutional neural network architecture for predicting protein-DNA binding. *IEEE/ACM transactions on computational biology and bioinformatics* 17(2), 679–689.

Zhang, Q., Zhu, L. and Huang, D.S. (2018) High-order convolutional neural network architecture for predicting DNA-protein binding sites. *IEEE/ACM Transactions on Computational Biology and Bioinformatics* 16(4), 1184–1192.

Zhang, Q., Shen, Z. and Huang, D.S. (2019) Predicting *in-vitro* transcription factor binding sites using DNA sequence+ shape. *IEEE/ACM transactions on computational biology and bioinformatics* 18(2), 667–676.

Zhang, Z., Zhao, Y., Feng, X., Luo, Z., Kong, S. *et al.* (2019) Genomic, molecular evolution, and expression analysis of NOX genes in soybean (Glycine max). *Genomics* 111(4), 619–628.

Zhou, T., Shen, N., Yang, L., Abe, N., Horton, J. *et al.* (2015) Quantitative modeling of transcription factor binding specificities using DNA shape. *Proceedings of the National Academy of Sciences* 112(15), 4654–4659.

Zhu, L.J., Christensen, R.G., Kazemian, M., Hull, C.J., Enuameh, M.S. *et al.* (2011) FlyFactorSurvey: A database of drosophila transcription factor binding specificities determined using the bacterial one-hybrid system. *Nucleic Acids Research* 39(suppl_1), D111–D117.

17 Deep Learning-Driven Smart and Sustainable Agriculture

Elshan Musazade[1], Isack Ibrahim Mrisho[2], and Xianzhong Feng[1]*

[1]Key Laboratory of Soybean Molecular Design Breeding, Northeast Institute of Geography and Agroecology, Chinese Academy of Sciences, Changchun, China; [2]College of Chinese Medicinal Material, Jilin Agricultural University, Changchun, China

Abstract

Deep learning (DL), a subfield of artificial intelligence (AI), is reshaping agriculture through data-driven, automated decision-making across the entire farming life cycle. Recent advancements in DL techniques, including convolutional neural networks (CNNs), recurrent neural networks (RNNs), generative adversarial networks (GANs), and Vision Transformers (ViTs), have demonstrated significant potential in soil analysis, crop monitoring, disease detection, and yield forecasting. DL models exhibit superior accuracy, scalability, and adaptability compared to traditional machine learning (ML) approaches. These capabilities are further enhanced by integration with enabling technologies, such as the Internet of Things (IoT), unmanned aerial vehicles (UAVs), and edge computing, facilitating real-time, field-deployable applications. Despite these innovations, several challenges persist, including data heterogeneity, limited rural infrastructure, and cybersecurity vulnerabilities. A critical examination of these challenges, along with a synthesis of recent technical progress and emerging trends, provides a foundation for developing smart and sustainable agricultural systems capable of addressing global food security and climate adaptation goals.

Keywords: artificial intelligence, deep learning, Internet of Things, precision agriculture, smart agriculture

17.1 Introduction

Agriculture remains one of the foundational pillars of the global economy, contributing approximately 6.4% to the global gross domestic product (GDP) and serving as a primary source of income, employment, and food security, mainly in developing countries (Naciones Unidas, 2019; Menaga and Vasantha, 2022; Mohyuddin *et al.*, 2024). However, this sector faces mounting challenges due to rapid urbanization, climate change, resource depletion, and population growth. Despite sufficient global food production, more than 821 million people remain undernourished, and nearly 500 million suffer from malnutrition (Naciones Unidas, 2019). Meeting the projected 70% increase in food demand by 2050 will require significant increases in cereal and meat production, alongside more sustainable use of land and water resources (Rahman, 2016; Trendov *et al.*, 2019; OECD-FAO, 2020; Canton, 2021).

To address these challenges, agriculture is undergoing a digital transformation through

*E-mail: fengxianzhong@iga.ac.cn

 Deep Learning in Plant Science and Agricultural Biotechnology (Jen-Tsung Chen ed.)
DOI: 10.1079/9781836991120.0017

the emergence of smart agriculture (SA), also known as digital, precision, or Agriculture 4.0. This data-driven paradigm integrates technologies such as wireless sensor networks (WSNs), Internet of Things (IoT), remote sensing, cloud computing, and big data analytics to enable real-time monitoring, adaptive control, and predictive decision-making (Fig. 17.1) (Ilyas and Ahmad, 2020; Goel *et al.*, 2021; Kashyap *et al.*, 2021; Wang *et al.*, 2022; Mohyuddin *et al.*, 2024). SA systems facilitate the continuous tracking of soil composition, crop health, irrigation efficiency, and environmental conditions, supporting site-specific interventions and improving both productivity and ecological sustainability (Azeem Ayaz Mirani *et al.*, 2023).

These technologies are supported by a layered infrastructure where sensing and data acquisition tools, such as Global Positioning System (GPS) modules, radio-frequency identification (RFID) tags, hyperspectral cameras, and environmental sensors, feed data into centralized cloud platforms or localized edge computing nodes (Matei *et al.*, 2017; Pratyush Reddy *et al.*, 2020). Edge computing, in particular, reduces latency and bandwidth consumption by enabling preliminary data processing at the source, facilitating timely and autonomous decision-making in the field (Foley, 2014; Mekala and Viswanathan, 2017). Applications range from GPS-guided tractors and drones to automated irrigation systems and robotics, coordinated through AI-powered farm management platforms for intelligent resource use and environmental protection (Zeraatpisheh *et al.*, 2018; Ghobadpour *et al.*, 2022; Sami *et al.*, 2022).

At the heart of this revolution is artificial intelligence (AI), particularly deep learning (DL), which, when combined with IoT and cloud computing, forms the backbone of intelligent agricultural systems (Prasad *et al.*, 2019; Sanath Rao *et al.*, 2021; Shaikh *et al.*, 2022). DL models, such as convolutional neural networks (CNNs) and recurrent neural networks (RNNs), can autonomously extract high-level features from raw agricultural data, outperforming classical ML models (e.g. support vector machines (SVMs), artificial neural networks (ANNs),

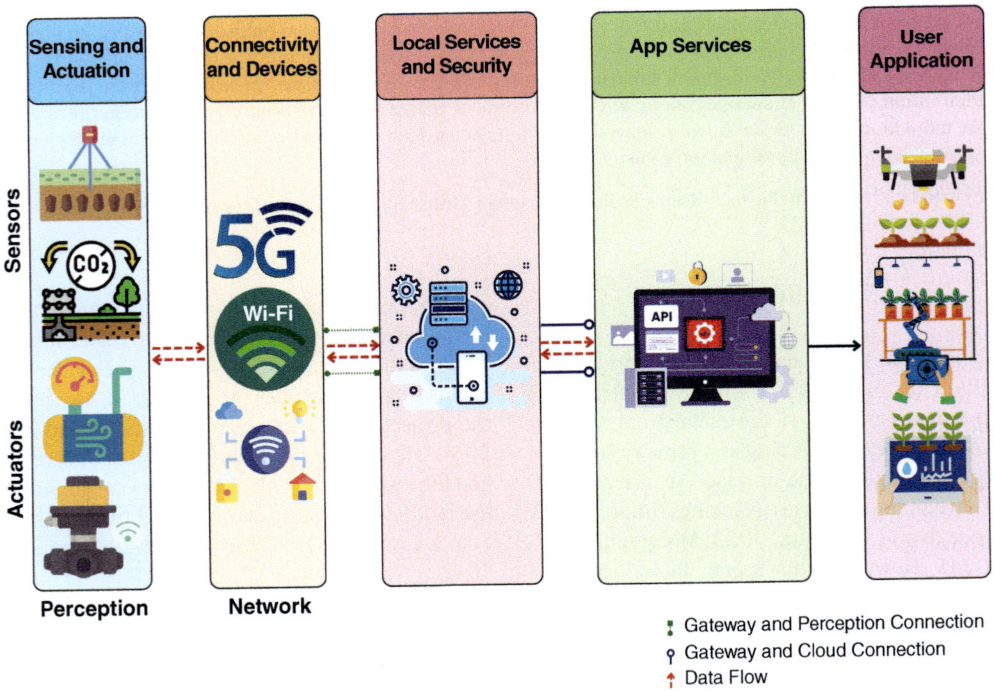

Fig. 17.1. Structure of smart agriculture system: integration of sensors, connectivity, local services, and app services for optimized agricultural operations. (Author's own image.)

k-nearest neighbors (KNNs)) in tasks such as disease detection, weed identification, and yield prediction (Meshram *et al.*, 2021; Durai and Shamili, 2022). For example, CNNs trained on the PlantVillage dataset have achieved over 99% accuracy in identifying tomato, potato, and rice leaf diseases, while transfer learning approaches, such as AlexNet, enhance model generalization across crops (Prasad *et al.*, 2019; Ayoub Shaikh *et al.*, 2022; Shaikh *et al.*, 2022; Mohyuddin *et al.*, 2024). Moreover, DL integrated with IoT platforms enables intelligent systems to monitor soil moisture, crop growth stages, and pest activity, thereby supporting early warning systems, optimizing pesticide use, and reducing postharvest loss (Hassan *et al.*, 2021; Naimi Marandi *et al.*, 2021; Yang *et al.*, 2022; Rahu *et al.*, 2023). These technologies also support predictive analytics in climate forecasting and risk assessment, which are critical to building climate-resilient agricultural systems.

However, despite their promise, the large-scale deployment of DL-driven agricultural systems faces technical and systemic challenges. These include system heterogeneity, limited data infrastructure in rural areas, privacy concerns, and difficulties in integrating diverse data sources across platforms (Varga *et al.*, 2017; Hassija *et al.*, 2019). Moreover, the effectiveness of DL models is contingent on the availability of large, labeled datasets, which are often difficult to obtain in real-world farming environments. Federated learning and edge computing have emerged as scalable solutions by reducing cloud dependency and enhancing real-time processing capabilities (Hassan *et al.*, 2021; Xiao *et al.*, 2021; Rajapaksha *et al.*, 2022). Ensuring interoperability, cybersecurity, and model transparency will be critical for the long-term sustainability of these systems (Varga *et al.*, 2017; Pathan *et al.*, 2020).

This review comprehensively synthesizes DL-enabled innovations across the agricultural life cycle, from preproduction planning to postharvest management. It highlights the intersection of DL with IoT, robotics, and automated systems and contrasts DL approaches with traditional ML methods. By examining recent advances, ongoing challenges, and future research directions, this chapter aims to inform the development of smart, scalable, and climate-adaptive agricultural systems that support global food security and environmental sustainability.

17.2 DL and Its Techniques in SA

DL, a subfield of machine learning (ML), employs ANNs to learn hierarchical and complex data representations through multiple layers of abstraction, enabling the automatic extraction of relevant features and eliminating the need for manual feature engineering (Lecun *et al.*, 2015). At the core of DL techniques are ANNs, with CNNs being particularly prominent in visual data processing tasks. CNN-based models can be trained directly on raw images, autonomously learning discriminative features through image-domain convolutions, thereby facilitating applications in image classification, object detection, and semantic segmentation (Liakos *et al.*, 2018). These vision-based capabilities are essential in automating and modeling critical agricultural processes, including crop monitoring, yield estimation, weed detection, and disease identification (Subeesh and Mehta, 2021). Recent advances in DL algorithms, computer vision, and hardware have significantly improved model performance in various domains, such as agriculture, robotics, automation, AI, and multimedia processing (Shaikh *et al.*, 2020). In addition to CNNs, other DL architectures, such as RNNs, generative adversarial networks (GANs), Vision Transformers (ViTs), and hybrid models, have demonstrated effectiveness across various agricultural applications, including time-series yield prediction, synthetic data generation, high-resolution crop trait modeling, and multimodal analysis (V. G. Dhanya *et al.*, 2022). DL systems iteratively inspect data to identify patterns, and their robustness to noise and capability for parallel data processing make them ideal for high-throughput and real-time applications. Nevertheless, DL models face challenges, including their "black-box" nature, which limits interpretability and transparency, and their reliance on computationally intensive hardware for parallel processing (Shaikh *et al.*, 2020). Despite these limitations, DL continues to advance precision agriculture (PA) and sustainable farming by enhancing automation, decision-making, and scalability in data-driven agricultural systems.

17.2.1 Convolutional neural networks (CNNs)

CNNs are a class of feedforward neural networks known for their effectiveness in computer vision tasks, including image classification, segmentation, and object detection. A standard CNN architecture includes an input layer and multiple hidden layers, typically comprising convolutional layers, activation functions such as rectified linear units (ReLU), pooling layers, and a fully connected output layer (Shaikh *et al.*, 2020). The input layer receives pixel data as multidimensional arrays, and the convolutional layers apply filters to extract spatial hierarchies and generate feature maps, while the ReLU function introduces nonlinearity by zeroing out negative values. Pooling layers subsequently reduce the spatial resolution, lowering computational complexity and enhancing translation equivariance, thus preserving spatial relationships despite linear shifts in input. Although CNNs are inherently parallelizable and suitable for large-scale data processing, they are limited by their vulnerability to adversarial attacks, sensitivity to geometric transformations, and lack of memory mechanisms for sequential data (Shaikh *et al.*, 2020). CNN-based architectures have gained prominence with the development of benchmark datasets, such as ImageNet, which enabled the training of robust models on millions of labeled images (Kriegeskorte and Tal, 2019; Miikkulainen *et al.*, 2019). Early CNN models, such as LeNet-5 (Lecun *et al.*, 2015), laid the foundation for subsequent innovations, such as AlexNet (Krizhevsky *et al.*, 2017), VGGNet (Simonyan and Zisserman, 2015), and ResNet (He *et al.*, 2016), each contributing to improved performance and reduced error rates in image recognition. Beyond classification, CNNs are central to image segmentation, which generates pixel-level masks for objects, and to object detection tasks that localize and classify multiple objects using bounding boxes. Popular object detection architectures, including single shot multibox detector (SSD) (Liu *et al.*, 2016), You Only Look Once (YOLO) (Redmon *et al.*, 2016), and Faster R-CNN (region-based convolutional neural networks) (Ren *et al.*, 2015), have been successfully applied across domains, such as automation and agriculture, highlighting the versatility and power of CNNs in modern DL applications. In agriculture, CNNs have been widely adopted for diverse tasks, including seed management, soil management, irrigation management, crop phenotyping, disease detection and management, pest detection and monitoring, weed management, abiotic stress detection, crop yield prediction, and various harvesting and postharvesting applications.

17.2.2 Recurrent neural networks (RNNs)

RNNs are a distinct class of neural networks designed to model temporal and sequential data by incorporating memory mechanisms, setting them apart from traditional feedforward neural networks. While feedforward networks (multilayer perceptrons (MLPs)) process data in a unidirectional manner, from input through hidden layers to output, each input is treated independently, lacking the capacity to retain contextual information from previous inputs. This absence of memory limits their effectiveness in tasks requiring temporal continuity or sequential dependency. In contrast, RNNs incorporate recurrent connections that allow information to persist over time steps, enabling the modeling of dependencies between successive elements in a sequence. This architectural feature makes RNNs particularly effective for modeling time series, natural language processing (NLP), and other tasks that demand contextual understanding over time. Furthermore, RNNs are often combined with convolutional layers to enhance spatiotemporal modeling, improving their performance in visual tasks where spatial and temporal dynamics are essential (Shaikh *et al.*, 2020). In agricultural contexts, RNNs have been applied to soil management, disease progression modeling, and crop yield prediction, where temporal data and sequential patterns are critical for accurate decision-making.

17.2.3 Advanced architectures in PA

Generative adversarial networks (GANs)

GANs are a special type of neural network used for unsupervised learning. GANs are a powerful approach to generative modeling, capable of learning to mimic input data distribution and producing realistic synthetic data samples. These models are designed to capture the underlying data structure, enabling them to generate new data points with varying properties while preserving the core characteristics of the original dataset. GANs have demonstrated state-of-the-art performance in numerous computer vision applications, including text-to-image synthesis (Xu *et al.*, 2018), image-to-image translation (Zhu *et al.*, 2017), and image super-resolution (Ledig *et al.*, 2017). In agriculture, GANs have also shown promising applications in areas such as plant phenotyping, where they are used to generate synthetic plant trait images for improved analysis, and postharvesting processes, including fruit quality assessment and defect detection.

The core architecture of GANs consists of two neural networks—a generator and a discriminator—trained in an adversarial framework. The generator aims to produce synthetic data that mimic the training distribution, while the discriminator evaluates whether the input data are real or generated. The generator improves by learning from the feedback of the discriminator, which penalizes it for generating implausible outputs (Reimers and Requena-Mesa, 2020). Through this iterative process, both networks evolve simultaneously, allowing the GAN to learn complex patterns in the data.

One of the major advantages of GANs is their ability to address data scarcity by generating realistic, synthetic images that can augment limited datasets. This makes GANs particularly valuable for data augmentation in DL-based computer vision, where they help reduce overfitting and improve model generalization (Hiriyannaiah *et al.*, 2020).

Vision transformers (ViTs)

While transformer models were initially developed for NLP, recent advancements have extended their application to computer vision through ViTs (Vaswani *et al.*, 2017; Dosovitskiy *et al.*, 2020). ViTs adapt the transformer architecture by modifying the input tokenization and embedding procedures to suit image data, allowing the processing of visual information using similar transformer blocks originally designed for text.

The typical architecture of ViTs includes three main components: patch embedding, feature extraction through stacked transformer encoders, and a classification head. The input image, typically represented as a three-dimensional (3D) array of height, width, and channels, is initially partitioned into smaller patches. These image patches are then flattened and linearly embedded into a sequence of vectors. A positional embedding is incorporated to preserve spatial information prior to inputting the sequence into the transformer encoder. The encoder comprises MLP and multiple multi-head self-attention (MHA) layers, each with layer normalization and residual connections (Wu *et al.*, 2021). The final MLP block, or classification head, produces the prediction output, often through a softmax function for classification tasks.

ViTs have been effectively applied to various vision tasks, including image classification, object detection, image segmentation, and multimodal applications, such as image-to-text and text-to-image generation (Bazi *et al.*, 2021; Li and Li, 2022). In agriculture, ViTs have shown promising results in tasks such as plant disease detection and management and weed identification and control by leveraging their ability to capture global contextual features for complex field conditions. Their ability to model global relationships through attention mechanisms and scalable architectures positions ViTs as a strong alternative to CNNs in many advanced vision tasks.

17.3 Applications of DL in SA

DL, particularly when integrated with IoT technologies, plays a pivotal role in enhancing SA by enabling real-time monitoring, detection, and diagnosis of key agricultural parameters, such as weather patterns, soil health, and plant diseases, thereby facilitating accurate yield estimation and timely decision-making (Fig. 17.2) (Hassan *et al.*, 2021). DL models find diverse applications throughout the agricultural life cycle, which can be broadly segmented into three major stages: preharvesting, harvesting, and postharvesting. Among these, the preharvesting phase is especially critical, as early interventions guided by DL-based predictive analytics can significantly influence crop performance. During harvesting and postharvesting, DL-driven automation and quality assessment tools maximize yield efficiency, minimize losses, and enhance product quality for improved market competitiveness.

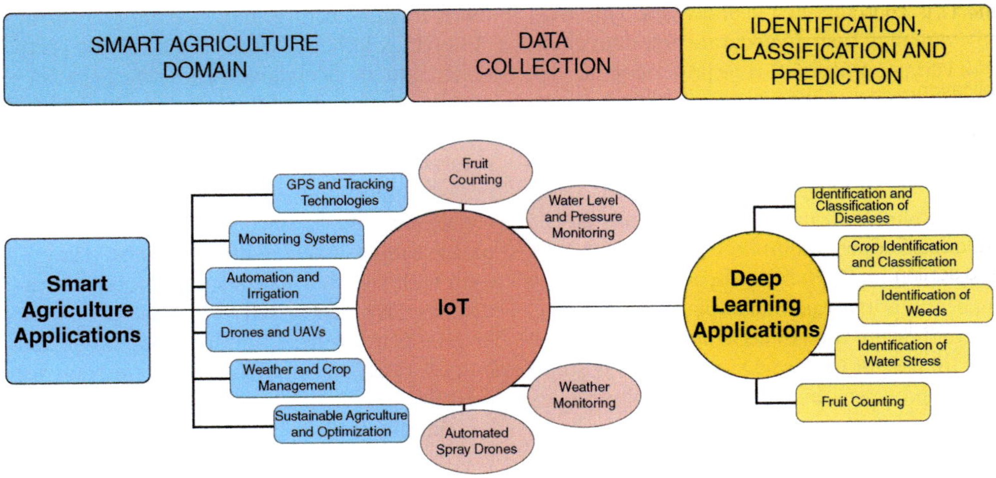

Fig. 17.2. Smart agriculture paradigm: a framework connecting IoT and DL for enhanced agricultural decision-making. DL, deep learning; IoT, Internet of Things (Author's own image.)

17.3.1 Preharvesting

The preharvesting phase is critical in agricultural management, as the real growth and yield potential of crops are strongly influenced by conditions during this stage. DL techniques, particularly CNNs and RNNs, are increasingly employed to monitor and analyze key variables, such as seed quality, soil health, fertilizer application, irrigation scheduling, and environmental factors. Accurate tracking and interpretation of these parameters enable farmers and agronomists to identify and implement optimal management practices, thereby reducing production losses and improving crop performance (Shaikh *et al.*, 2022; Mohyuddin *et al.*, 2024). By capturing complex, non-linear interactions among preharvest inputs and translating them into actionable insights, DL empowers precise, data-driven interventions that enhance decision-making and promote more innovative and sustainable agricultural systems.

Seed management

Seed management in smart farming has been significantly enhanced by advancements in DL-based computer vision, which have improved upon traditional seed analysis methods. DL techniques, particularly CNNs, have overcome the limitations of earlier approaches, enabling more accurate and efficient seed variety identification. Qiu *et al.* (2018) demonstrated that these methods, when combined with spectral imaging, enhance the precision of seed sorting, which is crucial for differentiating high-quality seeds from low-quality ones. Traditional seed sorting methods are labor-intensive and require sophisticated equipment, making them time-consuming and costly (Kannur *et al.*, 2011). However, DL-based computer vision allows for faster seed quality testing, enabling the extraction of morphological information from seed lots and grading them according to international standards (Bao and Deborah, 2021). The technology ensures the accurate assessment of key attributes, such as seed shape, length, size, and the presence of foreign bodies, supporting the evaluation of physical purity, genetic purity, seed health, and vigor (Granitto *et al.*, 2005). Table 17.1 summarizes some of the significant previous studies investigating the applications of DL in seed management.

Numerous studies have demonstrated the effectiveness of DL in seed classification and quality assessment. Veeramani *et al.* (2018) employed a CNN-based deep neural network (DNN) model for soybean seed analysis, focusing on estimating seed numbers per pod and classifying haploid seeds. This approach highlights the potential of DL in high-precision seed characterization, supporting advancements in crop breeding and genetic analysis (Veeramani *et al.*, 2018). Similarly, Tu *et al.* (2018) utilized an MLP neural network to classify high- and low-quality pepper

Table 17.1. Applications of DL in seed management. Author's own table.

Application	Plant species	Methodology	Results	References
Seed classification and sorting	Soybean	BPNN	Accuracy: 97.25% for seed size, germination, pod count	Liu *et al.*, 2015
Seed classification and sorting	Corn	DeepSort, SVM, LR, RF	Best performance after fivefold cross-validation	Veeramani *et al.*, 2018
Seed classification and sorting	Corn	Ensemble: KNN, LR, SVM, SURF, GoogLeNet, VGG19	Accuracy: 95.00%	Huang *et al.*, 2019
Seed classification and sorting	Cotton	CNN, ResNet + SVM, PLS-DA, LR	Accuracy: 80.00%	Zhu *et al.*, 2019
Seed classification and sorting	Multiple plant species	VGG16	Accuracy: 99.9% on test set (234 images)	Gulzar *et al.*, 2020
Seed classification and sorting	Sunflower	AlexNet, GoogLeNet, ResNet	GoogLeNet achieved 95% accuracy on 4800 seeds	Kurtulmuş, 2020
Seed classification and sorting	Corn	CNN + classifiers: ANN, SVM, KNN, boosted/bagged tree, LDA	CNN–ANN model: Accuracy 98.1%, Precision 98.2%, Recall 98.1%, F1-score 98.1%	Javanmardi *et al.*, 2021
Seed defect detection	Soybean	CNN (AlexNet, VGG19, GoogLeNet, ResNet50, MobileNetV2, ShuffleNet, SqueezeNet)	MobileNetV2 accurately classified defective seeds across sizes	Zhao *et al.*, 2021
Seed grading	Corn	DCNN	Accuracy: 98.2% on 408 corn test images	Ni *et al.*, 2019
Variety identification	Barley	CNN	Improved accuracy by 0.6%, sensitivity by 2.3%	Dolata and Reiner, 2018
Variety identification	Rice	KNN, SVM, CNN	CNN achieved 89.6% training and 87% testing accuracy	Qiu *et al.*, 2018
Variety identification	Oats	DCNN	Accuracy: 99.19% on test set	Wu *et al.*, 2019

ANN, artificial neural networks; BPNN, back-propagation neural network; DCNN, deep convolutional neural network; DL, deep learning; KNN, k-nearest neighbors; LDA, linear discriminant analysis; LR, logistic regression; PLS-DA, partial least squares discriminant analysis; RF, random forest; SVM, support vector machine; SURF, speeded-up robust features; VGG19, Visual Geometry Group 19-Layer Network.

seeds, showing that MLP networks, combined with binary logistic regression, can significantly improve germination rates and selection efficiency. Their research demonstrated the superior performance of deeper neural network architectures, such as GoogLeNet, over shallower networks, such as VGG19, in extracting seed features (Tu *et al.*, 2018).

DL has also been successfully integrated with spectroscopy for seed variety identification and classification. Zhu *et al.* (2019) demonstrated that combining CNN and ResNet models with spectroscopy achieved over 80% accuracy in cataloging cotton seeds, emphasizing the effectiveness of this integration in improving seed classification (Zhu *et al.*, 2019). Additionally,

Li *et al.* (2021) introduced SeedSortNet, a CNN-based model that achieved remarkable accuracy in sorting corn (97.33%) and sunflower (99.56%) seeds, further highlighting the potential of DL to automate seed sorting processes (Li *et al.*, 2019). Ma *et al.* (2020) demonstrated that CNN models could achieve 90% accuracy in predicting seed viability, showcasing the utility of DL in enhancing seed quality evaluation (Ma *et al.*, 2020).

The growing importance of seed quality in improving agricultural productivity has driven the development of advanced technologies for seed sorting. Taheri-Garavand *et al.* (2021) used a modified VGG16 model to identify chickpea varieties, demonstrating the role of computer vision in plant variety classification based on seed characteristics (Taheri-Garavand *et al.*, 2021). Zhao *et al.* (2021) tested various computer vision models to detect surface flaws in seeds, a crucial step for enhancing seed quality. Among the models tested, MobileNetV2 exhibited exceptional accuracy in detecting flaws in soybean seeds, significantly improving seed sorting efficiency (Zhao *et al.*, 2021). These studies collectively highlight the increasing role of DL and computer vision in automating seed sorting and classification, ultimately enhancing productivity in the seed industry and breeding programs.

Soil management

Soil is essential for plant growth, serving as a key medium for nutrients, water, and oxygen necessary for crop production. However, continuous cultivation has contributed to soil degradation, reducing fertility and crop yields. To address these challenges, soil health assessments, including analyzing inorganic nutrients and microbiological conditions, are critical for understanding soil limitations and informing remediation strategies (Suchithra and Pai, 2020; Kushwaha *et al.*, 2022). Traditional soil texture analysis methods, such as sieving and hydrometer techniques, remain fundamental but are increasingly being complemented by innovative technologies. Image-based soil analysis, enabled by the integration of computer vision and DL, provides a non-invasive and rapid approach to evaluating soil texture and particle size, enhancing both the efficiency and accuracy of soil assessments (Sudarsan *et al.*, 2016; Kushwaha *et al.*, 2022). Table 17.2 summarizes some of the significant

Table 17.2. Applications of DL in soil management. Author's own table.

Application	Methodology	Results	References
Soil classification	LCTF, 3D-CNN	Accuracy: 99.59%	Yu *et al.*, 2019
Soil contaminants	CNN, autoencoders	R^2 As: 0.86, Cu: 0.74, Pb: 0.82	Pyo *et al.*, 2020
Soil properties	Multitask DCNN	Outperformed single-task DCNN	Zhong *et al.*, 2016
Soil properties	CNN	Shallow CNN was the most effective	Riese and Keller, 2019
Soil properties	BiLSTM, RF, linear regression	BiLSTM outperformed all models	Li *et al.*, 2020
Soil properties	CNN	Outperformed traditional models	Yu *et al.*, 2021
Soil properties	CNN-LSTM, LSTM, CNN, RF, BiLSTM	BiLSTM achieved the best performance	Alibabaei *et al.*, 2021
Soil properties	DCNN (LucasResNet16)	Multitask DCNN outperformed single-task models	Zhong *et al.*, 2021
Soil properties	CNN, ANN, SVM, RF, KNN	Accuracies: 99.89%, 99.81%, and 99.58% at 20, 40, and 60 cm	Azadnia *et al.*, 2022
Soil properties	DCNN	5–26% improvement in sand, silt, and clay predictions	Lilburne *et al.*, 2022

ANN, artificial neural networks; BiLSTM, bidirectional long short-term memory; CNN, convolutional neural network; DCNN, deep convolutional neural network; KNN, k-nearest neighbors; LSTM, long short-term memory; RF, random forest; SVM, support vector machine.

previous studies investigating the applications of DL in soil management.

DL models, particularly CNNs, have also been widely adopted for automated soil quality assessments. Zhu *et al.* (2019) demonstrated the efficacy of CNNs in automating seed classification and estimating critical soil parameters, emphasizing the role of pattern recognition in improving agricultural efficiency (Zhu *et al.*, 2019). This approach reduces dependency on manual inspections, enabling higher-throughput and more accurate soil quality assessments. CNNs have thus become essential tools in agricultural monitoring, facilitating rapid and precise data collection that supports better decision-making.

In addition to soil texture and quality analysis, DL models have been employed to predict dynamic soil properties, such as soil temperature. Yu *et al.* (2021) utilized CNN architectures, including Conv2D, Conv3D, and ConvLSTM, to forecast soil temperature values, demonstrating the potential of these models in environmental monitoring (Yu *et al.*, 2021). Building on this, Shi *et al.* (2023) advanced these techniques by applying ConvLSTM networks with spatiotemporal data, incorporating empirical mode decomposition (EMD) to improve prediction accuracy (Shi *et al.*, 2015). Their models, particularly the EEMD-Conv3D architecture, showed significant performance improvements, underscoring the value of integrating historical and meteorological data to enhance prediction reliability. These innovations exemplify how DL can replace traditional soil analysis methods, offering more efficient and accurate alternatives.

Moreover, DL has proven valuable in enhancing soil and environmental parameter estimations using temporal and spatial models. Alibabaei *et al.* (2021) employed various RNN architectures, including BiLSTM, CNN-LSTM, and LSTM, to model soil-water content and evapotranspiration at multiple sites in Portugal. Their study found that BiLSTM outperformed all other models, achieving a mean squared error (MSE) as low as 0.014–0.056 (Alibabaei *et al.*, 2021). Similarly, Riese and Keller (2019) explored using one-dimensional (1D) CNNs for soil classification, demonstrating that simple architectures, such as LucasCoordConv, outperform more complex models, such as random forest (RF), in classification accuracy (Riese and Keller, 2019). These findings highlight the growing effectiveness of CNNs in improving soil classification and environmental monitoring.

Finally, applying DL in segmentation further contributes to agricultural monitoring. Zhang *et al.* (2020) introduced a mask-refined R-CNN for object segmentation, improving the integration of high- and low-level features in remote sensing applications. Their model, validated on the Cityscapes and COCO (common objects in context) datasets, demonstrated its versatility and effectiveness, offering significant potential for agricultural applications, such as crop monitoring and field analysis (Zhang *et al.*, 2020). These advancements in DL techniques, ranging from RNNs and CNNs to instance segmentation models, demonstrate their broad applicability in optimizing soil data analysis and supporting diverse agricultural practices.

Irrigation management

Efficient irrigation management is critical due to the increasing depletion of water resources, and optimizing irrigation processes has become a focal point in SA. Accurate assessment of crop water status is essential for maintaining a balance among hydrological, climatological, and agronomic factors, which directly impact crop yield and resource use (Elbeltagi *et al.*, 2023). To achieve this, researchers have explored the biophysical processes involved in water uptake by the root zone of soil and the transpiration process through plant canopies to determine the precise water requirements for crops (Kushwaha *et al.*, 2016; Vishwakarma *et al.*, 2022). In recent years, integrating image processing and DL techniques has significantly advanced smart irrigation practices. These technologies have enabled the automation of irrigation management, making it more efficient and responsive to real-time crop needs. Table 17.3 summarizes some of the significant previous studies investigating the applications of DL in irrigation management.

Goldstein *et al.* (2018) pioneered a novel calibration method for near-infrared (NIR) image data using CNNs with a streamlined architecture comprising a single convolutional and pooling layer. This CNN model was designed to improve the accuracy of irrigation-related predictions by learning spatial patterns within spectral imagery. Additionally, a decision tree

Table 17.3. Applications of DL in irrigation management. Author's own table.

Application	Methodology	Results	References
Irrigation efficiency and management	CNN on UAV images	Accuracy: 95.85%, precision: 93.3%	Zhang *et al.*, 2018
Irrigation efficiency and management	PVANET + Hough transform	Precision: 95%, recall: 95.5% using Sentinel-2 images	Tang *et al.*, 2021
Irrigation efficiency and management	SFAO-DeepCNN	Accuracy: 92%, specificity: 91.2%, sensitivity: 94.1%	Kumbi Mahantesh N., 2021
Irrigation system monitoring	Mask R-CNN	Performed well across various datasets	Gupta *et al.*, 2021
Irrigation system monitoring	Mask R-CNN on UAV images	Satisfactory results based on dataset size	Albuquerque *et al.*, 2020
Irrigation scheduling and optimization	DQN, CNN	Reduced water usage significantly vs traditional methods	M. Chen *et al.*, 2021
Irrigation scheduling and optimization	LSTM	RMSE: 2.35%, indicating high accuracy	Chantry *et al.*, 2021
Irrigation scheduling and optimization	DQN, CNN	Increased productivity by 11%; reduced water use by 20–30%	Alibabaei *et al.*, 2021
Irrigation system monitoring	PVANET	Precision: 95%, recall: 95.5% (Sentinel-2 images)	Tang *et al.*, 2021
Irrigation system monitoring	ResNet50 + U-Net on CCTV images	Dice score: 0.998; water-level prediction R^2: 0.97	Kim *et al.*, 2022
Water pollution detection	Shallow CNN + decision tree on NIR data	RMSEV: 25.47, R^2: 0.914	H. Chen *et al.*, 2020

CNN, convolutional neural network; DQN, deep q-network; LSTM, long short-term memory; PVANET, pyramid vision attention network; R-CNN, region-based convolutional neural network; RMSE, root mean square error; SFAO-DeepCNN, sunflower atom optimization-based deep convolutional neural network; UAV, unmanned aerial vehicle.

algorithm was employed to identify key features, which were then fed into the CNN, enhancing both model performance and interpretability. This hybrid approach demonstrates the potential of DL in sensor data calibration and smart irrigation systems (Goldstein *et al.*, 2018). Other studies have similarly explored the role of computer vision in optimizing irrigation. For example, Zhang *et al.* (2018) applied CNNs to monitor and allocate irrigation water in center pivot irrigation systems, achieving high precision (95.85%) and recall (93.33%) rates, further demonstrating the effectiveness of DL in irrigation management (Zhang *et al.*, 2018).

Moreover, innovations in autonomous agricultural systems have combined computer vision and multitasking capabilities to improve both irrigation and crop management. Chang and Lin (2018) developed a compact, intelligent agricultural machine that performs real-time plant and weed classification, enabling autonomous weeding and watering while maintaining soil moisture levels and optimizing herbicide application (Chang and Kuan-Ming, 2018). Kamyshova *et al.* (2022) proposed a computer vision-based system that uses a phyto-indication system for optimizing crop watering. This system, trained with a neural network using the resilient propagation method, achieved up to 93% accuracy in plant identification and 92% accuracy in growth stage detection (Kamyshova *et al.*, 2022). These developments highlight the growing role of DL and computer vision technologies in enhancing irrigation efficiency, ultimately contributing to more sustainable agricultural practices and increased crop productivity.

Crop phenotyping

Phenotyping, which involves the analysis of observable traits in plants, plays a critical role in plant breeding, especially in efforts to improve crop yield and develop resilient cultivars. Traditionally, phenotyping has been a manual, time-consuming process, but recent advancements in DL have revolutionized this field by enabling scalable and precise trait evaluation (Guo *et al.*, 2017). Early automated systems, such as the platform developed by Dee and French (2015), were pivotal in automating the detection and quantification of plant features without human intervention, marking a significant step forward in integrating DL in phenotyping (Dee and French, 2015). This foundational work paved the way for more sophisticated DL platforms, such as Deep Phenomics, which use pretrained neural networks for tasks such as leaf counting, mutant classification, and age regression, offering efficient, high-throughput solutions for plant scientists (Deery *et al.*, 2014; Shafiekhani *et al.*, 2017). Table 17.4 summarizes some of the significant previous studies investigating the applications of DL in crop phenotyping.

Further enhancing the capabilities of phenotyping, integrating autonomous ground vehicles (AGVs) and advanced sensing technologies has expanded the scope of DL-powered plant phenotyping. These systems enable non-invasive, large-scale trait extraction across various developmental stages, improving throughput and reducing the manual effort traditionally involved (Ubbens and Stavness, 2017). Moreover, the concept of "affordable phenotyping" has gained momentum, driven by the availability of low-cost imaging devices and mobile sensors compatible with DL frameworks. For instance, Feng *et al.* (2021) demonstrated the viability of mobile phone-embedded sensors for phenotypic data collection, making DL-driven phenotyping more accessible to the broader research community (Feng *et al.*, 2021).

These developments underscore the transformative impact of DL in high-throughput phenotyping, automating the extraction of complex plant traits and facilitating real-time data collection. The growing use of DL in phenotyping supports the efficiency of data collection and the advancement of precision breeding objectives, making it an indispensable tool for crop improvement. DL is helping bridge the gap between traditional phenotyping methods and modern plant breeding and genetic research demands by enabling scalable, accurate, cost-effective trait evaluation.

Disease and pest management

DISEASE DETECTION AND MONITORING. Plants are inherently vulnerable to infections due to their constant exposure to environmental factors, and their susceptibility to pathogens plays a significant role in disease outbreaks. Timely and accurate disease identification prevents significant yield losses and maintains crop quality. Advances in DL, particularly through the integration of CNNs, have enabled the development of precise disease detection models. These hybrid models offer scalable and efficient solutions for monitoring plant health across various parts, such as leaves, stems, and fruits, making them essential for early intervention and effective management. They have proven particularly useful in large-scale agricultural diagnostics, capable of handling diverse datasets from multiple crops. As these technologies continue to advance, they offer the potential to improve disease detection speed, minimize pesticide usage, and support more sustainable agricultural practices (Khirade and Patil, 2015). Table 17.5 summarizes some of the significant previous studies investigating the applications of DL in disease management.

CNNs have become a fundamental tool for disease detection, demonstrating impressive effectiveness across various crops. These models have shown scalability, efficiency, and accuracy in identifying crops such as rice, potato, corn, and grapevine diseases. For instance, Mohanty *et al.* (2016) developed a deep CNN capable of

Table 17.4. Applications of DL in crop phenotyping. Author's own table.

Application	Plant species	Methodology	Results	References
Plant species recognition and classification	Legumes (white bean, red bean, soybean)	CNN (five-layer)	Accuracy: 90.2% (white bean), 93.7% (red bean), 98.8% (soybean)	Grinblat *et al.*, 2016
Plant species recognition and classification	Multiple plant species	ResNet-based CNN	SR Precision: 91.84%, recall: 91.67%, F1-score: 91.49%; IHIL precision: 84%, recall: 83.14%, F1-score: 83.19%	Hati and Singh, 2021

CNN, convolutional neural network; IHIL, identification of healthy and infected leaves; SR, species recognition.

Table 17.5. Applications of DL in disease management. Author's own table.

Application	Plant species	Methodology	Results	References
Disease detection	Apple	AlexNet, VGG19, Inception, ResNet, DenseNet, BPNN, PlantDiseaseNet	Accuracy: 97.62%	Liu *et al.*, 2018
Disease detection	Multiple plant species	DNN, CNN	Accuracy: 99.53%	Ferentinos, 2018
Disease detection	Multiple plant species	GoogLeNet (CNN)	Accuracy: 96.17%	Arnal Barbedo, 2019
Disease detection	Multiple plant species	KNN, SVM, ELM, VGG16, GoogLeNet, AlexNet	Accuracy: 98%	Turkoglu and Hanbay, 2019
Disease detection	Rice	MobileNetV2, NASNetMobile, EfficientNet-b0, DenseNet121	MobileNetV2 showed superior accuracy	J. Chen *et al.*, 2020
Disease detection	Tomato	DCNN (with Adam optimizer)	Outperformed SGD and RMSprop in accuracy	Thangaraj *et al.*, 2021
Disease detection	Tomato and cotton	DenseNet + SVM (vs Inception, ResNet, MobileNet)	F1-score: 99.29%, precision ≥ 95% (in repeated tests)	Espejo-Garcia *et al.*, 2020
Disease detection	Coconut tree	2D-CNN with segmented images	Accuracy: 96.94%, Kappa: 0.91	Singh *et al.*, 2020
Disease detection	Tomato	DenseNet121	Accuracies: 99.51% (5 classes), 98.65% (7), and 97.11% (10)	Abbas *et al.*, 2021
Disease detection	Black gram	CNN with VirLeafNet-1, -2, -3	Accuracies: 91.23%, 96.42%, and 97.40%	Joshi *et al.*, 2021
Disease detection	Multiple plant species	ResTS (Residual Architecture)	F1-score: 0.991	Shah *et al.*, 2021
Disease detection	Corn	AlexNet, GoogLeNet, VGG16, VGG19, Softmax	GoogLeNet + Softmax: accuracy 99.94%	Pan *et al.*, 2022
Disease detection	Citrus plants	KNN, DNN	DNN: accuracy 99.89%; KNN: 89.89%	Barman and Choudhury, 2022
Disease detection	Grape	DeepLabV3 with ResNet50 backbone	Overall accuracy: 97.5% on a hold-out dataset	Ji Zhibin, 2022
Disease identification and classification	Citrus plants	Two-stage deep CNN	Detection accuracy: 94.37%, average precision: 95.8%	Syed-Ab-Rahman *et al.*, 2022
Disease identification and classification	Apple	CoaT-ViT (Transformer-based lightweight model)	Accuracy: 96.85% on apple leaf disease dataset	Li and Li, 2022
Disease identification and classification	Rice and cassava	Lightweight CNN with Inception and Residual connections	Accuracy: 99.39% (rice), 99.66% (PlantVillage rice), 76.59% (cassava)	Hassan and Maji, 2022

Continued

Table 17.5. Continued.

Application	Plant species	Methodology	Results	References
Disease identification and classification	Not specified	ACO-CNN, CNN, CGAN, SGD	CNN and ACO-CNN: Accuracy 100%, F1-score: 100; CGAN: accuracy 98%, F1-score: 97%; SGD: accuracy 85%, F1-score: 86%	Abd Algani *et al.*, 2022

ACO-CNN, ant colony optimization with convolution neural network; ANN, artificial neural networks; BPNN, back-propagation neural network; CGAN, conditional generative adversarial networks; CNN, convolutional neural network; DCNN, deep convolutional neural network; KNN, k-nearest neighbors; SVM, support vector machine.

identifying 14 crop species and 26 plant diseases, achieving an impressive accuracy of 99.35% on a held-out test set. They also applied pretrained models, such as AlexNet and GoogLeNet, achieving accuracy rates of 99.35% and 98.21%, respectively, using a public dataset containing healthy and infected leaves (Mohanty *et al.*, 2016). Similarly, Cruz *et al.* (2017) applied a CNN for detecting red grapevine diseases, accurately identifying frog eye, downy mildew, and septoria leaf blight using color images of leaf clippings (Cruz *et al.*, 2017). These studies demonstrate the scalability and effectiveness of CNNs in large-scale agricultural diagnostics.

Lu *et al.* (2017) applied CNNs for rice disease detection, achieving 95.48% accuracy using images of healthy and infected rice stems and leaves (Lu *et al.*, 2017). Oppenheim and Shani (2017) used CNNs to detect potato diseases, achieving accuracy rates ranging from 83% to 96%, depending on the amount of training data used (Oppenheim and Shani, 2017). Additionally, Durmus *et al.* (2017) evaluated SqueezeNet and AlexNet for tomato disease identification, reporting similar accuracy levels between the two models, further demonstrating the potential of CNNs for various crops (Durmus *et al.*, 2017). Rangarajan *et al.* (2018) applied VGG16 for tomato leaf disease classification and achieved an accuracy of 97.29%, slightly lower than the 97.50% accuracy obtained using AlexNet (Rangarajan *et al.*, 2018).

Picon *et al.* (2019) enhanced CNN models by integrating non-image metadata and crop information, achieving a balanced accuracy of 98% and reducing misclassifications by 71% compared to previous methods (Picon *et al.*, 2019). Similarly, Polder *et al.* (2019) proposed a deep CNN (DCNN) for the automatic detection of the Tulip Breaking Virus (TBV), achieving results comparable to earlier RGB-based methods (Polder *et al.*, 2019). Furthermore, Jiang *et al.* (2021) developed a CNN-based system for detecting infections in apple fruit, achieving an 88% recognition rate, demonstrating the effectiveness of CNNs in postharvest disease monitoring and containment (H. Jiang *et al.*, 2021).

CNN-based models have demonstrated strong potential for real-time, in-field detection of plant diseases. Mishra *et al.* (2020) developed a real-time disease detection system for corn plants using a deep CNN model deployed on a Raspberry Pi 3, achieving an accuracy of 88.46%, enabling timely interventions to reduce crop losses (Mishra *et al.*, 2020). Similarly, Sharma *et al.* (2020) developed the S-CNN model using segmented image data, achieving 98.6% accuracy in classifying ten different diseases, such as target spot disease in tomato plants (Sharma *et al.*, 2020). These advancements reinforce the potential of CNNs to improve the speed and accuracy of disease detection in practical agricultural settings.

DL models, particularly DNNs, have also demonstrated significant advancements in plant disease detection and classification. DNNs can identify more complex patterns in large datasets, making them particularly useful for analyzing multiple forms of data, including image and sensor data. Durai and Shamili (2022) achieved an impressive 99.53% accuracy rate in identifying plant diseases, highlighting the effectiveness of DNNs in precise and reliable disease classification (Durai and Shamili, 2022). This further emphasizes the growing role of DNNs in automating plant disease diagnostics, providing more accurate, scalable, and efficient solutions compared to traditional methods.

Hybrid DL models that combine CNNs with other techniques, such as RNNs and long short-term memory (LSTM) networks, have proven highly effective in enhancing plant disease detection. These hybrid models leverage the strengths of multiple neural network architectures to improve detection accuracy. For example, Sharif *et al.* (2018) proposed a CNN-based method for detecting bean diseases and recommended a hybrid CNN approach for citrus disease detection, achieving an accuracy of 97%. This highlights the power of CNNs in automating plant disease diagnostics, offering more precise and scalable solutions compared to traditional ML methods (Sharif *et al.*, 2018). Namin *et al.* (2018) combined LSTM networks with CNNs to predict plant phenotypes, reducing sensor reading errors by utilizing the ability of CNNs to process time-series data (Taghavi Namin *et al.*, 2018). Furthermore, Karthik *et al.* (2020) demonstrated the effectiveness of combining residual networks with CNN classifiers for tomato leaf disease detection, resulting in improved accuracy in disease identification (Karthik *et al.*, 2020).

Hybrid approaches have also benefited from the use of transfer learning. Wu *et al.* (2020) employed a CNN (FRCNN) with transfer learning to classify wheat grains, achieving an accuracy of 91% with a model loss of less than 0.5, demonstrating the robustness of hybrid models in agricultural settings (Wu *et al.*, 2021). Similarly, Bedi and Gole (2021) proposed the CAE-CNN model, which utilized high-resolution images of peach plant leaves to detect bacterial spot disease, achieving training and testing accuracies of 99.35% and 98.38%, respectively (Bedi and Gole, 2021). These results further underscore the effectiveness of CNN-based hybrid models for early disease detection.

Integrating attention mechanisms into hybrid models has also proven effective, particularly in enhancing spatiotemporal feature learning for disease detection tasks. Pandey and Jain (2022) proposed the DADCNN-5 model, which integrates CNNs with attention dense learning (ADL), achieving an impressive classification accuracy of 99.93% on a dataset of 54,305 images. This model demonstrated high diagnostic precision, with a sensitivity of 96.57%, specificity of 99.94%, and a very low false-positive rate of just 0.06%, highlighting the efficacy of hybrid CNN models in plant disease detection (Pandey and Jain, 2022). RNN-based approaches have also been incorporated into hybrid models to improve the detection of temporal dependencies in image sequences. Nandhini *et al.* (2022) developed a gated recurrent CNN (G-RCNN), which combines CNNs for feature extraction with RNNs to capture temporal dependencies. This hybrid model supports real-time monitoring of banana trees, enabling early disease detection through continuous image updates and providing farmers with timely intervention tools (Nandhini *et al.*, 2022). These studies highlight the growing role of hybrid DL models, particularly CNNs combined with transfer learning, RNNs, and attention mechanisms, in enhancing plant disease detection accuracy, efficiency, and scalability.

Moreover, advanced CNN architectures, such as ResNet, Inception, and VGG, have successfully detected plant disease by enabling deeper networks and more detailed feature extraction from high-resolution images. These architectures can handle complex datasets, making them ideal for large-scale agricultural applications. For example, Ramcharan *et al.* (2017) applied GoogLeNet (Inception) for cassava disease detection, yielding promising results that highlight the broad applicability of CNNs in plant pathology (Ramcharan *et al.*, 2017). Similarly, Fuentes *et al.* (2017) compared ResNet50 and ResNeXt-50 for tomato disease detection, finding ResNet50 to be more effective, thus reinforcing its robustness in agricultural applications. Their research also recommended CNN architectures such as R-FCN, R-CNN, VGGNet, SSD, and ResNet for distinguishing between nine tomato disease categories, showcasing the ability of CNN models to process high-resolution images and improve crop management (Fuentes *et al.*, 2017).

Zhang *et al.* (2019) further demonstrated the utility of advanced CNN architectures by utilizing GPDCNN for plant disease recognition, contributing to improved detection accuracy (Zhang *et al.*, 2019). Additionally, Shin *et al.* (2021) employed ResNet50 to detect powdery mildew in strawberries, achieving an impressive classification accuracy of 98.11%. Meanwhile, AlexNet was noted for its faster processing capabilities, handling 2320 images in just 40.73 seconds with a classification accuracy of 95.59% (Shin *et al.*, 2021). In citrus plant disease detection, Sujatha *et al.* (2021) evaluated various DL architectures, including VGG16, VGG19, and Inception v3, and found that DL methods significantly outperformed traditional ML techniques (Sujatha *et al.*, 2021). These studies emphasize the importance of integrating high-resolution imaging with advanced CNN architectures for reliable disease identification across multiple categories, enhancing the accuracy, efficiency, and scalability of plant disease detection systems.

Finally, transfer learning has become a critical technique for improving plant disease detection, especially when labeled data is scarce. Leveraging pretrained models enables the development of high-performing models with minimal data. Ramcharan *et al.* (2017) used Inception v3 for cassava disease detection, demonstrating its potential for in-field diagnosis with minimal visual data (Ramcharan *et al.*, 2017). Ferentinos (2018) trained a model on 87,848 images across 58 disease classes, achieving 99.53% accuracy on unseen images, showing its ability to generalize across unobserved data (Ferentinos, 2018). Similarly, Tetila *et al.* (2020) evaluated DL architectures such as ResNet50, VGG16, VGG19, Inception v3, and Xception for classifying soybean pests. Transfer learning improved model accuracy, with the best models achieving 93.82%, surpassing traditional methods (Tetila *et al.*, 2020).

Jiang *et al.* (2021) advanced transfer learning by combining ResNetV2 with multitask learning, achieving 95.67% accuracy in detecting rice leaf diseases. Using pretrained VGG16, their model achieved 97.22% accuracy for wheat and rice leaf diseases, outperforming ResNet50 and DenseNet121 (Z. Jiang *et al.*, 2021). Additionally, integrating transfer learning with handcrafted features has enhanced disease detection. Fan *et al.* (2022) combined these methods for apple and coffee leaves, achieving a high classification accuracy of up to 99.79% (Fan *et al.*, 2022).

These studies demonstrate that combining transfer learning with CNNs, hybrid models, and tailored features substantially enhances the accuracy, scalability, and efficiency of plant disease detection, supporting advancements in agricultural diagnostics and real-time field applications.

PEST DETECTION AND MONITORING. Pest control remains a critical challenge in both open-field and greenhouse agriculture, and DL techniques are increasingly integrated into PA to address this issue. These techniques enable the spatial and temporal targeting of agrochemical inputs, enhancing efficiency and reducing environmental harm. For instance, Ebrahimi *et al.* (2017) used DL methods in an image-processing system to classify pests such as thrips in strawberry greenhouses, facilitating real-time pest detection and improved control measures (Ebrahimi *et al.*, 2017). Similarly, Albanese *et al.* (2021) developed an embedded system for pest detection in fruit orchards, leveraging a low-power sensing platform and neural accelerator for image processing within pheromone-based traps (Albanese *et al.*, 2021). Recent advancements show that DL can enable fully automated, real-time pest monitoring without human involvement (Segalla *et al.*, 2020). Table 17.6 summarizes some of the significant previous studies investigating the applications of DL in pest management.

DL has demonstrated its potential in pest identification, with several studies utilizing CNNs for improved classification accuracy. Karar *et al.* (2021) introduced a mobile application powered by DL to classify agricultural pests, achieving a high accuracy of 99.0% across five pest groups (Cicadellidae, aphids, flea beetles, flax budworm, and red spider mite), offering real-time support to farmers (Karar *et al.*, 2021). In a related study, Butera *et al.* (2022) evaluated the effectiveness of advanced CNN models for detecting beetle-like pests in complex agricultural environments, demonstrating the robustness of these models in handling non-homogeneous image datasets (Butera *et al.*, 2022). Oliveri Conti *et al.* (2017) explored CNN-based approaches for classifying visually similar insect pests under challenging conditions, showcasing the versatility of DL in pest management (Oliveri Conti *et al.*, 2017). Similarly, Sanga *et al.* (2020) developed a banana disease detection system using five CNN architectures, further demonstrating the capability of DL in identifying crop pathologies (Sanga *et al.*, 2020). These studies highlight the growing integration of CNN-based object detection techniques for automated pest monitoring in PA.

Several studies also illustrate the growing integration of DL techniques in automated pest monitoring. Dawei *et al.* (2019) developed a diagnostic system using transfer learning to classify ten types of pests, achieving an average accuracy of 93.84% (Dawei *et al.*, 2019), while Chen *et al.* (2021) fine-tuned the VGG16 network

Table 17.6. Applications of DL in pest management. Author's own table.

Application	Plant species	Methodology	Results	Reference
Agrochemical optimization	Not specified	CAE	Accuracy: 86.78%	Khamparia *et al.*, 2020
Agrochemical optimization	Not specified	ResNet152	Accuracy: 99.2%	Sanga *et al.*, 2020
Pest classification and identification	Tomato	VGG16, VGG19, ResNet50	VGG16 achieved the highest accuracy: 91.9%	Mkonyi *et al.*, 2020
Pest classification identification	Multiple plant species	AlexNet, ResNet, GoogLeNet, VGGNet	Accuracies: 96.75%, 97.47%, and 95.97%	Thenmozhi and Reddy, 2019
Pest classification identification	Forest plants	Pretrained CNN models	Enhanced classification performance on the new dataset	Liu *et al.*, 2022
Pest detection	Multiple plant species	KNN, SVM, ELM, VGG16, VGG19, AlexNet	Accuracy: 98%	Turkoglu and Hanbay, 2019
Pest detection	Coconut tree	2D-CNN with segmented images	Accuracy: 96.94%, Kappa: 0.91	Singh *et al.*, 2020
Pest detection	Sugarcane	Inception v3, DenseNet121, ResNet50, Xception	Xception outperformed other models	Grijalva *et al.*, 2023

CAE, convolutional autoencoders; CNN, convolutional neural network; ELM, extreme learning machine; KNN, k-nearest neighbors; VGG, Visual Geometry Group network.

for tea pest classification, reaching an impressive accuracy of 97.75% (J. Chen *et al.*, 2021). In tomato plant pest detection, Mkonyi *et al.* (2020) applied DL algorithms, including VGG16, VGG19, and ResNet50, to identify the tomato leaf miner insect (*Tuta absoluta*), with VGG16 achieving the highest classification accuracy of 91.9% for tomato plant leaves (Mkonyi *et al.*, 2020). These studies collectively demonstrate the increasing role of DL in automating pest and disease diagnostics and management, offering a sustainable, data-driven approach to agriculture.

Weed management

Weeds significantly reduce agricultural productivity by competing with crops for vital resources such as water, light, and nutrients, necessitating effective management strategies to ensure optimal yields. Precision weed control, particularly through targeted herbicide application, is essential to minimize chemical use and environmental impact. Traditional robotic weeders often rely on detecting crop row patterns, yet their efficiency diminishes under dense weed conditions, where occlusion disrupts row visibility. In response, computer vision technologies have gained traction for enabling accurate weed identification and localization, forming the foundation for site-specific management practices. Recent studies have demonstrated the adaptability of vision-based models in agronomic tasks such as differentiating crop species from weeds or off-types (Dhanya *et al.*, 2022; Sau *et al.*, 2019). Table 17.7 summarizes some of the significant previous studies investigating the applications of DL in weed management.

The advent of DL has further advanced weed detection capabilities, outperforming traditional image-processing approaches that rely on manual feature extraction (e.g. color, texture, shape) and ML classifiers such as SVM. DNNs such as AlexNet, VGGNet, and ResNet have shown superior performance by autonomously learning multi-scale spatial and semantic features from raw images (Krizhevsky *et al.*, 2017). CNNs, in particular, have demonstrated high efficacy in plant species classification. For example, Hall *et al.* (2015) used CNNs to classify leaves from 32 crops and weed species (Hall *et al.*, 2015), while Kussul *et al.* (2017) distinguished weed species across multiple crops, including

Table 17.7. Applications of DL in weed management. Author's own table.

Application	Plant species	Methodology	Results	References
Crop–weed segmentation	Sugar beet	GAN + segmentation (U-Net, BONNET, SEGNET, U-Net-RESNET)	IoU improved significantly with synthetic data; all models performed well	Fawakherji *et al.*, 2020
Weed classification	Corn, wheat, and sugar beet	Hybrid network with VGGNet, AlexNet	Accuracy: 98.23%	Chavan Abhijeet, 2018
Weed classification	Multiple plant species	VGG16, SVM	VGG16 outperformed SVM	Sunil *et al.*, 2022
Weed detection	Spinach and beans	ResNet18 + linear iterative clustering	Accuracy: 94.84% (spinach), 95.70% (beans) with supervised labeling	Bah *et al.*, 2018
Weed detection	Corn	NVDI, YOLOv3 Detector	Accuracy: 98.00%	Etienne Dharmendra, 2019
Weed detection	Sugar beet and oilseed plants	Encoder–decoder CNN with RGB input	Accuracy: 96.06% and 96.12% across two datasets	A. Wang *et al.*, 2020
Weed detection	Lettuce	Compared Mask R-CNN with HOG SVM and YOLOv3	Mask R-CNN: accuracy 98%	Osorio *et al.*, 2020
Weed detection	Sugar beet and oilseed plants	FCN	Accuracy: 88.91% (pixel-wise), 96.12% (object-wise) segmentation	Huang *et al.*, 2020
Weed detection	Rice	VGG-based FCN	Achieved the highest segmentation accuracy	Veeranampalayam Sivakumar *et al.*, 2020
Weed detection	Soybean	SSD, Faster R-CNN	Faster R-CNN had the best detection and inference time	Asad and Bais, 2020
Weed detection	Canola	U-Net, VGG16, ResNet50	High-resolution images improved performance	Peng *et al.*, 2022
Weed detection	Rice	WeedDet (ResNet34-based)	Accuracy: 94.1%, outperformed YOLOv3	Chavan Abhijeet, 2018
Weed identification	Rice	SqueezeNet vs FCN and U-Net	SqueezeNet: accuracy 92.7%	Fawakherji *et al.*, 2020
Weed identification	Canola, corn, and radish	K-FLBPCM with VGG16, VGG19, ResNet18, Inception	K-FLBPCM outperformed other CNN models	Le *et al.*, 2020

FCN, fully convolutional network; GAN, generative adversarial network; HOG, histogram of oriented gradients; K-FLBPCM, filtered local binary patterns with contour masks and coefficient *k*; NDVI, normalized difference vegetation index; R-CNN, region-based convolutional neural network; RGB, red, green, blue (color model); SEGNET, segmentation network; SSD, single shot multibox detector; SVM, support vector machine; U-Net, U-shaped convolutional neural network; U-Net-RESNET, U-Net with residual network; VGG, Visual Geometry Group network; YOLO, You Only Look Once.

wheat, corn, and sunflower (Kussul *et al.*, 2017). Mortensen *et al.* (2016) applied a modified VGG16 to differentiate barley, grasses, oil crops, and weeds (Mortensen *et al.*, 2016), and dos Santos Ferreira *et al.* (2019) successfully categorized soybean field weeds into broadleaf and grass types using CNNs (dos Santos Ferreira *et al.*, 2019). Similarly, Moshia and Newete (2019) demonstrated crop–weed discrimination via row-guided robots equipped with DL networks (Moshia and Newete, 2019).

Bah *et al.* (2018) developed field-deployable systems integrating UAVs and CNNs to enable real-time weed detection and efficient dataset generation (Dian Bah *et al.*, 2018). Transfer learning techniques have been effectively employed for weed classification by Kounalakis *et al.* (2019) (Kounalakis *et al.*, 2019), while Partel *et al.* (2019) demonstrated the capabilities of smart sprayers integrated with AI and machine vision for selective herbicide application (Partel *et al.*, 2019). Milioto *et al.* (2017) introduced an RGB + NIR CNN model to distinguish sugar beet from surrounding weeds, later enhancing performance through a sequential encoder–decoder Fully convolutional network (FCN) using 3D convolution, which yielded an 11–14% improvement in F1-score. Incorporating vegetation indices further improved model generalization across varying field conditions (Milioto *et al.*, 2017). Garibaldi-Márquez *et al.* (2022) expanded this work by developing a CNN system that identified crops, broadleaf weeds (BLW), and narrow-leaf weeds (NLW) across multiple growth stages using a dataset of 15,240 images captured in natural cornfield environments (Garibaldi-Márquez *et al.*, 2022). Collectively, these advancements underscore the increasing robustness and scalability of DL-based weed management systems, offering practical solutions for PA in complex field conditions.

Abiotic stress detection

Identifying abiotic stress in plants, such as drought, salinity, and nutrient deficiencies, is essential for minimizing yield losses and ensuring crop health. These stressors often manifest visually through changes in the plant canopy, such as color, texture, and structure alterations. DL and computer vision technologies have become powerful tools for detecting these stress-induced symptoms. These technologies enable early stress detection by analyzing visual cues such as changes in leaf color and temperature, offering a non-invasive and high-throughput alternative to traditional, labor-intensive diagnostic methods (V. G. Dhanya *et al.*, 2022). For instance, drought causes stomatal closure, reducing photosynthesis and transpiration, which leads to measurable changes in leaf color and temperature (Nilsson, 1995). Similarly, nutrient deficiencies alter leaf texture and pigmentation, which can be captured and analyzed using image-based techniques (Xu *et al.*, 2011). These advancements have paved the way for more efficient decision-making in PA by providing timely and accurate stress detection. Table 17.8 summarizes some of the significant previous studies investigating the applications of DL in stress detection.

CNNs, a key DL technique, have proven particularly effective in identifying abiotic stress

Table 17.8. Applications of DL in stress detection. Author's own table.

Application	Plant species	Methodology	Results	References
Drought and water stress detection	Corn	DCNN	Accuracy: 98.14% and 95.95% for identification and classification	An *et al.*, 2019
Drought and water stress detection	Corn	CNN	Accuracy: 98% for the classification of three treatment types	Soffer *et al.*, 2021
Drought and water stress detection	Corn, okra, and soybean	AlexNet, GoogLeNet, Inception v3	Accuracy: 98.3% (AlexNet), 97.9% (GoogLeNet), 94.1% (Inception v3)	Chandel *et al.*, 2021
Nutrient deficiency detection	Sorghum	Custom DL (23 layers)	Accuracy: 8.25% better than traditional ML techniques	Azimi *et al.*, 2021
Nutrient deficiency detection	Rice	DS-CNN, ND-CNN	ND-CNN performed better and avoided overfitting	Zhu *et al.*, 2022

CNN, convolutional neural network; DCNN, deep convolutional neural network; DS-CNN, de-striping convolutional neural network; ND-CNN, nitrogen diagnosis convolutional neural network.

conditions such as water and nutrient stress. CNNs utilize layered convolution operations to extract deep features from input images, transforming them into actionable outputs that differentiate between healthy and stressed plants (Bauer *et al.*, 2019). Numerous studies have demonstrated the efficacy of CNNs in this area, highlighting their ability to detect stress-induced changes in plant morphology and physiology (Anami *et al.*, 2020; Abdalla *et al.*, 2021). Additionally, optical imaging techniques such as fluorescence, digital imaging, LIDAR, thermography, and multispectral and hyperspectral imaging have further improved stress detection capabilities. For example, thermal imaging can capture temperature shifts caused by altered transpiration and stomatal behavior, offering a reliable method for identifying stress (Lindenthal *et al.*, 2005).

To support real-time, large-scale stress monitoring, thermal and spectral sensors are increasingly deployed on unmanned aerial vehicles (UAVs) and ground automated vehicles (GAVs). These systems enable field-level surveillance, providing detailed, real-time data for stress detection and management (Wernick *et al.*, 2010). Integrating supervised ML algorithms, such as shape segmentation, feature extraction, and classification, further improves stress detection accuracy by identifying optimal decision boundaries in high-dimensional feature spaces. These advancements in sensing and DL technologies represent a significant leap forward in developing efficient and scalable abiotic stress detection systems, contributing to smart and sustainable agriculture practices.

Crop yield prediction

Crop yield prediction is a complex task influenced by genetic, environmental, and climatic factors, often requiring the integration of large, heterogeneous datasets. DL approaches have become increasingly prevalent due to their ability to model intricate non-linear interactions and unobserved variability. One such approach involves a semiparametric DNN architecture, combining the flexibility of non-parametric DL modeling with the interpretability of parametric structures, which enables more accurate yield forecasting under real-world conditions (Shaikh *et al.*, 2022; Mohyuddin *et al.*, 2024). Additionally, several studies have highlighted the effectiveness of DL models in yield prediction across various crops, utilizing different data sources such as multispectral imagery, time-series data, and computer vision techniques. Table 17.9 summarizes some of the significant previous studies investigating the applications of DL in crop yield prediction.

For instance, Torres-Tello and Ko (2021) proposed a DL-based method for spring wheat yield prediction using multispectral imagery, identifying late-heading, late-flowering, dough development, and harvesting stages as key periods for accurate forecasting. Their model emphasized the importance of red and red-edge bands in enhancing model performance (Torres-Tello and Ko, 2021). Similarly, Yang *et al.* (2019) demonstrated that CNNs trained on UAV-captured RGB and multispectral images effectively estimated rice grain yield at the ripening stage, outperforming traditional regression models based on vegetation indices (M. Yang *et al.*, 2019).

Shifting focus from cereals to horticultural crops, Zhou *et al.* (2021) developed an autonomous strawberry flower detection system using a Faster region-based CNN (Faster R-CNN) for counting strawberry blossoms and classifying mature and immature fruits. The model achieved a counting accuracy of 84.1%, demonstrating its potential to reduce labor costs and improve yield prediction efficiency (Zhou *et al.*, 2021).

For more dynamic yield prediction, De Alwis *et al.* (2019) integrated an attention mechanism with an LSTM network to identify influential factors affecting tomato yield in smart farming environments, capturing temporal dependencies relevant to crop growth and yield dynamics (De Alwis *et al.*, 2019). Similarly, Jeong *et al.* (2022) applied a hybrid DL architecture combining LSTM and 1D-CNN layers for yield prediction in North and South Korea, capturing regional yield variability and forecasting up to 2 months before harvest with strong performance metrics (R^2 = 0.859, Nash–Sutcliffe efficiency = 0.858, RMSE = 0.605 Mg/ha) (Jeong *et al.*, 2022). Moreover, Reddy and Kumar (2021) developed the BBI model for rice yield prediction, an end-to-end hybrid architecture combining two back-propagation neural networks (BPNNs) with an independently recurrent neural network (IndRNN). Using a six-layer structure, the model achieved the root mean square error (RMSE) and lowest mean absolute error (MAE) values of

Table 17.9. Applications of DL in crop yield prediction. Author's own table.

Application	Plant species	Methodology	Results	References
Flower-based yield estimation	Strawberry	Faster R-CNN	DL counting accuracy: 84.1%, occlusion rate: 13.5%	Chen *et al.*, 2019
Flower-based yield estimation	Grape	SegNet + VGG19 encoder	R^2 = 0.91 between actual and detected flowers	Palacios *et al.*, 2020
Fruit detection and counting	Apple	CNN + watershed algorithm	Count estimates using CNN and WS with R^2 = 0.826	Stein *et al.*, 2016
Fruit detection and counting	Tomato	Modified Inception-based model	Average test accuracy: 91%	Rahnemoonfar and Clay, 2017
Fruit detection and counting	Orchard	DL models	DL outperformed traditional models	A. Koirala *et al.*, 2019
Fruit detection and counting	Kiwi	SSD + MobileNetV2, Inception v3 (quantized versions)	TDR: 90.8%, 89.9%, 87.6%; recall: 72.8%	Zhou *et al.*, 2020
Fruit detection and counting	Apple	R-CNN, YOLO	R-CNN outperformed YOLO in fruit counting accuracy	Lu *et al.*, 2022
General crop yield prediction and mapping	Soybean	3D-CNN	Case study on soybean yield estimation using 3D CNN	Terliksiz and Altylar, 2019
General crop yield prediction and mapping	Strawberry	Faster R-CNN	DL counting accuracy: 84.1%, Occlusion: 13.5%	Chen *et al.*, 2019
General crop yield prediction and mapping	Soybean	CNN-LSTM	CNN-LSTM outperformed standalone CNN and LSTM in yield prediction	Sun *et al.*, 2019
General crop yield prediction and mapping	Rice	CNN with RGB + multispectral	MAPE: 20.4%, RMSE: 0.658, R^2: 0.585	Q. Yang *et al.*, 2019
General crop yield prediction and mapping	Corn	DNNs	RMSE: 12% of avg. yield; reduced to 11% with perfect weather data	Khaki and Wang, 2019
General crop yield prediction and mapping	Maize and soybean	CNN-RNN	RMSE: 9% (corn), 8% (soybean); outperformed other methods	Khaki *et al.*, 2020
General crop yield prediction and mapping	Soybean	DNN	Accuracy: R^2 = 0.720, RMSE = 15.9%	Maimaitijiang *et al.*, 2020
General crop yield prediction and mapping	Cotton	Faster R-CNN, SSD, SSD Lite	Mean percentage error: 8.54%	Tedesco-Oliveira *et al.*, 2020
General crop yield prediction and mapping	Kiwi	MobileNetV2, Quantized MobileNetV2, Inception v3	TDR: 90.8%, 89.7%, 72.8% across three models	Zhou *et al.*, 2020
General crop yield prediction and mapping	Wheat	CNN	R^2 = 0.75, RMSE = 732 kg/ha; also achieved= 0.77 and RMSE = 721 kg/ha	X. Wang *et al.*, 2020

General crop yield prediction and mapping	Soybean	LSTM	MAE = 0.24 Mg/ha (DOY 64); LSTM outperformed all models except OLS on DOY 16; MAE rose to 0.42 Mg/ha when forecast made earlier (DOY 16)	Schwalbert *et al.*, 2020
General crop yield prediction and mapping	Wheat	DL models	Outperformed ridge regression and RF; enabled interpretation of yield-related variables	Wolanin *et al.*, 2020
General crop yield prediction and mapping	Rice	BPNN + IndRNN	MAE/RMSE for summer rice: 0.0044/0.0057; for winter rice: 0.0074/0.0192 (best performance with six network layers)	Chu and Yu, 2020
General crop yield prediction and mapping	Coffee	SVM, GBR, RFR, PLSR, NEAT	December datasets identified as suitable for yield prediction	Barbosa *et al.*, 2021
General crop yield prediction and mapping	Not specified	CNN, RNN	Accuracy: 92%; model preserved original data distribution and outperformed others	Elavarasan and Durai Raj Vincent, 2021
General crop yield prediction and mapping	Strawberry	YOLOv3	UAV: mAP = 0.88 (2 m), AP = 0.93 (fully matured); Ground camera: mAP = 0.89, AP = 0.94 (fully matured)	Zhou *et al.*, 2021
General crop yield prediction and mapping	Not specified	GA-assisted DL	Outperformed gradient-based methods in convergence speed and accuracy	Bi and Hu, 2021
General crop yield prediction and mapping	Rice	LSTM, 1D-CNN	R^2 = 0.86, RMSE = 0.61 Mg/ha, demonstrating reliable early prediction	Jeong *et al.*, 2022

BPNN, back-propagation neural network; CNN, convolutional neural network; CNN-LSTM, convolutional neural network-long short-term memory; DL, deep learning; DNN, deep neural network; GA, genetic algorithm; GBR, gradient boosting regression; IndRNN, independently recurrent neural network; LSTM, long short-term memory; NEAT, NeuroEvolution of augmenting topologies; PLSR, partial least squares regression; R-CNN, region-based convolutional neural network; RFR, random forest regression; SSD, single shot multibox detector; SVM, support vector machine; VGG, Visual Geometry Group network; YOLO, You Only Look Once.

0.0057 and 0.0044 for summer rice and 0.0192 and 0.0074 for winter rice, respectively, demonstrating strong seasonal adaptability and high predictive performance (Reddy and Kumar, 2021).

Lastly, Bi and Hu (2021) proposed a genetic algorithm (GA)-assisted DL approach for agricultural yield prediction, addressing the limitations of conventional optimization methods and achieving superior convergence speed and predictive accuracy (Bi and Hu, 2021). These studies underscore the growing role of DL in advancing crop yield prediction, showcasing its versatility across diverse agricultural systems and its ability to automate and enhance the accuracy of yield estimation.

17.3.2 Harvesting

Harvesting is a critical phase in fruit and vegetable production, where factors such as fruit size, skin color, firmness, and maturity directly influence outcomes. Recent developments highlight the growing role of DL in optimizing this process. DL-driven vision systems, particularly autonomous harvesting robots, have improved harvesting precision and reduced labor costs. DL-based models also provide early and accurate yield predictions, essential for postharvest planning, policymaking, and crop management (Al-Gaadi *et al.*, 2016; Chlingaryan *et al.*, 2018; Wei *et al.*, 2020). Table 17.10 summarizes some of the significant previous studies investigating the applications of DL in harvesting.

DL models, particularly CNNs, have significantly advanced fruit detection, revolutionizing PA and enhancing the efficiency of fruit recognition and yield estimation. Early research by Habaragamuwa *et al.* (2018) employed a region-based CNN (R-CNN) with AlexNet to differentiate ripe from unripe strawberries in greenhouse environments, achieving an average precision of 82.61% (Habaragamuwa *et al.*, 2018). Bresilla *et al.* (2019) developed a CNN model using the YOLO algorithm for real-time fruit detection on trees. Trained on over 5000 annotated real and synthetic images of apple and pear trees, the model achieved over 90% accuracy, demonstrating the potential of DL-based object detection frameworks to enhance harvesting efficiency and support data-driven automation in agriculture (Bresilla *et al.*, 2019).

Computer vision has been extensively applied to automate fruit counting and size estimation in orchard crops, particularly apples, to improve yield forecasting accuracy and reduce labor demands. Cheng *et al.* (2017) developed two BPNN models for early and late-season apple yield prediction by analyzing canopy images of "Gala" and "Pinova" cultivars. Their method extracted features such as the total cross-sectional area of fruits and foliage using image segmentation, achieving R^2 values of up to 0.83 and mean absolute percentage errors (MAPE) as low as 8.9% near harvest, demonstrating strong predictive performance across different phenological stages (Cheng *et al.*, 2017). Complementarily, Kanwal *et al.* (2019) proposed a non-learning-based visual method for yield estimation in complex orchard environments, addressing challenges such as overlapping fruits and background clutter. Their approach combined color-based segmentation in the hue and saturation (HS) color space with morphological filtering, Gaussian smoothing, and Hough Transform-based detection of individual apple centers. This method effectively localized and counted tightly clustered apples, outperforming contour-based techniques and proving effective in real-world orchard conditions (Kanwal *et al.*, 2019). Collectively, these studies underscore the potential of both learning-based and learning-free computer vision techniques to enhance yield estimation under variable and noisy field conditions, supporting precision harvesting applications.

As the field evolved, several models emerged to improve fruit recognition and yield estimation. Koirala *et al.* (2019) introduced Mango-YOLO, based on YOLO-v2 and YOLO-v3, which outperformed Faster R-CNN, SSD, and earlier YOLO versions in memory utilization, speed, and accuracy (Anand Koirala *et al.*, 2019). Liang *et al.* (2018) enhanced mango and almond fruit detection by applying the SSD framework with data augmentation and smaller default boxes, yielding better results than the original SSD (Liang *et al.*, 2018). Stein *et al.* (2016) used a Faster R-CNN with a VGG16 backbone to identify fruits in mango orchards, achieving $R^2 = 0.90$ when trained on images from multiple perspectives (Stein *et al.*, 2016). Tian *et al.* (2019) integrated DenseNet with YOLO-v3 to detect apples at different developmental stages, with the model trained on a single-stage dataset outperforming one trained

Table 17.10. Applications of DL in harvesting. Author's own table.

Application	Plant species	Methodology	Results	References
Fruit detection	Mango	Faster R-CNN with VGG16	$R^2 = 0.90$	Stein *et al.*, 2016
Fruit detection	Mango, almond	SSD with data augmentation	Higher accuracy than the original SSD	Liang *et al.*, 2018
Fruit detection	Strawberry	R-CNN with AlexNet	82.61% average precision	Habaragamuwa *et al.*, 2018
Fruit detection	Apple, pear	YOLO	Over 90% accuracy	Bresilla *et al.*, 2019
Fruit detection	Mango	Mango-YOLO (YOLOv2/ YOLOv3)	Improved speed, memory, and accuracy	A. Koirala *et al.*, 2019
Fruit detection	Apple	DenseNet with YOLOv3	Higher F1-score with single-stage training	Tian *et al.*, 2019
Fruit recognition	Multiple plant species	CNN	95.68% accuracy	Leonardo *et al.*, 2018
Fruit detection	Apple	LedNet + ResNet110	86% accuracy	Kang and Chen, 2020
Fruit count and yield estimation	Apple	–	$R^2 = 0.83$, MAPE = 8.9% near harvest, RMSE ≈ 2.3–2.6 kg/tree	Cheng *et al.*, 2017
Fruit count and yield estimation	Apple	–	Successfully counted overlapping apples; outperformed contour-based methods in complex orchard environments	Kanwal *et al.*, 2019
Fruit count and yield estimation	Multiple plant species	Faster R-CNN + LSTM	SE: 6.59%, per tree yield error: 4.53%	Apolo-Apolo *et al.*, 2020
Fruit classification	Multiple plant species	VGG16	Outperformed 6-layer CNN	Hossain *et al.*, 2019
Fruit classification	Multiple plant species	CNN with feature fusion	Improved generalizability	Kirk *et al.*, 2020
Fruit maturity	Date	AlexNet, VGGNet	99.01% accuracy	Altaheri *et al.*, 2019
Maturity classification	Date	IHDS	99.4% accuracy, 99.4% F1-score, 99.7% sensitivity and precision	Faisal *et al.*, 2020
Autonomous harvesting	Apple	SSD + stereo cameras	90% detection, 16 s per cycle	Onishi *et al.*, 2019

CNN, convolutional neural networks; IHDS, intelligent harvesting decision system; R-CNN, region-based convolutional neural networks; SE, standard error; SSD, single shot multibox detector; YOLO, You Only Look Once.

on a multistage dataset in F1-score (Tian *et al.*, 2019). Similarly, Leonardo *et al.* (2018) applied a CNN architecture for fruit recognition, achieving an accuracy of 95.68% (Leonardo *et al.*, 2018). These studies demonstrate the diverse applications of CNN-based models, including YOLO, SSD, and Faster R-CNN variations, in advancing fruit detection and yield estimation.

Recent studies have introduced hybrid models for more comprehensive fruit analysis. Kang and Chen (2020) developed C-RCNN and LedNet for apple detection, incorporating advanced backbones such as ResNet110 and Darknet-53. LedNet with ResNet110 achieved the highest detection accuracy at 86%, while the lightweight network (LW-Net) showed superior computational efficiency

(Kang and Chen, 2020). Apolo-Apolo *et al.* (2020) developed a DL-based pipeline combining Faster R-CNN with an LSTM model to automate citrus fruit detection, counting, and size estimation. The model achieved a standard error (SE) of 6.59% for fruit counting and 4.53% for per-tree yield estimation, with a standard deviation (SD) of 0.97 kg per tree. At the orchard level, it maintained strong performance, achieving an SE of 7.22% and an SD of 4083.58 kg, significantly outperforming technician estimates, which had an SE of 13.74%, thereby demonstrating the effectiveness of DL in enhancing agricultural yield prediction (Apolo-Apolo *et al.*, 2020). These hybrid models highlight the growing complexity and accuracy of DL approaches in fruit detection and yield prediction, demonstrating the effectiveness of YOLO and SSD in precision horticulture.

Beyond detection, DL models have shown strong potential in fruit classification, offering improved accuracy and generalizability in agricultural settings. Shamim Hossain *et al.* (2019) compared two architectures: a six-layer CNN and a fine-tuned VGG16 model pretrained on a large dataset. Using a dataset of 2633 color images and a custom set of 5946 images across 10 fruit classes, VGG16 outperformed the compact CNN, achieving higher classification accuracy (Shamim Hossain *et al.*, 2019). Kirk *et al.* (2020) proposed a feature fusion strategy to improve DL model performance on unseen data by incorporating bioinspired features, enhancing generalizability. Their dataset comprised 6189 images, with 150 manually annotated for evaluation (Kirk *et al.*, 2020).

DL techniques have been increasingly applied to fruit maturity assessment and yield prediction, aiding harvesting decisions. Altaheri *et al.* (2019) developed a classification system for date fruit ripeness using transfer learning with AlexNet and VGGNet. Their dataset included 8072 images of five date varieties captured under varying conditions. VGG16 achieved the highest accuracy of 99.01% with a processing speed of 20.6 ms, effectively categorizing fruits by maturity and harvestability (Altaheri *et al.*, 2019). Similarly, Faisal *et al.* (2020) proposed an intelligent harvesting decision system (IHDS) using DL methods and computer vision to accurately classify date fruits into seven distinct maturity stages. Trained on datasets from the Center of Smart Robotics Research, the IHDS achieved accuracy, F1-score, sensitivity, and precision of 99.4%, 99.4%, 99.7%, and 99.7%, respectively (Faisal *et al.*, 2020). These advancements demonstrate the effectiveness of DL in automating maturity detection and facilitating precise harvesting decisions in agriculture.

Recent advances in DL have enabled the development of more accurate and efficient autonomous harvesting systems. Onishi *et al.* (2019) proposed a fruit harvesting system using robotic arms integrated with SSDs and stereo cameras for fruit detection and retrieval. The system achieved a 90% fruit detection rate in "Fuji" apple orchards and completed individual harvesting cycles in about 16 seconds by aligning the hand axis of the robotic arm with the identified fruit (Onishi *et al.*, 2019). These advancements highlight the growing significance of DL models in streamlining fruit detection, classification, and autonomous harvesting, making them invaluable tools in modern PA.

17.3.3 Postharvesting

Postharvest management represents a critical phase in the agricultural value chain, and inadequate attention to this stage can negate prior crop production and harvesting efforts, resulting in substantial financial losses. Fruit grading standards vary internationally, reflecting country-specific norms and regulatory frameworks (Mohyuddin *et al.*, 2024). As part of efforts to reduce postharvest losses and improve profitability in horticultural production, DL technologies have been increasingly applied to automate and enhance the accuracy of crop grading systems. Ucat and Cruz (2019) developed a classification model for Cavendish bananas utilizing Python, OpenCV, and TensorFlow, a DL framework. Their system categorized bananas into three distinct classes—Class A (big-hand), Class B (small-hand), and Cluster (partial-hand)—and achieved a classification accuracy exceeding 90% (Ucat and Dela Cruz, 2019).

Meshram *et al.* (2021) proposed a DL-based framework titled "MNet: Merged Net" to classify postharvest fruit images. Leveraging a self-constructed dataset of 12,000 images across six major Indian fruit categories, the framework aimed to improve classification accuracy through a merged network architecture, indicative of a DL approach (Meshram *et al.*, 2021). These studies underscore the growing relevance of DL models in automating quality assessment tasks in the postharvest phase of agriculture. Table 17.11 summarizes some of the significant previous studies investigating the applications of DL in postharvesting.

17.4 Challenges and Limitations of DL in SA

17.4.1 Limited availability and quality of training data

One of the most critical challenges in deploying DL in agriculture is the limited availability and quality of labeled training data, directly impacting model accuracy, generalizability, and real-world applicability. Agricultural datasets are often sparse, noisy, and imbalanced due to the variability in sensor outputs, imaging conditions, and environmental factors across different regions (Li and Chao, 2021). Moreover, collecting and annotating high-quality data is time-consuming, labor-intensive, and costly, especially when expert knowledge is required to label subtle distinctions in crop health, pest presence, or disease symptoms (Shaikh *et al.*, 2022). This scarcity and imbalance in data severely hinder DL tasks such as disease detection, yield prediction, and object classification, leading to biased models and suboptimal decisions. Synthetic data generation techniques, including GANs and conditional GANs (CGANs), have emerged as promising solutions to augment data and enhance diversity (Olatunji *et al.*, 2020; Zhu *et al.*, 2020; Cui *et al.*, 2021). However, challenges remain in ensuring the realism, consistency, and domain-specific relevance of these artificially generated datasets. Furthermore, the absence of standardized data collection protocols across agricultural systems exacerbates the problem, limiting the scalability and adaptability of DL models to new environments and use cases.

17.4.2 Domain adaptation and environmental variability

DL models in agriculture face significant challenges in generalizing across diverse environmental conditions due to domain adaptation issues. Agricultural environments are inherently dynamic, varying with region, season, crop type, soil composition, and local farming practices, drastically affecting model performance (Tian *et al.*, 2021). Models trained on data from a specific geographic or temporal context often fail when applied to different regions or subsequent growing seasons, where variations in climate, lighting, and crop phenology alter the visual and contextual features of input data (V. Dhanya *et al.*, 2022). For instance, a model trained on crop images in one location may underperform in another due to differences in soil or weather conditions. Additionally, seasonal fluctuations introduce temporal variability that further challenges model consistency; features learned in one season may not be applicable in another, necessitating frequent retraining with up-to-date, location-specific data (Sinha and Lee, 2024). These limitations hinder the scalability of DL technologies in agriculture and underscore the need for developing more robust and adaptive models capable of transferring learned representations across varied agricultural domains.

17.4.3 Model interpretability and transparency

A critical barrier to the widespread adoption of DL in agriculture is the lack of model interpretability

Table 17.11. Applications of DL in postharvesting. Author's own table.

Application	Plant species	Methodology	Results	References
Postharvest grading	Banana	OpenCV + TensorFlow	Over 90% classification accuracy	Ucat and Dela Cruz, 2019
Postharvest grading	Multiple plant species	MNet (Merged Net)	High accuracy with 12,000-image dataset	Meshram *et al.*, 2021

Mnet, merged net.

and transparency, often referred to as the "black-box" problem. Most DL models, particularly DNNs, operate with limited visibility into their internal decision-making processes, making it difficult for farmers, agronomists, and other stakeholders to trust and act upon their outputs (V. Dhanya *et al.*, 2022). This opacity becomes especially problematic in high-stakes applications such as pest management, disease diagnosis, or yield forecasting, where inaccurate or unexplained recommendations can result in substantial economic or environmental consequences (Kundu *et al.*, 2021). For instance, without a clear rationale behind a pesticide suggestion or disease classification, users may hesitate to implement the recommendations, undermining the practical utility of the technology. As a result, researchers are increasingly developing explainable AI (XAI) approaches that provide insights into model predictions through mechanisms such as attention layers, saliency maps, or visual decision paths (Sinha and Lee, 2024). Nonetheless, balancing the trade-off between interpretability and model complexity remains a significant challenge, and advancing transparent DL systems is essential for fostering user trust and facilitating the responsible deployment of AI in agriculture.

17.4.4 Computational complexity and resource constraints

The computational complexity and resource demands of DL models pose significant challenges to their widespread adoption in agriculture, particularly in resource-constrained or rural settings. DL-based computer vision applications typically require substantial processing power, memory, and energy for both training and inference, often necessitating high-performance computing infrastructure or cloud-based platforms (V. Dhanya *et al.*, 2022). While cloud computing offers scalability, it introduces latency issues that hinder real-time decision-making, an essential feature for field-based tasks such as pest detection or climate adaptation. This is particularly problematic for small-scale farms or remote regions lacking reliable internet connectivity and advanced computational infrastructure (Zeng *et al.*, 2022). Edge computing has emerged as a potential solution by enabling on-device inference, but edge devices such as IoT sensors and embedded systems often have limited processing capacity and memory. Running complex DL models on these devices without compromising performance remains a significant challenge. Approaches such as model compression, algorithm optimization, and lightweight architectures are being explored to improve efficiency. However, further research and development are needed to ensure that DL systems are accessible and functional in real-world agricultural environments (Sinha and Lee, 2024).

17.4.5 Practical challenges in real-world environments

Deploying DL models in real-world agricultural environments presents a range of practical challenges arising from the complex and dynamic nature of field conditions. Visual disturbances such as occlusion by leaves or branches, variations in illumination due to weather or time of day, changes in camera viewpoint, and cluttered or heterogeneous backgrounds can significantly degrade the accuracy and robustness of DL applications. These issues are particularly problematic for tasks such as fruit counting, plant segmentation, and weed identification, where precise visual recognition is essential (V. Dhanya *et al.*, 2022). To overcome these limitations, there is a growing need to develop more resilient model architectures and advanced image preprocessing techniques that can adapt to environmental variability and maintain consistent performance under field conditions.

17.4.6 Hyperparameter optimization

The performance and reliability of DL models in agriculture are highly dependent on effective hyperparameter optimization, which involves selecting suitable values for learning rates, batch sizes, activation functions, and network architectures. This process is inherently complex and typically demands expert knowledge and considerable trial-and-error experimentation to identify optimal configurations (V. Dhanya *et al.*, 2022). Inadequate hyperparameter tuning can lead to underfitting, overfitting, or inefficient training, compromising model performance in agricultural tasks. Although research into automated optimization techniques, such as Bayesian optimization, grid search, and reinforcement learning-based methods, is ongoing, these approaches often remain computationally intensive due to extensive search spaces and a lack of generalizability across diverse agricultural datasets and DL architectures. Further advancements are needed to streamline tuning for accessible and efficient deployment in smart farming applications.

17.4.7 Security, privacy, and data integrity

Integrating DL with IoT devices, cloud platforms, and edge computing in agricultural systems introduces significant security, privacy, and data integrity challenges. As these technologies increasingly handle sensitive geospatial, production, and environmental data, ensuring secure data transmission and system reliability becomes critical (V. Dhanya *et al.*, 2022). Without robust cybersecurity frameworks, DL-based agricultural systems remain vulnerable to threats such as unauthorized access, data tampering, and operational disruptions, which can compromise decision-making and lead to economic or environmental losses. Addressing these vulnerabilities requires the implementation of end-to-end encryption, secure communication protocols, and real-time monitoring systems to protect the integrity and confidentiality of agricultural data throughout the DL pipeline.

17.4.8 Economic and maintenance constraints

Economic and maintenance constraints present substantial barriers to adopting DL technologies in agriculture, particularly for smallholder farmers and those in low-income regions. The initial implementation of DL systems involves significant costs related to hardware procurement, software acquisition, sensor installation, data collection infrastructure, and model training—expenses that can be prohibitive for those operating on limited financial margins (Zeng *et al.*, 2022). Moreover, integrating DL solutions into existing farm management systems adds to the complexity and cost of deployment. Beyond these upfront investments, the long-term sustainability of DL systems requires continuous maintenance, including regular data updates and model retraining to maintain accuracy amid evolving environmental conditions (Sinha and Lee, 2024). These ongoing resource demands highlight the need for cost-effective, scalable solutions to ensure that DL technologies remain accessible and viable for various agricultural stakeholders.

17.5 Future Scope

17.5.1 Disease stage identification

One of the most critical aspects of plant disease identification is determining the stage of the disease, as each disease typically progresses through multiple phases. While most existing research has concentrated on disease type identification, relatively few studies have addressed identifying disease stages. Accurate stage recognition enables systems to recommend targeted interventions based on disease progression. Furthermore, incorporating disease forecasting into such systems can assist farmers and agriculturists in implementing timely preventive measures, thereby reducing the extent of damage and mitigating associated losses.

17.5.2 Quantification of a disease

The quantification of plant diseases represents an important and promising area for further research. Although substantial progress has been made, only a few studies have accurately measured the extent of disease impact. Disease quantification is particularly valuable, as it enables the implementation of corrective measures proportional to the severity of infection. This approach also supports the responsible use of pesticides by identifying the actual fraction of a crop affected, thereby minimizing excessive chemical application. Currently, many farmers resort to pesticide use without proper disease analysis or quantification, posing serious health risks. The development of practical image-processing applications can significantly aid in assessing the necessity and extent of chemical intervention, thus promoting more sustainable and health-conscious agricultural practices.

17.5.3 Mobile and online applications

Numerous solutions for plant disease identification have been proposed in the literature, yet only a few publicly accessible portals and mobile applications are currently available online. Among these, Leaf Doctor and Assess Software are two notable examples that are available for free. These applications, however, function optimally only with images captured against a flat, black background, which restricts their applicability in real-world agricultural settings. Despite this limitation, such online systems and mobile apps are critical tools for identifying plant diseases. They enable farmers to detect specific illnesses and facilitate the generation of analysis reports, which can be forwarded to agricultural experts for professional evaluation and recommendations, thereby enhancing disease management practices.

17.5.4 Exploring transfer learning to increase data size

Current trends in computer vision, particularly the shift toward DL approaches, remain somewhat limited in their effectiveness for plant disease diagnosis. This is mainly due to the complexity of data, especially during training, where transfer learning offers the most practical solution. Employing a heterogeneous domain strategy can support knowledge transfer in such contexts. Methods such as LSTM networks, temporal pooling, optical flow frames, and 3D convolution present promising directions for advancing automated plant disease identification. However, the field still requires more robust and well-designed methodologies, with data augmentation warranting further investigation.

17.5.5 Incorporating new technologies in agriculture

Agri-food supply chain traceability is a critical domain that warrants increased attention and demonstrates significant potential for integration with IoT technologies. The most commonly utilized IoT communication technologies in existing literature include ZigBee, LoRa, and Wi-Fi, while emerging high-speed communication technologies such as NB-IoT and 5G are anticipated to contribute substantially to the modernization and intelligence of agricultural production. With continuous technological advancements, IoT-based agriculture has considerable scope for innovation and the implementation of more effective solutions. In particular, there is a growing demand for low-cost systems characterized by low maintenance, autonomous operation, energy efficiency, robust architectural design, and ease of use.

17.6 Conclusion

AI, particularly DL, is redefining agricultural systems by enabling real-time, data-driven decision-making and intelligent automation. DL models have demonstrated significant potential to enhance productivity and environmental sustainability through plant disease detection, precision irrigation, crop phenotyping, and yield forecasting applications. Techniques such as CNNs, transfer learning, and image-based recognition have enabled high-accuracy diagnostics using datasets such as PlantVillage and ImageNet. However, generalizability remains a challenge due to domain-specific variability.

When integrated with IoT platforms, robotics, and autonomous systems, DL provides robust tools for addressing critical challenges in agriculture, including labor shortages, resource inefficiencies, and climate variability. Although classical ML models such as SVM, ANN, and KNN remain effective for tasks involving crop, soil, and water management, DL methods consistently outperform them in handling complex, high-dimensional, and unstructured data such as images, audio, and sensor streams.

However, widespread adoption is constrained by factors including limited dataset availability, system heterogeneity, deployment costs, and infrastructure gaps in rural areas. To unlock the full potential of DL in agriculture, future efforts must prioritize scalable model development, investment in data-sharing platforms, integration with real-world farming practices, and supportive policy frameworks. A multidisciplinary approach linking data science, agronomy, and engineering is essential to building resilient, intelligent, and sustainable agricultural systems for the future.

Acknowledgments

We thank the editors and reviewers for their critical comments on the manuscript.

Author Contributions

All authors have read and agreed to the published version of the manuscript.

Funding

This research was supported by the National Natural Science Foundation of China (U21A20215 and 32488102) and the Biological Breeding-National Science and Technology Major Project of China (2023ZD0403203).

Institutional Review Board Statement

Not Applicable.

Informed Consent Statement

Not Applicable.

Data Availability Statement

Not Applicable.

Conflicts of Interest

The authors declare no conflicts of interest.

References

Abbas, A., Jain, S., Gour, M., Vankudothu, S., 2021. Tomato plant disease detection using transfer learning with C-GAN synthetic images. *Comput. Electron. Agric.* 187, 106279. https://doi.org/10.1016/j.compag.2021.106279

Abd Algani, Y., Caro, O., Bravo, L., Kaur, C., Al Ansari, Dr. Mohammed Saleh, Bala, B., 2022. Leaf disease identification and classification using optimized deep learning. *Meas. Sensors* 25, 100643. https://doi.org/10.1016/j.measen.2022.100643

Abdalla, A., Cen, H., Wan, L., Mehmood, K., He, Y., 2021. Nutrient Status Diagnosis of Infield Oilseed Rape via Deep Learning-Enabled Dynamic Model. *IEEE Trans. Ind. Informatics* 17, 4379–4389. https://doi.org/10.1109/TII.2020.3009736

Al-Gaadi, K.A., Hassaballa, A.A., Tola, E., Kayad, A.G., Madugundu, R., Alblewi, B., Assiri, F., 2016. Prediction of Potato Crop Yield Using Precision Agriculture Techniques. *PLoS One* 11, e0162219. https://doi.org/10.1371/journal.pone.0162219

Albanese, A., Nardello, M., Brunelli, D., 2021. Automated Pest Detection with DNN on the Edge for Precision Agriculture. IEEE *J. Emerg. Sel. Top. Circuits Syst.* 11, 458–467. https://doi.org/10.1109/JETCAS.2021.3101740

Albuquerque, C.K.G., Polimante, S., Torre-Neto, A., Prati, R.C., 2020. Water spray detection for smart irrigation systems with Mask R-CNN and UAV footage, in: 2020 IEEE International Workshop on Metrology for Agriculture and Forestry (MetroAgriFor). *IEEE*, pp. 236–240.

Alibabaei, K., Gaspar, P.D., Lima, T.M., 2021. Modeling Soil Water Content and Reference Evapotranspiration from Climate Data Using Deep Learning Method. *Appl. Sci.* 11, 5029. https://doi.org/10.3390/app11115029

Altaheri, H., Alsulaiman, M., Muhammad, G., 2019. Date Fruit Classification for Robotic Harvesting in a Natural Environment Using Deep Learning. *IEEE Access* 7, 117115–117133. https://doi.org/10.1109/ACCESS.2019.2936536

An, J., Li, W., Li, M., Cui, S., Yue, H., 2019. Identification and classification of maize drought stress using deep convolutional neural network. *Symmetry (Basel).* 11, 256.

Anami, B.S., Malvade, N.N., Palaiah, S., 2020. Deep learning approach for recognition and classification of yield affecting paddy crop stresses using field images. *Artif. Intell. Agric.* 4, 12–20.

Apolo-Apolo, O.E., Martínez-Guanter, J., Egea, G., Raja, P., Pérez-Ruiz, M., 2020. Deep learning techniques for estimation of the yield and size of citrus fruits using a UAV. *Eur. J. Agron.* 115, 126030. https://doi.org/10.1016/j.eja.2020.126030

Arnal Barbedo, J.G., 2019. Plant disease identification from individual lesions and spots using deep learning. *Biosyst. Eng.* 180, 96–107. https://doi.org/10.1016/j.biosystemseng.2019.02.002

Asad, M.H., Bais, A., 2020. Weed detection in canola fields using maximum likelihood classification and deep convolutional neural network. *Inf. Process. Agric.* 7, 535–545.

Ayoub Shaikh, T., Rasool, T., Rasheed Lone, F., 2022. Towards leveraging the role of machine learning and artificial intelligence in precision agriculture and smart farming. *Comput. Electron. Agric.* 198, 107119. https://doi.org/10.1016/j.compag.2022.107119

Azadnia, R., Jahanbakhshi, A., Rashidi, S., Khajehzadeh, M., Bazyar, P., 2022. Developing an automated monitoring system for fast and accurate prediction of soil texture using an image-based deep learning network and machine vision system. *Measurement* 190, 110669. https://doi.org/10.1016/j.measurement.2021.110669

Azeem Ayaz Mirani, Zaman, S., Chohan, R., Siraj, S., Lakho, S., 2023. A Review of the Internet of Things and Deep Learning in Agriculture: A Smart Agriculture Prespective. *J. Hunan Univ. Nat. Sci.* 50. https://doi.org/10.55463/issn.1674-2974.50.10.18

Azimi, S., Kaur, T., Gandhi, T.K., 2021. A deep learning approach to measure stress level in plants due to Nitrogen deficiency. *Measurement* 173, 108650. https://doi.org/10.1016/j.measurement.2020.108650

Bah, M.D., Hafiane, A., Canals, R., 2018. Deep Learning with Unsupervised Data Labeling for Weed Detection in Line Crops in UAV Images. *Remote Sens.* 10, 1690. https://doi.org/10.3390/rs10111690

Bao and Deborah, F.B., 2021. Applicability of computer vision in seed identification: deep learning, random forest, and support vector machine classification algorithms. *Acta Bot. Brasilica* 35, 17–21. https://doi.org/10.1590/0102-33062020abb0361

Barbosa, B.D.S., Costa, L., Ampatzidis, Y., Vijayakumar, V., dos Santos, L.M., 2021. UAV-based coffee yield prediction utilizing feature selection and deep learning. *Smart Agric. Technol.* 1, 100010.

Barman, U., Choudhury, R.D., 2022. Smartphone assist deep neural network to detect the citrus diseases in agri-informatics. *Glob. Transitions Proc.* 3, 392–398.

Bauer, A., Bostrom, A.G., Ball, J., Applegate, C., Cheng, T., Laycock, S., Rojas, S.M., Kirwan, J., Zhou, J., 2019. Combining computer vision and deep learning to enable ultra-scale aerial phenotyping and precision agriculture: A case study of lettuce production. *Hortic. Res.* 6, 70. https://doi.org/10.1038/s41438-019-0151-5

Bazi, Y., Bashmal, L., Rahhal, M.M. Al, Dayil, R. Al, Ajlan, N. Al, 2021. Vision Transformers for Remote Sensing Image Classification. *Remote Sens.* 13, 516. https://doi.org/10.3390/rs13030516

Bedi, P., Gole, P., 2021. Plant disease detection using hybrid model based on convolutional autoencoder and convolutional neural network. *Artif. Intell. Agric.* 5, 90–101. https://doi.org/10.1016/j.aiia.2021.05.002

Bi, L., Hu, G., 2021. A genetic algorithm-assisted deep learning approach for crop yield prediction. *Soft Comput.* 25, 10617–10628. https://doi.org/10.1007/s00500-021-05995-9

Bresilla, K., Perulli, G.D., Boini, A., Morandi, B., Corelli Grappadelli, L., Manfrini, L., 2019. Single-shot convolution neural networks for real-time fruit detection within the tree. *Front. Plant Sci.* 10, 611.

Butera, L., Ferrante, A., Jermini, M., Prevostini, M., Alippi, C., 2022. Precise Agriculture: Effective Deep Learning Strategies to Detect Pest Insects. *IEEE/CAA J. Autom. Sin.* 9, 246–258. https://doi.org/10.1109/JAS.2021.1004317

Canton, H., 2021. Food and agriculture organization of the United Nations—FAO, in: The Europa Directory of International Organizations 2021. Routledge, pp. 297–305.

Chandel, N., Chakraborty, S., Rajwade, Y., Dubey, K., Tiwari, M.K., Jat, D., 2021. Identifying crop water stress using deep learning models. *Neural Comput. Appl.* 33. https://doi.org/10.1007/s00521-020-05325-4

Chang and Kuan-Ming, C.-L.L., 2018. Smart Agricultural Machine with a Computer Vision-Based Weeding and Variable-Rate Irrigation Scheme. *Robotics* 7, 38-NA. https://doi.org/10.3390/robotics7030038

Chantry, M., Hatfield, S., Düben, P., Polichtchouk, I., Palmer, T., 2021. Machine Learning Emulation of Gravity Wave Drag in Numerical Weather Forecasting. *J. Adv. Model. Earth Syst.* 13. https://doi.org/10.1029/2021MS002477

Chavan Abhijeet V., T.R. N., 2018. AgroAVNET for crops and weeds classification: A step forward in automatic farming. *Comput. Electron. Agric.* 154, 361–372. https://doi.org/10.1016/j.compag.2018.09.021

Chen, H., Chen, A., Xu, L., Xie, H., Qiao, H., Lin, Q., Cai, K., 2020. A deep learning CNN architecture applied in smart near-infrared analysis of water pollution for agricultural irrigation resources. *Agric. Water Manag.* 240, 106303. https://doi.org/10.1016/j.agwat.2020.106303

Chen, J., Liu, Q., Gao, L., 2021. Deep convolutional neural networks for tea tree pest recognition and diagnosis. *Symmetry (Basel).* 13, 2140.

Chen, J., Chen, Jinxiu, Zhang, D., Sun, Y., Nanehkaran, Y.A., 2020. Using deep transfer learning for image-based plant disease identification. *Comput. Electron. Agric.* 173, 105393. https://doi.org/10.1016/j.compag.2020.105393

Chen, M., Cui, Y., Wang, X., Xie, H., Liu, F., Luo, T., Zheng, S., Luo, Y., 2021. A reinforcement learning approach to irrigation decision-making for rice using weather forecasts. *Agric. Water Manag.* 250, 106838.

Chen, Y., Lee, W.S., Gan, H., Peres, N., Fraisse, C., Zhang, Y., He, Y., 2019. Strawberry Yield Prediction Based on a Deep Neural Network Using High-Resolution Aerial Orthoimages. *Remote Sens.* 11, 1584. https://doi.org/10.3390/rs11131584

Cheng, H., Damerow, L., Sun, Y., Blanke, M., 2017. Early Yield Prediction Using Image Analysis of Apple Fruit and Tree Canopy Features with Neural Networks. *J. Imaging* 3, 6. https://doi.org/10.3390/jimaging3010006

Chlingaryan, A., Sukkarieh, S., Whelan, B., 2018. Machine learning approaches for crop yield prediction and nitrogen status estimation in precision agriculture: A review. *Comput. Electron. Agric.* 151, 61–69. https://doi.org/10.1016/j.compag.2018.05.012

Chu, Z., Yu, J., 2020. An end-to-end model for rice yield prediction using deep learning fusion. *Comput. Electron. Agric.* 174, 105471-NA. https://doi.org/10.1016/j.compag.2020.105471

Cruz, A.C., Luvisi, A., De Bellis, L., Ampatzidis, Y., 2017. X-FIDO: An effective application for detecting olive quick decline syndrome with deep learning and data fusion. *Front. Plant Sci.* 8, 1741. https://doi.org/10.3389/fpls.2017.01741

Cui, X., Ying, Y., Chen, Z., 2021. CycleGAN based confusion model for cross-species plant disease image migration. *J. Intell. Fuzzy Syst.* 41, 6685–6696. https://doi.org/10.3233/JIFS-210585

Dawei, W., Limiao, D., Jiangong, N., Jiyue, G., Hongfei, Z., Zhongzhi, H., 2019. Recognition pest by image-based transfer learning. *J. Sci. Food Agric.* 99, 4524–4531. https://doi.org/10.1002/jsfa.9689

De Alwis, S., Zhang, Y., Na, M., Li, G., 2019. Duo Attention with Deep Learning on Tomato Yield Prediction and Factor Interpretation, in: *Lecture Notes in Computer Science (Including Subseries Lecture Notes in Artificial Intelligence and Lecture Notes in Bioinformatics).* Springer International Publishing, pp. 704–715. https://doi.org/10.1007/978-3-030-29894-4_56

Dee, H., French, A., 2015. From image processing to computer vision: Plant imaging grows up. *Funct. Plant Biol.* 42, iii–v. https://doi.org/10.1071/FPv42n5_FO

Deery, D., Jimenez-Berni, J., Jones, H., Sirault, X., Furbank, R., 2014. Proximal remote sensing buggies and potential applications for field-based phenotyping. *Agronomy* 4, 349–379. https://doi.org/10.3390/agronomy4030349

Dhanya, V., Subeesh, A., Vishwakarma, D., Kumar T, N., Gaba, R., Singh, A., 2022. Deep learning based computer vision approaches for smart agricultural applications. *Artif. Intell. Agric.* 6, 211–229. https://doi.org/10.1016/j.aiia.2022.09.007

Dhanya, V.G., Subeesh, A., Kushwaha, N.L., Vishwakarma, D.K., Nagesh Kumar, T., Ritika, G., Singh, A.N., 2022. Deep learning based computer vision approaches for smart agricultural applications. *Artif. Intell. Agric.* 6, 211–229. https://doi.org/10.1016/j.aiia.2022.09.007

Dian Bah, M., Hafiane, A., Canals, R., 2018. Deep learning with unsupervised data labeling for weed detection in line crops in UAV images. *Remote Sens.* 10, 1690-NA. https://doi.org/10.3390/rs10111690

Dolata, P., Reiner, J., 2018. Barley variety recognition with viewpoint-aware double-stream convolutional neural networks, in: *2018 Federated Conference on Computer Science and Information Systems (FedCSIS).* IEEE, pp. 101–105.

dos Santos Ferreira, A., Freitas, D.M., da Silva, G.G., Pistori, H., Folhes, M.T., 2019. Unsupervised deep learning and semi-automatic data labeling in weed discrimination. *Comput. Electron. Agric.* 165, 104963. https://doi.org/10.1016/j.compag.2019.104963

Dosovitskiy, A., Beyer, L., Kolesnikov, A., Weissenborn, D., Zhai, X., Unterthiner, T., Dehghani, M., Minderer, M., Heigold, G., Gelly, S., 2020. An image is worth 16x16 words: Transformers for image recognition at scale. arXiv Prepr. arXiv2010.11929.

Durai, S.K.S., Shamili, M.D., 2022. Smart farming using Machine Learning and Deep Learning techniques. *Decis. Anal. J.* 3, 100041. https://doi.org/10.1016/j.dajour.2022.100041

Durmus, H., Gunes, E.O., Kirci, M., 2017. Disease detection on the leaves of the tomato plants by using deep learning. 2017 6th Int. Conf. Agro-Geoinformatics, Agro-Geoinformatics 2017. https://doi.org/10.1109/Agro-Geoinformatics.2017.8047016

Ebrahimi, M.A., Khoshtaghaza, M.H., Minaei, S., Jamshidi, B., 2017. Vision-based pest detection based on SVM classification method. *Comput. Electron. Agric.* 137, 52–58. https://doi.org/10.1016/j.compag.2017.03.016

Elavarasan, D., Durai Raj Vincent, P.M., 2021. Fuzzy deep learning-based crop yield prediction model for sustainable agronomical frameworks. *Neural Comput. Appl.* 33, 13205–13224. https://doi.org/10.1007/s00521-021-05950-7

Elbeltagi, A., Kumar, M., Kushwaha, N.L., Pande, C.B., Ditthakit, P., Vishwakarma, D.K., Subeesh, A., 2023. Drought indicator analysis and forecasting using data driven models: case study in Jaisalmer, *India. Stoch. Environ. Res. Risk Assess.* 37, 113–131. https://doi.org/10.1007/s00477-022-02277-0

Espejo-Garcia, B., Mylonas, N., Athanasakos, L., Fountas, S., Vasilakoglou, I., 2020. Towards weeds identification assistance through transfer learning. *Comput. Electron. Agric.* 171, 105306. https://doi.org/10.1016/j.compag.2020.105306

Etienne Dharmendra, A.S., 2019. *Machine learning approaches to automate weed detection by UAV based sensors*. Auton. Air Gr. Sens. Syst. Agric. Optim. Phenotyping IV. https://doi.org/10.1117/12.2520536

Faisal, M., Alsulaiman, M., Arafah, M., Mekhtiche, M.A., 2020. IHDS: Intelligent Harvesting Decision System for Date Fruit Based on Maturity Stage Using Deep Learning and Computer Vision. *IEEE Access* 8, 167985–167997. https://doi.org/10.1109/ACCESS.2020.3023894

Fan, X., Luo, P., Mu, Y., Zhou, R., Tjahjadi, T., Ren, Y., 2022. Leaf image based plant disease identification using transfer learning and feature fusion. *Comput. Electron. Agric.* 196, 106892.

Fawakherji, M., Potena, C., Prevedello, I., Pretto, A., Bloisi, D.D., Nardi, D., 2020. Data augmentation using GANs for crop/weed segmentation in precision farming, in: *2020 IEEE Conference on Control Technology and Applications (CCTA)*. IEEE, pp. 279–284.

Feng, L., Chen, S., Zhang, C., Zhang, Y., He, Y., 2021. A comprehensive review on recent applications of unmanned aerial vehicle remote sensing with various sensors for high-throughput plant phenotyping. *Comput. Electron. Agric.* 182, 106033-NA. https://doi.org/10.1016/j.compag.2021.106033

Ferentinos, K.P., 2018. Deep learning models for plant disease detection and diagnosis. *Comput. Electron. Agric.* 145, 311–318. https://doi.org/10.1016/j.compag.2018.01.009

Foley, J., 2014. A five-step plan to feed the world. *Natl. Geogr. Mag.* 225, 26–58.

Fuentes, A., Yoon, S., Kim, S.C., Park, D.S., 2017. A robust deep-learning-based detector for real-time tomato plant diseases and pests recognition. *Sensors (Switzerland)* 17, 2022-NA. https://doi.org/10.3390/s17092022

Garibaldi-Márquez, F., Flores, G., Mercado-Ravell, D.A., Ramírez-Pedraza, A., Valentín-Coronado, L.M., 2022. Weed Classification from Natural Corn Field-Multi-Plant Images Based on Shallow and Deep Learning. *Sensors* 22, 3021. https://doi.org/10.3390/s22083021

Ghobadpour, A., Monsalve, G., Cardenas, A., Mousazadeh, H., 2022. Off-Road Electric Vehicles and Autonomous Robots in Agricultural Sector: Trends, Challenges, and Opportunities. *Vehicles* 4, 843–864. https://doi.org/10.3390/vehicles4030047

Goel, R., Yadav, C., Rastogi, R., 2021. Smart Agriculture –Urgent Need of the Day in Developing Countries. *Sustain. Comput. Informatics Syst.* 30, 100512. https://doi.org/10.1016/j.suscom.2021.100512

Goldstein, A., Fink, L., Meitin, A., Bohadana, S., Lutenberg, O., Ravid, G., 2018. Applying machine learning on sensor data for irrigation recommendations: revealing the agronomist's tacit knowledge. *Precis. Agric.* 19, 421–444. https://doi.org/10.1007/s11119-017-9527-4

Granitto, P.M., Verdes, P.F., Ceccatto, H.A., 2005. Large-scale investigation of weed seed identification by machine vision. *Comput. Electron. Agric.* 47, 15–24. https://doi.org/10.1016/j.compag.2004.10.003

Grijalva, I., Spiesman, B.J., McCornack, B., 2023. Image classification of sugarcane aphid density using deep convolutional neural networks. *Smart Agric. Technol.* 3, 100089.

Grinblat, G.L., Uzal, L.C., Larese, M.G., Granitto, P.M., 2016. Deep learning for plant identification using vein morphological patterns. *Comput. Electron. Agric.* 127, 418–424. https://doi.org/10.1016/j.compag.2016.07.003

Gulzar, Y., Hamid, Y., Soomro, A.B., Alwan, A.A., Journaux, L., 2020. A Convolution Neural Network-Based Seed Classification System. *Symmetry (Basel).* 12, 2018. https://doi.org/10.3390/sym12122018

Guo, D., Juan, J., Chang, L., Zhang, J., Huang, D., 2017. Discrimination of plant root zone water status in greenhouse production based on phenotyping and machine learning techniques. *Sci. Rep.* 7, 8303. https://doi.org/10.1038/s41598-017-08235-z

Gupta, D., Sharma, P., Choudhary, K., Gupta, K., Chawla, R., Khanna, A., Albuquerque, V.H.C. de, 2021. Artificial plant optimization algorithm to detect infected leaves using machine learning. *Expert Syst.* 38, e12501.

Habaragamuwa, H., Ogawa, Y., Suzuki, T., Shiigi, T., Ono, M., Kondo, N., 2018. Detecting greenhouse strawberries (mature and immature), using deep convolutional neural network. *Eng. Agric. Environ. Food* 11, 127–138. https://doi.org/10.1016/j.eaef.2018.03.001

Hall, D., McCool, C., Dayoub, F., Sunderhauf, N., Upcroft, B., 2015. Evaluation of Features for Leaf Classification in Challenging Conditions, in: *2015 IEEE Winter Conference on Applications of Computer Vision*. IEEE, pp. 797–804. https://doi.org/10.1109/WACV.2015.111

Hassan, S., Maji, A., Jasiński, M., Leonowicz, Z., Jasińska, E., 2021. Identification of Plant-Leaf Diseases Using CNN and Transfer-Learning Approach. *Electronics* 10, 1388. https://doi.org/10.3390/electronics10121388

Hassan, S.M., Maji, A.K., 2022. Plant Disease Identification Using a Novel Convolutional Neural Network. *IEEE Access* 10, 5390–5401. https://doi.org/10.1109/ACCESS.2022.3141371

Hassija, V., Chamola, V., Saxena, V., Jain, D., Goyal, P., Sikdar, B., 2019. A Survey on IoT Security: Application Areas, Security Threats, and Solution Architectures. *IEEE Access* 7, 82721–82743. https://doi.org/10.1109/ACCESS.2019.2924045

Hati, A.J., Singh, R.R., 2021. Artificial Intelligence in Smart Farms: Plant Phenotyping for Species Recognition and Health Condition Identification Using Deep Learning. *AI* 2, 274–289. https://doi.org/10.3390/ai2020017

He, K., Zhang, X., Ren, S., Sun, J., 2016. Deep Residual Learning for Image Recognition, in: *2016 IEEE Conference on Computer Vision and Pattern Recognition (CVPR)*. IEEE, pp. 770–778. https://doi.org/10.1109/CVPR.2016.90

Hiriyannaiah, S., Srinivas, A.M.D., Shetty, G.K., GM, S., Srinivasa, K.G., 2020. A computationally intelligent agent for detecting fake news using generative adversarial networks, in: *Hybrid Computational Intelligence*. Elsevier, pp. 69–96.

Hossain, M.S., Al-Hammadi, M., Muhammad, G., 2019. Automatic Fruit Classification Using Deep Learning for Industrial Applications. *IEEE Trans. Ind. Informatics* 15, 1027–1034. https://doi.org/10.1109/TII.2018.2875149

Huang, H., Lan, Y., Yang, A., Zhang, Y., Wen, S., Deng, J., 2020. Deep learning versus Object-based Image Analysis (OBIA) in weed mapping of UAV imagery. *Int. J. Remote Sens.* 41, 3446–3479. https://doi.org/10.1080/01431161.2019.1706112

Huang, S., Fan, X., Sun, L., Shen, Y., Suo, X., 2019. Research on Classification Method of Maize Seed Defect Based on Machine Vision. *J. Sensors* 2019, 1–9. https://doi.org/10.1155/2019/2716975

Ilyas, Q., Ahmad, M., 2020. Smart Farming: An Enhanced Pursuit of Sustainable Remote Livestock Tracking and Geofencing Using IoT and GPRS. *Wirel. Commun. Mob. Comput.* 2020, 1–12. https://doi.org/10.1155/2020/6660733

Javanmardi, S., Miraei Ashtiani, S.-H., Verbeek, F.J., Martynenko, A., 2021. Computer-vision classification of corn seed varieties using deep convolutional neural network. *J. Stored Prod. Res.* 92, 101800. https://doi.org/10.1016/j.jspr.2021.101800

Jeong, S., Ko, J., Yeom, J.M., 2022. Predicting rice yield at pixel scale through synthetic use of crop and deep learning models with satellite data in South and North Korea. *Sci.Total Environ.* 802, 149726-NA. https://doi.org/10.1016/j.scitotenv.2021.149726

Ji Zhibin, M.W., 2022. Automatic detection and severity analysis of grape black measles disease based on deep learning and fuzzy logic. *Comput. Electron. Agric.* 193, 106718. https://doi.org/10.1016/j.compag.2022.106718

Jiang, H., Li, X., Safara, F., 2021. WITHDRAWN: IoT-based Agriculture: Deep Learning in Detecting Apple Fruit Diseases. *Microprocess. Microsyst. NA*, 104321. https://doi.org/10.1016/j.micpro.2021.104321

Jiang, Z., Dong, Z., Jiang, W., Yang, Y., 2021. Recognition of rice leaf diseases and wheat leaf diseases based on multi-task deep transfer learning. *Comput. Electron. Agric.* 186, 106184.

Joshi, R.C., Kaushik, M., Dutta, M.K., Srivastava, A., Choudhary, N., 2021. VirLeafNet: Automatic analysis and viral disease diagnosis using deep-learning in Vigna mungo plant. *Ecol. Inform.* 61, 101197. https://doi.org/https://doi.org/10.1016/j.ecoinf.2020.101197

Kamyshova, G., Osipov, A., Gataullin, S., Korchagin, S., Ignar, S., Gataullin, T., Terekhova, N., Suvorov, S., 2022. Artificial Neural Networks and Computer Vision's-Based Phytoindication Systems for Variable Rate Irrigation Improving. *IEEE Access* 10, 8577–8589. https://doi.org/10.1109/ACCESS.2022.3143524

Kang, H., Chen, C., 2020. Fruit detection, segmentation and 3D visualisation of environments in apple orchards. *Comput. Electron. Agric.* 171, 105302. https://doi.org/10.1016/j.compag.2020.105302

Kannur, Anil, Kannur, Asha, Rajpuohit, V.S., 2011. Classification and Grading of Bulk Seeds Using Aritificial Neural Network. *Int. J. Mach. Intell.* 3, 62–73. https://doi.org/10.9735/0975-2927.3.2.62-73

Kanwal, Z., Basit, A., Jawad, M., Ullah, I., Ali, A., 2019. Overlapped apple fruit yield estimation using pixel classification and hough transform. *Int. J. Adv. Comput. Sci. Appl* 10, 567–573.

Karar, M.E., Alsunaydi, F., Albusaymi, S., Alotaibi, S., 2021. A new mobile application of agricultural pests recognition using deep learning in cloud computing system. *Alexandria Eng. J.* 60, 4423–4432. https://doi.org/10.1016/j.aej.2021.03.009

Karthik, R., Hariharan, M., Anand, S., Mathikshara, P., Johnson, A., Menaka, R., 2020. Attention embedded residual CNN for disease detection in tomato leaves. *Appl. Soft Comput. J.* 86, 105933-NA. https://doi.org/10.1016/j.asoc.2019.105933

Kashyap, P.K., Kumar, S., Jaiswal, A., Prasad, M., Gandomi, A.H., 2021. Towards Precision Agriculture: IoT-Enabled Intelligent Irrigation Systems Using Deep Learning Neural Network. *IEEE Sens. J.* 21, 17479–17491. https://doi.org/10.1109/JSEN.2021.3069266

Khaki, S., Wang, L., 2019. Crop yield prediction using deep neural networks. *Front. Plant Sci.* 10, 621-NA. https://doi.org/10.3389/fpls.2019.00621

Khaki, S., Wang, L., Archontoulis, S. V., 2020. A CNN-RNN Framework for Crop Yield Prediction. *Front. Plant Sci.* 10, 1750-NA. https://doi.org/10.3389/fpls.2019.01750

Khamparia, A., Saini, G., Gupta, D., Khanna, A., Tiwari, S., de Albuquerque, V.H.C., 2020. Seasonal Crops Disease Prediction and Classification Using Deep Convolutional Encoder Network. Circuits, *Syst. Signal Process.* 39, 818–836. https://doi.org/10.1007/s00034-019-01041-0

Khirade, S.D., Patil, A.B., 2015. Plant disease detection using image processing. Proc. - 1st Int. Conf. Comput. Commun. Control Autom. ICCUBEA 2015. https://doi.org/10.1109/ICCUBEA.2015.153

Kim, K.-H., Kim, M.-G., Yoon, P.-R., Bang, J.-H., Myoung, W.-H., Choi, J.-Y., Choi, G.-H., 2022. Application of CCTV image and semantic segmentation model for water level estimation of irrigation channel. *J. Korean Soc. Agric. Eng.* 64, 63–73.

Kirk, R., Cielniak, G., Mangan, M., 2020. L*a*b*Fruits: A Rapid and Robust Outdoor Fruit Detection System Combining Bio-Inspired Features with One-Stage Deep Learning Networks. *Sensors* 20, 275. https://doi.org/10.3390/s20010275

Koirala, A., Walsh, K.B., Wang, Z., McCarthy, C., 2019. Deep learning for real-time fruit detection and orchard fruit load estimation: benchmarking of 'MangoYOLO.' *Precis. Agric.* 20, 1107–1135. https://doi.org/10.1007/s11119-019-09642-0

Koirala, Anand, Walsh, K.B., Wang, Z., McCarthy, C., 2019. Deep learning – Method overview and review of use for fruit detection and yield estimation. *Comput. Electron. Agric.* 162, 219–234. https://doi.org/10.1016/j.compag.2019.04.017

Kounalakis, T., Triantafyllidis, G.A., Nalpantidis, L., 2019. Deep learning-based visual recognition of rumex for robotic precision farming. *Comput. Electron. Agric.* 165, 104973. https://doi.org/10.1016/j.compag.2019.104973

Kriegeskorte and Tal, N.G., 2019. Neural network models and deep learning. *Curr. Biol.* 29, R231–R236. https://doi.org/10.1016/j.cub.2019.02.034

Krizhevsky, A., Sutskever, I., Hinton, G.E., 2017. ImageNet classification with deep convolutional neural networks. *Commun. ACM* 60, 84–90. https://doi.org/10.1145/3065386

Kumbi Mahantesh N., A.A. B., 2021. Deep CNN based Sunflower Atom Optimization Method for Optimal Water Control in IoT. *Wirel. Pers. Commun.* 122, 1221–1246. https://doi.org/10.1007/s11277-021-08946-7

Kundu, N., Rani, G., Dhaka, V.S., Gupta, K., Nayak, S.C., Verma, S., Ijaz, M.F., Woźniak, M., 2021. IoT and Interpretable Machine Learning Based Framework for Disease Prediction in Pearl Millet. *Sensors* 21, 5386. https://doi.org/10.3390/s21165386

Kurtulmuş, F., 2020. Identification of sunflower seeds with deep convolutional neural networks. *J. Food Meas. Charact.* 15, 1024–1033. https://doi.org/10.1007/s11694-020-00707-7

Kushwaha, N., Elbeltagi, A., Mehan, S., Malik, A., Yousuf, A., 2022. Comparative study on morphometric analysis and RUSLE-based approaches for micro-watershed prioritization using remote sensing and GIS. *Arab. J. Geosci.* 15, 564. https://doi.org/10.1007/s12517-022-09837-2

Kushwaha, S.K., Chauhan, P., Hedlund, K., Ahren, D., 2016. NBSPred: a support vector machine-based high-throughput pipeline for plant resistance protein NBSLRR prediction. *Bioinformatics* 32, 1223–1225.

Kussul, N., Lavreniuk, M., Skakun, S., Shelestov, A., 2017. Deep Learning Classification of Land Cover and Crop Types Using Remote Sensing Data. *IEEE Geosci. Remote Sens. Lett.* 14, 778–782. https://doi.org/10.1109/LGRS.2017.2681128

Le, V.N.T., Ahderom, S., Alameh, K., 2020. Performances of the LBP Based Algorithm over CNN Models for Detecting Crops and Weeds with Similar Morphologies. *Sensors* 20, 2193. https://doi.org/10.3390/s20082193

Lecun, Y., Bengio, Y., Hinton, G., 2015. Deep learning. *Nature* 521, 436–444. https://doi.org/10.1038/nature14539

Ledig, C., Theis, L., Huszar, F., Caballero, J., Cunningham, A., Acosta, A., Aitken, A., Tejani, A., Totz, J., Wang, Z., Shi, W., 2017. Photo-Realistic Single Image Super-Resolution Using a Generative Adversarial Network, in: *2017 IEEE Conference on Computer Vision and Pattern Recognition (CVPR)*. IEEE, pp. 105–114. https://doi.org/10.1109/CVPR.2017.19

Leonardo, M.M., Carvalho, T.J., Rezende, E., Zucchi, R., Faria, F.A., 2018. Deep feature-based classifiers for fruit fly identification (Diptera: Tephritidae), in: *2018 31st SIBGRAPI Conference on Graphics, Patterns and Images (SIBGRAPI)*. IEEE, pp. 41–47.

Li, C., Zhang, Y., Ren, X., 2020. Modeling Hourly Soil Temperature Using Deep BiLSTM Neural Network. *Algorithms* 13, 173. https://doi.org/10.3390/a13070173

Li, X., Li, S., 2022. Transformer Help CNN See Better: A Lightweight Hybrid Apple Disease Identification Model Based on Transformers. *Agriculture* 12, 884. https://doi.org/10.3390/agriculture12060884

Li, Y., Chao, X., 2021. Toward Sustainability: Trade-Off Between Data Quality and Quantity in Crop Pest Recognition. *Front. Plant Sci.* 12. https://doi.org/10.3389/fpls.2021.811241

Li, Y., Randall, C.J., Woesik, R. van, Ribeiro, E., 2019. Underwater video mosaicing using topology and superpixel-based pairwise stitching. *Expert Syst. Appl.* 119, 171–183. https://doi.org/10.1016/j.eswa.2018.10.041

Liakos, K., Busato, P., Moshou, D., Pearson, S., Bochtis, D., 2018. Machine Learning in Agriculture: A Review. *Sensors* 18, 2674. https://doi.org/10.3390/s18082674

Liang, Q., Zhu, W., Long, J., Wang, Y., Sun, W., Wu, W., 2018. A Real-Time Detection Framework for On-Tree Mango Based on SSD Network. pp. 423–436. https://doi.org/10.1007/978-3-319-97589-4_36

Lilburne, L., Licorish, S., MacDonell, S., 2022. Soil Texture Prediction with Automated Deep Convolutional Neural Networks and Population Based Learning.

Lindenthal, M., Steiner, U., Dehne, H.-W., Oerke, E.-C., 2005. Effect of Downy Mildew Development on Transpiration of Cucumber Leaves Visualized by Digital Infrared Thermography. *Phytopathology®* 95, 233–240. https://doi.org/10.1094/PHYTO-95-0233

Liu, B., Zhang, Y., He, D.J., Li, Y., 2018. Identification of apple leaf diseases based on deep convolutional neural networks. *Symmetry (Basel).* 10, 11-NA. https://doi.org/10.3390/sym10010011

Liu, D., Ning, X., Li, Z., Yang, D., Li, H., Gao, L., 2015. Discriminating and elimination of damaged soybean seeds based on image characteristics. *J. Stored Prod. Res.* 60, 67–74. https://doi.org/10.1016/j.jspr.2014.10.001

Liu, W., Anguelov, D., Erhan, D., Szegedy, C., Reed, S., Fu, C.-Y., Berg, A.C., 2016. Ssd: Single shot multibox detector, in: *Computer Vision–ECCV 2016: 14th European Conference, Amsterdam, The Netherlands, October 11–14, 2016, Proceedings*, Part I 14. Springer, pp. 21–37.

Liu, Y., Liu, S., Xu, J., Kong, X., Xie, L., Chen, K., Liao, Y., Fan, B., Wang, K., 2022. Forest pest identification based on a new dataset and convolutional neural network model with enhancement strategy. *Comput. Electron. Agric.* 192, 106625.

Lu, S., Chen, W., Zhang, X., Karkee, M., 2022. Canopy-attention-YOLOv4-based immature/mature apple fruit detection on dense-foliage tree architectures for early crop load estimation. *Comput. Electron. Agric.* 193, 106696. https://doi.org/10.1016/j.compag.2022.106696

Lu, Y., Yi, S., Zeng, N., Liu, Y., Zhang, Y., 2017. Identification of rice diseases using deep convolutional neural networks. *Neurocomputing* 267, 378–384. https://doi.org/10.1016/j.neucom.2017.06.023

Ma, T., Tsuchikawa, S., Inagaki, T., 2020. Rapid and non-destructive seed viability prediction using near-infrared hyperspectral imaging coupled with a deep learning approach. *Comput. Electron. Agric.* 177, 105683. https://doi.org/10.1016/j.compag.2020.105683

Maimaitijiang, M., Sagan, V., Sidike, P., Hartling, S., Esposito, F., Fritschi, F.B., 2020. Soybean yield prediction from UAV using multimodal data fusion and deep learning. *Remote Sens. Environ.* 237, 111599. https://doi.org/10.1016/j.rse.2019.111599

Matei, O., Rusu, T., Petrovan, A., Mihuţ, G., 2017. A Data Mining System for Real Time Soil Moisture Prediction. *Procedia Eng.* 181, 837–844. https://doi.org/10.1016/j.proeng.2017.02.475

Mekala, M.S., Viswanathan, P., 2017. A Survey: Smart agriculture IoT with cloud computing. 2017 Int. Conf. Microelectron. Devices, Circuits Syst. ICMDCS 2017. https://doi.org/10.1109/ICMDCS.2017.8211551

Menaga, A., Vasantha, S., 2022. Smart Sustainable Agriculture Using Machine Learning and AI: A Review. pp. 447–458. https://doi.org/10.1007/978-981-16-7952-0_42

Meshram, V.A., Patil, K., Ramteke, S.D., 2021. MNet: A framework to reduce fruit image misclassification. *Ing. des Syst. d'Information* 26, 159–170. https://doi.org/10.18280/isi.260203

Miikkulainen, R., Liang, J., Meyerson, E., Rawal, A., Fink, D., Francon, O., Raju, B., Shahrzad, H., Navruzyan, A., Duffy, N., Hodjat, B., 2019. Evolving Deep Neural Networks, in: *Artificial Intelligence in the*

Age of Neural Networks and Brain Computing. Elsevier, pp. 293–312. https://doi.org/10.1016/B978-0-12-815480-9.00015-3

Milioto, A., Lottes, P., Stachniss, C., 2017. REAL-TIME BLOB-WISE SUGAR BEETS VS WEEDS CLASSIFICATION FOR MONITORING FIELDS USING CONVOLUTIONAL NEURAL NETWORKS. ISPRS Ann. Photogramm. Remote Sens. Spat. *Inf. Sci.* IV-2/W3, 41–48. https://doi.org/10.5194/isprs-annals-IV-2-W3-41-2017

Mishra, S., Sachan, R., Rajpal, D., 2020. Deep Convolutional Neural Network based Detection System for Real-time Corn Plant Disease Recognition. *Procedia Comput. Sci.* 167, 2003–2010. https://doi.org/10.1016/j.procs.2020.03.236

Mkonyi, L., Rubanga, D., Richard, M., Zekeya, N., Sawahiko, S., Maiseli, B., Machuve, D., 2020. Early identification of Tuta absoluta in tomato plants using deep learning. *Sci. African* 10, e00590. https://doi.org/10.1016/j.sciaf.2020.e00590

Mohanty, S.P., Hughes, D.P., Salathé, M., 2016. Using Deep Learning for Image-Based Plant Disease Detection. *Front. Plant Sci.* 7, 1419. https://doi.org/10.3389/fpls.2016.01419

Mohyuddin, G., Khan, M.A., Haseeb, A., Mahpara, S., Waseem, M., Saleh, A.M., 2024. Evaluation of Machine Learning Approaches for Precision Farming in Smart Agriculture System: A Comprehensive Review. *IEEE Access* 12, 60155–60184. https://doi.org/10.1109/ACCESS.2024.3390581

Mortensen, A.K., Dyrmann, M., Karstoft, H., Jørgensen, R.N., Gislum, R., 2016. *Semantic segmentation of mixed crops using deep convolutional neural network*.

Moshia, M.E., Newete, S.W., 2019. Mexican poppy (Argemone mexicana) control in cornfield using deep learning neural networks: a perspective. Acta Agric. Scand. Sect. B—Soil *Plant Sci.* 69, 228–234. https://doi.org/10.1080/09064710.2018.1536225

Naciones Unidas, 2019. World Population Prospects 2019: Highlights [WWW Document]. URL https://www.un.org/es/desa/world-population-prospects-2019-highlights (accessed 4.5.25).

Naimi Marandi, S., Ayoubi, S., Demattê, J.A., Zeraatpisheh, M., Amorim, M., Alcantara, F., 2021. Spatial Prediction of Soil Surface Properties in an Arid Region Using Synthetic Soil Image and Machine Learning. *Geocarto Int.* 37, 1–22. https://doi.org/10.1080/10106049.2021.1996639

Nandhini, M., Kala, K.U., Thangadarshini, M., Verma, S.M., 2022. Deep Learning model of sequential image classifier for crop disease detection in plantain tree cultivation. *Comput. Electron. Agric.* 197, 106915.

Ni, C., Wang, D., Vinson, R., Holmes, M., Tao, Y., 2019. Automatic inspection machine for maize kernels based on deep convolutional neural networks. *Biosyst. Eng.* 178, 131–144. https://doi.org/10.1016/j.biosystemseng.2018.11.010

Nilsson, H.-E., 1995. Remote sensing and image analysis in plant pathology. *Annu. Rev. Phytopathol.* 33, 489–527. https://doi.org/10.1146/annurev.py.33.090195.002421

OECD-FAO, 2020. OECD-FAO Agricultural Outlook 2020-2029. *OECD.* https://doi.org/10.1787/1112c23b-en

Olatunji, J.R., Redding, G.P., Rowe, C.L., East, A.R., 2020. Reconstruction of kiwifruit fruit geometry using a CGAN trained on a synthetic dataset. *Comput. Electron. Agric.* 177, 105699. https://doi.org/10.1016/j.compag.2020.105699

Oliveri Conti, G., Heibati, B., Kloog, I., Fiore, M., Ferrante, M., 2017. A review of AirQ Models and their applications for forecasting the air pollution health outcomes. *Environ. Sci. Pollut. Res.* 24, 6426–6445. https://doi.org/10.1007/s11356-016-8180-1

Onishi, Y., Yoshida, T., Kurita, H., Fukao, T., Arihara, H., Iwai, A., 2019. An automated fruit harvesting robot by using deep learning. *ROBOMECH J.* 6, 13. https://doi.org/10.1186/s40648-019-0141-2

Oppenheim, D., Shani, G., 2017. Potato Disease Classification Using Convolution Neural Networks. *Adv. Anim. Biosci.* 8, 244–249. https://doi.org/10.1017/S2040470017001376

Osorio, K., Puerto, A., Pedraza, C., Jamaica, D., Rodríguez, L., 2020. A Deep Learning Approach for Weed Detection in Lettuce Crops Using Multispectral Images. *AgriEngineering* 2, 471–488. https://doi.org/10.3390/agriengineering2030032

Palacios, F., Bueno, G., Salido, J., Diago, M.P., Hernández, I., Tardaguila, J., 2020. Automated grapevine flower detection and quantification method based on computer vision and deep learning from on-the-go imaging using a mobile sensing platform under field conditions. *Comput. Electron. Agric.* 178, 105796. https://doi.org/10.1016/j.compag.2020.105796

Pan, S., Qiao, J., Wang, R., Yu, H., Wang, C., Taylor, K., Pan, H., 2022. Intelligent diagnosis of northern corn leaf blight with deep learning model. *J. Integr. Agric.* 21, 1094–1105. https://doi.org/10.1016/S2095-3119(21)63707-3

Pandey, A., Jain, K., 2022. A robust deep attention dense convolutional neural network for plant leaf disease identification and classification from smart phone captured real world images. *Ecol. Inform.* 70, 101725.

Partel, V., Charan Kakarla, S., Ampatzidis, Y., 2019. Development and evaluation of a low-cost and smart technology for precision weed management utilizing artificial intelligence. *Comput. Electron. Agric.* 157, 339–350. https://doi.org/10.1016/j.compag.2018.12.048

Pathan, M., Patel, N., Yagnik, H., Shah, M., 2020. Artificial cognition for applications in smart agriculture: A comprehensive review. *Artif. Intell. Agric.* 4, 81–95. https://doi.org/10.1016/j.aiia.2020.06.001

Peng, H., Li, Z., Zhou, Z., Shao, Y., 2022. Weed detection in paddy field using an improved RetinaNet network. *Comput. Electron. Agric.* 199, 107179. https://doi.org/10.1016/j.compag.2022.107179

Picon, A., Seitz, M., Alvarez-Gila, A., Mohnke, P., Ortiz-Barredo, A., Echazarra, J., 2019. Crop conditional Convolutional Neural Networks for massive multi-crop plant disease classification over cell phone acquired images taken on real field conditions. *Comput. Electron. Agric.* 167, 105093. https://doi.org/10.1016/j.compag.2019.105093

Polder, G., Van De Westeringh, N., Kool, J., Khan, H.A., Kootstra, G., Nieuwenhuizen, A., 2019. Automatic detection of tulip breaking virus (TBV) using a deep convolutional neural network. *IFAC-PapersOnLine* 52, 12–17.

Prasad, R., Deo, R.C., Li, Y., Maraseni, T., 2019. Weekly soil moisture forecasting with multivariate sequential, ensemble empirical mode decomposition and Boruta-random forest hybridizer algorithm approach. *CATENA* 177, 149–166. https://doi.org/https://doi.org/10.1016/j.catena.2019.02.012

Pratyush Reddy, K.S., Roopa, Y.M., Kovvada Rajeev, L.N., Nandan, N.S., 2020. IoT based Smart Agriculture using Machine Learning. Proc. 2nd Int. Conf. Inven. Res. Comput. Appl. ICIRCA 2020. https://doi.org/10.1109/ICIRCA48905.2020.9183373

Pyo, J., Hong, S.M., Kwon, Y.S., Kim, M.S., Cho, K.H., 2020. Estimation of heavy metals using deep neural network with visible and infrared spectroscopy of soil. *Sci. Total Environ.* 741, 140162. https://doi.org/10.1016/j.scitotenv.2020.140162

Qiu, Z., Chen, J., Zhao, Y., Zhu, S., He, Y., Zhang, C., 2018. Variety Identification of Single Rice Seed Using Hyperspectral Imaging Combined with Convolutional Neural Network. *Appl. Sci.* 8, 212. https://doi.org/10.3390/app8020212

Rahman, M.H., 2016. Exploring sustainability to feed the world in 2050. *J. Food Microbiol.* 1, 7–16.

Rahnemoonfar and Clay, M.S., 2017. Deep Count: Fruit Counting Based on Deep Simulated Learning. *Sensors (Basel).* 17, 905-NA. https://doi.org/10.3390/s17040905

Rahu, M., Chandio, A., Khursheed, K., Karim, S., Alhussein, M., Anwar, M., 2023. Toward Design of Internet of Things and Machine Learning-Enabled Frameworks for Analysis and Prediction of Water Quality. *IEEE Access* PP, 1. https://doi.org/10.1109/ACCESS.2023.3315649

Rajapaksha, T.T., Alexander, A., Fernando, L., Than, A., Nguyen, H. Le, 2022. Real-Time Weather Monitoring and IoT-Based Palmtop Device for Smart Agriculture. *SN Comput. Sci.* 3, 91. https://doi.org/10.1007/s42979-021-00961-6

Ramcharan, A., Baranowski, K., McCloskey, P., Ahmed, B., Legg, J., Hughes, D.P., 2017. Deep learning for image-based cassava disease detection. *Front. Plant Sci.* 8, 1852. https://doi.org/10.3389/fpls.2017.01852

Rangarajan, A.K., Purushothaman, R., Ramesh, A., 2018. Tomato crop disease classification using pre-trained deep learning algorithm. *Procedia Comput. Sci.* 133, 1040–1047. https://doi.org/10.1016/j.procs.2018.07.070

Reddy, D.J., Kumar, M.R., 2021. Crop yield prediction using machine learning algorithm. Proc. - 5th Int. Conf. Intell. Comput. Control Syst. ICICCS 2021. https://doi.org/10.1109/ICICCS51141.2021.9432236

Redmon, J., Divvala, S., Girshick, R., Farhadi, A., 2016. You Only Look Once: Unified, Real-Time Object Detection, in: *2016 IEEE Conference on Computer Vision and Pattern Recognition (CVPR)*. IEEE, pp. 779–788. https://doi.org/10.1109/CVPR.2016.91

Reimers, C., Requena-Mesa, C., 2020. Deep learning–an opportunity and a challenge for geo-and astrophysics, in: *Knowledge Discovery in Big Data from Astronomy and Earth Observation*. Elsevier, pp. 251–265.

Ren, S., He, K., Girshick, R., Sun, J., 2015. *Faster r-cnn: Towards real-time object detection with region proposal networks*. Adv. Neural Inf. Process. Syst. 28.

Riese, F.M., Keller, S., 2019. SOIL TEXTURE CLASSIFICATION WITH 1D CONVOLUTIONAL NEURAL NETWORKS BASED ON HYPERSPECTRAL DATA. ISPRS Ann. Photogramm. Remote Sens. Spat. *Inf. Sci.* IV-2/W5, 615–621. https://doi.org/10.5194/isprs-annals-IV-2-W5-615-2019

Sami, M., Khan, S., Khurram, M., Farooq, M., Anjum, R., Aziz, S., Qureshi, R., Sadak, F., 2022. A Deep Learning-Based Sensor Modeling for Smart Irrigation System. *Agronomy* 12. https://doi.org/10.3390/agronomy12010212

Sanath Rao, U., Swathi, R., Sanjana, V., Arpitha, L., Chandrasekhar, K., Chinmayi, Naik, P.K., 2021. Deep Learning Precision Farming: Grapes and Mango Leaf Disease Detection by Transfer Learning. *Glob. Transitions Proc.* 2, 535–544. https://doi.org/https://doi.org/10.1016/j.gltp.2021.08.002

Sanga, S.L., Machuve, D., Jomanga, K. (2020) Mobile-based Deep Learning Models for Banana Disease Detection. *Eng. Technol. Appl. Sci. Res.* 10, 5674–5677. https://doi.org/10.48084/etasr.3452

Sau, S., Ucchesu, M., D'hallewin, G., Bacchetta, G., 2019. Potential use of seed morpho-colourimetric analysis for Sardinian apple cultivar characterisation. *Comput. Electron. Agric.* 162, 373–379. https://doi.org/10.1016/j.compag.2019.04.027

Schwalbert, R.A., Amado, T., Corassa, G., Pott, L.P., Prasad, P.V.V., Ciampitti, I.A., 2020. Satellite-based soybean yield forecast: Integrating machine learning and weather data for improving crop yield prediction in southern Brazil. *Agric. For. Meteorol.* 284, 107886. https://doi.org/10.1016/j.agrformet.2019.107886

Segalla, A., Fiacco, G., Tramarin, L., Nardello, M., Brunelli, D., 2020. Neural networks for Pest Detection in Precision Agriculture. 2020 IEEE Int. Work. Metrol. Agric. For. MetroAgriFor 2020 - Proc. https://doi.org/10.1109/MetroAgriFor50201.2020.9277657

Shafiekhani, A., Kadam, S., Fritschi, F.B., Desouza, G.N., 2017. Vinobot and vinoculer: Two robotic platforms for high-throughput field phenotyping. *Sensors (Switzerland)* 17, 214-NA. https://doi.org/10.3390/s17010214

Shah, W.H., Rasool, A., Saleem, S., Mushtaq, N.U., Tahir, I., Hakeem, K.R., Rehman, R.U., 2021. Understanding the Integrated Pathways and Mechanisms of Transporters, Protein Kinases, and Transcription Factors in Plants under Salt Stress. *Int. J. Genomics* 2021. https://doi.org/10.1155/2021/5578727

Shaikh, T.A., Ali, R., Beg, M.M.S., 2020. Transfer learning privileged information fuels CAD diagnosis of breast cancer. *Mach. Vis. Appl.* 31, 9. https://doi.org/10.1007/s00138-020-01058-5

Shaikh, T.A., Mir, W.A., Rasool, T., Sofi, S., 2022. Machine Learning for Smart Agriculture and Precision Farming: Towards Making the Fields Talk. *Arch. Comput. Methods Eng.* 29, 4557–4597. https://doi.org/10.1007/s11831-022-09761-4

Shamim Hossain, M., Al-Hammadi, M., Muhammad, G., 2019. Automatic Fruit Classification Using Deep Learning for Industrial Applications. *IEEE Trans. Ind. Informatics* 15, 1027–1034. https://doi.org/10.1109/TII.2018.2875149

Sharif, M., Khan, M.A., Iqbal, Z., Azam, M.F., Lali, M.I.U., Javed, M.Y., 2018. Detection and classification of citrus diseases in agriculture based on optimized weighted segmentation and feature selection. *Comput. Electron. Agric.* 150, 220–234. https://doi.org/10.1016/j.compag.2018.04.023

Sharma, P., Berwal, Y.P.S., Ghai, W., 2020. Performance analysis of deep learning CNN models for disease detection in plants using image segmentation. *Inf. Process. Agric.* 7, 566–574.

Shi, X., Chen, Z., Wang, H., Yeung, D.-Y., Wong, W.-K., Woo, W., 2015. *Convolutional LSTM network: A machine learning approach for precipitation nowcasting*. Adv. Neural Inf. Process. Syst. 28.

Shin, J., Chang, Y.K., Heung, B., Nguyen-Quang, T., Price, G.W., Al-Mallahi, A., 2021. A deep learning approach for RGB image-based powdery mildew disease detection on strawberry leaves. *Comput. Electron. Agric.* 183, 106042. https://doi.org/10.1016/j.compag.2021.106042

Simonyan, K., Zisserman, A., 2015. Very deep convolutional networks for large-scale image recognition. 3rd Int. Conf. Learn. Represent. ICLR 2015 - *Conf.* Track Proc.

Singh, D., Jain, N., Jain, P., Kayal, P., Kumawat, S., Batra, N., 2020. PlantDoc: A dataset for visual plant disease detection, in: Proceedings of the 7th ACM IKDD CoDS and 25th COMAD. pp. 249–253.

Sinha, S., Lee, Y.M., 2024. Challenges with developing and deploying AI models and applications in industrial systems. *Discov. Artif. Intell.* 4, 55. https://doi.org/10.1007/s44163-024-00151-2

Soffer, M., Hadar, O., Lazarovitch, N., 2021. Automatic Detection of Water Stress in Corn Using Image Processing and Deep Learning, in: *Lecture Notes in Computer Science*. Springer International Publishing, pp. 104–113. https://doi.org/10.1007/978-3-030-78086-9_8

Stein, M., Bargoti, S., Underwood, J., 2016. Image Based Mango Fruit Detection, Localisation and Yield Estimation Using Multiple View Geometry. *Sensors* 16, 1915. https://doi.org/10.3390/s16111915

Subeesh, A., Mehta, C., 2021. Automation and digitization of agriculture using artificial intelligence and internet of things. *Artif. Intell. Agric.* 5. https://doi.org/10.1016/j.aiia.2021.11.004

Suchithra, M.S., Pai, M.L., 2020. Improving the prediction accuracy of soil nutrient classification by optimizing extreme learning machine parameters. *Inf. Process. Agric.* 7, 72–82. https://doi.org/10.1016/j.inpa.2019.05.003

Sudarsan, B., Ji, W., Biswas, A., Adamchuk, V., 2016. Microscope-based computer vision to characterize soil texture and soil organic matter. *Biosyst. Eng.* 152, 41–50.

Sujatha, R., Chatterjee, J.M., Jhanjhi, N.Z., Brohi, S.N., 2021. Performance of deep learning vs machine learning in plant leaf disease detection. *Microprocess. Microsyst.* 80, 103615-NA. https://doi.org/10.1016/j.micpro.2020.103615

Sun, J., Di, L., Sun, Z., Shen, Y., Lai, Z., 2019. County-Level Soybean Yield Prediction Using Deep CNN-LSTM Model. *Sensors* 19, 4363. https://doi.org/10.3390/s19204363

Sunil, G.C., Zhang, Y., Koparan, C., Ahmed, M.R., Howatt, K., Sun, X., 2022. Weed and crop species classification using computer vision and deep learning technologies in greenhouse conditions. *J. Agric. Food Res.* 9, 100325.

Syed-Ab-Rahman, S.F., Hesamian, M.H., Prasad, M., 2022. Citrus disease detection and classification using end-to-end anchor-based deep learning model. *Appl. Intell.* 52, 927–938. https://doi.org/10.1007/s10489-021-02452-w

Taghavi Namin, S., Esmaeilzadeh, M., Najafi, M., Brown, T.B., Borevitz, J.O., 2018. Deep phenotyping: Deep learning for temporal phenotype/genotype classification. *Plant Methods* 14. https://doi.org/10.1186/s13007-018-0333-4

Taheri-Garavand, A., Nasiri, A., Fanourakis, D., Fatahi, S., Omid, M., Nikoloudakis, N., 2021. Automated In Situ Seed Variety Identification via Deep Learning: A Case Study in Chickpea. *Plants* 10, 1406. https://doi.org/10.3390/plants10071406

Tang, J., Arvor, D., Corpetti, T., Tang, P., 2021. Mapping Center Pivot Irrigation Systems in the Southern Amazon from Sentinel-2 Images. *Water* 13, 298. https://doi.org/10.3390/w13030298

Tedesco-Oliveira, D., Pereira da Silva, R., Maldonado, W., Zerbato, C., 2020. Convolutional neural networks in predicting cotton yield from images of commercial fields. *Comput. Electron. Agric.* 171, 105307. https://doi.org/10.1016/j.compag.2020.105307

Terliksiz, A.S., Altylar, D.T., 2019. Use Of Deep Neural Networks For Crop Yield Prediction: A Case Study Of Soybean Yield in Lauderdale County, Alabama, USA, in: *2019 8th International Conference on Agro-Geoinformatics (Agro-Geoinformatics)*. IEEE, pp. 1–4. https://doi.org/10.1109/Agro-Geoinformatics.2019.8820257

Tetila, E.C., Machado, B.B., Astolfi, G., de Souza Belete, N.A., Amorim, W.P., Roel, A.R., Pistori, H., 2020. Detection and classification of soybean pests using deep learning with UAV images. *Comput. Electron. Agric.* 179, 105836.

Thangaraj, R., Anandamurugan, S., Kaliappan, V.K., 2021. Automated tomato leaf disease classification using transfer learning-based deep convolution neural network. *J. Plant Dis. Prot.* 128, 73–86. https://doi.org/10.1007/s41348-020-00403-0

Thenmozhi, K., Reddy, U.S., 2019. Crop pest classification based on deep convolutional neural network and transfer learning. *Comput. Electron. Agric.* 164, 104906.

Tian, H., Wang, P., Tansey, K., Han, D., Zhang, J., Zhang, S., Li, H., 2021. A deep learning framework under attention mechanism for wheat yield estimation using remotely sensed indices in the Guanzhong Plain, PR China. *Int. J. Appl. Earth Obs. Geoinf.* 102, 102375. https://doi.org/10.1016/j.jag.2021.102375

Tian, Y., Yang, G., Wang, Z., Wang, H., Li, E., Liang, Z., 2019. Apple detection during different growth stages in orchards using the improved YOLO-V3 model. *Comput. Electron. Agric.* 157, 417–426. https://doi.org/10.1016/j.compag.2019.01.012

Torres-Tello, J., Ko, S.-B., 2021. Identifying useful features in multispectral images with deep learning for optimizing wheat yield prediction, in: *2021 IEEE International Symposium on Circuits and Systems (ISCAS)*. IEEE, pp. 1–5.

Trendov, N.M., Varas, S., Zeng, M., 2019. *Digital technologies in agriculture and rural areas*. FAO;

Tu, K., Li, L., Yang, L., Wang, J., Sun, Q., 2018. Selection for high quality pepper seeds by machine vision and classifiers. *J. Integr. Agric.* 17, 1999–2006. https://doi.org/10.1016/S2095-3119(18)62031-3

Turkoglu, M., Hanbay, D., 2019. Plant disease and pest detection using deep learning-based features. *TURKISH J. Electr. Eng. Comput. Sci.* 27, 1636–1651. https://doi.org/10.3906/elk-1809-181

Ubbens, J.R., Stavness, I., 2017. Deep plant phenomics: A deep learning platform for complex plant phenotyping tasks. *Front. Plant Sci.* 8, 1190. https://doi.org/10.3389/fpls.2017.01190

Ucat, R.C., Dela Cruz, J.C., 2019. Postharvest Grading Classification of Cavendish Banana Using Deep Learning and Tensorflow, in: *2019 International Symposium on Multimedia and Communication Technology (ISMAC)*. IEEE, pp. 1–6. https://doi.org/10.1109/ISMAC.2019.8836129

Varga, P., Plosz, S., Soos, G., Hegedus, C., 2017. Security threats and issues in automation IoT, in: *2017 IEEE 13th International Workshop on Factory Communication Systems (WFCS)*. IEEE, pp. 1–6. https://doi.org/10.1109/WFCS.2017.7991968

Vaswani, A., Shazeer, N., Parmar, N., Uszkoreit, J., Jones, L., Gomez, A.N., Kaiser, Ł., Polosukhin, I., 2017. Attention is all you need. Adv. Neural Inf. Process. Syst. 2017-Decem, 5999–6009.

Veeramani, B., Raymond, J.W., Chanda, P., 2018. DeepSort: deep convolutional networks for sorting haploid maize seeds. *BMC Bioinformatics* 19, 289. https://doi.org/10.1186/s12859-018-2267-2

Veeranampalayam Sivakumar, A.N., Li, J., Scott, S., Psota, E., J. Jhala, A., Luck, J.D., Shi, Y., 2020. Comparison of Object Detection and Patch-Based Classification Deep Learning Models on Mid- to Late-Season Weed Detection in UAV Imagery. *Remote Sens.* 12, 2136. https://doi.org/10.3390/rs12132136

Vishwakarma, D.K., Pandey, K., Kaur, A., Kushwaha, N.L., Kumar, R., Ali, R., Elbeltagi, A., Kuriqi, A., 2022. Methods to estimate evapotranspiration in humid and subtropical climate conditions. *Agric. Water Manag.* 261, 107378. https://doi.org/10.1016/j.agwat.2021.107378

Wang, A., Xu, Y., Wei, X., Cui, B., 2020. Semantic Segmentation of Crop and Weed using an Encoder-Decoder Network and Image Enhancement Method under Uncontrolled Outdoor Illumination. *IEEE Access* 8, 81724–81734. https://doi.org/10.1109/ACCESS.2020.2991354

Wang, Q., Xia, M., Chen, Y., Yu, Y., 2022. Design and Implementation of WSN-based Warehouse Temperature Sensor Monitoring System. *J. Phys. Conf. Ser.* 2246, 12006. https://doi.org/10.1088/1742-6596/2246/1/012006

Wang, X., Huang, J., Feng, Q., Yin, D., 2020. Winter Wheat Yield Prediction at County Level and Uncertainty Analysis in Main Wheat-Producing Regions of China with Deep Learning Approaches. *Remote Sens.* 12, 1744. https://doi.org/10.3390/rs12111744

Wei, M.C.F., Maldaner, L.F., Ottoni, P.M.N., Molin, J.P., 2020. Carrot Yield Mapping: A Precision Agriculture Approach Based on Machine Learning. *AI* 1, 229–241. https://doi.org/10.3390/ai1020015

Wernick, M., Yang, Y., Brankov, J., Yourganov, G., Strother, S., 2010. Machine Learning in Medical Imaging. *IEEE Signal Process. Mag.* 27, 25–38. https://doi.org/10.1109/MSP.2010.936730

Wolanin, A., Mateo-García, G., Camps-Valls, G., Gómez-Chova, L., Meroni, M., Duveiller, G., Liangzhi, Y., Guanter, L., 2020. Estimating and understanding crop yields with explainable deep learning in the Indian Wheat Belt. *Environ. Res. Lett.* 15, 024019. https://doi.org/10.1088/1748-9326/ab68ac

Wu, H., Xiao, B., Codella, N., Liu, M., Dai, X., Yuan, L., Zhang, L., 2021. Cvt: Introducing convolutions to vision transformers, in: *Proceedings of the IEEE/CVF International Conference on Computer Vision.* pp. 22–31.

Wu, N., Zhang, Y., Na, R., Mi, C., Zhu, S., He, Y., Zhang, C., 2019. Variety identification of oat seeds using hyperspectral imaging: investigating the representation ability of deep convolutional neural network. *RSC Adv.* 9, 12635–12644. https://doi.org/10.1039/C8RA10335F

Xiao, Z., Xu, X., Xing, H., Song, F., Wang, X., Zhao, B., 2021. A federated learning system with enhanced feature extraction for human activity recognition. *Knowledge-Based Syst.* 229, 107338. https://doi.org/10.1016/j.knosys.2021.107338

Xu, G., Zhang, F., Shah, S.G., Ye, Y., Mao, H., 2011. Use of leaf color images to identify nitrogen and potassium deficient tomatoes. *Pattern Recognit. Lett.* 32, 1584–1590. https://doi.org/10.1016/j.patrec.2011.04.020

Xu, T., Zhang, P., Huang, Q., Zhang, H., Gan, Z., Huang, X., He, X., 2018. Attngan: Fine-grained text to image generation with attentional generative adversarial networks, in: *Proceedings of the IEEE Conference on Computer Vision and Pattern Recognition.* pp. 1316–1324.

Yang, B., Zhang, Z., Wu, H., 2022. Detection and attribution of changes in agricultural eco-efficiency within rapid urbanized areas: A case study in the Urban agglomeration in the middle Reaches of Yangtze River, *China. Ecol. Indic.* 144, 109533. https://doi.org/https://doi.org/10.1016/j.ecolind.2022.109533

Yang, M., Xu, D., Chen, S., Li, H., Shi, Z., 2019. Evaluation of machine learning approaches to predict soil organic matter and pH using vis-NIR spectra. *Sensors (Switzerland)* 19, 263-NA. https://doi.org/10.3390/s19020263

Yang, Q., Shi, L., Han, J., Zha, Y., Zhu, P., 2019. Deep convolutional neural networks for rice grain yield estimation at the ripening stage using UAV-based remotely sensed images. *F. Crop. Res.* 235, 142–153. https://doi.org/10.1016/j.fcr.2019.02.022

Yu, F., Hao, H., Li, Q., 2021. An Ensemble 3D Convolutional Neural Network for Spatiotemporal Soil Temperature Forecasting. *Sustainability* 13, 9174. https://doi.org/10.3390/su13169174

Yu, Y., Xu, T., Shen, Z., Zhang, Y., Wang, X., 2019. Compressive spectral imaging system for soil classification with three-dimensional convolutional neural network. *Opt. Express* 27, 23029. https://doi.org/10.1364/OE.27.023029

Zeng, C., Zhang, F., Luo, M., 2022. A deep neural network-based decision support system for intelligent geospatial data analysis in intelligent agriculture system. *Soft Comput.* 26, 10813–10826. https://doi.org/10.1007/s00500-022-07018-7

Zeraatpisheh, M., Ayoubi, S., Jafari, A., Tajik, S., Finke, P., 2018. *Digital Mapping of Soil Properties Using Multiple Machine Learning in a Semi-arid Region*, Central Iran. Geoderma. https://doi.org/10.1016/j.geoderma.2018.09.006

Zhang, C., Yue, P., Di, L., Wu, Z., 2018. Automatic Identification of Center Pivot Irrigation Systems from Landsat Images Using Convolutional Neural Networks. *Agriculture* 8, 147. https://doi.org/10.3390/agriculture8100147

Zhang, Shanwen, Zhang, Subing, Zhang, C., Wang, X., Shi, Y., 2019. Cucumber leaf disease identification with global pooling dilated convolutional neural network. *Comput. Electron. Agric.* 162, 422–430. https://doi.org/10.1016/j.compag.2019.03.012

Zhang, Y., Chu, J., Leng, L., Miao, J., 2020. Mask-Refined R-CNN: A Network for Refining Object Details in Instance Segmentation. *Sensors* 20, 1010. https://doi.org/10.3390/s20041010

Zhao, G., Quan, L., Li, H., Feng, H., Li, S., Zhang, S., Liu, R., 2021. Real-time recognition system of soybean seed full-surface defects based on deep learning. *Comput. Electron. Agric.* 187, 106230. https://doi.org/10.1016/j.compag.2021.106230

Zhong, G., Wang, L.-N., Ling, X., Dong, J., 2016. An overview on data representation learning: From traditional feature learning to recent deep learning. *J. Financ. Data Sci.* 2, 265–278. https://doi.org/10.1016/j.jfds.2017.05.001

Zhong, L., Guo, X., Xu, Z., Ding, M., 2021. Soil properties: Their prediction and feature extraction from the LUCAS spectral library using deep convolutional neural networks. *Geoderma* 402, 115366. https://doi.org/10.1016/j.geoderma.2021.115366

Zhou, X., Lee, W.S., Ampatzidis, Y., Chen, Y., Peres, N., Fraisse, C., 2021. Strawberry Maturity Classification from UAV and Near-Ground Imaging Using Deep Learning. *Smart Agric. Technol.* 1, 100001. https://doi.org/10.1016/j.atech.2021.100001

Zhou, Z., Song, Z., Fu, L., Gao, F., Li, R., Cui, Y., 2020. Real-time kiwifruit detection in orchard using deep learning on Android™ smartphones for yield estimation. *Comput. Electron. Agric.* 179, 105856. https://doi.org/10.1016/j.compag.2020.105856

Zhu, F., He, M., Zheng, Z., 2020. Data augmentation using improved cDCGAN for plant vigor rating. *Comput. Electron. Agric.* 175, 105603. https://doi.org/10.1016/j.compag.2020.105603

Zhu, J.-Y., Park, T., Isola, P., Efros, A.A., 2017. Unpaired image-to-image translation using cycle-consistent adversarial networks, in: *Proceedings of the IEEE International Conference on Computer Vision.* pp. 2223–2232.

Zhu, S., Zhou, L., Gao, P., Bao, Y., He, Y., Feng, L., 2019. Near-Infrared Hyperspectral Imaging Combined with Deep Learning to Identify Cotton Seed Varieties. *Molecules* 24, 3268. https://doi.org/10.3390/molecules24183268

Zhu, Y., Abdalla, A., Tang, Z., Cen, H., 2022. Improving rice nitrogen stress diagnosis by denoising strips in hyperspectral images via deep learning. *Biosyst. Eng.* 219, 165–176. https://doi.org/10.1016/j.biosystemseng.2022.05.001

18 Deep Learning and Neural Network Modeling to Optimize Tissue Culture of Medicinal Plants

Akshita Patial[1,3], Rakhi Chandoliya[2,3], Khushbu Kumari[3,4], Vikas Sharma[1], and Rohit Joshi[3,4*]

[1]Department of Molecular Biology and Genetic Engineering, School of Bioengineering and Biosciences, Lovely Professional University, Jalandhar, Punjab, India; [2]Department of Biotechnology, School of Bioengineering and Biosciences, Lovely Professional University, Jaladhar, Punjab, India; [3]CSIR-Institute of Himalayan Bioresource Technology, Palampur, Himachal Pradesh, India; [4]Academy of Scientific and Innovative Research (AcSIR), CSIR-HRDC Campus, Ghaziabad, Uttar Pradesh, India

Abstract

Medicinal plants face significant threats of extinction due to habitat destruction, overexploitation, and environmental degradation. Effective conservation and sustainable utilization of these species require efficient propagation techniques. Plant tissue culture offers a promising method by allowing the regulated and swift increase of medicinal plants. As a key biotechnological tool, it possesses significant potential for progressing herbal medicine and therapeutic medical practices. This chapter aims to explore the application of deep learning (DL) methods in the *in vitro* development of medicinal plants. DL methods, especially convolutional neural networks (CNNs), have transformed image-based analysis in the processes of plant tissue culture. This chapter evaluates the processes involved in plant tissue induction and emphasizes the incorporation of DL in tissue culture techniques. Specifically, we discuss the development of predictive models that relate factors such as callus type, plant growth regulator (PGR) type, and PGR concentration to the success of indirect shoot regeneration and secondary metabolite production. The goal of these insights is to improve tissue culture techniques, which will, in turn, increase both the availability and yield of natural metabolites derived from medicinal plants. Despite extensive research on DL applications in plant biotechnology, a lack of empirical studies offering meaningful comparisons persists, primarily due to the use of diverse datasets in evaluations. This chapter addresses these challenges and proposes solutions to advance the automation of plant tissue culture systems. It also provides a comprehensive resource for plant biotechnologists and tissue culturists, fostering innovation in the conservation and sustainable production of endangered medicinal plant species.

Keywords: convolutional neural networks, deep learning, *in vitro* organogenesis, medicinal plant, plant growth regulator

18.1 Introduction

The rising global demand for medicinal plants, coupled with their expanding applications and the shrinking availability of arable land, has intensified the depletion of plant resources. In response to this problem, widespread cultivation of plant cells and tissues has risen as a viable

*E-mail: rohitjoshi@ihbt.res.in

DOI: 10.1079/9781836991120.0018

approach. However, progress in this field has historically been slow due to species-specific differences, underdeveloped cultivation techniques, and rigid research methodologies (Chandran *et al.*, 2020). Although plant tissue growth can be monitored at various stages *in vitro*, the growth conditions are highly complex, non-linear, and influenced by numerous external factors (Espinosa-Leal *et al.*, 2018). Traditional statistical approaches often struggle to optimize these intricate systems, necessitating more advanced computational methods. Recent technological advancements across multiple disciplines—such as information technology, engineering, analytical techniques, and material science—have significantly contributed to innovations in plant tissue culture (PTC). A deeper understanding of plant biological mechanisms, combined with these cutting-edge technologies, has accelerated progress in this field. Modern approaches integrate theoretical knowledge with advanced methodologies to refine plant cell and tissue culture techniques. These advancements seek to improve the extensive production of plant biomass and the creation of important secondary metabolites, thereby promoting the sustainable utilization of plant resources.

The process of indirect shoot regeneration follows three fundamental stages (Mikovski *et al.*, 2019). The first stage involves the careful selection of plant material, a critical step that significantly influences the success of the regeneration process (Fernando *et al.*, 2007). Choosing unsuitable explants can negatively impact the outcome (Hesami *et al.*, 2023). Ideally, explants should be sourced from young, vigorous plants that are growing under optimal environmental conditions during a favorable season (Hesami *et al.*, 2021). Furthermore, it is essential to collect explants from plant regions containing meristematic cells, ensuring the potential for continued growth and development (Hesami *et al.*, 2023). Another vital aspect of this phase is sterilization, which is necessary to maintain contamination-free cultures. This procedure typically involves treating the plant material with sterilizing agents such as sodium, calcium, or potassium hypochlorite, or ethanol, followed by multiple rinses with sterile water (Jafari *et al.*, 2016). The second phase focuses on establishing the culture. In this step, the properly prepared explants are transferred to a nutrient medium designed to support *in vitro* plant growth. This medium contains essential micro- and macronutrients, carbohydrates, and externally supplied phytohormones that regulate development and physiological activities (Hesami *et al.*, 2020). Moreover, the spatial layout and placement of explants inside the culture vessel are essential in influencing the quality of regenerated shoots. Elements like the closeness of explants and their contact with the medium can notably affect developmental results, impacting the overall success of regeneration (Mikovski *et al.*, 2019). During the early stages, callus formation is typically observed as a precursor to shoot development. The third and final stage involves the continued cultivation of callus tissue under conditions that promote *de novo* shoot formation. This process requires the incorporation of specific plant growth regulators (PGRs) into the medium, along with considerations of environmental factors such as light exposure and temperature control (Fernando *et al.*, 2007). Several variables, including callus type and hormonal balance, directly influence shoot regeneration efficiency (Lombardi *et al.*, 2007; Pacheco *et al.*, 2012; Hesami *et al.*, 2016; Rosa *et al.*, 2016). Fine-tuning these elements is essential for optimizing the regeneration process. However, conventional statistical approaches often prove cumbersome, as they rely on labor-intensive, sequential analyses of individual factors (Jafari *et al.*, 2022). To overcome these limitations, advanced computational methods such as deep learning (DL) have emerged as powerful tools for improving predictive accuracy and analytical efficiency in optimizing indirect shoot regeneration (Hesami *et al.*, 2020). By integrating these modern technologies, researchers can refine protocols more effectively and enhance the overall success of PTC techniques.

Advancements in intelligent technologies, particularly in the domains of machine learning (ML) and DL, have greatly influenced PTC and agricultural research. These innovations are transforming plant science by enhancing our understanding of complex biological processes, automating disease detection, and refining nutrient optimization strategies, thereby revolutionizing plant breeding (Hamazaki and Iwata, 2024). The introduction of DL, especially with the implementation of convolutional neural networks (CNNs), has revolutionized how plant imagery is interpreted and processed. In a groundbreaking research effort, Davidson *et al.* (2024) explored the utilization of CNNs for assessing mature somatic embryos, an essential process in the propagation of plants. Their research demonstrated the effectiveness of neural networks in performing semantic and instance segmentation on conifer embryos. By accurately distinguishing morphological features and counting cotyledons, this approach outperformed traditional techniques in both precision and efficiency. The use of image analysis based on artificial intelligence (AI) not only enhances the evaluation of embryos but also lays the groundwork for boosting crop production and promoting sustainability in agriculture and forestry. Additionally, AI-driven models are proving valuable in optimizing farming techniques and improving crop management. One persistent issue in wheat farming is seedling loss, which negatively impacts grain yield and economic returns. To address this challenge, Feng *et al.* (2024) developed MssiapNet, a DL model designed for wheat seedling variety identification. Their study underscored the significance of AI in streamlining seedling classification and ensuring better agricultural outcomes. By leveraging advanced computational tools, researchers and farmers can enhance efficiency, reduce losses, and promote more sustainable agricultural practices. As AI progresses further, its role in plant research and crop cultivation is expected to expand, offering innovative solutions for improving agricultural productivity and resilience. Sun *et al.* (2023) introduced SCGNet, an advanced DL model designed to improve data processing and feature integration. This model is specifically optimized for the fast and precise classification of wheat grain varieties. Their research demonstrated the effectiveness of DL in accelerating plant breeding by enabling accurate varietal identification. Similarly, Li *et al.* (2023) utilized a DL framework combining a one-dimensional convolutional neural network (1D-CNN) and hyperspectral imaging to assess cottonseed viability. Through preprocessing techniques and feature extraction methods, they successfully identified key patterns related to seed vitality. Their findings emphasized the capability of the 1D-CNN model in predicting seed viability, paving the way for automated quality control systems in cottonseed assessment. The integration of spectral and image-based features further enhanced predictive performance, providing a robust approach for evaluating seed quality. This chapter examines how DL is utilized for the classification of medicinal plant species, highlighting its capabilities in identifying and organizing different species. By examining these applications, researchers can gain insights into how AI can be leveraged for medicinal plant classification, ultimately advancing research in plant sciences and biotechnology.

18.2 DL for the Classification of Medicinal Plant Species

A significant portion of the population in developing countries—over 80%, according to the World Health Organization (WHO)—relies on traditional medicine for healthcare. Herbal remedies have been used for centuries to alleviate pain and treat various diseases (Abera, 2014). Medicinal plants are also valuable sources of new antimicrobial agents, offering promising prospects for biocompatible drug development. A thoroughly studied example is *Withania somnifera*, recognized for its various therapeutic advantages, such as alleviating stress and anxiety, anti-inflammatory effects, immune system modulation, tumor inhibition, and enhancement of sexual health (Anand *et al.*, 2019; Subhabrata *et al.*, 2021). However, human activities—both direct and indirect—are accelerating the extinction of medicinal plant species, posing a serious threat to biodiversity (Murphy and Romanuk, 2014). Given this, efficient identification and classification of medicinal plants are crucial for conservation and research efforts. Advanced technologies, such as DL, are revolutionizing this field. DL, a branch of ML, employs multilayered artificial neural networks (ANNs) to analyze and autonomously interpret intricate data (Chanyal *et al.*, 2022). This technology has considerable promise for improving biodiversity research and promoting the sustainable utilization of medicinal plants.

DL employs two main classification approaches: supervised learning, which relies on labeled datasets, and unsupervised learning, which identifies patterns without predefined labels (Pushpa *et al.*, 2021). To tackle computer vision challenges, researchers integrate these techniques—unsupervised learning for extracting features from unlabeled data, and supervised learning for precise classification using labeled datasets (Pham *et al.*, 2020). This synergy is particularly valuable in medicinal plant classification, where DL enhances automation and accuracy in species recognition. By leveraging advanced algorithms, researchers can streamline botanical identification, bridging the gap between computational intelligence and plant sciences. A recent study reviewed the role of DL in this domain, highlighting its potential to revolutionize species classification through innovative methodologies (Chanyal *et al.*, 2022). DL has shown significant potential in medicinal plant species classification due to its ability to extract and learn hierarchical features directly from raw data. Below are some ways DL is applied in this domain.

18.2.1 Image-based classification

DL models, particularly CNNs, can classify plant species based on visual features from images of leaves, flowers, fruits, or bark (Arun *et al.*, 2025). Example architectures include ResNet, Inception, VGG, and EfficientNet. Image-based classification of medicinal plants leverages advanced technologies like ML and computer vision to identify plant species using their visual characteristics. This method utilizes images of leaves, flowers, stems, or seeds captured via digital cameras or mobile devices. Features such as shape, color, texture, and venation patterns are extracted and analyzed using algorithms like CNNs. The technique enhances accuracy in species identification, aiding research, conservation, and the pharmaceutical industry. It provides a fast, scalable, and cost-effective alternative to traditional taxonomy, which relies on expert knowledge. This approach is especially valuable for documenting biodiversity and discovering novel medicinal plants.

18.2.2 Analysis of molecular data

Molecular data analysis of medicinal plants using DL, particularly recurrent neural networks (RNNs), revolutionizes the exploration of bioactive compounds and genetic patterns. RNNs, with their ability to process sequential data, are ideal for analyzing complex datasets such as DNA sequences, transcriptomes, and metabolomes (Askr *et al.*, 2023). These networks detect patterns and predict relationships between molecular structures and biological activities, enhancing drug discovery and phytochemical profiling. By integrating omics data, RNNs aid in identifying novel therapeutic compounds, optimizing pharmacological properties, and understanding gene expression dynamics. Their application accelerates precision in medicinal plant research, paving the way for advancements in natural product-based drug development and healthcare.

18.2.3 Multimodal learning

Combining different types of data (e.g. image, molecular, and environmental data) can improve classification accuracy. Multimodal networks can integrate these inputs effectively. Multimodal learning integrates diverse data types, such as images, text, and genomic information, to enhance the understanding of medicinal plants using DL techniques (Montesinos-Lopez *et al.*, 2024). Utilizing CNNs for analyzing images, natural language processing (NLP) for handling textual information, and graph neural networks (GNNs) for understanding molecular interactions allows researchers to gain diverse insights. This approach improves the identification of plant species, prediction of bioactive compounds, and understanding of therapeutic properties. Multimodal models enable more accurate and robust predictions by combining features across modalities, aiding drug discovery and biodiversity conservation. DL adaptability drives innovation in exploring the vast potential of medicinal plants.

18.2.4 Transfer learning

Models that have been pretrained (such as those trained on ImageNet) can be adapted for classifying medicinal plants. This method is particularly advantageous when dealing with small datasets. Transfer learning utilizes pretrained DL frameworks to speed up the analysis of medicinal plants. By employing models that have been trained on extensive datasets, such as images of plants or genetic information, researchers can customize these models for specific tasks related to medicinal plants, such as identifying species or predicting bioactive compounds (Patil *et al.*, 2022). This technique minimizes the requirement for large-scale data gathering and lengthy training periods, making it especially beneficial for rare or less-studied plants. Potential applications include drug discovery, conservation of biodiversity, and validation of traditional medicine. Integrating transfer learning with specialized datasets boosts model precision, allowing researchers to more efficiently and effectively harness the therapeutic benefits of medicinal plants.

18.2.5 Feature interpretation

Using techniques like Grad-CAM or SHAP (Shapley Additive Explanations) to visualize which features (e.g. leaf veins, flower patterns) the model uses for classification makes the results more interpretable (Mostafa *et al.*, 2023). Feature interpretation methods like Grad-CAM and SHAP are crucial for understanding DL models used in medicinal plant research. Grad-CAM (Gradient-weighted Class Activation Mapping) visualizes the regions in plant images that most influence model predictions, highlighting key features such as leaf patterns or plant morphology (Quach *et al.*, 2023). SHAP provides a detailed breakdown of how each feature contributes to a model's output, allowing for transparency in predictions. Utilizing these methods for classifying medicinal plants or predicting compounds enables researchers to better understand the rationale behind the model's decisions, enhancing the trust and clarity of AI-driven botanical research.

18.3 Current Developments in ML Utilizing Data from Various Sources Related to Medicinal Plants

18.3.1 Machine learning (ML)

ML is especially well suited to addressing learning challenges with small sample sizes. There are many distinct methods in the area of classical ML (Fig. 18.1). For instance, medicinal plants' quality assessment has made important use of both supervised and unsupervised learning (Zhu *et al.*, 2021).

Exploratory analysis

Exploratory analysis utilizes statistical principles to uncover connections between samples, variables, and the relationships between these samples and variables. Each of these three categories of information holds distinct significance. The data between samples can help detect anomalies and illustrate trends within the sample distribution. The correlations between the variables help identify information that is similar, redundant, and complementary. Furthermore, the important contribution of factors to sample categorization is reflected in the sample–variable relationships (Zhang *et al.*, 2023).

The importance of exploratory analysis is often underestimated, even though it is essential for understanding data relationships prior to classification and regression analysis. Through an improved knowledge of the interactions

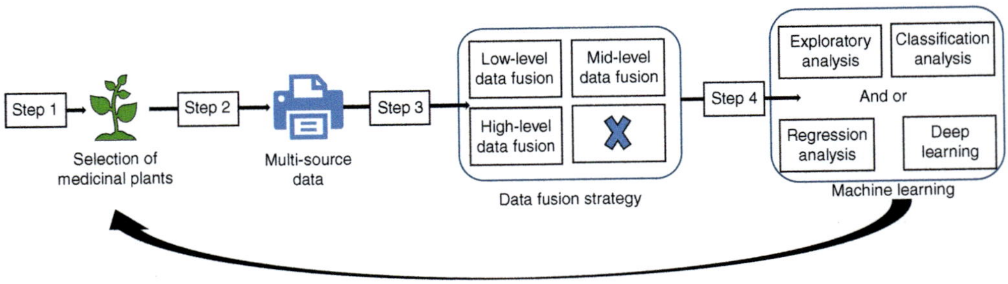

Fig. 18.1. Techniques used in machine learning include three basic categories of analysis. (Author's own image.)

between different data elements, it indirectly leads to the selection of models.

Regression analysis

Advanced mathematical models are integral to regression-based approaches for investigating how various features of medicinal plant data relate to key continuous variables, such as chemical composition and levels of adulteration. These techniques are crucial for accurate prediction and quantification, both of which are vital to ensuring the quality and authenticity of medicinal herbs (Li *et al.*, 2020). Widely used regression methods in this domain include Multivariate Linear Regression (MLR), Principal Component Regression (PCR), Partial Least Squares Regression (PLSR), and Support Vector Regression (SVR). Of these, PLSR is particularly favored for determining the concentration of active components and adulterants in herbal formulations. However, traditional regression models are typically limited to analyzing unimodal datasets. In contrast, data generated from medicinal plant studies are often not only multidimensional but also derived from multiple sources and modalities. To address this complexity, multimodal predictive modeling is gaining momentum. This emerging strategy significantly improves the accuracy of predictive outcomes and allows for a deeper understanding of the interrelated and unique contributions of diverse data types.

Classification analysis

Classification models are crucial for qualitatively assessing the quality of medicinal products. They aid in identifying varieties, detecting adulteration, and ensuring traceability across different regions. These models fall under the category of supervised learning models, meaning they require prior knowledge to be trained on known data before they can identify unknown instances. Classification models can be categorized into two types based on the specific issues they address. The first type, known as discriminant analysis, focuses on evaluating target compounds within defined categories, while the second type, class modeling, determines whether a target belongs to a specific class (Granato *et al.*, 2018). Classification analysis techniques largely rely on manual labeling, and to stay competitive in market applications, they must improve their efficiency in classifying large datasets. Although there have been recommendations to improve these techniques, they often involve only partial optimizations. Enhancing their capabilities in big data analysis and increasing their market appeal are the most critical considerations.

Classification using ML models

The domains of ecology, drug discovery, and conservation biology have achieved significant advancements thanks to improvements in automated identification and classification of plant species. Numerous initiatives have been launched to automatically classify plant species, tackling the issues arising from variations within a class and similarities between different classes, using leaf images taken with smartphones and high-resolution cameras.

In a 2016 study by Naresh and Nagendraswamy, a technique was introduced for recognizing species of Indian medicinal plants. They employed a K-NN classifier along with a chi-square dissimilarity measure, using a modified local binary pattern approach for feature extraction. Their models were evaluated using benchmark datasets as well as a self-assembled dataset that included 32 species. Various ML models were developed using features obtained from Oriented FAST, rotated BRIEF, histogram of oriented gradients, and scale-invariant feature transform (Roopashree *et al.*, 2022). Their experiments made use of a benchmark dataset in conjunction with a customized dataset consisting of 2515 samples from 40 distinct plant species. The smartphone application effectively identified the species through ML classification methods. Mostajer Kheirkhah and Asghari (2019) approached plant species recognition using the structure of GIST and trained it with various ML classifiers. Local Binary Patterns-Histogram Fourier (LBP-HF) was utilized by Bambil *et al.* (2020) for classifying plant species. Additionally, only a few studies (Muneer and Fati, 2020; Yang, 2021) have examined the identification of different medicinal plant species based on texture- and shape-related features from a limited selection of classes. The research conducted by Turkoglu and Hanbay (2019) centered on the classification of plant species by combining morphological features with non-parametric classifiers.

18.3.2 Deep learning (DL)

DL represents a branch of ML that emphasizes the training of multilayer ANNs to autonomously create data representations (Chanyal *et al.*, 2022). Techniques in DL have yielded outstanding outcomes in computer vision, with uses such as image recognition and enhancement being prevalent across various industries, including healthcare, agriculture, education, and manufacturing (Qiuosen *et al.*, 2022). There are essentially two classification methods in DL: supervised learning, which uses labeled data to guide the learning, and unsupervised learning, which groups similar patterns without labels in the beginning (Pushpa *et al.*, 2021). Researchers are merging supervised DL using labeled data with unsupervised DL using unlabeled data to overcome computer vision problems. The collaboration between contemporary botanical research and DL techniques emphasizes the ability of DL to enhance and automate the classification and recognition of medicinal plant species (Mulugeta *et al.*, 2024).

Categorization utilizing DL models

DL models have demonstrated their effectiveness in recognizing plant leaves due to their ability to handle a wide variety of species or classes. Numerous studies have investigated this area. To categorize plant species based on leaf characteristics, global features are fed into CNN in conjunction with DL techniques that employ transfer learning (Gladston and Sucithra, 2023). The identification of plant species, which was trained and tested using nine publicly accessible datasets, is accomplished by integrating form features with convolution features to different degrees (Wu *et al.*, 2023). In their study, Quach and associates (2023) presented CNN encoders that capture various combinations of shapes, colors, vein patterns, textures, and Fourier representations in preprocessed images. Yang *et al.* (2022) introduced a bilinear CNN utilizing transfer learning for the identification of plant species using standard datasets. A pretrained DL model, EfficientNet-B1, was both trained and evaluated on public and custom plant species datasets, achieving accuracies of 87% and 84% (Malik *et al.*, 2022). Azadnia *et al.* (2022) also developed a CNN model to identify various plant species. Their approach involved a box model that was subjected to several preprocessing techniques, such as image resizing, segmentation, and data augmentation, to expand the dataset. This process yielded 750 image samples representing five different medicinal plants. Similarly, Malik *et al.* (2022) introduced a DL-based approach aimed at automating the classification of medicinal plant species native to the Borneo region.

To improve the precision of real-time classification systems, it is crucial to thoroughly tackle the challenges posed by real-time image capture conditions. The DL models created by Wang *et al.* (2022) and Sennan *et al.* (2022) focus on detecting various disease variants within a single crop. In general, DL models have shown significant promise in identifying plant leaves and diagnosing diseases, and their use can enhance harvesting practices and result in increased yields. The findings indicate that much of the research is centered on standard datasets, disease and weed identification (Quach *et al.*, 2023), recognition of crop species (Wang *et al.*, 2022), and the classification of non-medical plants (Yang *et al.*, 2022). Contributions to the research of medicinal plant species that are unique to India have been exceptionally limited; studies concentrate on a narrow range of plant species datasets, and the images captured typically feature only a single leaf in a constrained environment against a plain backdrop. While ML feature extraction models have been incorporated, background research reveals that no attempts have been made to explore a wider array of medicinal plant species from India. Despite considering a large number of characteristics, the accuracy achieved remains constrained. Furthermore, the pursuit of identifying plant species with high intra-class variability and inter-class similarities has become outdated. To establish an effective recognition system across diverse datasets, it is essential to develop a structured approach that includes a limited set of features. Additionally, to evaluate the robustness of the proposed model, it is necessary to test it using real-time images.

18.4 Multisource Data of Medicinal Plants

Data from multiple sources about medicinal plants can be classified into two primary categories. The first emphasizes the use of various analytical techniques to identify the same portion of medicinal herbs. The second applies the same scientific approach to analyze different parts of medicinal plants (Table 18.1). These two categories of multisource data will be discussed in the next section.

18.4.1 Various detection methods

1. Spectroscopy and spectrometry
2. Chromatography
3. Elemental analysis
4. Artificial sensing methods

18.4.2 Utilization of ML integrated with diverse data sources pertaining to medicinal plants

Identification of adulterate

The presence of adulterants poses a major challenge to the purity and effectiveness of medicinal herbs. Since these adulterants typically do not impact the taste or alter the chemical profile in noticeable ways, they often go undetected through simple visual or sensory inspection. Identifying such contamination is a complex task, and existing analytical techniques are generally restricted to evaluating a limited number of known markers (Zhang *et al.*, 2023).

The integration of data from multiple sources has proven to significantly enhance the precision of adulteration detection, making it a widely adopted strategy in both food safety and medicinal plant research (Fig. 18.2). As a result, applying ML models to these fused datasets offers a powerful and promising approach for identifying fraudulent practices. This approach offers a valuable solution for detecting fraudulent activities in the pharmaceutical sector. However, it should be recognized that this method has its limits and cannot facilitate the detection of rapid and large-scale market fraud in the field of medicinal products. Therefore, further improvements and refinements of this technique are needed to promote wider adoption. These developments are crucial for the efficient management of the quality of pharmaceutical products.

Geographical tracking

Tracking the geographical origin of medicinal plants is essential for ensuring their quality, as environmental factors such as climate, soil type, and topography significantly influence their chemical composition. However, pinpointing the exact origin based on a few chemical markers is often unreliable due to the complex, overlapping, and interactive nature of phytochemicals. A single analytical technique is typically insufficient to capture the complete chemical landscape of a medicinal product (Wang *ét al.*, 2015). Designing robust origin-tracing frameworks remains a considerable challenge. Nonetheless, integrating diverse datasets can enhance compatibility between analytical methods and help construct a more holistic chemical fingerprint of medicinal plants, aiding in accurate origin determination (Zhong *et al.*, 2020). Recent advancements suggest that combining spectral analysis with ML tools offers promising solutions for this purpose. The fusion of multisource data with advanced algorithms has proven more effective in determining plant origin than traditional single-method approaches, delivering improved classification performance. Thus, it presents an innovative approach for tracking the geographical origins of pharmaceutical products. Additionally, the geographical tracking of medicines has applications in various scenarios, including characteristics of chromatophysics. As a result, spectroscopy has emerged as a favored method in this field.

Content forecasting

The therapeutic value of medicinal products largely depends on the levels of their bioactive constituents. As such, precise estimation of these compounds is essential for evaluating the overall quality of medicinal plant materials. To achieve this, calibration models are employed, drawing on different spectral regions and preprocessing strategies to enhance prediction accuracy. The performance of these models varies depending on the spectral input and data treatment techniques used, and optimizing these aspects can significantly improve prediction outcomes. While recent

Table 18.1. Various types of detection techniques along with applications, as well as advantages and disadvantages. Author's own table.

Types (techniques)	Advantages	Disadvantages	Applications
Spectroscopy and spectrometry (NMR, HSI, Raman, FT-IR, UV-Vis, MIR, NIR)	NMR: excellent reproducibility; no complicated preprocessing is required	Significant overlap in the signals	Rarely used to evaluate the quality of medicinal herbs
	HSI: acquiring both spectral and spatial data at the same time, across a broad range of scanning wavelengths	Unnecessary information is easily generated and time-consuming	A promising approach to assessing the quality of medicinal herbs
	Raman: moisture resistant, high sensitivity, and simple to operate	Fluorescence susceptibility	Less used in assessing the quality of medicinal herbs
	FT-IR: increased sensitivity, sharpness, and accuracy, along with ample spectral information	Tablet compression is highly affected by humidity and necessitates an extended detection period	Frequently employed in the qualitative assessment of medicinal herbs
	UV-Vis: quick identification, simple operation, and high sensitivity	Limited selectivity; prone to interference from multiple sources	Generally employed for the quantitative assessment of medicinal plants
	MIR: elevated sensitivity, elevated selectivity	Responsive to humidity, demonstrating a gradual reaction time	Frequently used in the qualitative examination of medicinal herbs
	NIR: quick identification, easy to use, non-invasive, and environmentally friendly	Poor sensitivity and significant signal interference	The most common method for evaluating medicinal plant quality
Chromatography (UHPLC, HPLC, GC)	UHPLC: low solvent use and quick analytical speed	Expensive equipment	Commonly used to ensure the quality of medicinal herbs
	HPLC: excellent separation efficiency, outstanding sensitivity	Expensive environmental contamination	Frequently utilized to assess the quality of herbal medicines
	GC: high specificity, sensitivity, fast processing speed, and excellent separation ability	Expensive equipment, limited qualitative ability, and the use of reference substances are necessary	Commonly employed for analyzing volatile or semi-volatile elements in medicinal herbs
Elemental analysis (IJBS, ICP-MS)	ICP-MS: great sensitivity and an extensive linear dynamic range	Pollution of the environment and intricate methods of sample preparation	Primarily employed for identifying components in medicinal plants
	IJBS: no preparation of samples is required. It is quick and instantaneous, while remaining cost-effective	Interference from the matrix, effect of self-absorption	Mainly utilized for identifying elements in medicinal plants
Artificial sensing technique (E-tongue, E-nose, CV)	E-tongue: simple and quick, responsive across various sensitivities, non-invasive, and environmentally friendly	Vulnerable to changes in humidity and temperature	Assessment of the quality of medicinal plants that are primarily utilized in liquid form
	E-nose: a quick analysis with a small sample size is required	Quantification is a significant challenge	The applications for evaluating the quality of medicinal plants are increasing
	CV: simple and quick, harmless, and eco-friendly	Only limited information is available.	The applications for classifying medicinal plants are on the rise

CV, Computer vision; FT-IR, Fourier transform infrared; HIS, hyperspectral imaging; MIR, mid-infrared; GC, gas chromatography; HPLC, high-performance liquid chromatography; ICP-MS, inductively coupled plasma mass spectrometry; NIR, near-infrared; NMR, nuclear magnetic resonance; UHPLC, ultra-high-performance liquid chromatography; UV-Vis, ultraviolet visible.

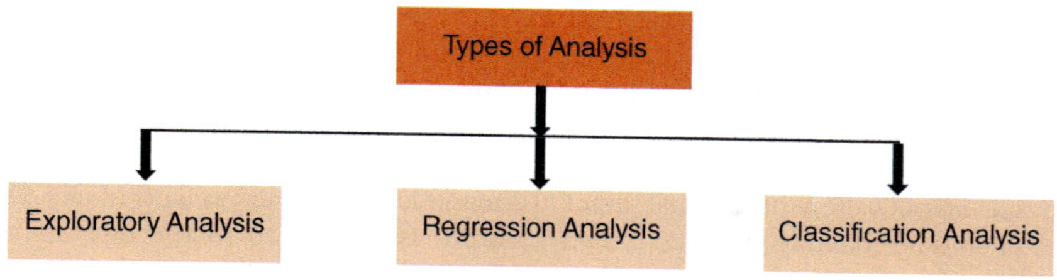

Fig. 18.2. Evaluation of the quality of herbal medicines. (Author's own image.)

studies have increasingly emphasized the geographic authentication of medicinal plants, the quantification of active ingredients has received comparatively less attention. Nonetheless, robust prediction of phytochemical content remains a key parameter for quality assurance and warrants more focused research.

Variety classification

Although medicinal plants belonging to the same genus often share morphological and therapeutic similarities, the composition of bioactive compounds can vary significantly across species, affecting their medicinal efficacy and quality. Traditional identification methods, which rely on visual assessment, are not only subjective but also demand skilled experts and substantial time investment. This underscores the pressing need for a faster, more objective, and scientifically grounded approach to differentiate among medicinal plant species. The classification and authentication of these plants is inherently complex and cannot be reliably achieved using a single analytical strategy. In recent years, notable progress has been made through the integration of diverse datasets—combining both qualitative traits and quantitative chemical profiles—alongside ML algorithms. Furthermore, breakthroughs in chemical imaging and electronic sensing technologies have opened up new avenues. These advanced tools offer promising, efficient alternatives to traditional identification methods based on physical features.

Other factors

The selection of medicinal drug product components and their growth duration are crucial for assessing the quality and market value of these products. Consequently, leveraging multiple data sources alongside ML is permitted in this domain. The integration of various data sources and ML is applied across different sectors. Although its application is restricted to specific areas, it still yields valuable insights for examining related subjects. Assessing the quality of pharmaceuticals is a multifaceted endeavor, and it is insufficient to concentrate solely on aspects like geographic origin or distinct counterfeits. Developing a comprehensive quality evaluation platform that incorporates blockchain technology is highly practical for overseeing the quality of medications.

18.5 Anticipation and Improvement of PTC through the Application of ML and Optimization Techniques

The success of *in vitro* plant cultivation is shaped by numerous variables, including nutrient formulation, plant genetic background, seed characteristics (such as type and age), and the concentration of PGRs and phytohormones (Hesami *et al.*, 2020). Due to the complex interactions and variability among these factors, conventional statistical tools such as ANOVA and linear regression often fall short in providing reliable insights. To overcome these analytical limitations, ML has emerged as a powerful solution, offering systems the ability to learn from data and improve predictions through experience (Ray *et al.*, 2019). ML encompasses several approaches: supervised learning, which uses labeled data to train models for accurate prediction and classification; unsupervised learning, which identifies hidden patterns within unlabeled datasets; and reinforcement learning, which optimizes decisions based on trial-and-error interactions with the environment (Alloghani *et al.*, 2020; Mahesh *et al.*, 2020). Recent advances in PTC have embraced these AI-driven techniques to interpret non-linear and multidimensional data. By integrating ML with optimization strategies, researchers have been able to explore the relationships between culture medium components and plant growth outcomes. This has facilitated the identification of optimal input conditions to enhance biomass production and overall plant development (Arab *et al.*, 2017).

18.5.1 AI models

Artificial neural networks (ANNs)

ANNs are designed to handle noisy, imperfect data and non-linear issues (Morris *et al.*, 1994). ANNs can identify and approximate any complicated non-linear system, allowing mathematical models to be developed quickly with limited experimental data (Ale-Dabbous *et al.*, 2017).

Generalized regression neural network (GRNN)

In 1991, Specht proposed the GRNN, a type of ANN designed for fast learning and statistical modeling (Fig. 18.3). Unlike conventional models, GRNN features a hidden layer with as many neurons as there are training samples. Its architecture is composed of four distinct layers: input, pattern, summation, and output. The pattern layer, fully connected to the input layer, processes incoming data and forwards signals to two types of neurons in the summation layer: the S-summation and D-summation neurons. These neurons serve different roles—S-summation neurons compute weighted sums where weights correspond to the target outputs, while D-summation neurons perform unweighted summations with fixed weights of one. The final output is calculated in the output layer by taking the ratio of the weighted sum (from the S-summation) to the unweighted sum (from the D-summation), yielding a predicted value for the given input vector. This design allows GRNN to generalize effectively with minimal training time.

Radial basis function (RBF)

Radial basis networks, which are structured statistical ANNs, are fundamentally based on the three-layer architecture of RBF networks, which includes an input layer, a hidden layer, and an output layer. Unlike biologically inspired traditional ANNs, statistical ANNs operate using regression techniques rather than mimicking natural neural pathways (Lin *et al.*, 2020). In the RBF model, each neuron's activation is determined by applying a transfer function to the Euclidean distance between the input vector and the neuron's center.

Probabilistic neural network (PNN)

It is a specialized form of ANN primarily built for classification tasks. Its structure and function resemble those of the RBF network. When an

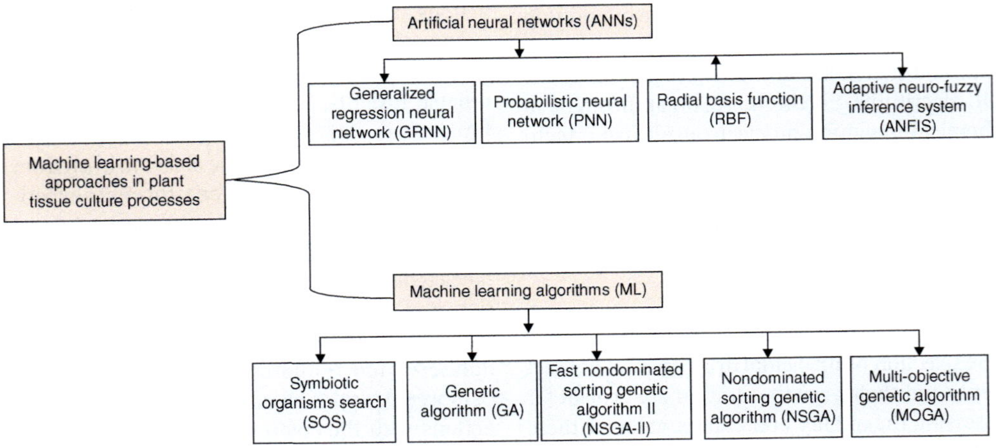

Fig. 18.3. Methods utilizing machine learning applied in plant tissue culture procedures. (Author's own image.)

input is provided, the initial layer computes its distance from stored training samples and transforms this into a likelihood distribution. From this probability set, the output layer selects the class with the highest probability through a decision function. Notably, PNNs are not suited for regression problems, as they lack the ability to model continuous output values. As such, their strength lies in offering reliable categorical predictions (Ying *et al.*, 2020).

Adaptive neuro-fuzzy inference systems (ANFIS)

Neuro-fuzzy logic merges the adaptive learning strength of ANNs with the reasoning power of fuzzy logic systems (Dorzhigulov and James, 2020). Unlike conventional approaches that depend heavily on numerical data, developing a neuro-fuzzy system emphasizes qualitative reasoning. This hybrid model translates observed patterns into interpretable statements, facilitating clearer insight into data relationships. Within these systems, ANNs play a critical role in fine-tuning fuzzy parameters and transforming raw numerical inputs into meaningful linguistic rules. As outlined by Hesami *et al.* (2020), the architecture of such models typically includes mechanisms for fuzzification, inference, and defuzzification. A prominent example is the ANFIS, which offers a structured and efficient framework for system modeling. According to Prado *et al.* (2020), ANFIS rapidly identifies optimal configuration parameters and effectively captures the dynamics of complex processes using Sugeno-type fuzzy inference rules.

18.5.2 Categories of ML algorithms

Utilizing ML algorithms in PTC can significantly reduce both costs and time when choosing the best options from existing alternatives. Consequently, embracing optimization techniques in PTC holds considerable importance (Osama *et al.*, 2015). Generally, optimization techniques can be categorized into two groups: classical algorithms and evolutionary algorithms.

Symbiotic organism search (SOS)

SOS is a highly promising metaheuristic optimization algorithm (OA) with cutting-edge global optimization capabilities. It relies on the symbiotic interaction between two organisms in an ecosystem to survive, and cycles to determine the optimal value of a particular function (Ezugwu *et al.*, 2019). The SOS algorithm comprises three phases: mutualism, commensalism, and parasitism. During the mutualism phase, each individual interacts with others in the population in order to benefit both individuals. During the commensalism phase, two organisms are randomly selected from the ecosystem to participate in symbiotic interactions. In the parasitism phase of the issue search, an artificial parasite vector is generated by mutating the organism and then altering its selected dimensions using a random interruption. The organism that will function as a host is chosen at random from the ecosystem's population. The evaluation is carried out in such a way that, if the fitness value is greater than that of the organism, the artificial host will replace the position of the mutating organism in the population; otherwise, if the fitness value of the mutating organism is greater, the mutating organism will develop immunity against the artificial parasite vector, after which the artificial parasite vector will be removed from the population list (Ezugwu *et al.*, 2018).

Genetic algorithm (GA)

The GA can potentially be used to optimize distinct *in vitro* culture stages in PTC (Fig. 18.3). Holland (1992) invented the GA for the first time and developed it as a powerful optimization tool. GA operates as a search technique inspired by evolutionary biology and the theory of natural selection. Drawing from Darwin's principle that organisms best adapted to their surroundings are more likely to endure, GAs simulate this survival mechanism to solve optimization problems (Yun *et al.*, 2020). To grasp how GAs function, it is essential to become familiar with several key components central to their framework. These include: (i) gene, the basic unit of information; (ii) chromosome, a sequence of genes; (iii) population, a collection of potential solutions; (iv) evolutionary operators that guide adaptation; (v) genetic operators such as mutation and crossover; (vi) elitism, which preserves top-performing solutions; and (vii) generation, representing each cycle of algorithmic evolution.

Non-dominated sorting genetic algorithm (NSGA-II)

Among the various evolutionary strategies for multi-objective optimization, NSGA-II stands out as one of the most influential and widely utilized algorithms (George and Amudha, 2020). Originally introduced by Srinivas and Deb in 1994 as NSGA, the algorithm was based on the principle of dominance and aimed to generate Pareto-optimal fronts with good solution diversity. However, its early version had limitations in terms of computational efficiency and diversity maintenance. To address these challenges, Deb *et al.* (2002) developed an improved version called NSGA-II, which has since become a benchmark in the field. Additionally, significant advancements were made in how non-dominated solutions were identified and preserved, greatly enhancing the algorithm's performance and scalability. As a result, NSGA-II has served as a foundation for numerous subsequent studies and algorithmic innovations in multi-objective optimization.

Multi-objective genetic algorithm (MOGA)

MOGA was implemented using MATLAB 2019a, following a sequence that mimicked evolutionary principles: selection, crossover, mutation, fitness evaluation, correction mechanisms, population renewal, and elitist preservation (Suha *et al.*, 2020). Unlike traditional GAs that optimize a single objective, multi-objective variants are designed to handle multiple competing objectives simultaneously.

18.5.3 ML and OAs in PTC

Laboratory cultivation encompasses complex and definitive developmental trajectories. The stages of laboratory culture are influenced by multiple variables, including the plant's genotype, the composition of the culture medium, and the various types and concentrations of PGRs. The information gathered during plant tissue cultivation includes (i) non-comparable production and two levels of binary entry, corresponding to embryonic and non-embryonic callus; (ii) variables that exceed two categories, such as roots, shoots, and embryos; (iii) persistent variables, like the length of the stem or root and the mass of the callus; (iv) time series data; (v) timing information; (v) characteristics such as vitrification degree, kernel color, and embryo developmental stages; and (vii) categorical variables like reaction types or the sources of plant phytohormones and carbohydrates (Osama *et al.*, 2015). Given the intricate nature of this issue and the interconnected aspects of the variables, it is challenging to optimize using traditional methods that isolate and assess individual variables in sequence. Recently, ML and optimization techniques have been employed to predict and enhance the stages of plant culture across various species.

In vitro sterilization

Initial micropropagation involves surface sterilization, which has a direct impact on PTC outcomes. The size, age, and shape of the explant, the physiological processes (reproductive or *in vitro*) of the parent plants, the physical conditions (temperature and light),the duration of the sterilization and immersion, and the concentration are some of the variables that can impact the effect of surface sterilization. Cuba-Díaz *et al.* (2020) discovered that using disinfectants at higher concentrations and for a longer period improves surface cleaning.

Somatic embryogenesis

Somatic cell embryonic production is the process of developing embryos in abnormally reproduced plant cells. This method removes the need for pollination. The presence of somatic cells suggests that plant cells are totipotent (Ibrahim, 2023). This somatic embryo is useful for a variety of applications, such as replicating better varieties, producing artificial seeds, improving genes, and producing secondary metabolites. When evaluating the influence of PGR on somatic cell embryos for *Chrysanthemum* (Hesami *et al.*, 2020), multiple layers of perceptron and SVR were compared for predicting accuracy. SVR outperforms multilayer perceptron (MLP) in all categories. In a study by Zielińska and Kępczyńska (2013), a medium containing kinetin, 2,4-dichlorophenoxyacetic acid, and sodium nitrate produced the highest number of somatic embryos per explant (57%) and the highest rate of embryogenesis (99%).

The study used image processing and an MLP model to predict the optimal combinations (2,4-dichlorophenoxyacetic acid, sucrose, kinetin concentrations, and explant age) to achieve high quality of phenotype. Where the process of selecting somatic embryos for embryo culture is difficult, costly, and time-consuming, ML models are used to automatically classify these embryos according to their stage of development. Image recognition techniques are used to rapidly identify somatic embryos that are biologically suitable for transfer to other growth media and to determine which embryos should be discarded for further growth. Similarly, a penalized logistic regression (PLR) was used to identify certain somatic embryos based on the prediction of the plant's development into an infectious, absorbed, reflected, excited, and healthy plant. This model has been shown to work well when applied to unprotected embryos with a different genetic background (Ibrahim *et al.*, 2023).

Rooting and acclimatization

In PTC, *in vitro* rooting and acclimatization are crucial processes. Rooting is a crucial aspect of plant growth (Ibrahim *et al.*, 2023). The process through which cultivated plants adapt to field or greenhouse conditions is known as acclimatization. Plantlets cultivated in regulated *in vitro* conditions are subjected to specific conditions designed to reduce stress and encourage plant growth. Plantlets have changed morphology, physiology, and anatomy as a result, requiring gradual exposure to environmental influences. Auxin and sucrose levels are the primary factors influencing rooting and adaptation. With the use of the multilayer perceptron-genetic algorithm (MLP-GA), the optimal composition was predicted and optimized to promote the optimal number of roots, nodes per plantlet, *ex vitro* leaves, and height during acclimatization. The model performed well, showing nearly perfect agreement between the predicted and observed values.

In a different study, extensive neuro-fuzzy logic was used to simulate the effects of light intensity and sugar concentration/content on acclimation (Gago *et al.*, 2010) (Table 18.2). The model successfully determined the ideal input levels and combinations for optimizing growth and development during *in vitro* rooting and

Table 18.2. Application of optimization algorithms and machine learning at various phases of plant tissue culture. Author's own table.

Plant species	Model	Purpose	References
Wrightia tinctoria	Neuro-fuzzy logic	Forecasting and improving *in vitro* acclimatization and rooting	Gago *et al.* (2010)
Wrightia tinctoria	GP	Forecasting and optimizing rooting *in vitro*	Mridula *et al.* (2018)
Medicago sativa	MLP	Forecasting and optimizing rooting *in vitro*	Tani *et al.* (1992)
Rauwolfia serpentina	ANN-HMM	Estimating and optimizing the growth of hairy roots	Mehrotra *et al.* (2013)
Artemisia annua	MLP	Estimating and optimizing the growth of hairy roots	Osama *et al.* (2013)
Glycyrrhiza glabra	MLP and RNN	Forecasting and maximizing hairy root culture	Prakash *et al.* (2010)
Pinus taeda	MLP	Forecasting and maximizing shoot organogenesis	Barone (2019)
Bryophyllum daigremontianum	Neuro-fuzzy logic	Forecasting and maximizing shoot organogenesis	García-Pérez *et al.* (2020)
Swertia paniculata	MLP	Forecasting and optimizing the production of secondary metabolites	Kaur *et al.* (2020)
Chrysanthemum × grandiflorum	MLP-NSGA-II	Estimating and improving the sterilization process	Hesami*et al.* (2019)
Daucus carota	MLP	Forecasting cell growth and making it optimal	Albiol *et al.* (1995)
Nicotiana tabacum	MLP	Identification of plant viruses in cultured cells	Frossyniotis *et al.* (2008)

acclimatization. To evaluate the effects of different mineral concentrations in a Hoagland mineral solution on three physiological disorders and growth parameters (hardening efficiency, newly formed shoot length, total leaf number, leaf area, and leaf chlorophyll content) during the experiment, a five-dimensional IV-design space was created using a design of experiments (DOE) methodology and a hybrid of fuzzy logic and ANN (Maleki *et al.*, 2021). The neuro-fuzzy logic's capacity to forecast all growth indicators and the frequency of leaf necrosis showed a strong correlation between observed and expected values. Additionally, the "IF-THEN" rule of the model showed that magnesium ions and calcium ions with a positive charge had opposite effects on the amount of chlorophyll in the leaves, but increased some growth parameters and decreased leaf necrosis. Several parameters were negatively affected by the addition of nitrate to the media, while some growth characteristics were positively impacted by the combination of ammonium and copper (II) ions or magnesium ions.

18.6 Application of Data Modeling in PTC

18.6.1 Optimizing tissue culture schemes with data-driven modeling technology

To achieve effective and efficient PTC, it is essential to scientifically enhance various cultural parameters, including light, temperature, humidity, as well as the suitable composition of the medium and the regulatory factors influencing plant growth. The process of experimenting to determine the most effective cultural environments can occasionally be time-consuming and ineffective. Moreover, inadequate analysis of the various types and amounts of several hormones and minerals can be challenging, difficult, expensive, and often result in poor quality. In order to optimize PTC media, several control trials are required because most optimization is based on previous experimental results. The effective design and optimization of new media are also essential. For the *in vitro* cultivation of plant tissues to be effectively beneficial, the interactions of minerals and hormones with other elements and different plant tissues are essential (Arab *et al.*, 2018).

The relationships between the numerous media components are very complex and involve many unknown relationships. It is quite challenging to adapt the medium to different plant species and genotypes. The development and implementation of a predictive modeling system for medium composition and cultural conditions may thus increase the effectiveness of culture optimization. Hesami *et al.* (2017) suggest that data-driven modeling holds promise for improving biological processes and resolving multimodel non-linearity. To improve performance, modeling systems are often coupled with optimization methods. Both the output and the likelihood of a software sensor can be predicted using appropriate modeling techniques. Modeling approaches have improved significantly in recent years. Some common approaches include combining gene expression programming (GEP) with GA (Jamshidi *et al.*, 2019) and integrating an AI model with an OA (Hesami *et al.*, 2020).

18.6.2 Innovative tools to support bioreactor engineering

The main source of secondary metabolites is still medicinal plant cell and tissue culture (Ramawat *et al.*, 2021). Plant cells and tissues can be produced on a largescale using bioreactors, which offer the following benefits: stability, high yield, efficiency, and affordability (Fig. 18.4). Microorganisms or plant cells and tissues can be cultivated in bioreactors under carefully regulated environmental conditions. A highly competitive solution to the current problems of high cultivation costs will be the large-scale production of secondary metabolites by bioreactor, especially for medicinal plants like *Panax ginseng* and *Panax notoginseng* that have slow growth, low bioactive components, and are easily harmed by the environment. By developing the optimal circumstances for the efficient growth and synthesis of bioactive compounds—which can then be applied to secondary metabolites, recombinant proteins, microbial fermentation, etc.—the culture outcomes can be greatly improved. In terms of economics, the bioreactor design aims to reduce the cost of plant production. To do this,

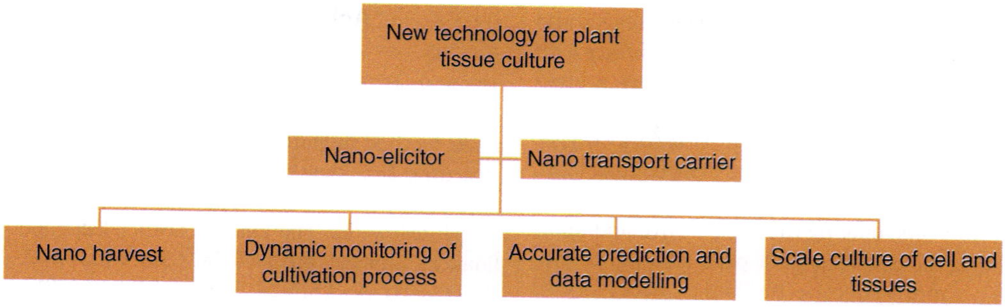

Fig. 18.4. Development of various new technologies for plant tissue culture. (Author's own image.)

all expenses related to equipment, substrate, time, and ultimate power must be considered (Khanahmadi *et al.*, 2017).

Glass or stainless steel is frequently used to make bioreactors for plant cell and tissue cultures (Su *et al.*, 2019). Bioreactors will be categorized as liquid-phase, gas-phase, gas-phase combination, or liquid-phase combination bioreactors based on their changing environment. The structural characteristics of the cultivation materials will determine whether they are categorized as tissue culture bioreactors or suspension cell culture bioreactors. Mechanical and gas-powered reactors are examples of liquid-phase reactors. An unstable reactor will damage the culture material. Both the oxygen flow and the shear force should be taken into account when designing gas-powered reactors. Reactors for plant tissue and cell culture, such as airlift, tubular membrane, silicone tube air flow, slug bubble, wavelet, orbitally shaken, spray, helical ribbon impellers, bioreactors, and stirred reactors, have been developed steadily in recent years (Valdiani *et al.*, 2019).

The stirred tank reactor (STR) is now the most used bioreactor for plant tissue cell growth. The airlift and bubble column systems were mainly developed to enhance the biological oxygen demand within the culture. For high-density plant suspension cultures, helical and double-helical ribbon impellers have demonstrated effective performance. The use of disposable bioreactors is one important advancement in the cultivation of plant cells and tissues. By replacing the stainless-steel components of the reactor with plastic ones, the disposable reactor significantly reduces the time and cost of reactor production.

18.6.3 Nanomaterials in the tissue culture of plants

In biotechnology, nanomaterials are used in many different domains, such as pharmaceuticals, agriculture, and medicine. Among their many applications are the creation of sensors, the breakdown of agricultural chemicals, soil remediation, and drug transportation. Lipids, polymers, magnetic materials, ceramics, silicates, metallic oxides, and semiconductor quantum dots (QDs) are among the materials used to create nanoparticles. Whereas metal-based nanomaterials are size-dependent, dendritic macromolecules and polymer nanoparticles encapsulated in emulsion have controlled and degraded release. Nanomaterials increase the system's activity by making the particles smaller, more soluble, and having a larger surface area (Ghormade *et al.*, 2011). Nanoparticles are frequently employed in PTC as elicitors to produce secondary metabolites (Lala *et al.*, 2021) or as molecules that selectively extract advantageous molecules from plant cells by transporting plant hormones and other substances (Khan *et al.*, 2021) (Fig. 18.4).

However, nanoparticles can successfully limit the pollution that plant growth produces in the environment by delivering certain chemicals into cells in a targeted manner. Currently, inorganic nanoparticles are being researched for potential application in drug delivery. A 20–30% increase in agricultural crop yields has been attributed to nano-fertilizers and nano-insecticides (Kah *et al.*, 2018). Examples of nanomaterials with multiple uses that are appropriate for cellular delivery include calcium phosphate, gold, carbon materials, silicon oxide, iron oxide, lactate dehydrogenase, which have broad availability, high surface function, high biocompatibility, and the ability to deliver drugs to specific targets with measurable release. At the right times, growth regulators can be added to plants to track their development. They can also be used to trigger other developmental reactions. Temperature, light, and other environmental factors can readily harm PGRs, making them ineffective and influencing the development of plant organs. Developing a more effective method is necessary to increase the growth of PTC material.

18.7 Conclusion

In conclusion, medicinal plants are at risk of extinction due to various threats, necessitating effective conservation and sustainable practices. This chapter emphasizes the role of PTC as a vital biotechnological tool for the rapid and controlled propagation of these species. By integrating DL approaches, particularly CNNs, researchers can enhance image-based analysis during tissue culture stages. The focus on predictive models relating callus type and PGRs to shoot regeneration success aims to improve methodologies and increase the yield of valuable natural metabolites. Addressing current challenges in empirical studies within plant biotechnology, this chapter provides valuable insights for advancing the sustainable production of endangered medicinal plants.

Acknowledgment

The authors thank the Director, CSIR-Institute of Himalayan Bioresource Technology, Palampur, India, for their support and for providing the infrastructure. K.K. thanks the CSIR-UGC (NTA Ref. No.: 191620158661) for providing a senior research fellowship. The authors acknowledge the financial support provided by CSIR (MLP-207, IHP002401 No. CSPS24/RDSF/IHBT/IHP24/01, and MMP025301 No.AB-CSIR/255/2024-TMD) and DBT (BT/PR45280/NER/95/1918/2022).

Author Contributions

A.P. and K.K. wrote the manuscript; R.C. prepared the figures and tables of the manuscript; V.S. reviewed the manuscript; R.J. conceived the idea, designed the study, and reviewed and edited the final manuscript. All authors have read and agreed to the published version of the manuscript. The study was approved by the Institutional Review Board. This manuscript represents CSIR-IHBT communication No. 5823.

Conflicts of Interest

The authors declare no conflict of interest.

References

Abera, B. (2014) Medicinal plants used in traditional medicine by Oromo people, Ghimbi District, Southwest Ethiopia. *Journal of Ethnobiology and Ethnomedicine.* 10 (1): 1–15.

Albiol, J., Campmajo, C., Casas, C., Poch, M. (1995) Biomass estimation in plant cell cultures: a neural network approach. *Biotechnology Progress.* 11(1): 88–92.

Ale-Dabbous, A. N., Kumar, P., & Khan, A. R. (2017) Prediction of airborne nanoparticles at roadside location using a feedforward artificial neural network. *Atmospheric Pollution Research.* 8: 446–454.

Alloghani, M., Al-Jumeily, D., Mustafina, J., Hussain, A., Aljaaf, A.J. (2020) A systematic review on supervised and unsupervised machine learning algorithms for data science. *Supervised Unsupervised Learn. Data Science.* 177: 3–21.

Anand, U., Jacobo-Herrera, N., Altemimi, A., and Lakhssassi, N. (2019) A comprehensive review on medicinal plants as antimicrobial therapeutics: potential avenues of biocompatible drug discovery. *Metabolites.* 9 (11): 258.

Arab, M.M.,Yadollahi, A., Ahmadi, H., Eftekhari, M., Maleki, M. (2017) Mathematical modelling and optimizing of in vitro hormonal combination for G× N15 vegetative rootstock proliferation using Artificial Neural Network-Genetic Algorithm (ANN-GA). *Frontiers Plant Science.* 8: 1853.

Arab, M.M., Yadollahi, A., Eftekhari, M., Ahmadi, H., Akbari, M., Khorami, S.S. (2018) Modelling and Optimizing a New Culture Medium for In Vitro Rooting of G× N15 Prunus Rootstock Using Artificial Neural Network-Genetic Algorithm. *Scientific Reports.* 8: 9977.

Arun, A., Sharma, S., Singh, B., Hazra, T. (2025).Identification of Plant Species Using Convolutional Neural Network with Transfer Learning. *Journal of Phytopathology*, 173(1): e70032.

Askr, H., Elgeldawi, E., Aboul Ella, H., Elshaier, Y. A., Gomaa, M. M., Hassanien, A. E. (2023) Deep learning in drug discovery: an integrative review and future challenges. *Artificial Intelligence Review.*56(7): 5975-6037.

Azadnia, R., Al-Amidi, M.M., Mohammadi, H., Cifci, M.A., Daryab, A., Cavallo, E. (2022) An AI based approach for medicinal plant identification using deep CNN based on global average pooling. *Agronomy* 12 (11): 2723.

Bambil, D., Pistori, H., Bao, F., Weber, V., Alves, F.M., Gonçalves, E.G., Bortolotto, I.M. (2020) Plant species identification using color learning resources, shape, texture, through machine learning and artificial neural networks. *Environment Systems and Decisions.* 40(4): 480–484.

Barone, J.O. (2019)Use of multiple regression analysis and artificial neural networks to model the effect of nitrogen in the organogenesis of *Pinustaeda* L. *Plant Cell Tissue Organ Culture.* 137(3):455–464.

Chandran, H., Meena, M.,Barupal, T., Sharma, K. (2020) Plant Tissue Culture as a Perpetual Source for Production of Industrially Important Bioactive Compounds. *Biotechnology Reports.* 26: e00450.

Chanyal, H., Yadav, R. K., Saini, D. K. J. (2022)Classification of medicinal plants leaves using deep learning technique: A review. *International Journal of Intelligent Systems and Applications in Engineering.* 10(4): 78–87.

Cuba-Díaz, M., Rivera-Mora, C., Navarrete, E. and Klagges, M. (2020) Advances of native and non-native Antarctic species to in vitro conservation: improvement of disinfection protocols. *Scientific Reports.* 10(1):3845.

Davidson, S. J., Saggese, T., & Krajňáková, J. (2024) Deep learning for automated segmentation and counting of hypocotyl and cotyledon regions in mature *Pinusradiata* D. Don. somatic embryo images. *Frontiers in Plant Science.*15: 1322920.

Deb, K., Pratap, A., Agarwal, S., Meyarivan, T. (2002) A fast and elitist multi-objective genetic algorithm: NSGA-II. *IEEE Transactions on Evolutionary Computation.* 6(2):182–197.

Dorzhigulov, A., James, A.P. (2020) Deep neuro-fuzzy architectures. In: *Deep learning classifiers with memristive networks. Springer*, Berlin, pp. 195–213.

Espinosa-Leal, C.A., Puente-Garza, C.A., García-Lara, S. (2018) *In Vitro* Plant Tissue Culture: Means for Production of Biological Active Compounds. *Planta.* 248: 1–18.

Ezugwu, A.E., Adeleke, O.J., Viriri, S. (2018) Symbiotic organisms search algorithm for the unrelated parallel machines scheduling with sequence-dependent setup times. *PLoS ONE.* 13(7): p.e0200030.

Ezugwu, A.E.,Prayogo, D. (2019) Symbiotic organisms search algorithm: Theory, recent advances and applications. *Expert Systems with Applications.*119: 184–209.

Feng, Y., Liu, C., Han, J., Lu, Q., Xing, X. (2024) Identification of wheat seedling varieties based on Mssiap Net. *Frontiers in Plant Science*, 14: 1335194.

Fernando, J.A., Vieira, M.L.C., Machado, S.R., Appezzato-da-Gloria, B. (2007) New insights into the in vitro organogenesis process: the case of Passiflora. *Plant Cell Tissue Organ Culture.* 91(1):37–44.

Frossyniotis, D., Moschopoulou, G., Yialouris, C. (2008) Artificial Neural Network Selection for the Detection of Plant Viruses.*World Journal of Agricultural Science.* 4(1):114–120.

Gago, J., Martinez-Nunez, L., Landin, M., Gallego, P. (2010) Artificial neural networks as an alternative to the traditional statistical methodology in plant research. *Journal of Plant Physiology.* 167(1):23–27.

Gago, J., Martinez-Nunez, L., Landin, M.,Flexas, J., Gallego, P.P. (2014) Modelling the effects of light and sucrose on *in vitro* propagated plants: A multiscale system analysis using artificial intelligence technology. *PLoS ONE.9:* e85989

García-Pérez, P., Lozano-Milo, E., Landín, M., Gallego, P.P. (2020) Combining medicinal plant in vitro culture with machine learning technologies for maximizing the production of phenolic compounds. *Antioxidants.*9(3):210.

George, T., Amudha, T. (2020) Genetic algorithm based multi-objective optimization framework to solve traveling salesman problem. *In: Advances in Intelligent Systems and Computing. Springer*, Berlin, pp. 141–151.

Ghormade, V., Deshpande, M.V.,Paknikar, K.M. (2011) Perspectives for Nano-Biotechnology Enabled Protection and Nutrition of Plants. *Biotechnology Advances.* 29: 792–803.

Gladston, A., Sucithra, B. (2023) Hand-crafted feature extraction and deep learning models for leaf image recognition. In: Handbook of Research on Computer Vision and Image Processing in the Deep Learning Era. *IGI Global, pp.* 161–178.

Granato, D., Putnik, P., Kovacevic, D.B. (2018) Trends in chemometrics: Food authentication, microbiology, and effects of processing. *Comprehensive Reviews in Food Science and Food Safety.* 17:663–677.

Hamazaki, K., and Iwata, H. (2024) AI-assisted selection of mating pairs through simulation-based optimized progeny allocation strategies in plant breeding. *Frontier Plant Science.* 15: 1361894.

Hesami, M., Jones, A.M.P. (2020) Application of artificial intelligence models and optimization algorithms in plant cell and tissue culture. *Applied Microbiology and Biotechnology.* 104(22): 9449-85.

Hesami, M., Adamek, K., Pepe, M., Jones, A.M. (2023) Effect of explant source on phenotypic changes of in vitro grown cannabis plantlets over multiple subcultures. *Biology.* 12(3):443.

Hesami, M., Baiton, A., Alizadeh, M., Pepe, M., Torkamaneh, D., Jones, A.M. (2021) Advances and perspectives in tissue culture and genetic Engineering of *Cannabis*. *International Journal of Molecular-Sciences.* 22(11):5671.

Hesami, M., Daneshvar, M.H. (2016) Development of a regeneration protocol through indirect organogenesis in Chenopodium quinoa Willd. *Indo-American Journal of Agricultural Veterinary Science.* 4(2):25–32.

Hesami, M., Jones, A.M.P. (2020) Application of artificial intelligence models and optimization algorithms in plant cell and tissue culture. *Applied Microbiology and Biotechnology.* 104(22):9449–85.

Hesami, M., Jones, A.M.P. (2023) Potential roles of epigenetic memory on the quality of clonal cannabis plants: content and profile of secondary metabolites. In: Medicinal Usage of Cannabis and Cannabinoids 2023 Jan 1 (pp. 91-104). Academic Press.

Hesami, M., Naderi, R., Tohidfar, M. (2019) Modelling and optimizing *in vitro* sterilization of *chrysanthemum* via multilayer perceptronnon-dominated sorting genetic algorithm-II (MLP-NSGAII).*Frontiers in Plant Science.* 10:282.

Hesami, M., Naderi, R., Tohidfar, M., Yoosefzadeh-Najafabadi, M. (2020) Development of support vector machine-based model and comparative analysis with artificial neural network for modelling the plant tissue culture procedures: Effect of plant growth regulators on somatic embryogenesis of chrysanthemum, as a case study. *Plant Methods.* 16: 112.

Hesami, M., Naderi, R., Yoosefzadeh-Najafabadi, M., Rahmati, M. (2017) Data-Driven Modelling in Plant Tissue Culture. *Journal of Applied Environmental Biology Sciences.* 7: 37–44.

Hesami, M., Tohidfar, M., Alizadeh, M., Daneshvar, M.H. (2020) Effects of sodium nitroprusside on callus browning of Ficusreligiosa: an important medicinal plant. *Journal of Forest y Research.* 31(3):789–96.

Hesami, M., Condori-Apfata, J.A., Valencia, M.V., Mohammadi, M. (2020c) Application of artificial neural network for modelling and studying *in vitro* genotype-independent shoot regeneration in wheat. *Applied Sciences.* 10: 5370.

Holland, J.H. (1992) Adaptation in natural and artificial systems: an introductory analysis with applications to biology, control, and artificial intelligence. *MIT Press*.

Ibrahim, N., Adedamola, A. M., Ibrahim, B., Ahmed, R. T., Raji, I. D., & Bello-Salau, H. (2023) Survey of Machine Learning and Optimization Algorithms in Plant Tissue Culture. *Engineering Proceedings.* 56(1): 122.

Jafari, M., Daneshvar, M.H., Jafari, S., Hesami, M. (2022) Machine learning-assisted *In Vitro* Rooting optimization in *Passifloracaerulea*. *Forests.* 13(12):2020.

Jafari, M., Daneshvar, M.H., Lotfi-Jalalabadi, A. (2016) Control of in vitro contamination of Passifloracaerulea by using of sodium hypochlorite. *Indo-American Journal of Agricultural Veterinary Science.* 4:8–15.

Jamshidi, S., Yadollahi, A., Arab, M.M., Soltani, M., Eftekhari, M., Sabzalipoor, H, Sheikhi, A., Shiri, J. (2019) Combining Gene Expression Programming and Genetic Algorithm as a Powerful Hybrid Modelling Approach for Pear Rootstocks Tissue Culture Media Formulation. *Plant Methods.* 15: 136.

Kah, M.,Kookana, R.S.,Gogos, A.,Bucheli, T.D. (2018) A Critical Evaluation of Nano pesticides and Nano fertilizers against Their Conventional Analogues. *Nature Nanotechnology* 13: 677–684.

Kaur, P., Gupta, R.C., Dey, A., Malik, T., Pandey, D.K. (2020) Optimization of salicylic acid and chitosan treatment for bitter secoiridoid and xanthone glycosides production in shoot cultures of *Swertiapaniculata* using response surface methodology and artificial neural network. *BMC Plant Biology.* 20(1):225.

Khan, M.A., Fugate, M., Rogers, D.T., Sambi, J., Littleton, J.M., Rankin, S.E., Knutson, B.L. (2021) Mechanism of Mesoporous Silica Nanoparticle Interaction with Hairy Root Cultures During Nano harvesting of Biomolecules. *Advances in Biology.* 5: 2000173.

Khanahmadi, M., Paek, K.Y. (2017) Bioreactor Technology for Sustainable Production of Valuable Plant Metabolites: Challenges and Advances. In: *Crop Improvement: Sustainability Through Leading-Edge Technology*; Springer: Cham, Switzerland, pp. 169–189.

Lala, S. (2021) Nanoparticles as Elicitors and Harvesters of Economically Important Secondary Metabolites in Higher Plants: A Review. *IET Nanobiotechnology.* 15: 28–57.

Li Y., Shen Y., Yao C. (2020) Quality assessment of herbal medicines based on chemical fingerprints combined with chemometrics approach: A review. *Journal of Pharmaceutical and Biomedical Analysis.* 185: 113215.

Li, Q., Zhou, W., Zhang, H. (2023) Integrating spectral and image information for prediction of cottonseed vitality.*Frontiers in Plant Science*, 14: 1298483.

Lin J, Zhao Y, Watson D, Chen C (2020) The radial basis function differential quadrature method with ghost points.*Mathematics and Computers in Simulation.*173:105–114.

Lombardi, S.P., Passos, IRdS.,Nogueira, M.C.S., Appezzato-da-Gloria B. (2007) *In vitro* shoot regeneration from roots and leaf discs of *Passifloracincinnata* Mast.*Brazilian Archives of Biology and Technology.*50:239–247.

Mahesh, B. (2020) Machine learning algorithms—A review. *International Journal of Scientific Research.*9: 381–386.

Maleki, S.,Zanjani, B.M.,Kohnehrouz, B.B., Landin, M., Gallego, P.P. (2021) Computer-based tools unmask critical mineral nutrient interactions in Hoagland solution for healthy kiwiberry plant acclimatization. *Frontier Plant Science.*12: 723992.

Malik, O.A., Ismail, N., Hussein, B.R., Yahya, U. (2022) Automated real-time identification of medicinal plants species in natural environment using deep learning models-A case study from Borneo region. *Plants* 11(15): 1952.

Mehrotra, S., Prakash, O., Khan, F., Kukreja, A. (2013) Efficiency of neural network-based combinatorial model predicting optimal culture conditions for maximum biomass yields in hairy root cultures. *Plant Cell Reports.* 32(2):309–317.

Mikovski, A.I., Silva,NTd S., Machado,CdS., Otoni, M.D., Carvalho, W.C., Rocha, I.F., Silva, D.I. (2019) Tissue culture and biotechnological techniques applied to passion fruit with ornamental potential: an overview. *Ornam Horticulturae.* 25:189–99.

Montesinos-Lopez, O. A., Chavira-Flores, M., Kismiantini, Crespo-Herrera, L., Saint Pierre, C., Li, H.,Crossa, J. (2024) A review of multimodal deep learning methods for genomic-enabled prediction in plant breeding. *Genetics*, 228(4): iyae161.

Morris, A.J., Montague, G.A., Willis, M.J. (1994) *Artificial neural networks: studies in process modeling and control.Transactions of the Institution of Chemical Engineers.* 72A: 3–19.

Mostafa, S., Mondal, D., Panjvani, K., Kochian, L., Stavness, I. (2023) Explainable deep learning in plant phenotyping. *Frontiers in Artificial Intelligence*, 6: 1203546.

Mostajer Kheirkhah, F., Asghari, H. (2019) Plant leaf classification using GIST texture features. *IET Computer Vision.*13(4): 369–375.

Mridula, M.R., Nair, A.S., Kumar, K.S. (2018) Genetic programming-based models in plant tissue culture: an addendum to traditional statistical approach. *PLoS Computational Biology* 14(2): e1005976.

Mujtaba, M., Khawar, K.M., Camara, M.C.,Carvalho, L.B., Fraceto, L.F.,Morsi, R.E., Elsabee, M.Z., Kaya, M.,Labidi, J., Ullah, H. (2020) Chitosan-Based Delivery Systems for Plants: A Brief Overview of Recent Advances and Future Directions. *International Journal of Biological Macromolecules.*154: 683–697.

Mulugeta, A.K., Sharma, D.P., Mesfin, A.H. (2024) Deep learning for medicinal plant species classification and recognition: a systematic review. *Frontier Plant Science.* 14:1286088.

Muneer, A., Fati, S.M. (2020) Efficient and automated herbs classification approach based on shape and texture features using deep learning. *IEEE Access* 8: 196747–196764.

Murphy, G. E., and Romanuk, T. N. (2014) A meta-analysis of declines in local species richness from human disturbances. *Ecology and Evolution.* 4(1): 91–103.

Naresh, Y.G., Nagendraswamy, H.S. (2016) Classification of medicinal plants: an approach using modified LBP with symbolic representation. *Neurocomputing.*173: 1789–1797.

Niazian, M.,Niedbała, G. (2020) Machine learning for plant breeding and biotechnology.*Agriculture.*10: 436.

Osama, K., Mishra, B.N., Somvanshi, P. (2015) Machine learning techniques in plant biology. In: Plant Omics: *The Omics of plant science. Springer*, Berlin, pp. 731–754.

Osama, K., Somvanshi, P., Pandey, A.K., Mishra, B.N. (2013) Modelling of nutrient mist reactor for hairy root growth using artificial neural network. *European Journal of Science Research.* 97(4):516–526.

Pacheco, G., Garcia, R., Lugato, D., Vianna, M., Mansur, E. (2012) Plant regeneration, callus induction and establishment of cell suspension cultures of *Passifloraalata* Curtis.*ScientiaHorticulturae.*144:42–47.

Patil, S. S., Patil, S. H., Pawar, A. M., Patil, N. S., &Rao, G. R. (2022) Automatic classification of medicinal plants using state-of-the-art pre-trained neural networks. *Journal of Advanced Zoology*, 43: 80-88.

Pham, T.N., Van Tran, L. and Dao, S.V.T. (2020) Early disease classification of mango leaves using feed-forward neural network and hybrid metaheuristic feature selection. *IEEE Access.* 8: 189960-189973.

Prado, F., Minutolo, M.C., Kristjanpoller, W. (2020) Forecasting based on an ensemble autoregressive moving average-adaptive neuro-fuzzy inference system–neural network-genetic algorithm framework. *Energy.* 197:117159.

Prakash, O., Mehrotra, S., Krishna, A., Mishra, B.N. (2010) A neural network approach for the prediction of in vitro culture parameters for maximum biomass yields in hairy root cultures. *Journal of Theoretical Biology.* 265(4):579–585.

Pushpa, B., AV S. H., Ashok A. (Eds.) (2021) "Diseased leaf segmentation from complex background using indices-based histogram," in 2021 6thInternational Conference on Communication and Electronics Systems (ICCES), Coimbatore, India (IEEE).

Qi, L., Zhong, F., Chen, Y. (2020) An integrated spectroscopic strategy to trace the geographical origins of emblic medicines: Application for the quality assessment of natural medicines, *Journal of Pharmaceutical Analysis.* 10: 356e364.

Qiaosen S., Fengsheng W., Dong C., Gang C., Chao L., Leyi W. (2022) Deep convolutional neural networks with ensemble learning and transfer learning for automated detection of gastrointestinal diseases. *Computers in Biologyand Medicine.* 150: 106054.

Quach, B.M., Dinh, V.C., Pham, N., Huynh, D., Nguyen, B.T. (2023) Leaf recognition using convolutional neural networks-based features. *Multimedia Tools and Applications.* 82(1): 777–801.

Quach, L. D., Quoc, K. N., Quynh, A. N., Thai-Nghe, N., Nguyen, T. G. (2023) Explainable deep learning models with gradient-weighted class activation mapping for smart agriculture. *IEEE Access.* 11*:* 83752-83762.

Ramawat, K.G. (2021) An Introduction to the Process of Cell, Tissue, and Organ Differentiation, and Production of Secondary Metabolites. In: *Plant Cell and Tissue Differentiation and Secondary Metabolites: Fundamentals and Applications*; *Springer*: Cham, Switzerland, pp. 1–22.

Ray, S.A. (2019) Quick review of machine learning algorithms. In Proceedings of the 2019 International Conference on Machine Learning, Big Data, Cloud and Parallel Computing (COMITCon), Faridabad, India, 14–16 February 2019; pp. 35–39.

Roopashree, S., Anitha, J., Mahesh, T.R., Kumar, V.V., Viriyasitavat, W., Kaur, A. (2022) An IoT based authentication system for therapeutic herbs measured by local descriptors using machine learning approach. *Measurement* 200: 111484.

Rosa, Y.B.C.J., Monte-Bello, C.C., Dornelas, M.C. (2016).*In vitro* organogenesis and efficient plant regeneration from root explants of *Passiflorasuberosa L.* (Passifloraceae). *InVitro Cell Dev Biology—Plant.* 52(1):64–71.

Sennan, S., Pandey, D., Alotaibi, Y., Alghamdi, S. (2022) A novel convolutional neural networks-based spinach classification and recognition system. *Computers, Materials & Continua* 73 (1).

Specht, D.F. (1991) A general regression neural network.*IEEE T Neur Net* 2(6):568–576.

Srinivas, N., Deb, K. (1994).Multi objective optimization using nondominated sorting in genetic algorithms.*Evolutionary Computation* 2(3): 221–248.

Su, R., Sujarani, M., Shalini, P., Prabhu, N. (2019) A Review on Bioreactor Technology Assisted Plant Suspension Culture.*Asian Journal of Biotechnology and Bioresources.Technology.*5*:* 1–13.

Subhabrata, P., Shreya, C., Uttpal, A., Swarnali, D., Samapika, N., Mimosa, G. (2021) *Withaniasomnifera* (L.) Dunal (Ashwagandha): A comprehensive review on ethnopharmacology, pharmacotherapeutics, bio medicinal and toxicological aspects.*Biomedicine & Pharmacotherapy.* 143: 112175.

Suha, E., Azmi, Y., Muhammad, R.M., Nazmi, M.N., Modather, M., Sami, M.M.E., Kheiralla, A.F. (2020) Applying multi-objective genetic algorithm (MOGA) to optimize the energy inputs and greenhouse gas emissions (GHG) in wetland rice production, *Energy Reports.*6: 2988-2998.

Sun, X., Li, Y., Li, G., Jin, S., Zhao, W., Liang, Z., & Zhang, W. (2023) SCGNet: efficient sparsely connected group convolution network for wheat grains classification. *Frontiers in Plant Science*, 14*:* 1304962.

Tani, A., Murase, H., Kiyota, M., Honami, N. (1992) Growth simulation of alfalfa cuttings *in vitro* by Kalman filter neural network.*ActaHorticulturae* 319:671–676.

Turkoglu, M., Hanbay, D. (2019) Leaf-based plant species recognition based on improved local binary pattern and extreme learning machine.*Physica A: Statistical Mechanics and Its Applications.* 527: 121297.

Valdiani, A., Hansen, O.K., Nielsen, U.B., Johannsen, V.K., Shariat, M., Georgiev, M.I., Omidvar, V., Ebrahimi, M., Dinanai, E.T.,Abiri, R. (2019) Bioreactor-Based Advances in Plant Tissue and Cell Culture: Challenges and Prospects. *Critical Review of Biotechnology.* 39*:* 20–34.

Wang, B., Li, H., You, J., Chen, X., Yuan, X., Feng, X. (2022) Fusing deep learning features of triplet leaf image patterns to boost soybean cultivar identification. *Computers and Electronics in Agriculture.* 197: 106914.

Wang, P., Yu, Z. (2015) Species authentication and geographical origin discrimination of herbal medicines by near infrared spectroscopy: A review, *Journal of Pharmaceutical Analysis.* 5: 277e284.

Wu, H., Fang, L., Yu, Q., Yuan, J. and Yang, C. (2023) Plant leaf identification based on shape and convolutional features. *Expert Systems with Applications.* 219:119626.

Yang, C. (2021) Plant leaf recognition by integrating shape and texture features. *Pattern Recognition.* 112: 107809.

Yang, Z., He, W., Fan, X., Tjahjadi, T. (2022) Plant Net: transfer learning-based fine-grained network for high-throughput plants recognition. *Soft Computing.* 26(20): 10581–10590.

Ying, X., Tang, B., Zhou, C. (2020) Nursing scheme based on back propagation neural network and probabilistic neural network in chronic kidney disease. *Journal of Medical Imaging and Health Informatics.* 10(2):416–421.

Yun, Y., Chuluunsukh, A., Gen, M. (2020) Sustainable closed-loop supply chain design problem: a hybrid genetic algorithm approach. *Mathematics* 8(1):84.

Zhang, Y., Wang, Y. (2023) Recent trends of machine learning applied to multi-source data of medicinal plants. *Journal of Pharmaceutical Analysis.* 13(12):1388–1407.

Zhu, L., Spachos, P., Pensini, E. (2021) Deep learning and machine vision for food processing: A survey. *Current Research in Food Science.* 4:233–249.

Zielińska, S., Kępczyńska, E. (2013) Neural modelling of plant tissue cultures: A review. *Biotechnologia,* 94: 253–268.

Index

Note: Page numbers in **bold** type refer to **figures.**
Page numbers in *italics* type refer to *tables.*

Printed and bound by CPI Group (UK) Ltd, Croydon, CR0 4YY

16/09/2026

14963051-0003